W0049814

Plant
Systematics
and
Evolution   Supplement 9

U. Jensen and J. W. Kadereit (eds.)

Systematics and Evolution of the Ranunculiflorae

Springer-Verlag Wien GmbH

Univ.-Prof. Dr. Uwe Jensen
Lehrstuhl für Pflanzenökologie und Systematik
Universität Bayreuth, Bayreuth, Bundesrepublik Deutschland

Univ.-Prof. Dr. Joachim W. Kadereit
Institut für Spezielle Botanik und Botanischer Garten
Johannes Gutenberg-Universität
Mainz, Bundesrepublik Deutschland

With 78 Figures

Library of Congress Cataloging-in-Publication Data

Systematics and evolution of the Ranunculiflorae/U. Jensen and J. W. Kadereit (eds.).
p. cm. – (Plant systematics and evolution. Supplement; 9)
Papers and posters contributed to a conference held September 10-11, 1994 at Bayreuth, Germany.
Includes bibliographical references and indexes.
ISBN 978-3-7091-7361-9   ISBN 978-3-7091-6612-3 (eBook)
DOI 10.1007/978-3-7091-6612-3
1. Ranunculales – Classification – Congresse. 2. Ranunculales – Evolution – Congresses.
I. Jensen, Uwe, 1931–   . II. Kadereit, J. W. (Joachim W.), 1956–   . III. Series: Plant systematics and evolution. Supplementum; 9. QK495.A12S97 1995 583'.111–dc20

ISSN 0172-6668
ISBN 978-3-7091-7361-9

Preface

This volume contains the papers and posters contributed to the International Conference "Systematics and Evolution of the *Ranunculiflorae*" held on 10/11 September 1994 at Bayreuth/Germany. These contributions deal, on different systematic levels and in different contexts, with virtually all taxa contained in the group, explore a wide array of characters employed for the classification of higher plants, and also discuss aspects of the biology of the plants in question. After the consideration of the entire *Ranunculiflorae*, emphasis is placed on the *Papaveraceae* s.l., *Lardizabalaceae*, *Ranunculaceae* and *Berberidaceae* including various segregate genera. Unfortunately little new information was presented for the *Menispermaceae* apart from their possible phylogenetic position within the group. This certainly illustrates our unbalanced knowledge of reasonably small temperate versus large tropical flowering plant families, and emphasizes the need for intensified research in the latter.

The particular strength of this volume is the combination of morphological and molecular evidence used for the understanding of the phylogeny of the *Ranunculiflorae* and its constituent families. On a theoretical basis the issue of data combination is addressed by BANDELT.

Following an introduction into the history of the concept of the *Ranunculiflorae* (*Ranunculidae*, *Ranunculanae*) and its problems by KUBITZKI, comparative interpretations of wood anatomy (CARLQUIST), sieve-element plastids and phloem proteins (BEHNKE), epicuticular wax structure (BARTHLOTT & THEISEN), floral structure and floral biology (ENDRESS), structure of the androecium (RONSE DECRAENE & SMETS), palynology (BLACKMORE & al.), seed structure (BRÜCKNER), secondary compounds (JENSEN) and chloroplast (*rbc*L, *atp*B) and nuclear DNA (18S) sequence data (HOOT & CRANE), and a cladistic analysis of various non-DNA characters are presented (LOCONTE & al.).

These studies result in often divergent opinions on the circumscription of the group, its affinities, the phylogenetic relationships among its constituent families, and its monophyletic or paraphyletic status.

The *Ranunculiflorae* are, with some exceptions, distinguished from the "woody Magnoliids" by tricolpate or tricolpate-derived (BLACKMORE & al.) pollen grains, absence of ethereal oil cells, benzylisoquinoline alkaloids of the berberine and morphine type (JENSEN, KUBITZKI), form-Ss sieve element plastids of unusually homogeneous and comparatively large size (BEHNKE), and mostly clustered epicuticular wax tubules which, however, are not exclusive to this group (BARTHLOTT & THEISEN). Consensus on the members of the group is broad, and *Pteridophyllum, Hypecoum, Fumariaceae, Papaveraceae, Sargentodoxa, Lardizabalaceae, Menisper-*

maceae, *Hydrastis, Circaeaster, Kingdonia, Ranunculaceae, Nandina* and *Berberidaceae* are included by all authors who address this question. There is, however, disagreement about the inclusion of *Glaucidium* and *Nelumbo*.

An affinity of the *Ranunculiflorae* to "woody Magnoliids" is discussed by KUBITZKI, BEHNKE and ENDRESS. ENDRESS describes similarities in flower structure between his basal *Menispermaceae/Lardizabalaceae* on the one hand and *Winteraceae* and *Schisandraceae* on the other hand. KUBITZKI, however, explicitly excludes *Schizandraceae* as a possible close relative of the *Ranunculiflorae*. An affinity to the "paleoherbs" is considered by KUBITZKI, and suggested by CARLQUIST on the basis of similarities in wood anatomy between *Ranunculiflorae* and *Piperales* (*Aristolochiaceae, Lactoridaceae, Piperaceae*). Finally, the possibility of a close relationship of *Euptelea* to the *Ranunculiflorae* (HOOT & CRANE, CARLQUIST) might indicate an affinity of the group to the "lower Hamamelids". Not much agreement was reached on interfamilial relationships or basal and derived taxa in the *Ranunculiflorae*. Virtually all major families (*Lardizabalaceae* and/or *Menispermaceae*: BEHNKE, BRÜCKNER, CARLQUIST, ENDRESS, LOCONTE & al.; *Papaveraceae* s.l.: HOOT & CRANE; *Ranunculaceae*: RONSE DECRAENE & SMETS) are considered as basal by the different authors, and diverse sister group relationships between families are proposed. Strong morphological similarities between, e.g., *Menispermaceae/Lardizabalaceae* and *Berberidaceae/Papaveraceae* (ENDRESS) are not reflected in some of the molecular phylogenies presented (HOOT & CRANE), although in part in others (OXELMAN & LIDÉN). The sequence of families presented in this volume follows the topology suggested by HOOT & CRANE.

The question raised by KUBITZKI, whether *Ranunculiflorae* are mono-or paraphyletic, also has been answered differently. Most contributions assume, often implicitly, monophyly. LOCONTE & al., however, regard the *Ranunculiflorae* as paraphyletic in relation to the *Dilleniales*. Without intending to formulate a phylogenetic hypothesis, RONSE DECRAENE & SMETS see the androecium of the *Ranunculiflorae* and notably the *Ranunculaceae* to foreshadow structures found in the Dilleniids, Rosids and Asterids.

Opinions about phylogenetic relationships within **Papaveraceae** s.l. (incl. *Pteridophyllum, Hypecoum, Fumariaceae*) differ widely. In their morphology-based phylogenetic analysis LOCONTE & al. obtain the topology *Platystemonoideae* (incl. the papaveroid *Canbya*), *Papaveroideae, Eschscholzioideae, Glaucioideae, Chelidonioideae, Pteridophylloideae, Hypecoideae* and *Fumarioideae*, whereas KADEREIT & al., also based on morphological characters, present and interpret morphologically, geographically and ecologically an almost inversed pattern of relationships (*Pteridophyllum, Fumariaceae + Hypecoaceae, Chelidonioideae, Eschscholzioideae, Glaucium/Dicranostigma, Papaveroideae* incl. *Platystemonoideae*). The somewhat separate position of the chelidonioid genera *Glaucium/Dicranostigma* found in both these studies is supported by the serological findings of SHNEYER & al. A position of *Pteridophyllum* entirely outside the *Papaveraceae* s.l. was found by OXELMAN & LIDÉN using 28S rDNA sequences. On the basis of both morphological and cpDNA data BLATTNER & KADEREIT document and compare three pairs of intercontinental disjuncts in subf. *Chelidonioideae*. A detailed analysis of subf. *Papaveroideae* based on cpDNA restriction site variation is presented by SCHWARZBACH & KADEREIT (New World taxa) and JORK & KADEREIT (Old World taxa). The New World genera are believed to have originated in a fast radiation event. Of the Old World genera, *Meconopsis* is hypothesized to represent a paraphyletic base group which gave rise to several lines of *Papaver* as well as *Roemeria* and the New World *Stylomecon*. A phylogenetic analysis of *Corydalis* based on the ITS region of nuclear ribosomal DNA leads LIDÉN & al. to the recognition of several well-supported clades within this genus which can be diagnosed with various seed characters. Another result of this study is the paraphyletic status of *Dicentra*.

Nuclear and chloroplast DNA sequences (HOOT & al.) do not provide unequivocal evidence about the position of *Sargentodoxa* in relation to the **Lardizabalaceae**. Together with *Sinofranchetia* and *Decaisnea* this genus forms the unresolved basal clade of the family. Beyond this, *Boquila, Lardizabala, Akebia, Holboellia* and *Stauntonia* are resolved as a well-supported group, of which the last three genera form an equally well-supported subclade.

It is still unclear whether **Hydrastis** should be included or excluded from the *Ranunculaceae.* Inclusion is favoured by TAMURA and KOSUGE & al., and a close relationship between these two is seen by OXELMAN & LIDÉN and HOOT and JOHANSSON. LOCONTE & al., however, assign a position remote from the *Ranunculaceae* to it, and CARLQUIST regards its wood anatomy as very different from that of the *Ranunculaceae*.

Equally, no consensus was reached on **Circaeaster** and **Kingdonia**. Evidence for a close relationship between these two genera, which is denied by TAMURA, is presented by HOOT & CRANE, OXELMAN & LIDÉN and LOCONTE & al. Only LOCONTE & al., however, include them in the *Ranunculaceae*. A sister group relationship to other families of the *Ranunculiflorae* is suggested by HOOT & CRANE and OXELMAN & LIDÉN.

Although many different characters have been considered for the classification and an appropriate description of the phylogenetic relationships within the **Ranunculaceae**, particularly the molecular data included in this volume further improve the understanding of this family. Both chloroplast (*atp*B, *rbc*L, restriction site variation) and nuclear (rDNA, *adh*, legumin) characters have been analysed by HOOT, JENSEN, JOHANSSON, KOSUGE & al. and LANG & FISCHER. To our knowledge no similarly broad molecular data set has ever been used for the reconstruction of the phylogenetic relationships within a family of flowering plants. The high degree of congruence among the molecular trees (JENSEN & al.), and between molecular and non-molecular data demonstrates the utility of molecular markers in this family. As a result, a more reliable classification of the *Ranunculaceae* is presented (JENSEN & al.). Several of the molecular data sets analysed suggest that the *Coptidoideae* and *Isopyroideae* (both with T-type chromosomes) are basal to the *Ranunculoideae* (R-type chromosomes). The same conclusion was drawn by LOCONTE & al. on the basis of non-molecular data. These authors regard T-type chromosomes and accumulation of benzylisoquinoline alkaloids as plesiomorphic characters. As far as investigated, the T-type chromosome groups are also characterized by the production of the highly unusual 18:3 – fatty columbinic acid (AITZETMÜLLER). The *Cimicifugeae* (with R-type chromosomes) share relatively small embryos with the T-type chromosome groups (ENGELL). For the R-type chromosome genera, the molecular data provide evidence for a position of *Eranthis* close to the

Cimicifuginae, and of *Nigella* – an account of the reproductive biology of *Nigella* is presented by WEBER – close to the *Delphiniinae*. JENSEN & al. also postulate a common origin for *Trollius* and *Adonis*. This latter relationship has not been detected by non-molecular data analysed either cladistically (LOCONTE & al.) or numerically (NIKOLIC).

Flower (ENDRESS, RONSE DECRAENE & SMETS) and pollen morphological (BLACKMORE & al.) characters vary considerably in the family, and their assessment as plesiomorphic or apomorphic remains difficult. Some suspected synapomorphies, however, can be used for the aggregation of taxa. Other characters studied are rather uniform. These include epicuticular wax tubules (BARTHLOTT & THEISEN) and sieve-element plastids (BEHNKE).

The reproductive biology of some species of *Ranunculus* s. str. has been studied by STEINBACH & GOTTSBERGER. DAHLGREN identifies polyploidy and hybridization, environmental adaptation, and many other processes to be responsible for the present-day pattern of variation in *Ranunculus* subg. *Batrachium*.

In *Anemone*, seed fat composition (AITZETMÜLLER) and morphological and molecular (HOOT, EHRENDORFER) variation was studied. The incongruence between molecular and morphological evidence found is hypothesized to be caused by the parallel evolution of very similar morphologies in adaptation to similar environmental requirements by EHRENDORFER.

New molecular and morphological evidence contributes to an improved classification and phylogenetic understanding of the **Berberidaceae** and confirs the recognition of four major chromosomal groups (x = 10, 8, 7, 6). *Rbc*L sequence and cpDNA restriction site data (KIM & JANSEN) as well as *gap*A sequence data (ADACHI & al.) largely support earlier morphology-founded classifications recognizing the tribes *Nandineae, Leonticeae, Berberideae* and *Epimedieae*. A phylogenetic interpretation of inflorescence structure and other morphological characters, however, results in a partly different cladogram (NICKOL). *Nandina*, often considered a distinct family, should be included in the *Berberidaceae* in a basal position, perhaps together with *Caulophyllum* (KIM & JANSEN, ADACHI & al., LOCONTE & al., CARLQUIST, NICKOL). The *Leonticeae* (*Caulophyllum, Leontice, Gymnospermium*; x = 8) are molecularly (KIM & JANSEN) and phytochemically (quinolizidine alkaloids) well-circumscribed. Also the *Berberideae* (x = 7) including the woody *Berberis* and *Mahonia* and the herbaceous *Ranzania* are well-supported by morphological (LOCONTE & al.; not inflorescence morphology: NICKOL) and molecular data (KIM & JANSEN, ADACHI & al.). Within the large *Epimedieae* (x = 6), *Jeffersonia* diverges in wood anatomy (CARLQUIST), and apparently occupies a basal position (KIM & JANSEN). *Diphylleia, Podophyllum* and other closely related genera are phytochemically peculiar by containing lignans instead of benzylisoquinoline alkaloids and probably form a separate clade within the *Epimedieae*. Their epicuticular waxes differ considerably from the remainder of the tribe (BARTHLOTT & THEISEN).

We believe that the 38 contributions shortly summarized above have improved our understanding of the systematics and evolution of the *Ranunculiflorae* significantly, and hope that the present volume will be of interest to a large part of the systematic community.

Finally, we wish to acknowledge financial support of the symposium by the Deutsche Forschungsgemeinschaft (DFG) and the government of the Freistaat

Bayern, thank all those at Bayreuth and Mainz who helped organize the symposium and complete this volume, and express our gratitude to the reviewers of the written versions of all contributions.

Bayreuth and Mainz, March 1995   *U. Jensen*
                                                                           J. W. Kadereit

Contents

Listed in Current Contents

Pl. Syst. Evol. [Suppl.] 9: 1–10 (1995)

Ranunculiflorae – delimitation, phylogeny, diversification

K. KUBITZKI

Received September 11, 1994

Key words: *Ranunculanae*–Ranalean families, eudicot hypothesis, tricolpate pollen.

Abstract: Within the Ranalean complex the *Ranunculiflorae* form a natural entity that in all modern classifications re-appears more or less in the same circumscription, although at different rank and under different names. The *Ranunculiflorae* are so distinct from woody *Ranales* (*Magnoliiflorae*) that a higher (superordinal) taxonomic rank appropriately expresses their relative isolation and individuality. On morphological and phytochemical grounds, several families that at one time have been assigned to this alliance, have to be excluded from it, such as the *Nymphaeaceae* (but not *Nelumbo*), *Nepenthaceae,* and *Paeoniaceae*. Phylogenetic links of the *Ranunculiflorae* with magnolialean families may exist but are probably less tight than often assumed; the *Illiciales* are no vital link between the two major groups. In contrast to the "eudicot" concept it is argued that tricolpate pollen in the *Ranunculiflorae* originated independently from non-magnolialean dicotyledons. Attention is called to different themes of diversification in several families of the *Ranunculiflorae*.

The superorder *Ranunculiflorae* is a very natural alliance, comprising the *Berberidaceae, Lardizabalaceae, Menispermaceae, Ranunculaceae, Papaveraceae*, and some segregate families. Among them, the *Ranunculaceae* have been favourite subjects of botanical research and teaching because they are widely distributed and readily available in the northern hemisphere. Due to their predominantly herbaceous growth they also have often been used in biosystematic experiments. The study of their showy flowers has yielded interesting insights into their reproductive biology; their incompatibility system has been analyzed by geneticists; and their chromosomes have been important criteria in the subdivision of the family. The wealth of biodynamically active principles of various families of the *Ranunculiflorae* has been used by indigenous people, studied by natural compound chemists, and evaluated by chemosystematists; their proteins have been used in serological studies; the wood anatomy of their ligneous members has been studied in depth; the pollen morphology of various families has been carefully investigated, and phylogenetic relationships within this complex have been analyzed.

In spite of all accumulated knowledge about this plant group, many questions connected with it have remained unresolved or have emerged from recent studies. The tools of macromolecular analysis now available open the way for answering some of them with much greater precision than conventional approaches permitted.

Therefore it is pertinent, as an introduction to this symposium, to formulate some of the systematic problems inherent in this plant group.

Circumscription of *Ranunculiflorae*

In DAHLGREN's (1980) classification the Ranunculiflorae form a superorder that includes the orders Ranunculales and Papaverales and the families mentioned above. Earlier (DAHLGREN 1975), and again later (see the posthumously published classification in DAHLGREN 1989), DAHLGREN preferred the name *Ranunculanae*. TAKHTAJAN (1969, 1980, 1987) assigned to this alliance the status of a subclass named *Ranunculidae*. CRONQUIST (1981), although applying the subclass concept as well, merged TAKHTAJAN's Ranunculidae into his subclass *Magnoliidae*. Similarly THORNE (1982) included the contents of the *Ranunculales* and *Papaverales* within a single order, *Berberidales*, which forms part of his superorder *Magnolianae*. Thus the contents of the *Ranunculidae* re-appears in all modern classifications, although at different rank and under different names, but in more or less the same circumscription.

A first approximation of the group of families in question can be found in the first volume of A. P. DE CANDOLLE's (1824) Prodromus. The sequence of families of his first subclass, *Thalamiflorae*, starts with the *Ranunculaceae* and is followed by various families of present-day *Magnoliales, Ranunculales, Papaverales*, and, obviously for their polystaminate and apocarpous condition, by the *Dilleniaceae*. BARTLING (1830) emphasized the apocarpous condition by coining for this alliance the name "*Polycarpicae*" that is still in use particularly in German textbooks. Later authors variously added the waterlilies and removed the *Papaveraceae*, but the decisive step forward was taken by H. HALLIER (1905), who made the fundamental distinction between the woody "*Polycarpicae*" with spherical ethereal oil cells and the predominantly herbaceous "*Ranales*" without such idioblasts. This division foreshadowed TAKHTAJAN's (1969) division of the Ranalean families into *Magnoliidae* and *Ranunculidae*. Apart from the *Ranunculaceae, Berberidaceae* and *Menispermaceae* HALLIER included also the *Papaveraceae, Nymphaeaceae, Ceratophyllaceae* and *Podostemonaceae* in his *Ranales*. Some years before HALLIER (1896) had been one of the first to defend the Ranalean hypothesis that for the first time seems to have been mentioned, although very briefly, by BESSEY (1893). This hypothesis received support by the influential papers of ARBER & PARKIN (1907) and of BESSEY (1915), and decades later by the work of BAILEY and his associates.

The Englerians, to which HALLIER stood in opposition, remained unaffected by these suggestions and up to the last edition of the "Syllabus" (MELCHIOR 1964) continued with a sequence of dicotyledonous orders ("Reihen") in which the Ranalean families were preceded by apetalous and homochlamydeic orders. While ENGLER had recognised the significance of ethereal oil cells as early as 1897 and had used them for characterising his suborder ("Unterreihe") *Magnoliineae*, WETTSTEIN's (1935) treatment of the *Polycarpicae* ignored this important character.

Another key character equally important in the classification of the Ranalean families, the distinction between unisulcate and tricolpate pollen grains, was pointed out by WODEHOUSE (1936) and BAILEY and his associates in the 1940s. In post-war classifications a broad consolidation of the concept of the *Ranunculanae* or

Ranunculidae took place, irrespective of differences in the name and rank of this alliance (as subclass, superorder, or order) and of the circumscription of component families.

All modern classifications of the *Ranunculanae* agree in the exclusion of the *Nymphaeaceae* in the strict sense (i.e., except *Nelumbo*). Nearly all characters the *Nymphaeaceae* have in common with the *Ranunculanae*, such as polymerous or trimerous flowers, varying degrees of apocarpy, and unisulcate(-derived) pollen grains, are plesiomorphic, but the geologically early appearance of the *Nymphaeaceae*, their peculiar seed characters and aberrant secondary chemistry place them apart from the whole magnolialean (Ranalean) clade (KUBITZKI 1972); even in the modern concept of palaeoherbs (DONOGHUE & DOYLE 1989) they are strange bedfellows. This does not apply to *Nelumbo*, whose ranunculalean affiliation is increasingly appreciated. It is true that the benzylisoquinoline alkaloids of *Nelumbo* are of a rather primitive type characteristic of woody magnoliids. The possession of nuclear endosperm development also points to magnoliids rather than ranunculids; in the latter group this endosperm character is restricted to the *Lardizabalaceae*. The tricolpate pollen of *Nelumbo*, however, is the characteristic ranunculalean advance and to my mind leaves little doubt about the correct inclusion of *Nelumbo* in the ranunculids. Otherwise, the unnecessary assumption of an twofold origin of tricolpates in the Ranalean complex had to be made.

Two other controversial genera included in TAKHTAJAN's (1987) *Ranunculidae* and placed in the neighbourhood of THORNE's (1992) *Berberidales*, respectively, are *Paeonia* and *Glaucidium*. From the time of DE CANDOLLE (1824), *Paeonia* has been considered a member of the *Ranunculaceae* by many authors until the growing understanding of multistaminate androecia and particularly the distinction between primary and secondary polyandry made this untenable (HIEPKO 1965). In *Glaucidium*, stamen development and embryology hardly appear compatible with a Ranalean affinity (TAMURA 1972, TOBE 1981). Although most authors presently place these two genera in the proximity of the *Dilleniaceae*, their relationship with the latter family does not appear particularly close.

LINDLEY (1846) may have been the first to compare the *Sarraceniaceae* with ranunculid families, certainly for their spirocyclic and multistaminate flowers and similarities with the *Papaveraceae*, and ENGLER (1897) created an order *Sarraceniales* to which he assigned the *Sarraceniaceae*, *Nepenthaceae*, and *Droseraceae*. WETTSTEIN (1935) included all insectivorous pitcher plants, *Sarraceniaceae*, *Nepenthaceae*, and *Cephalotaceae*, in his *Polycarpicae*; it is beyond dispute, however, that the latter are saxifragalean. The two remaining families, *Sarraceniaceae* and *Nepenthaceae*, have gone through a real odyssey during the last decades. In TAKHTAJAN's (1987) and DAHLGREN's (1989) classification this odyssey ended with the assignment of the two families (and of the *Droseraceae*, which often, and with justification, have been considered as closely related with *Nepenthes*) to three different superorders. THORNE (1976, 1992) attached the *Sarraceniaceae* and *Nepenthaceae* to the *Theales*, although not in close proximity, while in 1976 he included the *Droseraceae* in the *Rosales* and 1992 in the *Saxifragales*. CRONQUIST (1981) classified all three families in the order *Nepenthales*.

In spite of these uncertainties about the relationships of these families all evidence now available confirms that none of them is ranunculalean. The *Sar-*

raceniaceae certainly are not related with *Nepenthaceae* and *Droseraceae*, as indicated both by morphological (DEBUHR 1975) and chemical (iridoids) evidence. Several important characters, known in part since the time of MARKGRAF (1955) and SCHMID (1964), link the *Droseraceae*, *Nepenthaceae*, *Ancistrocladaceae*, and *Dioncophyllaceae*. This is corroborated by the presence of naphthoquinones of the plumbagin and 7-methyl juglone type in all four families, and the presence of ancistrocladine alkaloids in the latter two (HEGNAUER 1990): where one of the four families goes, the others must follow.

Seemingly all out of the blue, CRONQUIST (1968) attached the families *Coriariaceae*, *Corynocarpaceae*, and *Sabiaceae* to his order *Ranunculales*. The problem with these families is that no close relatives are known for them. Therefore, their affiliation with the *Rutales*, *Celastrales*, and *Sapindales*, respectively, has never been fully satisfactory. In 1981 CRONQUIST followed other authors in transferring the *Corynocarpaceae* to the *Celastrales* (not supported by NOWICKE & SKVARLA 1983), while the *Coriariaceae* and *Subiaceae* remained assigned to his *Ranunculales*. The evidence for this placement of *Coriaria* is very weak, and the *Sabiaceae*, which have been shifted from the *Menispermaceae* to the *Sapindales* and back, may finally prove to be allied to neither family; unfortunately, their chemistry, which in view of their vague floral characters would be important, is nearly unknown.

SMITH (1972) and TAKHTAJAN (1969) in earlier versions of his system included the *Illiciaceae* and *Schisandraceae* in subclass *Ranunculidae*. Their Ranalean nature is undisputed but most authors now place them in close proximity to the *Winteraceae*. Indeed their mixture of magnolialean and ranunculalean characters, i.e., the presence of ethereal oil cells and allegedly tricolpate pollen, invited phylogenetic speculations, in which the *Illiciales* acted as a vital link between the two groups, an idea that now has been abandoned.

Phylogeny

When HALLIER (1905) made the distinction between *Polycarpicae* and Ranales, he was very explicit about the phylogeny of the latter. As ancestors of the *Berberidaceae* he proposed the *Schisandraceae*, and the *Menispermaceae* were thought as being derived from the *Lardizabalaceae* or *Schisandraceae*. I was unable to find out on which grounds these statements were made, but it is known that HALLIER tended to rely strongly on exomorphic similarity. Later, when through the work of WODEHOUSE (1935) the pollen morphology of the *Illiciales* as syn-tricolpate to syn-hexacolpate had become known, HALLIER's hypothesis seemed to make sense. Thus TAKHTAJAN (1969) stated that the *Ranunculales* were "evidently derived from the ancestors of the *Illiciales*". Recent work by LIU & YANG (1989) and DOYLE & al. (1990) has shown, however, that the pollen of the *Illiciales* very likely is an elaboration of the trichotomosulcate condition, which in turn is derived from an anasulcate stage by an extension of the radii of a trichotomous aperture. Such a transformation had been suggested by STRAKA (1963) as one possible mode for the origin of tricolpates from unisulcates (Fig. 1). Today it appears to us as a rather abnormal way leading to pseudo-tri-(or in the case of the *Illiciales* hexa-)colpate pollen elsewhere only known from palms. "True" tricolpates, in contrast, are very

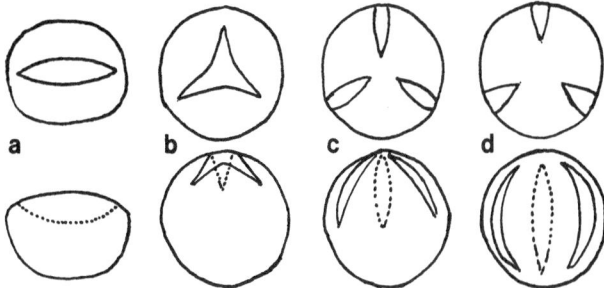

Fig. 1. Hypothetical origin of (pseudo-) tricolpate pollen (*d*) from unisulcate (*a*) pollen, with intermediate trichotomosulcate (*b*) and anacolpate (*c*) stages. Pollen in upper row seen from the distal pole, in lower low in equatorial view. From STRAKA (1963), with permission

likely to have originated from unisulcates through a rapid change in the symmetry of the tetrad (HUYNH 1976).

In view of these findings the *Illiciales* can hardly be considered as transitional to the tricolpates; they are unisulcate-derived, i.e., specialized magnoliids. Nevertheless, a derivation of ranunculids from magnoliids is generally assumed and seems acceptable. It is less certain, however, whether the woody magnoliids or the "paleoherbs" are closer to the ancestor of the ranunculids (DOYLE & DONOGHUE 1993). The strongest argument in favour of a magnolialean relationship of the *Ranunculanae* seems to me the uninterrupted structural continuity of the benzylisoquinoline alkaloids from magnoliids to ranunculids (see Table 3 in GOTTLIEB & al. 1993). Otherwise the similarity between the two groups seems to boil down to plesiomorphic, or seemingly plesiomorphic, traits such as seeds with small embryos, numerous floral parts in spiral arrangement, conduplicate carpels developing into polyspermous follicles, etc. It could be, however, that the leaf-like structure of the carpel in *Ranunculaceae* is not plesiomorphic. The increased number of floral parts both in *Magnoliaceae* and *Ranunculaceae* may be due to secondary multiplication, and some authors favour the trimerous condition, which is very frequent in magnoliids and also predominant in *Lardizabalaceae* and *Menispermaceae*, as a more basal character state. Yet the spirocyclic floral organization of *Sargentodoxa*, in which the carpels are helically arranged, would contradict such a view (see also KUBITZKI 1987b) and rather point to a spiral beginning.

The phylogenetic relationship between the *Ranunculanae* and the remaining dicotyledons with tricolp(or)ate pollen is a particularly critical issue. Are the ranunculids a blind-ending branch, or are they paraphyletic in relation to the other dicotyledons? The transition from unisulcate to tricolpate pollen grains may have been a rare event, but phylogeneticists keep silent about how often it has happened. TAKHTAJAN's (1987) scheme implies three or four independent transitions to the tricolpate condition, one on the way to the hamamelids, a second to ranunculids/caryophyllids (*Nelumbo* separate or not), and another to dilleniids/rosids.

In contrast, the transition to the tricolpate pollen appears as a unique event in the cladistic scenarios based on parsimony studies employing morphological (DONOGHUE & DOYLE 1989, LOCONTE & STEVENSON 1991, DOYLE & DONOGHUE 1993) and molecular (QIU & al. 1993) evidence; consequently, all tricolpates are

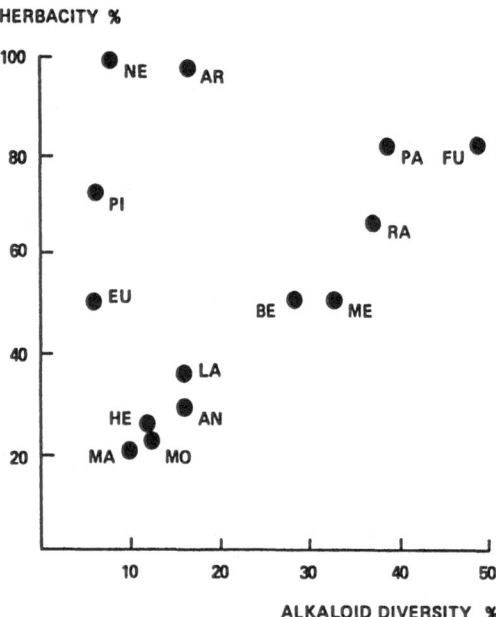

Fig. 2. Correlation between herbacity and alkaloid diversity in magnolialean families. For calculation of herbacity, see GOTTLIEB (1982, fig. 3.2). "Alkaloid diversity" refers to number of different skeletal types of benzylisoquinoline types. *AN Annonaceae, AR Aristolochiaceae, BE Berberidaceae, EU Eupomatiaceae, FU Fumariaceae, HE Hernandiaceae, LA Lardizabalaceae, MA Magnoliaceae, MO Monimiaceae, NE Nelumbonaceae, ME Menispermaceae, PA Papaveraceae, PI Piperaceae, RA Ranunculaceae.* From GOTTLIEB (1982), with permission

considered as forming the monophyletic "eudicot" branch (Fig. 3). This concept does not relate well to the view of HUBER (1990) and myself (for a summary see KUBITZKI 1993), according to which *Ranunculanae* are considered a blind-ending branch which is rooted in the magnoliids. This latter concept implies a different origin for the trochodendralean/hamamelid branch, which may have been basal to all non-ranunculalean tricolpates, and very early in angiosperm evolution was represented by tricolpates (*Trochodendrales, Platanaceae,* or still earlier representatives) (Fig. 2B). Indeed, an early divergence of non-Ranalean tricolpates from an unknown angiospermous ancestor is increasingly appreciated by palaeobotanists and may have taken place at a time when "angiosperm diversity was still low ···, and probably prior to any extensive angiosperm radiation at the magnoliid grade" (FRIIS & CRANE 1989).

Evolution

CRONQUIST (1981) has raised the interesting point that the *Ranunculales* might be primitively herbaceous and that their woody members are only secondarily woody. He was impressed by the observation that the woody members of the *Ranunculaceae* have broad medullary rays and the *Menispermaceae* have anomalous secondary thickening; he also believed that the more primitive members of the *Berberidaceae* were herbaceous rather than woody. For various reasons, however, a woody ancestry of the *Ranunculales* seems more likely to me. First, it is true that the *Menispermaceae* are generally woody or herbaceous climbers. In two African genera, however, *Penianthus* and *Sphenocentron*, the arborescent habit correlates with seemingly unspecialized traits such as pinnately veined leaves and free floral parts (DEKKER 1983), although the vessels have simple perforations. Second, in the *Ranunculaceae* and *Berberidaceae* derivatives of cinnamate including caffeic acid esters and lignans are found. This is typical of herbaceous derivatives from woody ancestors, because cinnamate, the building stone of lignin, becomes available for syntheses other than lignin in the metabolism of herbs of woody ancestry (KUBITZKI 1987a). This does not imply that evolution in the *Ranunculiflorae* followed the one direction from woody to herbaceous growth, and it is not proven that *Clematis, Bocconia,* or *Romneya* are woody relicts in otherwise herbaceous families.

Within the herbaceous lifestyle, which is so prominent among the *Ranunculanae*, there is much room for variation of characters or complexes of characters. Such variation concerns the duration of the life cycle, growth forms and pollination systems, but also more subtle traits such as the reproductive system and secondary metabolites. In the *Fumariaceae*, e.g., enormous variation affects the underground organs, which range from fleshy to fibrous, from unbranched to branched, monopodial to sympodial, tuberous and rhizomatous (RYBERG 1960), and this variation can clearly be understood in an ecogeographic context. Among secondary metabolites, in the predominantly herbaceous families benzylisoquinoline alkaloids have undergone an enormous diversification (Fig. 2). Also diterpenes, recently recognised as characteristic attributes of herbaceous taxa (FIGUEIREDO & al. 1995), are prominent in the *Ranunculaceae*, in which in addition to diterpenes of common types a great variety of

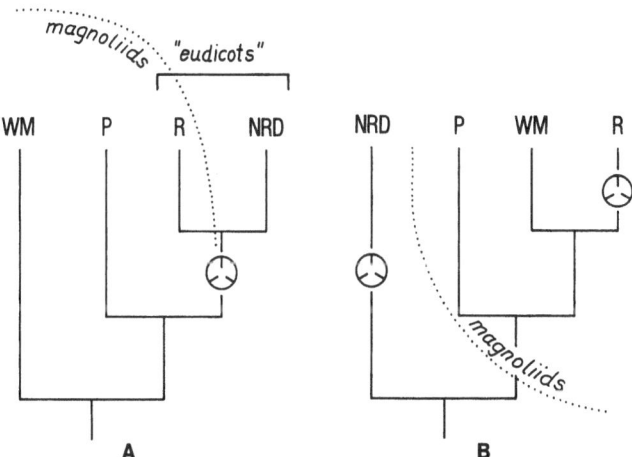

Fig. 3. Hypothetical phylogenetic relationships between ranunculalean and non- ranunculalean dicotyledons, with indication of postulated origin of tricolpate pollen. *A* one of the scenarios supported by numerical parsimony analyses based on morphological and DNA sequence data (alternatively, the eudicot branch may be closer to woody magnoliids); *B* Scenario implicit in HUBER's (1990) and KUBITZKI's (1993) proposal. For further explanation see text. WM woody magnoliids (*Magnolianae*); P paleoherbs; R ranunculids (*Ranunculanae*) NRD (tricolpate) non-ranunculalean dicotyledons

complex atisane and aconane based diterpene alkaloids is found. It seems that the herbaceous lifestyle in compensation for the loss of lignin and condensed tannins requires an increasing investment in other allelochemical substances.

A most notable diversification is also known from micromorphological characters, which has been documented for the exine structure and sculpture of pollen grains in *Berberidaceae* (NOWICKE & SKVARLA 1981), Menispermaceae (THANIKAIMONI 1986), and other families. We remain ignorant about how far this variation is the product of natural selection, how far it follows an inner, "nomothetic" trend (VAN CAMPO 1976), and how far it is mere "evolutionary noise".

It is certain, however, that most of these different morphological, reproductive, and developmental characteristics are not distributed at random but are linked in meaningful combinations. Their analysis would permit the recognition of broader relationships such as evolutionary patterns and life strategies, an important, though neglected task for botanical research. The approach needed would have to be highly integrated and certainly would have to transcend the purely analytic procedure that is employed in search for genealogical relationships, which at present attracts most attention of botanists.

References

ARBER, E. A. N., PARKIN, J., 1907: On the origin of angiosperms. – J. Linn. Soc. Bott. **38**: 29–80.

BARTLING, F. G., 1830: Ordines naturales plantarum. – Göttingen: Dieterich.

BESSEY, C. E., 1893: Evolution and classification. – Bot. Gaz. **18**: 329–333.

– 1915: The phylogenetic taxonomy of flowering plants. – Ann. Missouri Bot. Gard. **2**: 109–164.

CANDOLLE, A. P. de, 1824: Prodromus systematis naturalis regni vegetabilis, **1**. – Paris.

CRONQUIST, A., 1968: The evolution and classification of flowering plants. – Boston: Houghton Mifflin.

– 1981: An integrated system of classification of flowering plants. – New York: Columbia University Press.

DAHLGREN, G., 1989: The last Dahlgrenogram. System of classification of the dicotyledons, pp. 249–260. – In KIT TAN, (Ed.): The DAVIS and HEDGE Festschrift. – Edinburgh: University Press.

DAHLGREN, R., 1975: A system of classification of the angiosperms to be used to demonstrate the distribution of characters. – Bot. Not. **128**: 119–147.

– 1980: A revised system of classification of the angiosperms. – Bot. J. Linn. Soc. **80**: 91–124.

DEBUHR, L. E., 1975: Phylogenetic relationships of the *Sarraceniaceae*. – Taxon **24**: 297–306.

DEKKER, A. J. F. M., 1983: A revision of the genera *Penianthus* MIERS and *Sphenocentrum* PIERRE, (*Menispermaceae*) of West and Central Africa. – Bull. Jard. Bot. Natl. Belg. **53**: 17–66.

DONOGHUE, M. J., DOYLE, J., 1989: Phylogenetic analysis of angiosperms and the relationships of *Hamamelidae*. – In CRANE, P. R., BLACKMORE, S., (Eds): Evolution, systematics and fossil history of the *Hamamelidae*, 1, pp. 17–45. – Oxford: Clarendon.

DOYLE, J. A., DONOGHUE, M. J., 1993: Phylogenies and angiosperm diversification. – Paleobiol. **19**: 141–167.

HOTTON, C. A., WARD, J. V., 1990: Early Cretaceous tetrads, zonasulculate pollen, and *Winteraceae*. II. Cladistic analysis and implications. – Amer. J. Bot. **77**: 1558–1568.

ENGLER, A., 1897: Übersicht über die Unterabteilungen, Klassen, Reihen, Unterreihen und Familien der *Embryophyta* siphonogamae. – In ENGLER, A., PRANTL, K., (Eds): Die natürlichen Pflanzenfamilien. Nachträge und Register zu Teil II–IV, pp. 341–357. – Leipzig: Engelmann.

FIGUEIREDO, M. R., KAPLAN, M. A., GOTTLIEB, O. R., 1995: Diterpenes, taxonomic markers? – Pl. Syst. Evol. **195** (3/4): 149–158.

FRIIS, E.-M., CRANE, P. R., 1989: Reproductive structures of Cretaceous *Hamamelidae*. – In CRANE, P. R., BLACKMORE, S., (Eds): Evolution, systematics, and fossil history of the *Hamamelidales*, **1**, pp. 155–174. – Oxford: Clarendon.

GOTTLIEB, O. R., 1982: Micromolecular evolution, systematics and ecology. – Berlin: Springer.

KAPLAN, M. A. C., ZOCHER, D. H. T., 1993: A chemosystematic overview of *Magnoliidae*, *Ranunculidae*, *Caryophylidae* and *Hamamelidae*. – In KUBITZKI, K., ROHWER, J. G., BITTRICH, V, (Eds): The families and genera of vascular plants, II, pp. 20–31. – Berlin: Springer.

HALLIER, H., 1896: Betrachtungen über die Verwandtschaftsbeziehungen der Ampelideen und anderer Pflanzenfamilien. – Natuurk. Tijdschr. Nederl. Indie **56**: 300–331.

– 1905: Ein zweiter Entwurf des natürlichen (phylogenetischen) Systems der Blütenpflanzen. Vorläufige Mitteilung. – Ber. Deutsch. Bot. Ges. **23**: 85–91.

HEGNAUER, R., 1990: Chemotaxonomie der Pflanzen, **9**. – Basel: Birkhäuser.

HIEPKO, P., 1965: Das zentrifugale Androeceum der *Paeoniaceae*. – Ber. Deutsch. Bot. Ges. **77**: 427–435.

HUBER, H., 1990: Angiospermen. Leitfaden durch die Ordnungen und Familien der Bedecktsamer. – Stuttgart: G. Fischer.

HUYNH, K. L., 1976: Arrangement of some monosulcate, trisulcate, dicolpate, and tricolpate pollen types in the tetrads, and some aspects of evolution in the angiosperms. – In FERGUSON, I. K., MULLER, J., (Eds): The evolutionary significance of the exine, pp. 101–124. London: Academic Press.

KUBITZKI, K., 1972: Probleme der Grossystematik der Blütenpflanzen. Ber. Deutsch. Bot. Ges. **85**: 259–277.

– 1987a: Phenylpropanoid metabolism in relation to land plant origin and diversification. – J. Pl. Physiol. **131**: 17–24.

– 1987b: Origin and significance of trimerous flowers. – Taxon **36**: 21–28.

– 1993: Introduction. – In KUBITZKI, K., ROHWER, J. G., BITTRICH, V., (Eds): The Families and Genera of Vascular Plants, II, pp. 1–12. Berlin: Springer.

LINDLEY, J., 1846: The vegetable kingdom. – London: Bradbur & Evans.

LIU, H., YANG, C.-S., 1989: Pollen morphology of *Illiciaceae* and its significance in systematics. – Chinese J. Bot. **1**: 104–115.

LOCONTE, H., STEVENSON, D. W., 1991: Cladistics of the *Magnoliidae*. – Cladistics **7**: 267–296.

MARKGRAF, F., 1955: Über Laubblatthomologien und verwandtschaftliche Zusammenhänge bei *Sarraceniales*. – Planta **46**: 414–446.

MELCHIOR, H., (Ed.), 1964: A. ENGLER's Syllabus der Pflanzenfamilien, II. – Berlin: Borntraeger.

NOWICKE, J., SKVARLA, J. J., 1981: Pollen morphology and phylogenetic relationships of the *Berberidaceae*. – Smithsonian Contrib. Bot. **50**.

– 1983: Pollen morphology and the relationships of the *Corynocarpaceae*. – Taxon **32**: 176–183.

QIU, Y.-L., CHASE, M. W., LES, D. H., PARKS, C. R., 1993: Molecular phylogenetics of the *Magnoliidae*: Cladistic analyses of nucleotide sequences of the plastid gene rbcL. – Ann. Missouri Bot. Gard. **80**: 587–606.

Ryberg, M., 1960: A morphological study of the *Fumariaceae* and the taxonomic significance of the characters examined. – Acta Horti Bergiani **19**, **4**: 121–248, pl. I–XII.

Schmid, R., 1964: Die systematische Stellung der Dioncophyllaceen. – Bot. Jahrb. Syst. **83**: 1–56.

Smith, A. C., 1972: An appraisal of the orders and families of primitive extant angiosperms. – J. Indian Bot. Soc., Golden Jubilee Vol. **50A**: 215–226.

Straka, H., 1963: Über die mögliche phylogenetische Bedeutung der Pollenmorphologie der madagassischen *Bubbia perrieri* R. Cap. (*Winteraceae*). – Grana Palynol. **4**: 355–360.

Takhtajan, A., 1969: Flowering plants, origin and dispersal. – Washington: Smithsonian Institution Press.

– 1980: Outline of the classification of flowering plants (*Magnoliophyta*). – Bot. Rev. **46**: 225–359.

– 1987: Systema Magnoliophytorum. – Leningrad: Nauka.

Tamura, M., 1972: Morphology and phyletic relationship of the *Glaucidiaceae*. – Bot. Mag. (Tokyo) **85**: 29–41.

Thanikaimoni, G., 1986: Evolution of *Menispermaceae*. – Canad. J. Bot. **64**: 3130–3133.

Thorne, R. F., 1976: A phylogenetic classification of the angiospermae. – Evol. Biol. **9**: 35–106.

– 1992: Classification and geography of the flowering plants. – Bot. Rev. **58**: 225–348.

Tobe, H., 1981: Embryological studies in *Glaucidium palmatum* Sieb. et Zucc. with discussion on the taxonomy of the genus. – Bot. Mag. (Tokyo) **94**: 207–224.

van Campo, M., 1976: Patterns of pollen morphological variation within taxa. – In Ferguson, I. K., Muller, J., (Eds): The evolutionary significance of the exine, pp. 125–137. – London: Academic Press.

Wettstein, R. V., 1935: Handbuch der systematischen Botanik, 4th edn. – Wien: Deuticke.

Wodehouse, R. P., 1935: Pollen grains. – New York: McGraw Hill.

– 1936: Evolution of pollen grains. – Bot. Rev. **2**: 67–84.

Address of the author: K. Kubitzki, Institut für Allgemeine Botanik und Herbarium der Universität Hamburg, Ohnhorststrasse 18, D-22609 Hamburg, Federal Republic of Germany.

Accepted December 14, 1994

Pl. Syst. Evol. [Suppl.] 9: 11–24 (1995)

Wood anatomy of *Ranunculiflorae*: a summary

SHERWIN CARLQUIST

Received October 20, 1994

Key words: *Ranunculiflorae, Berberidaceae, Glaucidiaceae, Hydrastis, Lardizabalaceae, Menispermaceae, Nandina, Papaveraceae, Ranunculaceae.* – Phylogenetic wood anatomy.

Abstract: Recent surveys of wood anatomy of *Bereridaceae, Glaucidiaceae, Lardizabalaceae, Menispermaceae, Papaveraceae*, and *Ranunculaceae* reveal that wood anatomy is more closely related to habits than to systematic distinctions. The families can, however, be ranked from primitive to specialized in terms of vessel element length and "F/V ratio" (length of imperforate tracheary elements divided by length of vessel elements) as well as in terms of morphology of imperforate tracheary elements. *Lardizabalaceae* – especially the genus *Decaisnea* – rank as the family of *Ranunculiflorae* with the most primitive wood features. Wood characters that are of prime importance in determining relationship of *Ranunculiflorae* to other families include vessel restriction patterns, storying, and ray type; characters of less value include helical sculpturing in vessels, axial parenchyma type, and mode of crystal occurrence. The families of *Piperales, (Aristolochiaceae, Lactoridaceae, Piperaceae)* show more numerous resemblances to the families of *Ranunculiflorae* than do families of other orders. *Paeoniaceae*, thought by some closely related to *Glaucidiaceae*, have wood unlike that of *Glaucidium* or other families of *Ranunculiflorae*. The wood of *Eupteleaceae* is similar to that of *Decaisnea*, but this resemblance may derive merely from similarity in phylogenetic level. Wood anatomy supports inclusion of *Hydrastis* in *Ranunculaceae* and *Nandina* in *Berberidaceae*.

In an attempt to provide new information on wood anatomy of the families of *Ranunculiflorae*, a series of papers leading to the present summary has been constructed. This series includes studies on *Lardizabalaceae* (CARLQUIST 1984), *Papaveraceae* (CARLQUIST & ZONA 1988, CARLQUIST & al. 1994), *Ranunculaceae* (including *Hydrastis*) and *Glaucidiaceae* (CARLQUIST, unpubl.), *Berberidaceae* (CARLQUIST, unpubl.), and *Menispermaceae* (CARLQUIST unpubl.). *Kingdonia* and *Circaeaster* have not been studied here because their xylem is too limited for comparison even to herbaceous *Ranunculiflorae* with some secondary growth. *Sargentodoxa* has not been studied in the series leading to this summary because material was not available to me. However, data on wood anatomy of *Sargentodoxa* (LEMESLE 1943) and information on pollen ultrastructure (NOWICKE & SKVARLA 1982) seem to place that genus in *Lardizabalaceae*. The series of papers on wood anatomy of *Ranunculiflorae* includes many species not hitherto studied, and includes new reports of anatomical features, such as those on storied cambia and axial parenchyma types. New information on wood anatomy of other families, such as *Piperales* (e.g., *Aristolochiaceae, Lactoridaceae, Piperaceae*) and *Chloranthales* (CARLQUIST 1992, 1993) provide data that prove vital in comparisons of *Ranunculiflorae* with other orders.

Wood is often valuable with relation to demonstration of systematic affinities of dicotyledonous genera and families, so a review of wood data of *Ranunculiflorae* is of value. If kinds of data other than those on wood give clear ideas of phylesis and relationships within *Ranunculiflorae*, however, those interpretations may help interpretations of patterns of wood evolution. The present paper is divided into three parts. The first deals with phyletic sequences of wood features within *Ranunculiflorae*. The second discusses wood features of greatest value for under-standing relationships of *Ranunculiflorae* to families of other orders, and indicates which of those relationships are most likely. In a third section, distributions of wood features that tend to define particular groups within *Ranunculiflorae* are discussed. In fact, relationships between wood anatomy and ecology or habit are often more persuasive than those that seem to link wood anatomy with systematic distinctions. Because detailed descriptions of habits and ecological preferences of taxa require much space, relationships at the generic and specific level between these features and wood character expressions are largely covered in individual papers earlier in this series.

Comparisons at the familial level with respect to habit are, however, of concern in this summary. For example, are *Ranunculiflorae* primitively lianoid, as OGANEZOWA (1975) claimed? The system of QIU & al. (1993) places the arboreal family *Eupteleaceae* near the base of the ranunculid clade, but *Papaveraceae* is lower and the remainder of the order higher, so that one might have to hypothesize a conversion from herbaceous to arboreal to lianoid; is this possible? *Ranunculiflorae* have a number of features regarded as primitive, and may be close to families that are predominantly herbaceous, such as *Piperales*; if the "paleoherb" hypothesis (TAYLOR & HICKEY 1992), based on groups such as the *Piperales* is valid, are *Ranunculiflorae* primitively herbaceous or relatively nonwoody? Even if these questions cannot be answered satisfactorily at present, the close relationship between wood anatomy and habit is evident. For example, *Cocculus trilobus* (THUNB.) DC. has only a single cambium, has libriform fibers, and has only upright and square cells in rays, in contrast to the successive cambia, tracheids, and predominantly procumbent ray cells in lianoid *Menispermaceae* (CARLQUIST, unpubl.). These contrasts are related to the sprawling subshrub habit of *Cocculus trilobus*.

Study of woods of *Ranunculiflorae* has highlighted distinctive modes of wood structure. The problem of *Jeffersonia*, which has wood features different from those of other *Berberidaceae*, is one of these (CARLQUIST, unpubl.). In some instances, (*Glaucidium, Hydrastis*), wood formulas can be used in helping to decide whether a genus should be segregated or recognized within a larger family. A distinctive wood pattern would support segregation.

The terminology, materials and methods, and raw data that support conclusions in the present paper are given in the series of papers cited above; space does not permit the inclusion of these details here. Literature is also cited more extensively in those papers.

Wood anatomy and phylogenetic sequences

Only one species of *Ranunculiflorae*, *Decaisnea fargesii* FRANCH. has long scalariform perforation plates in vessels of the secondary xylem (CARLQUIST 1984). The primitiveness of these perforation plates in underlined by the fact that lateral walls of vessels of *Decaisnea* have a primitive pitting configuration, scalariform to transitional, on intervascular contacts. The remainder of *Ranunculiflorae* have much more specialized vessel elements, and the only occurrences of primitive character states that can be cited are much more advanced than those of *Decaisnea*. Occasional perforation plates with one of three bars occur in secondary xylem of *Epimedium* and *Glaucidium*, but most perforation plates are simple. In *Hydrastis*, only simple plates plus a very few with a single bar can be seen in secondary xylem.

There are reports of scalariform perforation plates in vessels of primary xylem of some *Ranunculiflorae*, however. BIERHORST & ZAMORA (1965) list scalariform perforation plates (mixed with simple plates) in species of *Menispermaceae* and *Ranunculaceae*. Primary xylem contains a scattering of scalariform perforation plates also in *Lardizabalaceae* (SOLEREDER 1908), *Hydrastis* (LEMESLE 1948, TOBE & KEATING 1985), and *Glaucidium* (TAMURA 1972). BAILEY (1944) thought of the primary xylem as a refugium for primitive vessel features phylogenetically, and this concept appears justified on the basis of numerous studies, so that one might expect more primitive perforation plates in primary xylem than in secondary xylem in particular species.

LEMESLE (1943) figures "tracheids with areolate pits" for vessel elements he observed in *Sargentodoxa* (*Sargentodoxaceae*; included in *Lardizabalaceae* by some authors); these should be called fibriform vessel elements. Narrow vessel elements of lianas are occasionally or commonly fibriform, so occurrence of such cells in *Sargentodoxa* is not surprising.

The length of vessel elements of families of *Ranunculiflorae* is given in Table 1. These mean lengths are based on averages of mean vessel elements lengths of all species studied in the respective families of the series of papers cited earlier. The families in Table 1 have been arranged in order of descending length of vessel elements. According to BAILEY & TUPPER (1918), long vessel elements (which have virtually the same length as the fusiform cambial initials from which they were

Table 1. Quantitative wood features of *Ranunculiflorae*[1]

	Vessel element length	F/V ratio
Lardizabalaceae	411	1.51
Lianoid *Menispermaceae*	318	3.20
Nandina	279	1.55
Papaveraceae	272	1.82
Berberis (including *Mahonia*)	246	1.74
Ranunculaceae	195	1.91
Glaucidiaceae	159	1.97

[1] Sources for data given in introduction, further explanation in text.

derived) are primitive and short ones are specialized. Of the families in Table 1, *Lardizabalaceae* have the longest vessel elements. In certain dicotyledons with particular growth forms (e.g., rosette trees that appear to have been derived from herbs), paedomorphosis occurs in wood, and one finds longer vessel elements than one would expect on the basis of degree of phyletic specialization of a particular plant (CARLQUIST 1962). However, the *Ranunculiflorae* studied, except for some with herbaceous characteristics (e.g., *Bocconia* of the *Papaveraceae* and relatively non-woody species in several families), do not possess indicators of paedomorphosis in wood to any appreciable extent. In fact, the herbaceous species of *Ranunculaceae* have shorter vessel elements than the woody species of the family. Lianoid species do not on the average have vessel elements longer than those of nonlianoid woody dicotyledons (CARLQUIST 1975). Therefore, one can use the vessel element lengths of Table 1 as accurate indicators of degree of phyletic advancement. The vessel elements of *Lardizabalaceae* are markedly longer than those of the other families, followed by *Menispermaceae*, then by *Papaveraceae, Berberidaceae,* and *Ranunculaceae.* This sequence is close to that in recent phylogenetic treatments (TAKHTAJAN 1987, THORNE 1992). The cladogram of QIU & al. (1993), based on *rbc*L evidence, is similar, except that *Papaveraceae* is accorded a basal position. Families are by no means uniform in vessel element length, and shortening of fusiform cambial initials has occurred polyphyletically. Nevertheless, the sequences of Table 1 are useful preliminary indicators of phyletic advancement.

The F/V ratio (Table 1) is another key to phyletic advancement. After origin of vessels in dicotyledons, there is progressive division of labor between vessel elements and imperforate tracheary elements. With phyletic advance, vessel elements become progressively shorter, as noted above, but intrusiveness of imperforate tracheary elements (before they form secondary walls) increases, so they become progressively longer than the vessel elements they accompany. This phenomenon of increasing division of labor, accompanied by length divergence between vessel elements (V) and the imperforate tracheary elements ("fibers," F) they accompany has been documented in detail earlier (CARLQUIST 1975). The F/V ratio does not increase indefinitely with phyletic advance: ratios above 4.0 are rare. Moreover, the increase in the ratio may be affected in some groups by factors other than phyletic advancement. The basal position of *Lardizabalaceae* in this series is notable. The high F/V ratio in *Menispermaceae* is perhaps related to the occurrence of successive cambia in that family. Except for *Menispermaceae*, the sequence of families is identical with the sequence for vessel element length. The families putatively more primitive for two categories given in Table 1 are woody, with a progressive increase in herbaceousness in the families lower on the list. This could be interpreted as evidence that *Ranunculiflorae* are primitively woody, but I do not regard the evidence as conclusive. In *Chloranthaceae* and *Piperales*, a sympodial habit appears to have preceded a monopodial habit (CARLQUIST 1992), and a similar phenomenon may have occurred in *Ranunculiflorae* as a whole or in parts of it. Certainly the habits of *Nandina* and *Mahonia* are sympodial, as are the habits of various herbaceous genera, and this growth form should be considered in phyletic analysis. The sympodial habit can occur in somewhat woody species, but is much more common in less woody *Chloranthales* and *Piperales* than in arboreal ones.

Wood anatomy and relationships of *Ranunculiflorae*

Vessel restriction patterns. Vessels in dicotyledons are not always distributed randomly in fascicular areas (areas of vertical cells) of wood. Vessels may be confined to the central portions of fascicular areas, with very few or no vessels touching rays. Such phenomena have been termed vessel restriction patterns (CARLQUIST & ZONA 1988). Vessel restriction patterns are not common in dicotyledons at large. Aside from *Ranunculiflorae*, they have been reported only from *Valeriana* of *Valerianaceae* (CARLQUIST 1983), *Launea* of *Asteraceae* (CARLQUIST 1988) and *Isomeris* of *Capparaceae* (new report). In addition, vessel restriction is newly reported for *Aristolochiaceae* and *Piperaceae* in Table 2. Vessel restriction patterns occur in all families of *Ranunculiflorae* (Table 2, column 1), although vessel restriction is not conspicuous in all species. In *Lardizabalaceae*, I was able to observe it clearly only in *Boquila trifoliata* DECAISNE. Although my materials of *Menispermaceae* did not reveal vessel restriction clearly, the phenomenon was evident in that family to METCALFE & CHALK (1950: 56), who state, "Vessels ⋯ seldom in contact with the rays." In *Glaucidium*, vessels are most conspicuously confined to central portions of fascicular areas in fiber-containing bands, less obviously confined in zones in which vessels occur within a ground tissue of axial parenchyma.

Helical sculpturing in vessels. In Table 2, column 2, forms of helical sculpturing in vessels are listed. Grooves interconnecting pit apertures (G) and helical thickenings (T) are sometimes interrelated (CARLQUIST 1988). In *Clematis* and other genera, one can see in some vessels that grooves interconnecting pit apertures are flanked by vessels. Helical sculpturing tends to be more pronounced in narrower vessels, less obvious or absent in wider vessels. Although these forms of helical sculpturing occur in numerous families of dicotyledons, their occurrence in all major families of *Ranunculiflorae* is noteworthy. Absence of helical sculpturing in vessels of *Glaucidium* is probably related to the highly mesic preference of that species; also, herbs tend to have less helical sculpturing (e.g., it is lacking in *Delphinium* and *Hydrastis*) than in woody representatives of a family. The widespread occurrence of helical sculpturing in vessels of *Ranunculiflorae* is probably a better indicator of ecology than of relationship, because this feature has been found mostly in species of habitats with marked dry or cold seasons (CARLQUIST 1975).

Imperforate tracheary elements. Table 2, column 3 lists distribution of types of imperforate tracheary elements in *Ranunculiflorae* and some other families (data on the latter from CARLQUIST 1990, 1992, 1993). One can see that tracheids occur in all but one species of *Menispermaceae*, in some *Lardizabalaceae*, and in *Jeffersonia* of the *Berberidaceae*. Tracheids are generally interpreted as the primitive type of imperforate tracheary element; fiber-tracheids are considered intermediate in specialization, and libriform fibers most specialized (BAILEY 1944; CARLQUIST 1975, 1988). Therefore, one might expect tracheid presence is a plesiomorphy in the genera that have them. There is another possibility, however; tracheids might result from an evolutionary diversification of fiber-tracheids, a process termed fiber-tracheid dimorphism (CARLQUIST 1988: 109). The fiber-tracheids themselves do not, of course, diverge: the genetic information for fiber-tracheid formation is modified so that a plant produces both fiber-tracheids and tracheids. As a result of fiber-tracheid dimorphism, a wood would contain both tracheids (which are vasicentric in this

Table 2. Qualitative wood features of *Ranunculiflorae*. [1] Data from CARLQUIST papers cited in text and from NAST & BAILEY (1946); further explanation in text. Understanding of the phyletic status of a character state (e.g., plesiomorphic, synapomorphic) is essential to use of that character state as an indicator of relationship.

Abbreviations: + or a letter: presence of a feature in at least some members of a family (for families with more than one genus, parentheses indicate the feature occurs in only one genus), *O* absence of feature, *G* grooves interconnecting pit apertures, *T* thickenings, *F* fiber-tracheids, *L* libriform fibers, *T* tracheids, *VT* vasicentric tracheids, *D* diffuse, *DA* diffuse-in-aggregates, *IV* intervascular, *P* pervasive, *VS* vasicentric scanty, *MP* multiseriate and paedomorphic with cells mostly upright, *MS* multiseriate with upright cells only as sheathing cells, *U* uniseriate rays present, *AP* rhomboidal crystals in axial parenchyma, *D* druses in rays, *R* rhomboidal crystals in rays

	Vessel restriction patterns	Helical sculpturing in vessels	Imperforate element types	Axial parenchyma types	Ray histology	Storied structure	Crystal occurrence
Lardizabalaceae	+	T	F, T	VS + D	MS (+U)	+	(R)
Menispermaceae	+	G	T, (L)	D, DA, VS	MS	+	AP, R
Nandina	+	T	L + T	0	MS	+	R
Other *Berberidaceae*	+	G, T	L	(VS), 0	MS	+	R
Papaveraceae	+	T	L, VT	VS	MP	+	0
Ranunculaceae	+	G, T	L, VT	P, IV, VS	MS	+	(R)
Glaucidiaceae	+	0	L, (T)	IV, P	MS	0	0
Eupteleaceae	0	0	F	D + DA	MS, (U)	0	0
Chloranthaceae	0	0	F, T	D, DA, VS	MS, MP + U	0	0
Aristolochiaceae	+	0	F, T	D, DA	MP	+	D
Lactoridaceae	0	0	F	VS	MP	+	F
Piperaceae	–	0	F, T	VS	MP	+	0

case) plus fiber-tracheids (distal from the vessels). This situation is actually represented in *Boquila* and *Lardizabala* (*Lardizabalaceae*). If such a phylad then progressed further and lost the fiber-tracheids, a wood containing tracheids only would be evolved as an apomorphy. This might have happened in the ancestors of *Akebia, Holboellia, Sinofranchetia,* and *Stauntonia.* Alternatively, these genera might have retained tracheids from tracheid-bearing ancestors as a plesiomorphy, in which case *Decaisnea* may have shifted from tracheid to fiber-tracheid presence in response to selection for greater mechanical strength in the wood of this self-supporting shrub. The difference is relatively minor, because the shift from tracheids to fiber-tracheids (or back again) is not a huge one. The presence of libriform fibers in *Berberidaceae, Glaucidiaceae, Papaveraceae,* and *Ranunculaceae* is definitely a specialization, however. The level of these four families in imperforate tracheary elements parallels the ranking of families in the features considered in Table 1.

Nandina is more primitive than other *Berberidaceae* in retaining fiber-tracheids – but this proves to be the only difference between *Nandina* and *Berberis* in wood anatomy. The presence of vasicentric tracheids in *Clematis* can best be explained as an apomorphy, the result of a process termed vessel dimorphism (CARLQUIST 1988: 109). In vessel dimorphism, both wide and very narrow vessels are formed; among the latter are sometimes so narrow that they lack perforation plates and thus are, by definition, tracheids. Wherever this process has happened, the tracheids are in contact with vessels and thus must be termed vasicentric tracheids. The presence of tracheids in *Jeffersonia* might be a plesiomorphy, or it might be explained by fiber-tracheid dimorphism or vessel dimorphism. Further investigation of *Jeffersonia* and genera close to it is necessary to understand this. The work of KIM & JANSEN (1995) shows that *Jeffersonia* is closer to a basal position than hitherto thought. KIM (pers. comm.) considers *Jeffersonia* "definitely not the terminal group of the *Berberidaceae*." *Vancouveria*, a genus close to *Jeffersonia*, has vessels (with scalariform or pseudoscalariform lateral wall pitting) and libriform fibers.

Axial parenchyma. Axial parenchyma types are given in Table 2, column 4. The generally accepted scheme of KRIBS (1937) indicates that diffuse parenchyma is the most primitive character state, followed by diffuse-in-aggregates; the only other type present in *Ranunculiflorae*, vasicentric scanty, is considered more specialized than diffuse-in-aggregates (KRIBS 1937). All three types can occur together in some species (notably Menispermaceae). Aggregation of diffuse cells into either diffuse-in-aggregates or vasicentric axial parenchyma is apparently about equally easy to achieve genetically. However, absence of the most primitive type, diffuse, in the families of *Ranunculiflorae* other than *Lardizabalaceae* and *Menispermaceae* accords with the rankings of Table 1. Intervascular parenchyma (axial parenchyma cells scattered among vessel elements, rather than surrounding groups of vessel elements) is a relatively recently recognized type, as is pervasive parenchyma (parenchyma that comprises the background tissue of a fascicular area, or at least large tangential bands of it, to the exclusion of imperforate tracheary elements. In the wood of *Clematis*, latewood contains numerous narrow vessels but few or no libriform fibers, so axial parenchyma is inevitably intervascular in distribution – although this is an expression related to vasicentric, since all axial parenchyma cells are in contact with vessels. Pervasive parenchyma is related to herbaceous habits: It likely serves for storage in stems that overwinter.

Berberidaceae have largely departed from the remainder of *Ranunculiflorae*; vasicentric parenchyma was observed in one species each of *Berberis* and *Jeffersonia*. In all other species of the family (including *Nandina*), axial parenchyma is absent but functionally replaced by living (nucleated) libriform fibers.

Rays. Wide, tall multiseriate rays little altered during ontogeny, relatively intact extensions of primary rays, are characteristic of the families of *Ranunculiflorae* (Table 2, column 5). These contrast with multiseriate rays in most dicotyledons, in which the large primary rays are rapidly broken into smaller segments during growth of the stem or root. In most *Ranunculiflorae*, the multiseriate rays are composed of procumbent cells except for one or two layers of sheathing cells, which are upright. These rays would come closes to Homogeneous Type II of Kribs, except that the presence of upright cells – at least where most abundant, as in *Papaveraceae* – is indicative of paedomorphosis, and rays with a large proportion of upright cells would be termed Paedomorphic Type II (Carlquist 1988). Some rays, therefore, can be considered intermediate between these two types. Paedomorphic rays, as in *Cocculus trilobus* (*Menispermaceae*) as well as *Epimedium* and *Jeffersonia* of the *Berberidaceae*, *Glaucidium*, and the more nearly herbaceous genera of *Ranunculaceae*, characterize plants with less woody habits. Uniseriate rays are abundant only in *Decaisnea* and *Jeffersonia*. Because presence of both multiseriate and uniseriate rays is more primitive than presence of multiseriate rays only (Kribs 1935), the ray configuration of *Decaisnea* may be one more indication of the primitive status of that genus. In *Jeffersonia*, the rays are paedomorphic and may have another explanation. In *Delphinium*, ray areas can be abruptly converted largely to zones of fibers, leaving a series of uniseriate and biseriate rays; this process may happen in *Jeffersonia* also. Uniseriate rays are uncommon in *Lardizabalaceae* other than *Decaisnea*.

Storying. Storying of fusiform cambial initials occurs in all families of *Ranunculiflorae* other than *Glaucidiaceae*. In *Glaucidiaceae*, as in herbaceous *Ranunculaceae*, accumulation of secondary xylem may be insufficient for storying to appear; where present in dicotyledons, it appears only after an appreciable number of divisions leading to increase in girth of the cambium.

Where present, storying is often not revealed in imperforate tracheary elements, because these elongate to various degrees compared with length of fusiform cambial initials from which they were derived, and the varied nature of the elongation prevents storying from becoming evident. If numerous vessels elements are present side by side tangentially, storying is evident in them, but this is not true in some *Ranunculiflorae* (e.g., *Lardizabalaceae*). Storying is, however, quite evident in secondary phloem of *Lardizabalaceae* (and *Menispermaceae*), because the numerous sieve tube elements lie side by side tangentially and reflect the storied nature of the fusiform cambial initials well. Given the absence of storying in herbaceous species, *Ranunculiflorae* as a whole do have storied cambia, and therefore varying degrees of storied wood structure.

Crystals. Rhomboidal calcium oxalate crystals are not rare in wood of dicotyledons, but such crystals still have been reported in a minority of the genera of woody dicotyledons (see Carlquist 1988). In *Ranunculiflorae*, calcium oxalate crystals were observed in rays in all families except *Glaucidiaceae* and *Ranunculaceae*. One must remember that only a small portion of each family was surveyed, so conceivably crystals might yet be found in wood of *Ranunculaceae*. Crystals were observed in ray cells of only one species of *Lardizabalaceae* (*Stauntonia hexaphylla* DECAISNE); this example shows that we should attach more significance to presence of a feature than absence. In the case of crystals in wood, one should also take into account phloem; crystals present in phloem rays of a species do not always occur in the xylem rays, or are less common in xylem rays.

Synthesis of features indicating relationship. Wood features are of highest value in systematic significance if they are not easily evolved with respect to ecological conditions, and if they characterize relatively few families. For example, vestured pits occur almost universally in woods of *Myrtales*, but also in a small number of woods from other orders (see listing in CARLQUIST 1988). If several features of restricted distribution in dicotyledon woods are common to a group of families, the likelihood that they are valid indicators of close relationship increases accordingly. One should also be able to find suites of features (e.g., pollen ultrastructure, embryology) that link families in which wood features seem to demonstrate close relationship.

The wood features that appear most promising as indicators of relationships of *Ranunculiflorae* by virtue of the above considerations are vessel restriction patterns, ray histology, and storying. To be indicators of relationship, the character states described should be synapomorphic, and all conceivably are in *Ranunculiflorae*. Features of subsidiary importance as indicators of relationships include axial parenchyma type and distribution, crystal occurrence in wood, imperforate tracheary element type, and helical sculpturing occurrence. The character state distributions outside of *Ranunculiflorae* for the features listed as most significant point primarily to *Piperales* (*Aristolochiaceae, Lactoridaceae, Piperaceae*).

Can similarities in these character state distributions result from similarity in habit or evolutionary level? Vessel restriction patterns appear unrelated to either of these considerations (the significance of vessel restriction patterns has not as yet been elucidated). Storying is more likely to occur in dicotyledons with short fusiform cambial initials (and thus could be related to evolutionary level), but there are many groups of dicotyledons with short fusiform cambial initials but no indications of storying. The peculiar nature of multiseriate rays in *Ranunculiflorae* (cells procumbent except for 1–2 layers of sheath cells; uniseriate rays absent in most species) is not matched in very many families, and few "woody herbs" have this ray type.

In addition to the wood features, there are some remarkable similarities in bark anatomy that link *Ranunculiflorae* and *Piperales*. In inner bark of some species, one finds a cylinder of large brachysclereids (in which phloem fibers may also be included); a single large rhomboidal crystal may be found in at least some of these sclereids. From the cylinder of sclereids, sclereid plates may extend radially into the outer xylem rays of the woody cylinder. Sclereid cylinders with all of the features mentioned have been observed and figured in *Lardizabalaceae* (CARLQUIST 1984), *Menispermaceae* (CARLQUIST, unpubl.), and, in *Piperales*, *Aristolochiaceae* (CARLQUIST 1993). To be sure, all of these families are lianoid, but bark sclereids corresponding to the formula cited are not present in bark of most lianoid families of dicotyledons.

The potential significance of the wood resemblances between *Ranunculiflorae* and *Piperales* is underlined by the cladogram of QIU & al. (1993), based on analysis of *rbc*L. In that cladogram, *Piperales* occur in a clade adjacent to and basal to that including the families of *Ranunculiflorae*. This placement is pertinent because the families of *Piperales* figure prominently in the paleoherb hypothesis of TAYLOR & HICKEY (1992).

Chloranthaceae, interpreted here as the sole family of *Chloranthales*, possess fewer similarities to *Ranunculiflorae* than do *Piperales*. The rays of *Chloranthaceae* are like those of the allegedly primitive *Decaisnea* of *Lardizabalaceae*. *Chloranthaceae* have relatively tall fusiform cambial initials, so one would not expect storying in that family. Vessel restriction patterns are not thus far evident in *Chloranthaceae*. The long scalariform perforation plates, scalariform to transitional wall pitting, and fiber-tracheids of *Decaisnea* are similar to those of *Chloranthaceae*.

Eupteleaceae are placed in the clade that includes *Ranunculiflorae* according to the results of QIU & al. (1993). Although this placement is unprecedented in phylogenies based primarily on macromorphology (e.g., TAKHTAJAN 1987, THORNE 1992), it should be given consideration. The wood of *Euptelea*, which has been described by NAST & BAILEY (1946), is actually rather similar to that of *Decaisnea* or *Chloranthaceae*: long scalariform perforation plates, scalariform to transitional lateral wall pitting of vessels, fiber-tracheids present as the imperforate tracheary element type, axial parenchyma diffuse to diffuse-in-aggregates, rays Heterogeneous Type II. All of these features occur in *Decaisnea*. *Eupteleaceae* do not have crystals in ray cells, nor do they have vessel restriction patterns.

Resemblances between wood of *Decaisnea* and that of *Euptelea* can be claimed to be based mostly on similarity in evolutionary level, and one should keep this possibility in mind. LOCONTE & ESTES (1989) accord a basal position to *Decaisnea* in their cladogram of the genera of *Lardizabalaceae*. Interestingly, NAST & BAILEY (1946) cite transitions between tricolpate and polycolpate pollen as characteristic not only of *Eupteleaceae*, but also *Berberidaceae* and *Ranunculaceae*. The possibility that *Eupteleaceae* might be related to and basal to *Ranunculiflorae* needs to be examined in detail; that idea stands in contradiction to the paleoherb hypothesis of TAYLOR & HICKEY (1992).

Most workers now exclude *Glaucidium* from *Ranunculaceae* (or other families). TOBE (1981) stresses similarity of *Glaucidium* to *Paeoniaceae*. In terms of wood anatomy, however, the resemblances are very few (data on wood of *Paeonia* based on KEEFE & MOSELEY 1978). Wood of *Paeonia* characteristically has scalariform perforation plates with 1–7 bars (only 37% of perforation plates are simple); wood of *Glaucidium* has mostly simple plates, and only occasionally plates with 1–3 bars. Lateral wall pitting on vessels of *Paeonia* is transitional to alternate; lateral wall pitting on vessels of *Glaucidium* appears in part scalariform, but the pattern is probably basically pseudoscalariform (a basically alternate pattern with pits much widened laterally). Imperforate tracheary elements of *Paeonia* are tracheids; those of *Glaucidium* are libriform fibers. Axial parenchyma of *Paeonia* is apotracheal and sparse; in *Glaucidium* it is paratracheal and pervasive. Rays in *Paeonia* are narrowly multiseriate (2–8 cells wide) plus uniseriate, and have lignified cell walls; those of *Glaucidium* are tall and wide, multiseriate only, and have thin nonlignified cell walls. Thus, the resemblances in wood between *Paeonia* and *Glaucidium* are minimal. On the contrary, wood of *Glaucidium* resembles that of *Delphinium* or *Hydrastis*. *Hydrastis* has been separated as a monogeneric family from *Ranunculaceae* or *Berberidaceae* by various authors (e.g., TOBE & KEATING 1985), but KEENER (1993), reviewing all available data, concludes that *Hydrastis* is best placed in

Ranunculaceae. Wood data show that *Hydrastis* agrees with herbaceous or minimally woody *Ranunculaceae* in all features, differing only in having in secondary xylem a very small number of scalariform perforation plates with a single bar – a minor difference when one considers that scalariform plates can be found in primary xylem of a number of *Ranunculaceae* (BIERHORST & ZAMORA 1965). The concepts of KEENER (1993) and the wood data cited here agree with the treatment of HOOT (1991), who, in a cladistic analysis of *Ranunculaceae*, accords *Hydrastis* a near-basal position in the family.

Wood anatomy and familial definitions. In many respects, wood anatomy of *Ranunculiflorae* is like a continuum, with distinctive modes of structure relating to particular habits and, to a lesser extent, ecological conditions, rather than to systematic distinctions. For example, *Cocculus trilobus* differs from lianoid *Menispermaceae* by having libriform fibers rather than tracheids, vasicentric axial parenchyma rather than diffuse and diffuse-in-aggregates, and relatively narrow rays with upright cells rather than very wide, tall rays with cells procumbent except for a few sheathing cells (CARLQUIST, unpubl.). These differences seem related to habit primarily: *Cocculus trilobus* is a sprawling subshrub. In addition, the mean vessel diameter of *C. trilobus* is smaller than that of any other *Menispermaceae* in the study cited, a fact related to the dryland tropical island sites where this species grows. Yet *C. trilobus* unquestionably belongs in the same genus as large upright shrubs and scandent shrubs that become true lianas with age.

Although one should note with caution the above example, there are instances where wood features are restricted to a particular family. For example, wood of *Lardizabalaceae* is like that of *Menispermaceae* in many respects, but the occurrence of successive cambia is restricted to *Menispermaceae*, and likely successive cambia occur in the vast majority of genera and species in that family.

The genus *Berberis* (including *Mahonia*) is distinctive in that true axial parenchyma is almost entirely absent – it has been reported in one species (CARLQUIST, unpubl.), although in view of the size of the genus, the possibility that axial parenchyma may occur in more than one species is considerable. Nucleated fibers substitute functionally for axial parenchyma; there even appears to be incipient dimorphism between the fibers located distal to vessel groups and those within vessels groups; the latter fibers are somewhat more parenchymalike in pits and wall thickness.

The wood of *Nandina* is very similar to that of *Berberis*. *Nandina* has fiber tracheids plus (adjacent to rays) nucleated libriform fibers; otherwise, *Nandina* wood has no features not also found widely in *Berberis*. Differences between the genera claimed by SHEN (1954) are likely the results of misinterpretations. *Nandina* has been listed separately in Tables 1 and 2. The data given there as well as other information cited elsewhere (CARLQUIST 1995c) show that *Nandina* does not warrant segregation into a monogeneric family, although some authors recognize such a family. The presence of fiber-tracheids in *Nandina*, and the fact its values in Table 1 rank as slightly more primitive than those of *Berberis* justify a basal position for the genus within the family *Berberidaceae*, as shown in the cladistic study of LOCONTE & ESTES (1989). One is tempted to characterize the family *Berberidaceae* on the basis of *Berberis*, because wood samples of that genus are so much more readily available than wood portions of the other genera, most of which are relatively nonwoody.

However, modes of wood structure other than that shown by *Berberis* occur in the family. In particular, the wood of *Jeffersonia* is distinctive, and more intense study of wood of the subfamily (*Epimedioideae*), to which *Jeffersonia* belongs, is recommended.

Our concepts of wood anatomy of *Ranunculaceae* are influenced by the fact that *Clematis* has been the source of most wood samples studied thus far for the family. When wood of herbaceous genera with some degree of woodiness (e.g., *Delphinium*, *Helleborus*) is added, a much more diverse picture emerges. As with *Berberidaceae*, we need a survey of wood of more of the herbaceous genera, particularly those species with some appreciable degree of woodiness in rhizomes. The likelihood that *Hydrastis* should be included in *Ranunculaceae* on the basis of wood anatomy is discussed above. Although wood anatomy of *Glaucidium* shares many features with wood of *Hydrastis* and *Delphinium*, those features may be due to similarity in habit and ecology, and other features of *Glaucidium* may well be of overriding significance in segregating it as a monogeneric family.

Woods of *Papaveraceae* are separable from those of other *Ranunculiflorae*– particularly those with more specialized xylem (*Papaveraceae* all have libriform fibers). The juvenilism in rays of *Papaveraceae* is much more pronounced than in other families of *Ranunculaceae*, although one can cite selected genera from those families that are like *Papaveraceae* in ray structure (e.g., *Epimedium*, *Xanthorhiza*). Adaptation of the various genera and species of *Papaveraceae* to particular habits and ecological conditions (especially xeric conditions) is of overriding importance in explaining the xylary character states of the component genera of the family.

References

BAILEY, I. W., 1994: The development of vessels in angiosperms and is significance in morphological research. – Amer. J. Bot. **31**: 421–428.

TUPPER, W. W., 1918: Size variation in tracheary cells. I. A. comparison between the secondary xylems of vascular cryptogamns, gymnosperms, and angiosperms. – Proc. Amer. Acad. Arts. Sci. **54**: 149–204.

BIERHORST, D. W., ZAMORA, P. M., 1965: Primary xylem elements and element associations of angiosperms. – Amer. J. Bot. **52**: 657–710.

CARLQUIST, S., 1962: A theory of paedomorphosis in dicotyledonous woods. – Phytomorphology **12**: 30–45.

– 1975: Ecological strategies of xylem evolution. – Berkeley, Los Angeles: University of California Press.

– 1983: Wood anatomy of *Calyceraceae* and *Valerianaceae*, with comments on aberrant perforation plates in predominantly herbaceous groups of dicotyledons. – Aliso **10**: 413–425.

– 1984: Wood and stem anatomy of *Lardizabalaceae*, with comments on the vining habit, ecology, and systematics. – Bot. J. Linn. Soc. **88**: 257–277.

– 1988: Comparative wood anatomy. – Berlin, Heidelberg: Springer.

– 1990: Wood anatomy and relationships of *Lactoridaceae*. – Amer. J. Bot. **77**: 1498–1505.

– 1992: Wood anatomy and stem of *Chloranthus*; summary of wood anatomy of *Chloranthaceae*, with comments on relationships, vessellessness, and the origin of monocotyledons. – Bull., n.s., **13**: 3–16.

– 1993: Wood and bark anatomy of *Aristolochiaceae*: systematic and habital correlations. – J. **14**: 341–357.

– ZONA, S., 1988: Wood anatomy of *Papaveraceae*. – Bull., n.s., **9**: 253–267.

– SCHNEIDER, E. L., MILLER, R. B., 1994: Wood and bark anatomy of *Argemone* (*Papaveraceae*). – J. **15**: 247–255.

HOOT, S. B., 1991: Phylogeny of the *Ranunculaceae* based on epidermal microcharacters and macromorphology. – Syst. Bot. **16**: 741–755.

KEEFE, J. M., MOSELEY, M. F. Jr., 1978: Wood anatomy and phylogeny of *Paeonia* section *Moutan*. – J. Arnold Arbor. **59**: 274–297.

KEENER, C. S., 1993: A review of the classification of the genus *Hydrastis* (*Ranunculaceae*). – Aliso **13**: 551–558.

KIM, Y., JANSEN, K., 1995: Phylogenetic implications of chloroplast DNA variation in the *Berberidaceae*. – Pl. Syst. Evol. [Suppl.] 9: 341–349.

KRIBS, D. A., 1935: Salient lines of structural specialization in the wood rays of dicotyledons. – Bot. Gaz. **96**: 547–557.

– 1937: Salient lines of structural specialization in the wood parenchyma of dicotyledons. – Bull. Torrey Bot. Club **64**: 177–186.

LEMESLE, R., 1943: Les trachéides à ponctuations aréolées du *Sargentodoxa cuneata* REHD. et. WILS. et leur importance dans la phylogénie des Sargentodoxacées. – Bull. Bot. Soc. France **90**: 104–107.

– 1948: Position phylogénétique de l'*Hydrastis canadensis* L. et du *Crossosoma californicum* NUTT., d'après les particularités histologiques du xylème. – Compt. Rend. Séances Acad. Sci. **227**: 221–223.

LOCONTE, H., ESTES, J. R., 1989: Phylogenetic systematics of *Berberidaceae* and *Ranunculales* (*Magnoliidae*). – Syst. Bot. **14**: 565–579.

METCALFE, C. R., CHALK, L., 1950: Anatomy of the dicotyledons. – Oxford: Clarendon.

NAST, C. G., BAILEY, I. W., 1946: Morphology of *Euptelea* and comparison with *Trochodendron*. – J. Arnold Arbsor. **27**: 186–192.

NOWICKE, J. W., SKVARLA, J. J., 1982: Pollen morphology and the relationships of *Circaeaster*, of *Kingdonia*, and of *Sargentodoxa* to the *Ranunculales*. – Amer. J. Bot. **69**: 990–998.

OGANEZOWA, G. G., 1975: On the evolution of life forms in the family *Berberidaceae* s.l. – Bot. Ž. **60**: 1665–1675.

QIU, Y.-L., CHASE, M. W., LES, D. H., PARKS, C. R., 1993: Molecular phylogenetics of the *Magnoliidae*: cladistic analyses of nucleotide sequences of the plastid gene *rbc*L. – Ann. Missouri Bot. Gard. **80**: 587–606.

SHEN, Y.-F., 1954: Phylogeny and wood anatomy of *Nandina*. – Taiwania **5**: 85–92.

SOLEREDER, H., 1908: Systematic anatomy of the dicotyledons [translated by L. A. BOODLE & F. E. FRITSCH]. – Oxford: Clarendon.

TAKHTAJAN, A., 1987: Systema Magnoliophytorum. – Leningrad: Nauka.

TAMURA, M., 1972: Morphology and phyletic relationship of the *Glaucidiaceae*. – Bot. Mag. (Tokyo) **85**: 29–41.

TAYLOR, D. W., HICKEY, L. J., 1992: Phylogenetic evidence for the herbaceous origin of angiosperms. – Pl. Syst. Evol. **180**: 127–156.

THORNE, R. F., 1992: Classification and geography of the flowering plants. – Bot. Rev. **58**: 225–348.

TOBE, H., 1981: Embryological studies in *Glaucidium palmatum* SIEB. et ZUCC. with a discussion on the taxonomy of the genus. – Bot. Mag. (Tokyo) **94**: 207–224.

– KEATING, R.C., 1985: The morphology and anatomy of *Hydrastis* (*Ranunculales*): systematic reevaluation of the genus. – Bot. Mag. (Tokyo) **98**: 291–316.

Note: The three papers cited as Carlquist (unpubl.) are being published concurrently in the journal Aliso.

Address of the author: Prof. SHERWIN CARLQUIST, 4539 Via Huerto, Santa Barbara, CA 93110-2323, USA.

Accepted December 20, 1994

Pl. Syst. Evol. [Suppl.] 9: 25–37 (1995)

Sieve-element plastids, phloem proteins, and the evolution of the *Ranunculanae*[1]

H.-Dietmar Behnke

Received September 30, 1994

Key words: *Ranunculanae, Papaverales, Ranunculales, Nelumbonales.* – Sieve-element plastids, phloem-proteins, ultrastructure, systematics.

Abstract: The sieve-element plastids of 85 species of the *Ranunculanae* were investigated by transmission electron microscopy. All contained form-Ss plastids. The average diameter of the sieve-element plastids from stem phloem is with 1.62 µm rather large, both compared to other superorders and to the dicotyledon average. Among the five families recognized in the *Ranunculales* (incl. *Papaverales*), *Lardizabalaceae* and *Menispermaceae* contain the largest average diameters (above 1.9 µm) while those of *Berberidaceae, Ranunculaceae,* and *Papaveraceae* are about 1.5 µm. Within the families there is considerable size variation which permits the inclusion of segregate taxa, even if their plastid diameter is much different from the family average (e.g., *Hydrastis*). *Nelumbo nucifera* (*Nelumbonales*) contains form-Ss plastids that are very similar to those of the *Ranunculales.* – Dispersive P-protein is a common feature found in the sieve elements of all taxa studied. Nondispersive protein bodies are recorded for *Clematis* and *Nelumbo* only.

The significance and reliability of sieve-element plastids as a character contributing to the classification of the flowering plants has in the past been established by the detection of specific P-type plastids confined to distinct higher taxa. Prominent examples may be found among almost every higher category such as represented by the *Monocotyledoneae, Caryophyllales,* or *Buxaceae* (see BEHNKE 1981a, 1982, 1994a). In addition, diameter measurements have been introduced to evaluate the different sizes of sieve-element plastids recorded in the more than 2500 studied angiosperm species (BEHNKE 1988, 1994a).

A preliminary ultrastructural study of the sieve-element plastids in the *Ranunculanae* (Behnke 1971) and a few additional references (see Table 1, D) covered 18 species only, all with S-type plastids. In view of their key position as one of the basic angiosperm taxa and the great diversity of sieve-element plastids in the *Magnoliidae* (BEHNKE 1988) a more complete study of the *Ranunculanae* was initiated.

---

[1] Sieve-element plastids, phloem proteins, and the evolution of the flowering plants, V. – For previous parts in this series see (IV.:) BEHNKE (1989) and (III.:) BEHNKE (1988).

Phloem (P-) proteins, dispersive and nondispersive, are a specific feature of angiosperm sieve elements not found in other plant classes. The presence of distinct P-protein bodies is often confined to clearly defined angiosperm taxa (Behnke 1991a).

Material and methods

85 species belonging to the five major families generally recognized in the *Ranunculales* and one of the two species in the unigeneric *Nelumbonales* were investigated in a total of some 100 fixations for transmission electron microscopy (TEM). Table 1 lists all species studied, their origin, and the sieve-element data used. Sieve-element plastid vouchers that are presently stored with the author, but will be placed in the Heid herbarium, contain a sample of the specimen collected, an electron micrograph of its typical plastid, and an embedded block of the respective fixation used.

Living material is a prerequisite for fixation of sieve elements and their investigation by TEM. Thin hand sections were made with a razor blade from different plant parts and immediately immersed into a sodium cacodylate buffered fixing solution containing 4% formaldehyde and 5% glutaraldehyde. Postfixation in buffered 1% OsO_4, stepwise dehydration with acetone, embedding and polymerization in epon-araldite or low viscosity epoxy mixtures were according to standard procedures (cf. Behnke 1982). Semithin sections were performed from embedded material and inspected for phloem parts, before ultrathin sections were cut with a diamond knife and subsequently stained with aqueous uranyl acetate and lead citrate.

Metaphloem sieve elements were used for the recording of plastid diameters. The measurements were taken directly from the TEM negatives and averages were calculated for each species or specific sampling from a minimum of ten different plastids. Material from different plant parts (e.g., stem, petiole, peduncle) was treated separately in the calculation of a total of some 1400 measurements for species, subfamily and family averages.

Sieve-element plastids

The investigated 85 species contain S-type plastids, all of which belong to form-Ss (for definitions see Behnke 1991b), i.e. with only starch grains, except the sieve-element plastids of *Helleborus lividus* Ait. which are almost devoid of starch (Fig. 1 m). Number, sizes and morphology of the starch grains are also rather homogeneous. Thus, the average sieve-element plastid in the stem phloem of the *Ranunculanae* has a diameter of 1.62 µm and contains between 5 and 10 medium-sized (much less than 1 µm), globular starch grains (see Figs. 1, 2) which may be surrounded by small particles (Figs. 1 b, e; 2 g, n) or, rarely, be particulate all over the plastid (Figs 1 p; 2 e, p).

Fig. 1. S-type sieve-element plastids of *Menispermaceae* (MNS), *Lardizabalaceae* (LAR) including *Sargentodoxaceae* (SRG), and *Ranunculaceae* (RAN) including *Hydrastidaceae* (HDS) from stem (*st*), petiole (*pt*), peduncle (*pd*) and rhizome (*rz*) phloem. *a Menispermum canadense* (st). *b Cocculus laurifolius* (st). *c Akebia trifoliata* (pd). *d. Decaisnea fargesii* (pt). *e. Clematis stanleyi* (st). *f Cimicifuga cordifolia* (st). *g Consolida regalis* (st). *h Myosurus minimus* (pd). *i Sargentodoxa cuneata* (st). *k Hepatica nobilis* (rz). *l Eranthis cilicica* (pd). *m. Helleborus lividus* (pt). *n Ranunculus bulbosus* (st). *o Callianthemum anemonoides* (pt). *p Hydrastis canadensis* (st); all × 20, 000.

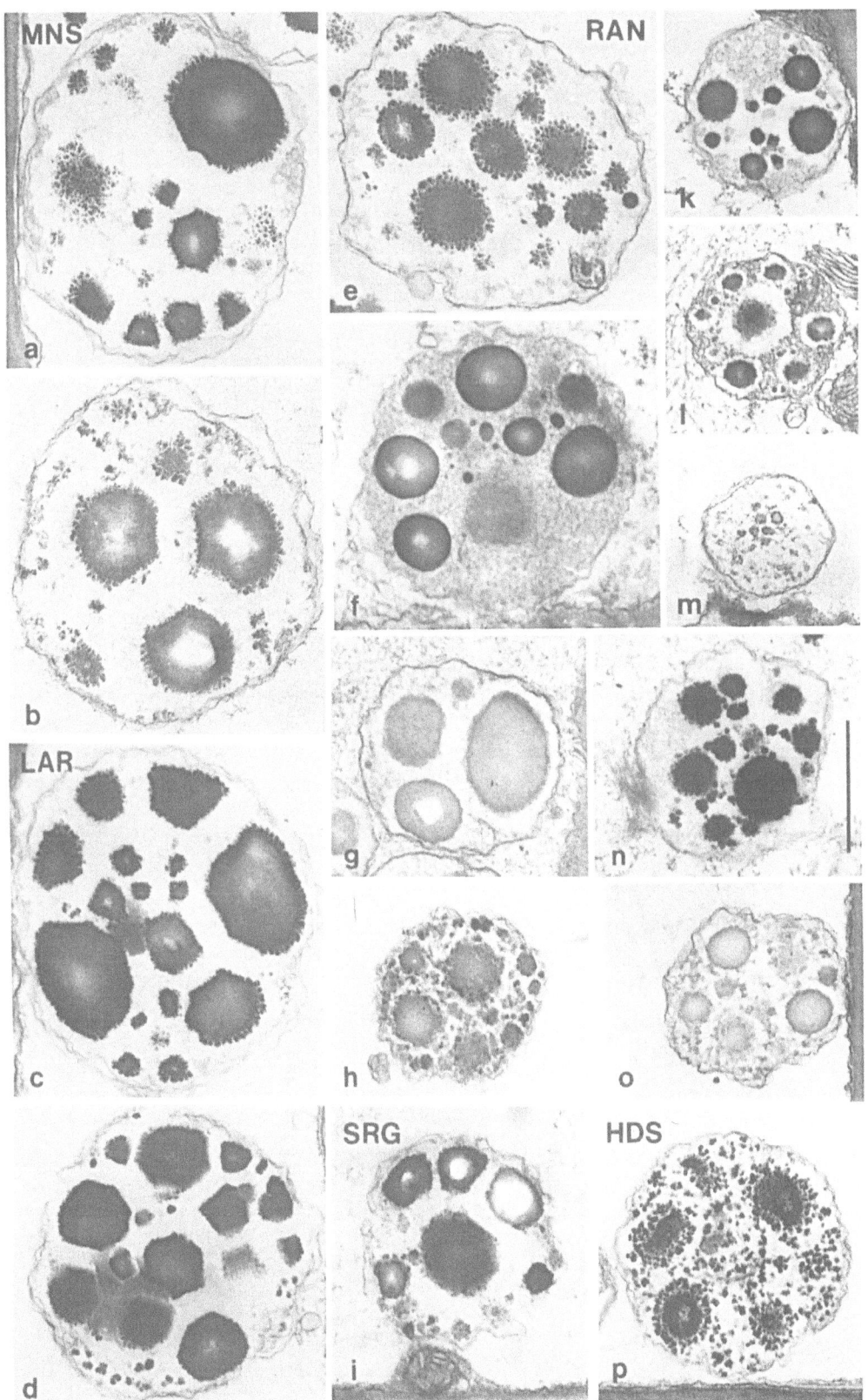

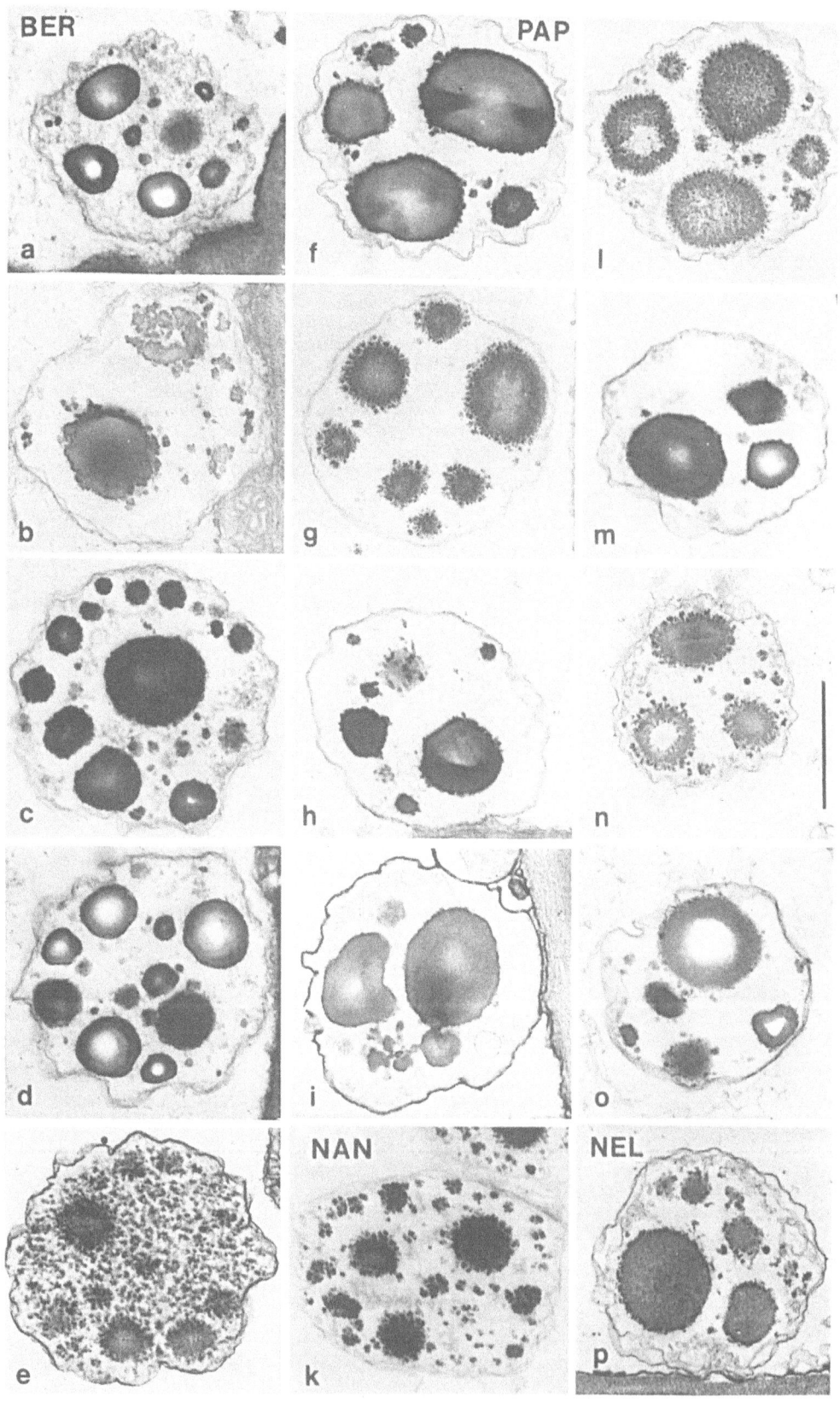

Differences are recorded for family and subfamily averages of plastid diameters. If sieve-element plastids from stem phloem are compared, those of the *Lardizabalaceae* and *Menispermaceae* have the largest and *Hydrastidoideae* the smallest diameters (Table 1). Where data are available, petiole and peduncle sieve-element plastids show the same trend.

Berberidaceae (BER; Fig. 2 a–e, k). 14 species (including *Nandina*) investigated, with rather inhomogeneous diameters of sieve-element plastids in both stem and petiole phloem. Therefore, plastid data obtained for *Nandina* would not strongly support its treatment as a separate family (TAKHTAJAN 1987).

Lardizabalaceae (LAR; Fig. 1c, d, i). 6 of the 8 genera (including *Sargentodoxa*) recognized are studied, all containing comparatively large sieve-element plastids (Table 1). Due to unsatisfactory preservation during several fixations, measurements for *Holboellia* most likely resulted in much increased values. The average diameter of the sieve-element plastids of *Sargentodoxa* (1.49 µm) is the smallest in the family, the family average (without *Sargentodoxa*) being higher than 2.0 µm.

Menispermaceae (MNS; Fig. 1a, b). The average diameters of the stem sieve-element plastids from the seven species studied are among the largest measured in the *Ranunculanae* and make the family average (1.94 µm) almost equal to that of the *Lardizabalaceae. Chasmanthera dependens* HOCHST. and *Dioscoreophyllum cummensii* DIELS do contain form-Ss plastids. Since, however, in repeated tests all plastids were burst, these taxa are not included in Table 1.

Papaveraceae (PAP; Fig. 2 f–i, l–o). 26 species from the three subfamilies are investigated. Within the same plant part, The sieve-element plastids are of rather uniform sizes, those of petioles of *Stylophorum lasiocarpum* FEDDE and *Papaver miyabeanum* TATEW. being at the extreme lower limit (see Table 1). Data from sieve-element plastids do not really favour family status of the *Fumariaceae* and *Hypecoaceae* (see e.g., TAKHTAJAN 1987, DAHLGREN 1989), actually the average size of the sieve-element plastids in the *Fumariaceae* is almost identical to that of the *Papaveraceae*.

Ranunculaceae (RAN; Fig. 1 e–h, k–p). The average plastid diameters of the 29 species studied are mainly taken from stem phloem, a few derive from peduncle and petiole phloem (see Table 1). The family average for stem and peduncle phloem is very similar (about 1.50 µm). Species averages vary from rather high values for *Ceratocephala, Cimicifuga,* and *Clematis* to others containing small plastids, like *Helleborus lividus* Ait., *Ranunculus ficaria* L. and *Hydrastis canadensis* L.

Pteridophyllum racemosum SIEB. & ZUCC. contains form-Ss sieve-element plastids with globular starch grains marginally fissured into small particles. Due to inadequate preservation prior to the fixation for TEM, diameter measurements could not be obtained.

◀──

Fig. 2. S-type sieve element plastids of *Berberidaceae* (BER) including *Nandinaceae* (NAN), *Papaveraceae* (PAP), and *Nelumbonaceae* (NEL) from stem (*st*), petiole (*pd*) and peduncle (*pd*) phloem. *a. Berberis thunbergii* (st). *b Caulophyllum thalictroides* (st). *c Podophyllum emodi* (pt). *d Mahonia aquifolium* (st). *e. Ranzania japonica* (st). *f Papaver glaucum* (st). *g Dicranostigma franchetianum* (pt). *h Bocconia frutescens* (pt). *i Argemone ochroleuca* (st). *k Nandina domestica* (pt). *l Meconopsis wallichii* (pt). *m Glaucium flavum* (st). *n Roemeria refracta* (pd). *o Adlumia fungosa* (st). *p Nelumbo nucifera* (pt), all × 20,000.

Table 1. Investigated species of the *Ranunculanae*, arranged into families and subfamilies and listing forms and average diameters (in μm) of sieve-element plastids from stem, petiole (Pet.), and peduncle (Ped.) phloem and origin of material; * refers to insufficient number of measurements, not included into family or subfamily averages. *V* Sieve-element plastid voucher; *D* Reference to previous publications (1 = Behnke 1971; 2 = Behnke 1972; 3 = Behnke 1975; 4 = Esau 1975; 5 = Behnke 1977; 6 = Behnke 1981b; 7 = Behnke 1988; 8 = Behnke 1991b)

Species	Form	Average diameters			Origin	V	D
		stem	pet.	ped.			
Lardizabalaceae	**Ss**	**1.98**					
Decaisnea fargesii Franch.	Ss		1.86		BG-BONN	V	1
Holboellia coriacea Diels	Ss	1.74			BG-K	V	
Holboellia latifolia Wall.	Ss	2.54			BG-K	V	
Stauntonia hexaphylla Decne.	Ss	1.93			BG-B	V	
Akebia quinata (Decne.) Koidz.	Ss	2.19			BG-HEID	V	
Akebia trifoliata (Thunb.) Koidz.	Ss			1.72	BG-BONN	V	1
Sinofranchetia chinensis (Franch.) Hemsl.	Ss	*			BG-MJG	V	
Sargentodoxa cuneata Rehder & Wilson	Ss	1.49			BG-NAS	V	
Menispermaceae	**Ss**	**1.94**					
Tinospora cf. tenera Miers	Ss	2.15			BG-HEID		
Stephania glabra Miers	Ss	1.74			BG-HEID		2, 5
Cocculus laurifolius DC.	Ss	2.08			BG-HEID	V	
Cocculus thunbergii DC.	Ss	2.09			BG-HEID	V	3
Menispermum canadense L.	Ss	2.19			BG-HEID		
Menispermum dauricum DC.	Ss	1.84			BG-BONN	V	1
Sinomenium acutum Rehder & Wilson	Ss	1.64			BG-BONN	V	

Taxon							
Berberidaceae	**Ss**	**1.53**	**1.72**				
Nandinoideae							
Nandina domestica THUNB.	Ss		1.70		BG-BONN	V	1
Berberidoideae	Ss	1.53	1.72				
Caulophyllum thalictroides (L.) MICHX.	Ss	1.76			BG-HEID	V	
Ranzania japonica T. ITO	Ss	1.79			BG-HEID	V	
Mahonia aquifolium NUTT.	Ss	1.28			BG-HEID	V	
Mahonia bealii (FORT.) CARR.	Ss			*	BG-BONN		1
Berberis candidula (SCHNEID.) SCHNEID.	Ss	1.31			BG-BONN		1
Berberis thunbergii DC.	Ss	1.24			BG-HEID	V	
Vancouveria hexandra (HOOK.) C. MORR. & DECNE.	Ss	1.67			BG-HEID	V	
Epimedium pinnatum FISCH.	Ss	*			BG-HEID	V	
Bongardia chrysogonum BOISS.	Ss	1.63			BG-HEID	V	
Achlys triphylla DC.	Ss		1.64		BG-MJG	V	6
Jeffersonia dubia (MAXIM.) BENTH. & HOOK. f.	Ss		1.56		BG-HEID	V	
Diphylleia cymosa MICHX.	Ss		1.76		BG-BONN	V	
Podophyllum emodi WALL.	Ss		1.92		BG-BONN	V	1
Ranunculaceae	**Ss**	**1.53**	**1.37**	**1.54**			
Helleboroideae	Ss	1.54	1.44				
Caltha palustris L.	Ss		1.75		BG-MJG	V	
Trollius europaeus L.	Ss	1.25			BG-HEID	V	
Helleborus lividus AIT.	Ss		1.12		BG-BONN	V	1

Table 1. (Continued)

Species	Form	Average diameters			Origin	V	D
		stem	pet.	ped.			
Eranthis cilicica Schott & Kotschy	Ss			1.18	BG-HEID		6
Cimicifuga cordifolia Pursh.	Ss	2.08			BG-HEID	V	
Actaea alba (L.) Mill.	Ss	1.60			BG-HEID	V	
Nigella damascena L.	Ss	*			horticultured, H.-D. Behnke 13-07-81		8
Aconitum napellus L.	Ss	1.30			BG-HEID	V	
Delphinium grandiflorum L.	Ss	1.61			BG-HEID	V	
Consolida regalis S. F. Gray	Ss	1.38	1.23		BG-HEID	V	
Ranunculoideae							
Callianthemum anemonoides Endl. ex Heynh.	Ss	1.62	1.23	1.63	BG-HEID	V	
Adonis annua L. emend. Huds.	Ss	1.53			BG-HEID	V	
Anemone baldensis Jacq.	Ss			1.61	BG-HEID	V	
Anemone hortensis L.	Ss			*	Metzovo, Greece, H.-D. Behnke 02-04-83		
Hepatica nobilis Mill.	Ss	1.34			BG-HEID	V	
Knowltonia vesicatoria Sims	Ss	1.51			BG-C	V	
Pulsatilla vulgaris Mill.	Ss			1.54	BG-HEID	V	
Clematis × jouiniana Schneid. cv. Mrs. Robert Brydon	Ss	2.03			BG-HEID	V	
Clematis recta L.	Ss	2.03			BG-BONN	V	1
Clematis stanleyi Hook. (= *Clematopsis*)	Ss	1.90			BG-U	V	
Clematis texensis Buckl.	Ss	*			BG-MJG	V	
Myosurus minimus L.	Ss			1.27	BG-HEID	V	
Ranunculus bulbosus L.	Ss	1.53			BG-HEID	V	
Ranunculus ficaria L.	Ss	1.09			Ortona, Italy, H.-D. Behnke 01-04-83	V	

Taxon	Ss				Source	
Ceratocephala falcata L.	Ss			2.10	BG-MJG	V [2]
Isopyroideae	Ss	1.40				
Semiaquilegia ecalcarata (MAXIM.) SPRAGUE & HUTCH.	Ss	1.24			BG-HEID	V
Aquilegia vulgaris L.	Ss	1.56			BG-HEID	V
Thalictroideae	Ss					
Thalictrum tuberosum L.	Ss	1.33			BG-HEID	V
Hydrastidoideae						
Hydrastis canadensis L.	Ss	1.28			BG-MJG	
Papaveraceae	**Ss**	**1.56**	**1.37**	**1.57**		
Chelidonioideae	Ss	1.63	1.39			
Chelidonium majus L.	Ss	1.81			BG-HEID	V
Eomecon chionantha HANCE	Ss		1.54		BG-MJG	V
Macleaya microcarpa (MAXIM.) PERS.	Ss	1.25			BG-MJG	V
Bocconia frutescens L.	Ss		1.62		BG-MJG	V
Stylophorum lasiocarpum FEDDE	Ss		1.01		BG-MJG	V
Dicranostigma franchetianum (PRAIN) FEDDE	Ss	1.78			BG-MJG	V
Glaucium flavum CRANTZ	Ss	1.67			Palavas, France, H.-D. BEHNKE 30-04-84	V
Eschscholzioideae	Ss	1.33				
Eschscholzia californica CHAM.	Ss	1.48			BG-B	
Hunnemannia fumariifolia SWEET	Ss	1.27			BG-C	
Papaveroideae	Ss	1.62	1.34	1.57		
Argemone mexicana L.	Ss	1.68			BG-HEID	V
Argemone ochroleuca SWEET.	Ss	1.52			BG-MJG	V
Romneya coulteri HARV.	Ss	*			BG-MJG	V

Table 1. (Continued)

Species	Form	Average diameters			Origin	V	D
		stem	pet.	ped.			
Papaver atlanticum (Ball.) Coss.	Ss			1.49	BG-BONN	V	1
Papaver californicum A. Gray	Ss			1.77	BG-MJG	V	
Papaver glaucum Boiss. & Hausskn.	Ss	1.49			BG-MJG	V	
Papaver miyabeanum Tatew.	Ss		1.07	1.67	BG-MJG	V	
Roemeria refracta DC.	Ss			1.36	BG-MJG	V	
Meconopsis wallichii Hook.	Ss		1.60		BG-MJG	V	
Stylomecon heterophylla G. Taylor	Ss	1.79			BG-MJG	V	
Fumarioideae							
Dicentra spectabilis (L.) Lem.	Ss	1.51 *			BG-HEID	V	
Adlumia fungosa (Ait.) Greene ex B.S.P.	Ss	1.47			BG-HBG	V	
Corydalis lutea L.	Ss	1.65			BG-BONN	V	1
Pseudofumaria alba (Miller) M. Lidén	Ss	1.28			BG-MJG	V	
Platycapnos spicata L.	Ss	1.46			BG-MJG	V	
Fumaria officinalis L.	Ss	1.70			BG-TUB	V	
Hypecoideae							
Hypecoum procumbens L.	Ss	1.71			Nea Michaniona, Greece H.-D. Behnke 03-04-83	V	
Nelumbonaceae							
Nelumbo nucifera Gaertn.	Ss		1.65		BG-BONN; BG-HEID	V	1,4,7

Nelumbonaceae (NEL; Fig. 2p). Previous studies have recorded form-Ss plastids in *Nelumbo nucifera* GAERTN. (BEHNKE 1971, 1988; ESAU 1975). Petiolar sieve-element plastids contain five and more globular starch grains very similar to those found in other *Ranunculanae*. The size of the plastids, corrected to 1.65 μm, also conforms with the average calculated for the superorder.

Phloem proteins

Dispersive P-protein is commonly found in the *Ranunculanae* and especially abundant in several of the investigated *Berberidaceae* and *Menispermaceae*. Nondispersive protein bodies (up to 3 in one sieve element) are only recorded for *Clematis* x *jouiniana* and *C. stanleyi* HOOK. (BEHNKE 1991a).

Nelumbo nucifera GAERTN. contains tubular P-protein and rod-shaped non-dispersive protein bodies (ESAU 1975).

P-protein is a characteristic of angiosperm sieve elements and only absent in a few families, e.g., *Poaceae*. Nondispersive protein bodies are found in only about 40 dicotyledon families (BEHNKE 1991a) and the monocotyledon *Zingiberaceae* (BEHNKE 1994b). While there are taxa (like *Malvanae*, *Violanae*, and *Rosanae*) where nondispersive protein bodies occur in most of their families, the *Ranunculanae* are among those superorders that only sporadically contain this character.

Evolution of the *Ranunculanae*

The strict homogeneity of the sieve-element plastids in the *Ranunculanae* (form-Ss plastids throughout) is paralleled by only few and their size (average of 1.62 μm within stem) unrivalled among the other dicotyledon superorders. Most of them are heterogeneous with respect to either the plastid types (i.e. contain P- and S-type plastids, e.g., *Magnolianae*, BEHNKE 1988) or plastid forms, e.g., *Caryophyllanae* (with several P-forms, BEHNKE 1994a) and *Violanae* (with form-Ss and form So-plastids, unpubl. data.).

Provided that large form-Ss plastids are the starting point from which all other plastid forms and smaller sizes are derived (see BEHNKE 1988 for supporting data and arguments), their sieve-element plastid characters place the *Ranunculanae* close to the centre of angiosperm evolution, next to or combined with the *Magnolianae* (cf. subclass *Magnoliidae* sensu CRONQUIST 1988). This view is supported by the following: (1) The superorder average of form-Ss plastids in the *Magnolianae* (about 1.5 μm) is not much different from that of the *Ranunculanae* and still significantly higher than the value for the overall dicotyledon average (about 1.3 μm); (2) six of the eight qualitatively distinct forms of sieve-element plastids are present in the two taxa combined, a diversity not found elsewhere.

Within the *Ranunculales* their large sieve-element plastids support the basic position of *Lardizabalaceae* and *Menispermaceae*. The size of form-Ss plastids in the other three families (provided that *Papaveraceae* is included and does not represent a separate order, cf. KUBITZKI & al. 1993) is much smaller and rather uniform (see Table 1).

The sieve-element plastid data of *Nelumbo* (see Table 1 and Fig. 2p) fit into either *Ranunculanae* or *Magnolianae* but do not favour an inclusion of *Nelumbonaceae* into

Nymphaeales (see Behnke 1988, but cf. Cronquist 1988).

This ultrastructural study was made possible through the expert help of Mrs Marianne Von Der Decken and Mrs Brigitte Moraw securing the success of the sophisticated TEM techniques and Mrs Doris Laupp developing and printing all the micrographs. The directors of the botanic gardens listed in Table 1 gave permission to collect the plant samples used. J. W. Kadereit and U. Hecker (Bot. Garten Mainz, FRG), K. Kramer (Bot. Garten Heidelberg, FRG), and Sheng Cheng-kui (HORTUS Bot. Nanjingensis, V. R. China) assisted the author with finding further plant material. This research was supported by the Deutsche Forschungsgemeinschaft. My sincere thanks go to all afore mentioned persons and institutions.

References

Behnke, H.-D., 1971: Sieve-tube plastids of *Magnoliidae* and *Ranunculidae* in relation to systematics. – Taxon **20**: 723–730.
– 1972: Sieve-tube plastids in relation to angiosperm systematics – an attempt towards a classification by ultrastructural analysis. – Bot. Rev. **38**: 155–197.
– 1975: The bases of angiosperm phylogeny: Ultrastructure. – Ann. Missouri Bot. Gard. **62**: 647–663.
– 1977: Transmission electron microscopy and systematics of flowering plants. – Pl. Syst. Evol. [Suppl.] **1**: 155–178.
– 1981a: Siebelement-Plastiden, Phloem-Protein und Evolution der Blütenpflanzen: II. Monokotyledonen. – Ber. Deutsch. Bot. Ges. **94**: 647–662.
– 1981b: Sieve-element characters. – Nordic J. Bot. **1**: 381–400.
– 1982: Sieve-element plastids, exine sculpturing and the systematic affinities of the *Buxaceae*. – Pl. Syst. Evol. **139**: 257–266.
– 1988: Sieve-element plastids, phloem protein and evolution of flowering plants: III. *Magnoliidae*. – Taxon **37**: 699–732.
– 1989: Sieve-element plastids, phloem-proteins and the evolution of flowering plants. IV. *Hamamelidae*. – In Blackmore, S., Crane, P. R., (Eds): Systematics, evolution and fossil history of the *Hamamelidae*, 1, pp. 105–128. – Syst. Ass. Spec. Vol. **40A**. Oxford: Clarendon.
– 1991a: Nondispersive protein bodies in sieve elements: a survey and review of their origin, distribution and taxonomic significance. – IAWA Bull. n.s. **12**: 143–175.
– 1991b: Distribution and evolution of forms and types of sieve-element plastids in the dicotyledons. – Aliso **13**: 167–182.
– 1994a: Sieve-element plastids: Their significance for the evolution and systematics of the order. – In Behnke, H.-D., Mabry, T. J., (Eds): *Caryophyllales*. Evolution and systematics, pp. 87–121. – Berlin, Heidelberg, New York: Springer.
– 1994b: Sieve-element plastids, nuclear crystals and phloem proteins in the *Zingiberales*. – Bot. Acta **107**: 3–11.
Cronquist, A., 1988: The evolution and classification of flowering plants. 2nd edn. – Bronx: New York Botanical Garden.
Dahlgren, G., 1989: The last Dahlgrenogram – system of classification of the dicotyledons. – In Kit Tan, (Ed.): The Davis and Hedge Festschrift, pp. 249–260. – Edinburgh: Edinburgh University Press.
Esau, K., 1975: The phloem of *Nelumbo nucifera* Gaertn. – Ann. Bot. (London) **39**: 901–913.

KUBITZKI, K., ROHWER, J. G., BITTRICH, V., (Eds), 1993: The families and genera of vascular plants, **2**: Flowering plants. Dicotyledons. – Berlin, Heidelberg, New York: Springer.

TAKHTAJAN, A., 1987: Systema Magnoliophytorum. – Leningrad: Nauka (in Russian).

Address of the author: H.-DIETMAR BEHNKE, Zellenlehre, Ruprecht-Karls-Universität Heidelberg, lm Neuenheimer Feld 230, D-69120 Heidelberg, Federal Republic of Germany.

Accepted November 30, 1994

Pl. Syst. Evol. [Suppl.] 9: 39–45 (1995)

Epicuticular wax ultrastructure and classification of *Ranunculiflorae*

Wilhelm Barthlott and Inge Theisen

Received November 11, 1994

Key words: Angiosperms, *Ranunculiflorae*. -Cuticle, wax, scanning electron microscopy, wax ultrastructure, systematics, taxonomy.

Abstract: Epidermal surfaces of about 199 species from 66 genera of ranunculiflorous families are examined by scanning electron microscopy. The micromorphology of their epicuticular wax crystals is presented and discussed under taxonomic aspects. All families of the *Ranunculiflorae* s. str. (*Ranunculales, Papaverales*) prove to be highly uniform: apart from a few exceptions they are characterized by the presence of clustered wax tubules (*Berberis* type), chemically dominated by the secondary alcohol nonacosan-10-ol in the species analysed. This is in marked contrast to the *Magnoliidae* s. str. (*Aristolochiales, Laurales* s. l., *Magnoliales*), which are almost uniformly defined by transversely ridged crystals of the *Aristolochia* type, chemically characterized by the presence of palmitone and the absence of nonacosan-10-ol. However, the *Canellaceae, Nelumbonaceae*, and *Winteraceae* produce the *Berberis* type tubules similar to *Ranunculiflorae*. This suggests a reconsideration of the position of these families and in particular for *Nelumbo* a close relation to the *Ranunculiflorae* also based on other chemical data. Within the polyphyletic *Hamamelididae* the "lower" families *Cercidiphyllaceae, Daphniphyllaceae*, and certain genera of *Hamamelidaceae* are also characterized by *Berberis* type waxes.

Over the last decades systematically relevant information on structure and composition of the plant cuticle based on SEM studies have been published (reviews in BEHNKE & BARTHLOTT 1983, JEFFREE 1986, BARTHLOTT 1990). The cuticles of the majority of higher plants are covered with epicuticular wax secretions. Epicuticular "waxes" occur throughout bryophytes, pteridophytes, gymno- and angiosperms. They are chemically diverse mixtures of lipophilic substances like long-chain fatty acids (true waxes), alkanes, ketones, esters, and cyclic compounds like pentacyclic triterpenes, phytosterols, and flavonoids. These substances occur usually as local projections of crystalline nature (JEFFREE & al. 1975) exhibiting surprisingly a high ultrastructural diversity. The micromorphology of epicuticular wax crystals revealed by SEM has provided valuable new criteria for the classification of angiosperms (BARTHLOTT 1993, DITSCH & BARTHLOTT 1994, FRÖLICH & BARTHLOTT 1988, HENNIG & al. 1994, THEISEN & BARTHLOTT 1994).

The monocots are characterized by the presence of two wax types: parallel oriented platelets (*Convallaria* type) within most liliiflorous families including *Triuridales*. The subclasses *Commelinidae*, *Zingiberidae*, and *Arecidae* are characterized by aggregated rodlets with longitudinal striations (*Strelitzia* type). The core group of *Magnoliidae* (*Aristolochiales, Annonales, Laurales, Magnoliales*) as well as *Paeoniaceae* are defined by the occurrence of transversely ridged rodlets (*Aristolochia* type) chemically containing a considerable amount of palmitone. *Canellaceae, Nelumbonaceae*, and *Winteraceae* share epicuticular wax features with the *Ranunculiflorae* and they are discussed below.

The data presented in the present paper are based on our comprehensive survey of *Magnoliidae, Ranunculidae*, and *Hamamelididae* (HENNIG & al. 1994), which we also refer to for material examined and methods of high resolution SEM. The conclusions are based on the examination of some 199 species of 66 genera of almost all families; they are interpreted in comparison with data of about 13.000 species of angiosperms analysed in the last 18 years. It became obvious, that both *Ranunculiflorae* s. str. and *Magnoliiflorae* s. str. each are characterized by highly specific uniform wax types – but share no common features under this aspect.

Results

Epicuticular wax crystal ultrastructure of *Ranunculiflorae*

Micromorphology of ranunculiflorous waxes. Thin wax films forming continous layers are present in many taxa: they are not detectable by SEM and under systematic aspects they are not relevant.

Within the subclass treated here only two types of epicuticular crystals (apart from a minor exception in *Podophyllum*) occur. The dominating *Berberis* type of hollow tubular waxes occurs throughout the order (Fig. 1c–h). They chemically consist in the few examples analysed (e.g., *Papaver*) of the secondary alcohol nonacosan-10-ol. The tubules are often branched, and they are frequently not uniformly arranged on the outer cuticular surface, but show a specific "clustered" distribution characteristic for this wax type (Fig. 1c–d). This is not easily detectable in some taxa.

The second wax type occurring only within certain taxa of few families are irregular platelets (Fig. 1 b) of differing size and shape, but without the specific parallel orientation of the liliiflorous monocotyledons (*Convallaria* type). These platelets occur throughout all major orders of angiosperms and they are of less systematic significance.

Systematic survey of waxes in the *Ranunculiflorae*. The circumscription of the families listed alphabetically follows CRONQUIST (1988); tentatively related families are added. Enlisted are several additional families of the *Magnoliidae* and *Hamamelididae* which produce *Berberis* type waxes. The numbers in brackets indicate the numbers of genera and species examined within the family (gen./spp.).

Berberidaceae (8/29) (incl. *Leonticaceae*): Uniformly characterized by the *Berberis* type apart from one *Epimedium* species with irregular platelets. *Podophyllum* differs significantly (see *Podophyllaceae*).

Canellaceae (2/2): *Berberis* type.
Cercidiphyllaceae (1/2): *Berberis* type.
Circaeasteraceae (1/1): *Berberis* type.
Coriariaceae (1/3): Wax crystals absent.
Daphniphyllaceae (1/2): *Berberis* type.
Fumariaceae (4/12): *Berberis* type
Glaucidiaceae (1/1): Few small irregular platelets.
Hamamelidaceae (7/12): Often wax crystals absent; rarely irregular platelets and two genera with tubules (*Berberis* type).
Hydrastidaceae (1/1): Few small irregular platelets.
Hypecoaceae (1/1): *Berberis* type.
Kingdoniaceae: No material examined.
Lardizabalaceae (2/4): *Berberis* type.
Menispermaceae (14/28): *Berberis* type.
Nandinaceae (1/1): *Berberis* type.
Nelumbonaceae (1/2): *Berberis* type.
Paeoniaceae (1/20): Transversely rigded rodlets (*Aristolochia* type).
Papaveraceae (10/21): *Berberis* type.
Platystemonaceae: No material examined.
Podophyllaceae (1/1): *Podophyllum* differs from *Berberidaceae* by exhibiting solid rodlets sometimes aggregated to larger units.
Pteridophyllaceae (1/1): Only eroded small platelets (herbarium artefact?).
Ranunculaceae (19/87): Often wax crystals absent. Irregular shaped platelets, but throughout the family also tubules of the *Berberis* type present.
Sabiaceae (1/5) (incl. *Meliosmaceae*): Wax crystals absent.
Sargentodoxaceae (1/1): *Berberis* type.
Winteraceae (3/6): *Berberis* type.

Discussion

Familiar relations and circumscriptions within the subclass. Apart from a few exceptions the *Ranunculiflorae* (*Ranunculales, Papaverales*) prove to be highly uniform, they are almost completely characterized by the presence of clustered wax tubules of the *Berberis* type.

A second wax type of irregular shaped platelets also appears in a few families which produce tubules (*Ranunculaceae, Berberidaceae*). Occurring less frequently in the *Ranunculidae*, but widespread in the angiosperms, they are without systematic significance. For that reason the wax data give no hint for remaining or segregation the *Glaucidiaceae* and *Hydrastidaceae* from the *Ranunculaceae*. THORNE (1992) and TAKHTAJAN (1987) see a close relationship between *Glaucidiaceae* and *Paeoniaceae*, the latter put both families into the *Ranunculidae*. The *Paeoniaceae* possess transversely ridged rodlets, a striking argument against a position within *Ranunculidae*. In a previous paper (DITSCH & BARTHLOTT 1994) we stated a close relationship between *Paeoniaceae* and *Magnoliidae* s. str. The monotypic *Glaucidiaceae* with its irregular platelets appear to be an independent taxon with doubtful affinities.

The relationship between *Sabiaceae, Coriariaceae*, and the *Ranunculiflorae* hypothesized by CRONQUIST (1988) could not be confirmed for lack of epicuticular waxes.

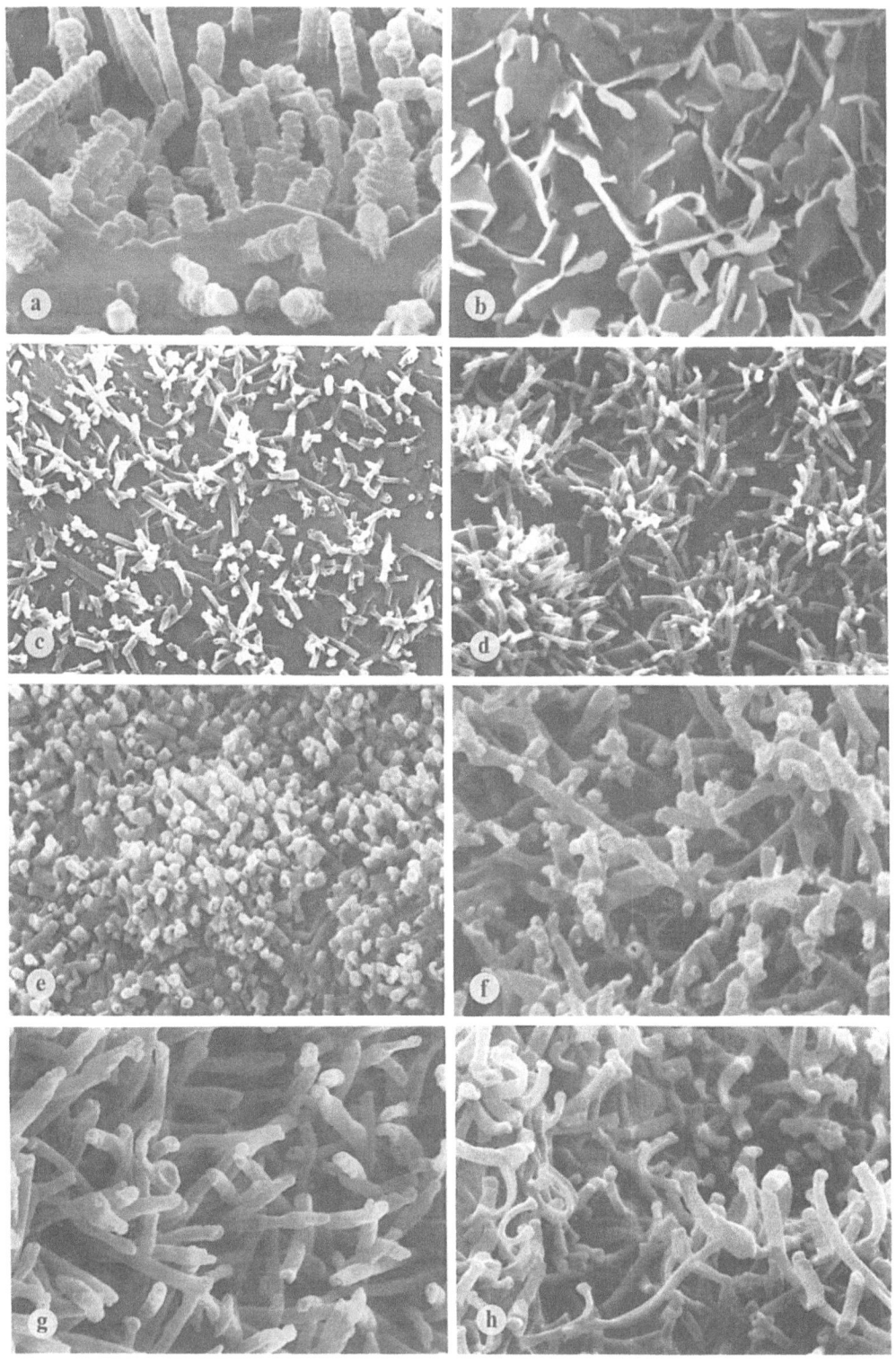

Relations and circumscription of the subclass. In contrast to the view of several authors, the results of wax ultrastructure strongly support the splitting of the *Magnoliidae* s. l. (in the sense of CRONQUIST 1988) into the predominantly woody "lower magnoliids" and the predominantly herbaceous "higher magnoliids" (corresponding TAKHTAJAN's 1987 subclasses *Magnoliidae* and *Ranunculidae*). The core group of ranunculiflorous angiosperms (*Ranunculales, Papaverales*) are almost uniformly characterized by clustered wax tubules of the *Berberis* type (Fig. 1c–d), which is also common in gymnosperms (JEFFREE 1986). Whereas the magnoliid families (*Aristolochiales, Magnoliales, Laurales*) are defined by transversely ridged rodlets(*Aristolochia* type; Fig. 1 a) and by the absence of clustered tubules (*Berberis* type). The holarctic *Nelumbonaceae* (BARTHLOTT & al., in print), long time stated as a member of the *Nymphaeales*, is characterized by the *Berberis* type. According to other characters (e.g., tricolpate pollen, serological data, alkaloids), it is consequently included on order level into the "eudicot" *Ranunculidae* adjacent to *Papaverales*.

Concerning the wax ultrastructure of the *Winteraceae* and the uncertain position of the *Canellaceae* both families show no relation to the *Magnoliidae*. Instead of the transversely ridged rodlets (*Aristolochia* type) of the magnoliids they both produce tubules (*Berberis* type). Under aspect of wax data the *Winteraceae* and *Canellaceae* could be considered as ranunculid relatives. A relationship of these two families has been previously discussed (LEROY 1977, SUH & al. 1992), and also *rbc*L analyses (CHASE & al. 1993, QIU & al. 1993) show the close connection of *Canellaceae* and *Winteraceae*. Perhaps they are a link between *Ranunculidae* and the "lower hamamelids". Whereas the hamamelid *Trochodendrales* and *Illiciales* are not characterized by their epicuticular wax (both show irregular platelets), the remaining "lower hamamelids" (*Cercidiphyllaceae, Daphniphyllaceae*, few species of *Hamamelidaceae*) are defined by more or less clustered tubules (*Berberis* type). The proposed sister-group relationship between *Illiciales* and *Winteraceae* (DONOGHUE & DOYLE 1989) are not supported by wax data.

We are particular indebted to the Royal Botanic Gardens Kew (Prof. G. PRANCE and Prof. G. LUCAS) for supply of important material over many years for our studies. Also we greatly acknowledge the help of Prof. W. GREUTER and Prof. P. HIEPKO (Botanischer Garten und Botanisches Museum Berlin-Dahlem).

◄————————————————————————————————

Fig. 1. Epicuticular wax on leaf surfaces. *a Annona squamosa* L. (*Annonaceae*), x 12.520; transversely ridged wax rodlets (*Aristolochia* type); characteristic for the core families of *Magnoliidae*. *b Ranunculus lingua* L. (*Ranunculaceae*), x 9.360; irregular platelets; occurring throughout all major orders of angiosperms. *c–h* Wax tubules (*Berberis* type): *c Berberis thunbergii* DC. (*Berberidaceae*), x 6.510. *d Corydalis cava* SCHWEIGG. & KORT. (*Fumariaceae*), x 7.720; tubules with a particular clustered arrangement are characteristic for the *Berberis* type. *e Nelumbo nucifera* GAERTN. (*Nelumbonaceae*), x 12.380. *f Drimys brasiliensis* MIERS (*Winteraceae*), x 18.140. *g Canella winterana* GAERTN. (*Canellaceae*), x 17.860. *h Cercidiphyllum japonicum* SIES. & ZUCC. (*Cercidiphyllaceae*), x 14.980

References

BARTHLOTT, W., 1990: Scanning electron microscopy of the epidermal surface in plants. – In CLAUGHER, D., (Ed.): Application of the scanning EM in taxonomy and functional morphology, pp. 69–94. – Syst. Assoc. Spec. Vol. **41**. Oxford: Clarendon Press.

– 1993: Epicuticular wax ultrastructure and systematics. – In BEHNKE, H.-D., MABRY, T. J., (Eds): Evolution and systematics of the *Caryophyllales*. – Berlin, Heidelberg, New York: Springer.

– NEINHUIS, C., JETTER, R., BOURAUEL, T., RIEDERER, M.: Waterliliy, poppy, or sycamore: on the systematic position of Nelumbo. – Flora (in print).

BEHNKE, H.-D., BARTHLOTT, W., 1983: New evidence from the ultrastructural and micromorphological fields in angiosperm classification. – Nordic J. Bot. **3**: 43–66.

CHASE, M. W., SOLTIS, D. E., OLMSTEAD, R. G., MORGAN, D., LES, D. H., MISHLER, B. D., DUVALL, M. R., PRICE, R., HILLS, H. G., QUI, Y.-L., KRON, K. A., RETTIG, J. H., CONTI, E., PALMER, J. D., MANHART, J. R., SYTSMA, K. J., MICHAELS, H. J., KRESS, W. J., DONOGHUE, M. J., CLARK, W. D., HEDRON, M., GAUT, B. S., JANSEN, R. K., KIM, K.-J., WIMPEE, C. F., SMITH, J. F., FURNIER, G. R., STRAUSS, S. H., XIANG, Q.-Y., PLUNKETT, G. M., SOLTIS, P. S. SWENSEN, S., EGUIARTE, L. E., LEARN, G. H., BARRETT, S. C. H., GRAHAM, S., DAYANANDAN, S., ALBERT, V. A., 1993: Phylogenetics of seed plants: an analysis of nucleotide sequences from the plastid gene *rbc*L. – Ann. Missouri Bot. Gard. **80**: 528–580.

CRONQUIST, A., 1988: The evolution and classification of flowering plants. 2nd edn. – New York: The New York Botanical Garden.

DITSCH, F., BARTHLOTT, W., 1994: Mikromorphologie der Eqicuticularwachse und die Systematik der *Dilleniales, Lecythidales, Malvales* und *Theales*. – Trop. Subtrop. Pflanzenwelt **88**: 1–74. – Akad, Wiss. Lit. Mainz. Stuttgart: F. Steiner.

DONOGHUE, M. J., DOYLE, J. A., 1989: Phylogenetic relationships of angiosperms and the relationships of *Hamamelidae*. – In CRANE, P. R., BLACKMORE, S., (Eds): Evolution, systematics, and fossil history of the *Hamamelidae* **1**: pp. 17–45. – Oxford: Clarendon Press.

FRÖLICH, D., BARTHLOTT, W., 1988: Mikromorphologie der epicuticularen Wachse und das System der Monokotylen. – Trop. Subtrop. Pflanzenwelt **63**: 1–74. – Akad. Wiss. Lit. Mainz. Stuttgart: F. Steiner.

HENNIG, S., BARTHLOTT, W., MEUSEL, I., THEISEN, I., 1994: Mikromorphologie der Epicuticularwachse und die Systematik der *Magnoliidae, Ranunculidae* und *Hamamelididae*. – Trop. Subtrop. Pflanzenwelt **90**: 1–60. – Akad. Wiss. Lit. Mainz. Stuttgart: F. Steiner.

JEFFREE, C. E., 1986: The cuticle, epicuticular waxes and trichomes of plants, with reference to their structure, functions and evolution. – In JUNIPER, B., SOUTHWOOD, R., (Eds): Insects and plant surfaces, pp. 23–64. – London: Edward Arnold.

– BAKER, E. A., HOLLOWAY, P. J., 1975: Ultrastructure and recrystallisation of plant epicuticular waxes. – New Phytol. **75**: 539–549.

LEROY, J.-F., 1977: A compound ovary with open carpels in *Winteraceae* (*Magnoliales*): evolutionary implications. – Science **196**: 977–978.

QIU, Y. -L., CHASE, M. W., LES, D. H., PARKS, C. R., 1993: Molecular phylogenetics of the *Magnoliidae*: cladistic analyses of nucleotide sequences of the plastid gene *rbc*L. – Ann. Missouri Bot. Gard. **80**: 587–606.

SUH, Y., THIEN, L. B., ZIMMER, E. A., 1992: Phylogeny of the basal flowering plants: evidence from the sequences of 26S ribosomal DNA. – 5th Int. Symp. Int. Organ. Pl. Biosyst. (Abstract): p. 37. – St. Louis, Missouri.

TAKHTAJAN, A. L., 1987: Systema Magnoliophytorum. – Leningrad: Nauka.

THEISEN, I., BARTHLOTT, W., 1994: Mikromorphologie der Epicuticularwachse und die Systematik der *Gentianales*, *Rubiales*, *Dipsacales* und *Calycerales*. – Trop. Subtrop. Pflanzenwelt **89**: 1–62. – Akad. Wiss. Lit. Mainz. Stuttgart: F. Steiner.

THORNE, R. F., 1992: Classification and geography of the flowering plants. – Bot. Rev. **58**: 225–348.

Addresses of the authors: WILHELM BARTHLOTT (correspondence) and INGE THEISEN, Botanisches Institut und Botanischer Garten der Universität Bonn, Meckenheimer Allee 170, D-53115 Bonn, Federal Republic of Germany.

Accepted December 19, 1994

Pl. Syst. Evol. [Suppl.] 9: 47–61 (1995)

Floral structure and evolution in *Ranunculanae*

PETER K. ENDRESS

Received September 11, 1994

Key words: *Ranunculanae, Menispermaceae, Lardizabalaceae, Ranunculaceae, Berberidaceae, Papaveraceae.* – Floral structure, floral evolution.

Abstract: Apomorphic tendencies of unusual features shared by two or more families of the *Ranunculanae* are: early caducous sepals, congenitally closed gynoecium up to the level of the stigma (but still apically gaping after ovule initiation!), protruding diffuse placentae, transverse fruit dehiscence (in *Berberidaceae* and *Papaveraceae*); excessive number of tepal whorls (in some *Menispermaceae* and *Berberidaceae*); unisexual flowers, synandry, wet stigmas forming an external compitum (in *Menispermaceae* and *Lardizabalaceae*); elaborate nectar-flowers, pollen with tricolpate-multiaperturate series, seeds with elaiosomes (in some *Ranunculaceae, Berberidaceae* and *Papaveraceae*). Other unusual features shared by *Menispermaceae* and *Lardizabalaceae* may be plesiomorphic or apomorphic: non-peltate carpels, fleshy fruits. Thus, floral structure suggests close relationships between *Menispermaceae* and *Lardizabalaceae* as well as *Berberidaceae* and *Papaveraceae*. A floral syndrome (probably myiophilous) consisting of small, flat, open, brownish or greenish flowers with short, spathulate, often bilobed staminodes (petals) that present open nectar from their apex occurs in (more or less basal) representatives of four families (except for *Papaveraceae*). It may be plesiomorphic or represent a basal apomorphic tendency in the *Ranunculanae.*

The *Ranunculanae (Ranunculiflorae)* (DAHLGREN 1980, TAKHTAJAN 1987) are circumscribed here as comprising five families taken in a wide sense and including some segregate families: *Menispermaceae, Lardizabalaceae* (incl. *Sargentodoxaceae*), *Ranunculaceae* (incl. *Hydrastidaceae, Circaeasteraceae, Glaucidiaceae*), *Berberidaceae* (incl. *Nandinaceae*), and *Papaveraceae* (incl. *Pteridophyllaceae, Hypecoaceae, Fumariaceae*).

Our knowledge on the five families is very unbalanced. While in *Ranunculaceae, Berberidaceae* and *Papaveraceae* floral structure and biology has been relatively well studied, because they contain many temperate ornamental plants with attractive flowers, the largely tropical *Menispermaceae* with inconspicuous flowers is very poorly known. *Menispermaceae* and *Lardizabalaceae* are almost completely unknown concerning floral biology and floral development.

In this survey, which also contains some new results, I am focusing on unusual characteristics, which are either synapomorphies or apomorphic tendencies

Menispermaceae   Lardizabalaceae   Ranunculaceae Berberidaceae   Papaveraceae

Floral organs commonly in trimerous whorls, including gynoecium		Floral organs never spiral, not even in polyandrous or polygynous flowers

Floral organs commonly in trimerous
   whorls, including gynoecium

Floral organs spiral or irregular
   in polyandrous or polygynous
   flowers

Petals, if present, small, often
   apically nectariferous

Synandry common

Carpels not peltate, even if
   gynoecium is unicarpellate

Fruits commonly fleshy

Floral organs never spiral, not even
   in polyandrous or polygynous
   flowers

Sepals early caducous

Gynoecium congenitally closed up to stigma

Placentae protruding-diffuse in some genera

Fruit dehiscence transverse, if dehiscent

Flowers bisexual

Elaborated nector-flowers present in a few genera

Pollen with tricolpate-multiaperturate series (van Campo
   & Vernier 1984)

Seeds with elaiosomes in some genera

Fig. 1. Floral features common to two or three of the five families of *Ranunculanae* (plesiomorphies, synapomorphies and apomorphic tendencies)

(Cantino 1985) shared by two or more families (Fig. 1). Not only synapomorphies but also shared apomorphic tendencies of outstanding characters may be indicative for close relationship. In contrast, the search for plesiomorphic features common to several families may help in revealing basic conditions for the *Ranunculanae*.

Structure

Gender distribution. Among *Ranunculanae* (*Ranunculiflorae*) as in the *Magnolianae* (*Magnoliiflorae*) there is a large proportion of taxa with unisexual flowers. This is not surprising in view of the low synorganisation level of the floral parts in both these groups, where strong architectural constraints on the presence of the organs of both sexes are lacking (Endress 1994). Unisexual flowers are exclusively present in *Menispermaceae* and *Lardizabalaceae*. On the contrary, it is surprising that there are not more unisexual flowers in the other families. This state is very rare in *Ranunculaceae* (*Xanthorhiza*, pers. obs.; species of *Pulsatilla, Anemone, Thalictrum*) and probably absent in *Berberidaceae* and *Papaveraceae*.

Floral phyllotaxis. In all five families of *Ranunculanae* flowers with whorled phyllotaxis occur, either almost exclusively or at least in a number of genera. Spiral phyllotaxis does not occur at all in *Berberidaceae* and *Papaveraceae*, and only in *Ranunculaceae* it is present in a relatively large proportion of the genera. In all five families flowers with the organs in trimerous whorls exist (in *Ranunculaceae* and *Papaveraceae* not in the androecium).

In four families flowers with polyandrous androecia exist. However, the development of polyandry is diverse. In *Menispermaceae* and *Ranunculaceae* the phyllotaxis in the androecium is whorled, spiral or irregular, with centripetal inception

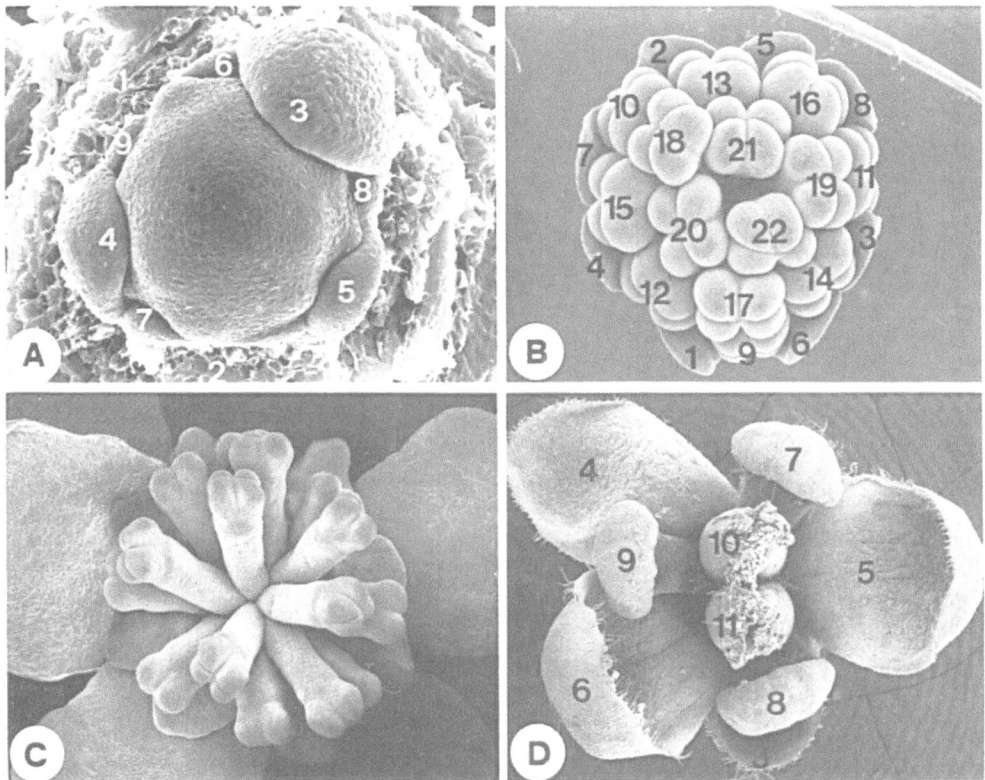

Fig. 2. Spiral floral phyllotaxis in flowers of *Menispermaceae*. (Collection numbers with "E" are by the author) *A, B Hypserpa laurina* Diels, male flowers, E 9169, *A* Young floral bud, inner perianth members numbered (x 200). *B* Old floral bud, petals and stamens numbered (x 25). *C* Anthetic flower (x 15). *H. decumbens* Diels, female anthetic flower, with 6 sepals, 3 petals and 2 carpels, E 9270 (x 13)

(Schöffel 1932; C. Wagner, pers. comm.; for *Menispermaceae*, pers. obs.; Fig. 2 A–C). In *Papaveraceae* it is whorled or irregular, with centripetal inception (Murbeck 1912, Karrer 1991). In *Berberidaceae* it is whorled, with centrifugal inception (*Podophyllum*, deMaggio & Wilson 1986) or irregular, with centripetal inception (*Achlys*, Endress 1989). The distribution of trimery and the disparity of polyandrous forms in *Ranunculanae* make it probable that polyandry and spiral floral phyllotaxis in *Ranunculanae* are not primitive but have evolved several times in parallel. The often cited spiral, pentamerous terminal flowers in *Berberis* inflorescences are not really spiral but stamens are epipetalous, an indication that this special case is derived from a whorled pattern (Endress 1987).

In *Menispermaceae* spiral phyllotaxis was also found in oligomerous flowers of *Hypserpa decumbens* (Fig. 2D).

Most likely a flexible situation with easy transition between different phyllotaxis patterns or trimery (and not spiral phyllotaxis) is plesiomorphic in *Ranunculanae* (see also Endress 1987).

A rare feature among angiosperms is the arrangement of an increased number of organs of the same category in regular isomerous whorls. Among *Ranunculanae* they

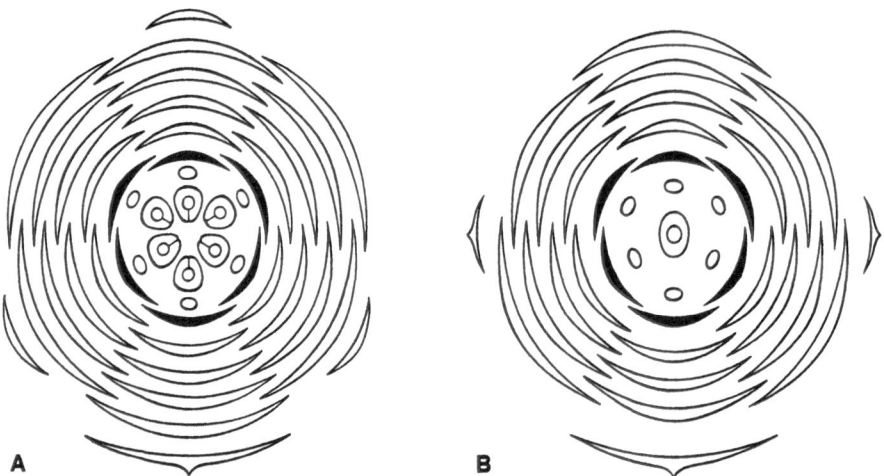

Fig. 3. Floral diagrams of flowers with excessive number of sepal whorls. Sepals white, petals black. *A Sciadotenia paraensis* DIELS, female flower (*Menispermaceae*) (after EICHLER 1864). *B Nandina domestica* THUNBERG, bisexual flower (*Berberidaceae*) (after EICHLER 1878 and TERABAYASHI 1985)

occur in two organ categories: (1) In two families polysepalous flowers with the sepals in trimerous whorls are formed: *Anisocycla, Carronia, Sciadotenia, Sphenocentrum* (*Menispermaceae*; EICHLER 1864, as *Sychnosepalum*; KESSLER 1993), *Nandina* (*Berberidaceae*; TERABAYASHI 1985) (Figs. 3, 4 A, B). In *Sciadotenia* tepal number is 16–36 (according to KESSLER 1993), in *Nandina* it is 21–30 (according to TERABAYASHI 1985). (2) In *Ranunculaceae* polyandrous flowers with all organs in pentamerous whorls occur in *Aquilegia* with 40 or more stamens (SCHÖFFEL 1932). In *Menispermaceae* the presence of whorled polyandry is suggested from some illustrations in the literature, but these taxa have not been critically studied (e.g., *Epinetrum*, TROUPIN 1962; *Albertisia*, FORMAN 1986). In *Ranunculaceae* (e.g. *Thalictrum*) and *Papaveraceae* (esp. *Eschscholzia* and relatives) polyandrous flowers with the stamens in tetramerous whorls occur, and by repeated "doubling" and "halfing" of the positions whorls of higher or lower numbers may arise (SCHÖFFEL 1932, ENDRESS 1987, KARRER 1991, RONSE DECRAENE & SMETS 1993).

Perianth. In all five families of *Ranunculanae* there are representatives with a double perianth showing sepals and (ontogenetically retarded) petals. This is in contrast to the magnoliids where in flowers with more than one perianth series, the inner series are not obviously ontogenetically retarded, even if they are petal-like (in size and showiness) at maturity (HIEPKO 1965). In *Menispermaceae, Lardizabalaceae,* the basal clades of *Ranunculaceae,* and in *Caulophyllum* (*Berberidaceae*) the petals, if present, are much smaller than the sepals and produce nectar in the apical region (epidermal nectaries without stomata; VOGEL 1993 for *Ranunculaceae*; ENDRESS, unpubl., for other families). In some *Ranunculaceae, Berberidaceae* and *Papaveraceae* the petals may be larger (and then the main or only optically attractive organs) and/or of complicated shapes with nectar produced more basally or not at all. *Nigella* (*Ranunculaceae*) is especially complicated with hidden nectar and, in addition, two pseudonectaries (glistening but dry parts that are optically attractive to insects) at the outside (see WEBER 1993). Pseudonectaries near the nectaries also occur in *Trollius* (*T. pumilus*) and *Xanthorhiza*. In *Menispermaceae, Lardizabalaceae, Ranunculaceae* and *Berberidaceae* there is a tendency for the petals to be bilobed (Fig. 7 A–C).

While in *Menispermaceae, Lardizabalaceae, Ranunculaceae* and *Berberidaceae* nectar is produced on petals (staminodes) in nectariferous flowers, in *Papaveraceae* it is produced at the stamen base (only fumarioid and hypecooid lineage). In *Corydalis* (*Papaveraceae*) species with long spurs it is secondarily transferred deep into the spur by intercalary growth. In *Lardizabalaceae*, when petals are lacking, the wet stigmas may exude large amounts of sweet secretions (*Akebia*), or nectar may also be produced on stamens (*Decaisnea*). The often cited carpellary nectaries of *Caltha* (*Ranunculaceae*) (e.g., PETERSON & al. 1979) are probably not nectaries but secrete lipophilic substances (PELLMYR, pers. comm.).

In *Berberidaceae* and *Papaveraceae* (especially fumarioids) there is a shared tendency for the (outer) sepals to remain small without protective function in later bud stages and to fall off before anthesis.

Synsepaly and sympetaly are almost absent. Synsepaly occurs in *Cyclea* (*Menispermaceae*) (KESSLER 1993) and some *Papaveraceae* (in both papaveroids and fumarioids) (KADEREIT 1993). Sympetaly is known from *Antizoma, Cyclea* and *Cissampelos* (*Menispermaceae*) (DIELS 1910, TROUPIN 1962, KESSLER 1993), and *Consolida* (*Ranunculaceae*) (HIEPKO 1965). It is clearly autapomorphic for each group.

Androecium (see also under floral phyllotaxis). Stamen structures in *Ranunculanae* are less diverse than in the *Magnolianae*. The most prominent forms are unique valvate anthers in *Berberidaceae*, where one narrowly hinged valve serves both pollen sacs of a theca (see ENDRESS & HUFFORD 1989). A somewhat similar dehiscence pattern is present in *Brassiodendron* and *Endiandra* (*Lauraceae*) (ROHWER 1994). More simple valvate anthers are also known from a few *Ranunculaceae* (ENDRESS & HUFFORD 1989, WEBER 1993). In *Menispermaceae, Lardizabalaceae* and *Papaveraceae* valvate anthers have not been found as yet, although an extended search has not been carried out. In *Mahonia* (*Berberidaceae*) stamen filaments may have two lateral appendages (HIEPKO 1965) which may again recall the situation in *Laurales* (see ROHWER 1994). However, in *Berberidaceae* these appendages are not nectariferous.

Synandry is prominent among *Menispermaceae* and *Lardizabalaceae* (MIERS 1864–1871, DIELS 1910, SCHAEPPI 1976) but absent in the other families (except for a slight expression in the elaborate flowers of fumarioids (*Papaveraceae*) with two phalanges each consisting of three united stamens; see, e.g., RONSE DECRAENE & SMETS 1992). There is a clear correlation of synandry with the presence of unisexual flowers. In male flowers the stamens are the innermost organs, and fusion is not prevented by constraints set by a gynoecium (ENDRESS 1990).

Slight basal fusion of petals and stamens in the same radius has been found in whorled flowers of *Berberidaceae* (e.g., HIEPKO 1965) and *Lardizabalaceae* (pers. obs.). In *Berberidaceae* this is associated with the presence of "common primordia" between petals and stamens of the same radius (e.g., BRETT & POSLUSZNY 1982).

Pollen. In all five families the basic pollen structure is tricolpate. However, *Menispermaceae* are outstanding among *Ranunculanae* because of the dominance of

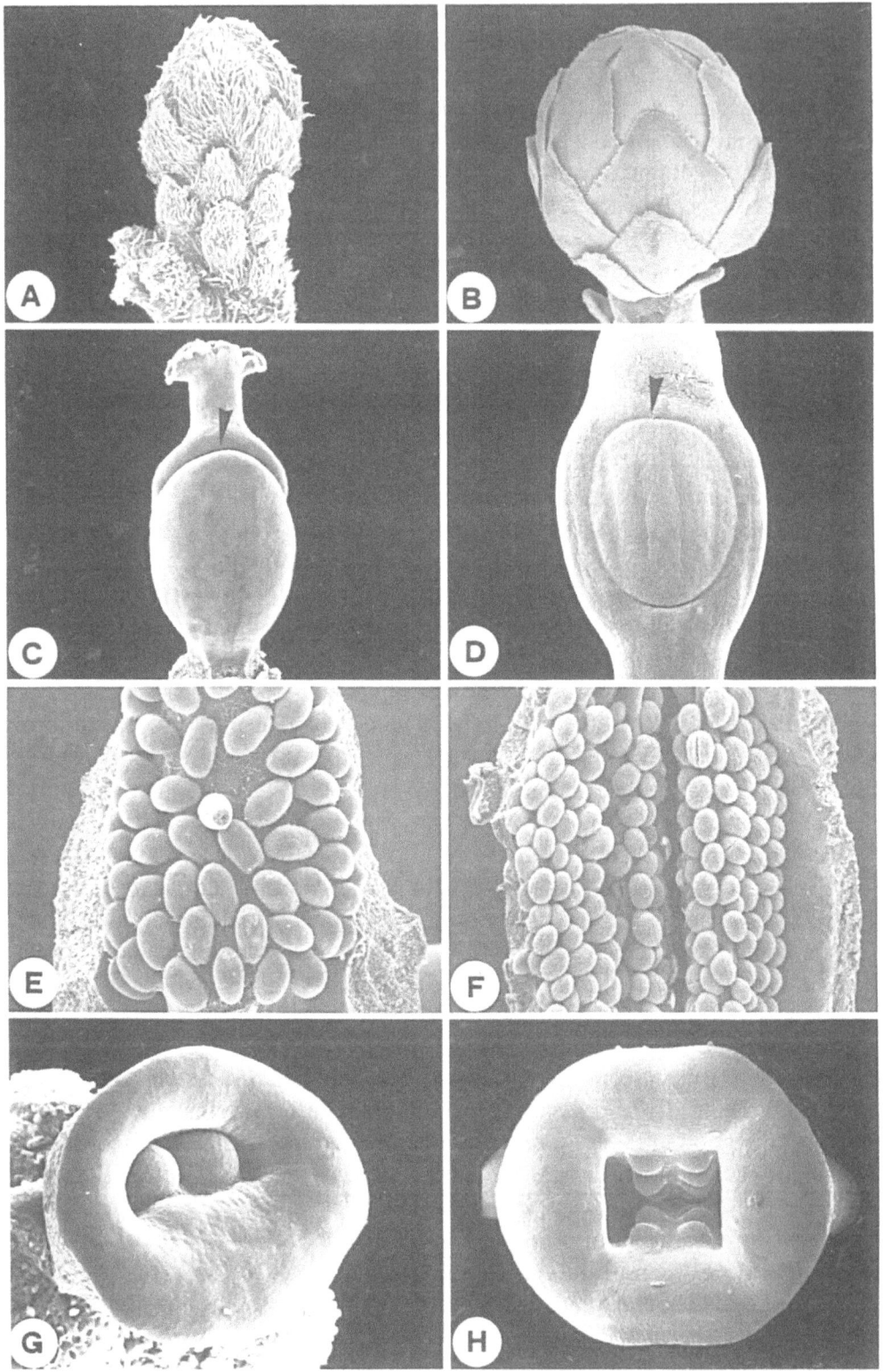

tricolporate pollen (FERGUSON 1975, THANIKAIMONI 1984, 1986), which is otherwise indicative of a higher evolutionary level among the dicots. On the other hand, *Ranunculaceae, Berberidaceae,* and *Papaveraceae* exhibit parallel series of aperture configurations, from tricolpate to multiaperturate (série successiforme after VAN CAMPO & VERNIER 1984) (for *Ranunculaceae,* see e.g., TAMURA 1993; for *Berberidaceae* NOWICKE & SKVARLA 1981; for *Papaveraceae* VAN CAMPO & VERNIER 1984). This series occurs in parallel several times also within the family *Papaveraceae* s. l. (see e.g., KADEREIT 1993).

Gynoecium. The carpel bauplan greatly varies within *Ranunculanae.* Carpels are not or only very slightly peltate in *Lardizabalaceae* (VAN HEEL 1983) and *Menispermaceae* (pers. obs.). This greatly contrasts with the constantly and strongly ascidiate carpel in *Berberidaceae* (e.g., KUMAZAWA 1938, GUÉDÈS 1977, SCHNYDER 1982). In *Ranunculaceae* (e.g., TAMURA 1965, SCHAEPPI & FRANK 1962, ROHWEDER 1967) and *Papaveraceae* (KARRER 1991) both non-peltate or weakly peltate and strongly ascidiate forms occur.

Some *Berberidaceae* and *Papaveraceae* share a protruding-diffuse placenta (concomitant with increased ovule number), a unique feature in the entire *Magnolianae – Nymphaeanae – Ranunculanae* complex (Fig. 4E, F). The only difference in the two families is that *Berberidaceae* uniformly have a single carpel, while *Papaveraceae* have two or more carpels. In the common morphological terminology, the unicarpellate gynoecium of *Berberidaceae* is completely ascidiate (LEINFELLNER 1956, GUÉDÈS 1977), while in the bi- or pluricarpellate gynoecium of *Papaveraceae* each carpel is only basally peltate (except for *Romneya,* KARRER 1991). Thus quite different! However, if we look at it from a different angle, the similarities become apparent. The basal condition in *Papaveraceae* is a completely syncarpous bicarpellate gynoecium having a unilocular ovary with two parietal placentae. There is a trend within the family to increase carpel number with concomitant increase of placenta number. The reverse trend would be decrease from two to one carpel with only one placenta. This is exactly what occurs in *Berberidaceae.* In this sense the gynoecia of *Berberidaceae* are unicarpellate with a single parietal placenta. Evolutionary changes may go in both directions. The completely syncarpous gynoecium of *Papaveraceae* may have evolved from a completely ascidiate carpel as in *Berberidaceae* by doubling of the placenta number, or, vice versa, the unicarpellate gynoecium of *Berberidaceae* may have evolved from a syncarpous bicarpellate condition (see ENDRESS 1994; Fig. 5). Ontogenetically, the gynoecium in both

---

Fig. 4. Apomorphic tendencies in pairs of families. *A,B* Floral buds with excessive number of sepal whorls. *A Carronia protensa* DIELS (*Menispermaceae*), E 9042 (x 15). *B Nandina domestica* THUNBERG (*Berberidaceae*), E 4751 (x 18). *C,D.* Transverse dehiscence lines in anthetic ovaries. *C Jeffersonia dubia* BENTHAM & HOOKER (*Berberidaceae*), E 4578 (x 15). *D Bocconia frutescens* L. (*Papaveraceae*), E 1285 (x 30). *E,F.* Protruding diffuse placentae, at anthesis. *E Podophyllum peltatum* L. (*Berberidaceae*), E 7334 (x 13). *F Papaver lecoqii* LAMOTTE (*Papaveraceae*), E 4660 (x 18). *G, H.* Gynoecia congenitally closed up to stigma but apically gaping still after ovule inception in floral bud. *G Berberis vulgaris* L. (*Berberidaceae*) (x 160); courtesy of NORBERT SCHNYDER (from SCHNYDER 1982: fig. 5). *H Hunnemannia fumariifolia* SWEET (*Papaveraceae*) (x 80); courtesy of ANDREA FETZ-KARRER (from KARRER 1991: fig. 92)

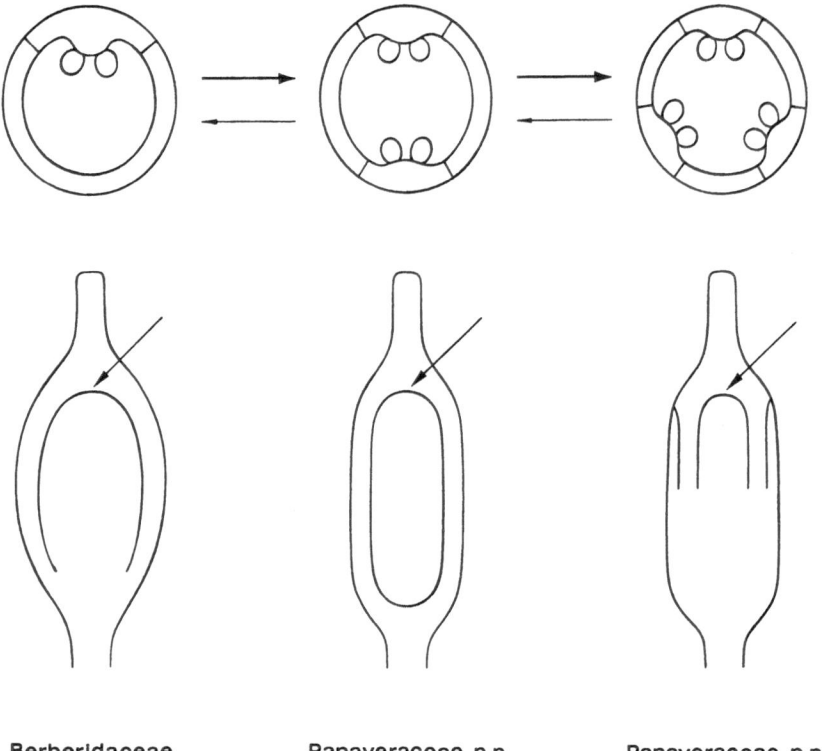

Berberidaceae   Papaveraceae p.p.   Papaveraceae p.p.

Fig. 5. Gynoecium structure in *Berberidaceae* and *Papaveraceae*. Upper row: Transverse sections of ovaries, showing placentae with ovules and dehiscence lines at fruit maturity. Arrows indicate possible evolutionary trends. Lower row: Gynoecia from the side with transverse dehiscence lines (arrows)

families arises as a tubular structure, and the thick placentae with the young ovules are sometimes visible from the outside long before the gynoecium closes (e.g. SCHNYDER 1982, KARRER 1991; Fig. 4G, H). Associated with this peculiar gynoecium structure in both *Berberidaceae* and *Papaveraceae* is the unusual pattern of fruit dehiscence. Each carpel has an upper transverse dehiscence line with two longitudinal extensions (Figs. 4C–D, 5). (These two longitudinal extensions were the reason for the often discussed interpretions of the gynoecium as pseudomonomerous in *Berberidaceae* and pseudodimerous in *Papaveraceae*.) In the bicarpellate *Papaveraceae* these two longitudinal extensions additionally join at the base. The result is that from each fruit two valves fall off at maturity. In *Berberidaceae* and in pluricarpellate *Papaveraceae* the resulting valves remain attached at the base (Fig. 5). Another shared trend associated with this fruit type in both families is to produce seeds with elaiosomes (among *Papaveraceae* s.l. in both *Fumariaceae* and *Papaveraceae* s. str.).

In contrast, many *Lardizabalaceae* have a laminar-diffuse placenta (PAYNE & SEAGO 1968, QIN 1989). This is different from the protruding-diffuse placenta by its position on each carpel flank along the postgenitally fused carpel margins (see ENDRESS 1994). It is associated with increased ovule number, hemitropy or almost orthotropy of the ovules and heavy mucilaginous secretion in the placental region by tufts of long hairs between the ovules, a character complex known also from a few other angiosperm groups (among basal angiosperms also from some *Nymphaeanae*). Incomplete fusion of these carpels has also been mentioned (PAYNE & SEAGO 1968, QIN 1989). However, the carpels are not really open but they are sealed by the abundant mucilage produced on the inner carpel surface. The fruits are large bluish berries. In *Akebia* (with laminar-diffuse placentation) the carpels are ventrally open at maturity in a follicular manner. In *Sinofranchetia* and *Decaisnea* (with linear placentation) they remain closed. At anthesis, heavy secretion also occurs at the stigma. In genera where the three stigmas are close together (*Decaisnea, Sinofranchetia*), the secretion may unite the carpels to form an external compitum (Fig. 6A, B; ENDRESS 1982). From its systematic distribution and its specialization laminar-diffuse placentation is most likely an autapomorphic character complex within the *Ranunculanae* and even within the *Lardizabalaceae*.

Ovules are basically bitegmic, anatropous and crassinucellar. In *Menispermaceae* and *Ranunculaceae* there is a tendency to form unitegmic ovules, in both families in more than one subclade (BOUMAN & CALIS 1977, SOEJIMA 1990).

Stigma differentiation is various. A noteworthy tendency is to form stigmas with two convoluted crests with multicellular protuberances, as in some *Menispermaceae*, some *Berberidaceae* and *Hydrastis* (basal clade of *Ranunculaceae*; LOCONTE & ESTES 1989, JOHANNSON & JANSEN 1993, KEENER 1993; HOOT 1994; Fig. 6C–F). Wet stigmas are prominent in *Lardizabalaceae, Berberidaceae* and some *Menispermaceae*, dry stigmas in *Ranunculaceae* and *Papaveraceae*.

Indehiscent, more or less fleshy fruits may be plesiomorphic in the *Ranunculanae*. They consistently occur in *Menispermaceae*, and they are also present in the putative basal clades of *Lardizabalaceae* (*Sinofranchetia, Decaisnea*), *Ranunculaceae* (*Hydrastis*) and *Berberidaceae* (*Nandina*). They are absent in *Papaveraceae*.

Biology and evolution

A curious simple flower form occurs in four of the five families of *Ranunculanae*: small, open (flat) brownish or greenish flowers with short, spathulate, often bilobed staminodes (petals) that present open nectar from their apex: many *Menispermaceae*, *Sinofranchetia* of *Lardizabalaceae* (possibly basal in the family, see HOOT & al. 1993; second basal, after *Decaisnea*, according to LOCONTE & ESTES 1989), *Xanthorhiza* and *Kingdonia* of *Ranunculaceae* (possibly basal genera in the family, if *Hydrastis* is set aside, HOOT 1994), and *Caulophyllum* of *Berberidaceae* (basal in the family next to *Nandina*, LOCONTE & ESTES 1989; Fig. 7A–D). This is an unspecialized pollination syndrome, possibly with special attraction to flies. Flies are an important pollinator group in primitive angiosperms (e.g., ENDRESS 1990). In view of its general primitive nature and its wide occurrence in the *Ranunculanae*, this syndrome may be seen as an especially primitive feature in the *Ranunculanae*. It is either plesiomorphic in the *Ranunculaceae* or a basal apomorphic tendency. the pollination biology of these genera has not actually been studied. Also *Trollius* as another primitive candidate in *Ranunculaceae* (JENSEN 1968 and pers. comm.) is fly-pollinated (PELLMYR 1992).

In contrast, more elaborated flowers with concealed nectar in spurs, sometimes of considerable length, have been evolved in parallel in *Ranunculaceae, Berberidaceae* and *Papaveraceae*. Furthermore, in all three families, there are genera with more than one spur per flower, which is otherwise rare in the angiosperms. Examples are

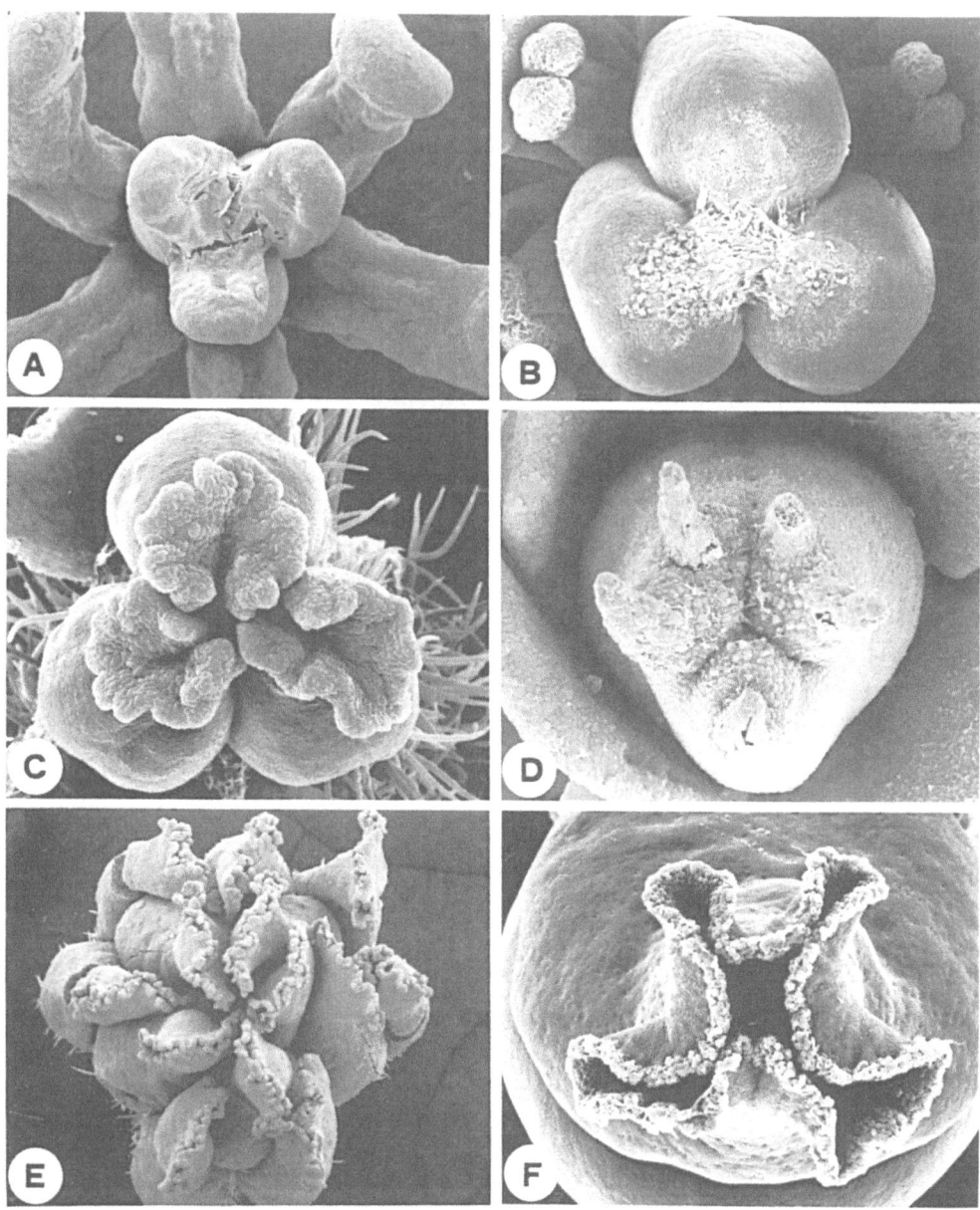

Fig. 6. Gynoecia from above, showing stigma differentiation at anthesis. *A,B.* Wet stigmas with external compitum in *Lardizabalaceae. A Decaisnea fargesii* Franch., E 5157 (x 18). *B Sinofranchetia chinensis* Hemsley, E 4669 (x 35). *C, D. Menispermaceae. C Hypserpa decumbens* Diels (shortly before anthesis), E 9270 (x 80). *D Stephania japonica* O. Kuntze, Gynoecium of a single lobed carpel, ventral side up, E 4397b (x 70). *E,F.* Convolute stigmas with protuberances in *Ranunculaceae* and *Berberidaceae. E Hydrastis canadensis* L., C. Wagner 201.10.3 (x 18). *F Jeffersonia dubia* Bentham & Hooker, E 4578 (x 50)

Fig. 7. Syndrome of small, open, brownish or greenish (myiophilous) flowers with short, cuneate, apically bilobed and nectariferous petals as occurring in four families of the *Ranunculanae* (*s* sepal, *p* petal). *A Sinomenium acutum* REHDER & WILSON (*Menispermaceae*), E 9522 (x 13). *B Sinofranchetia chinensis* HEMSLEY (*Lardizabalaceae*), E 4669 (x 13). *C Xanthorhiza simplicissima* MARSH. (*Ranunculaceae*), E 7206 (x 15). *D Caulophyllum thalictroides* MICHAUX (*Berberidaceae*), E 5131 (x 13)

Aquilegia (*Ranunculaceae*), *Epimedium* (*Berberidaceae*; e.g., SUZUKI 1984), and *Dicentra* (*Papaveraceae*; e.g., MACIOR 1970). Pollination by bumblebees is common in these flowers. In a few *Ranunculaceae* also hummingbird- and hawkmoth-pollination occurs.

Nectar-producing staminodes or petals are probably not basically peltate as proposed by, e.g., LEINFELLNER (1959), but nectar-bearing grooves have evolved in different groups from simple, flat organs. The variability is often high within a genus and can not always be clearly reconciled with peltate structures (e.g., HIEPKO 1965, DAHLGREN 1992, KOSUGE 1994). This appears even more so, if a comparison among all four families with nectariferous petals is made. If a container is needed (advantageous), such as in a nectariferous region to hold the nectar, concave parts may be readily selected as soon as they are formed on an organ by differential growth. This may or may not involve a peltate development. If such organs develop like peltate structures, this peltate nature is most probably evolutionarily secondary. Petals have probably evolved several times in the *Ranunculanae* from outer (nectar-producing) staminodes (see Hiepko 1965; Tamura 1965, 1984, 1993; Kosuge & Tamura 1989; Kosuge 1994). They may also have been lost and evolved again in some groups.

Relationships between the families

In most systems the deepest split within the *Ranunculanae* is between *Papaveraceae* and all other families, whereby the *Papaveraceae* is the most advanced group (Dahlgren 1980, Takhtajan 1987, Cronquist 1988, Thorne 1992) or the basal-most group (Drinnan & al. 1994). Is this because the total syncarpy of *Papaveraceae* seemingly contrasts so much with the apocarpy of the other families? However, other groupings have been proposed. Loconte & Stevenson (1991) put the deepest split between *Lardizabalaceae* and other *Ranunculanae*. From the point of view of floral structure there are two pairs of families: *Menispermaceae/Lardizabalaceae* and *Berberidaceae/Papaveraceae*. *Ranunculaceae* are close to *Berberidaceae/ Papaveraceae*. Whether there are sister group relationships between the families of each pair or whether one is nested within the other cannot be decided with the data available. Integration with other kinds of features, especially from DNA studies, should throw more light on this question. As is so often the case, the distinction between synapomorphy and shared apomorphic tendencies in sister groups (e.g., multiple petal evolution) is difficult but crucial and is the most delicate problem in attempts of phylogenetic reconstruction. Although both indicate close relationship, the primitive situation cannot be revealed, without a distinction between the two.

I am grateful for discussions to P. R. Crane, S. Hoot and C. Wagner. I thank A. Fetz-Karrer and N. Schnyder for providing SEM micrographs from their theses, R. Siegrist for skilful microtome work, U. Jauch (Institute of Plant Biology) for expert help with SEM micrographs, and A. Zuppiger for careful photographic work.

References

Bouman, F., Calis, J. I. M., 1977: Integumentary shifting – a third way to unitegmy. – Ber. Deutsch. Bot. Ges. **90**: 15–28.

Brett, J. F., Posluszny, U., 1982: Floral development in *Caulophyllum thalictroides* (*Berberidaceae*). – Canad. J. Bot. **60**: 2133–2144.

Van Campo, M., Vernier, D., 1984: Les grains de pollen. Séries morphologiques et structure de l'exine. – G. Bot. Ital. **118**: 255–270.

Cantino, P. D., 1985: Phylogenetic inference from nonuniversal derived character states. – Syst. Bot. **10**: 119–122.

Cronquist, A., 1988: The evolution and classification of flowering plants, 2nd edn. – Bronx: New York Botanical Garden.

Dahlgren, G., 1992: *Ranunculus* subgenus *Batrachium* on the Aegean islands and adjacent areas: nectary types and breeding system. – Nordic J. Bot. **12**: 299–310.

Dahlgren, R. M. T., 1980: A revised system of classification of the angiosperms. – Bot. J. Linn. Soc. **80**: 91–124.

deMaggio, A. E., Wilson, C. L., 1986: Floral structure and organogenesis in *Podophyllum peltatum* (*Berberidaceae*). – Amer J. Bot. **73**: 21–32.

Diels, L., 1910: *Menispermaceae*. – In Engler, A., (Ed.): Das Pflanzenreich IV, 94. –Leipzig: Engelmann.

DRINNAN, A. N., CRANE, P. R., HOOT, S. B., 1994: Patterns of floral evolution in the early diversification of non-magnoliid dicotyledons (eudicots). – Pl. Syst. Evol. [Suppl.] **8**: 93–122.

EICHLER, A. W., 1864: Versuch einer Charakteristik der natürlichen Pflanzenfamilie *Menispermaceae*. – Denkschr. Königl. Bayer. Bot. Ges. Regensburg **5**: 1–42.

– 1878: Blüthendiagramme. – Leipzig: Engelmann.

ENDRESS, P. K., 1982: Syncarpy and alternative modes of escaping disadvantages of apocarpy in primitive angiosperms. – Taxon **31**: 48–52.

– 1987: Floral phyllotaxis and floral evolution. – Bot. Jahrb. Syst. **108**: 417–438.

– 1989: Chaotic floral phyllotaxis and reduced perianth in *Achlys* (*Berberidaceae*). – Bot. Acta **102**: 159–163.

– 1990: Evolution of reproductive structures and functions in primitive angiosperms. – Mem. New York Bot. Gard. **55**: 5–34.

– 1994: Diversity and evolutionary biology of tropical flowers. – Cambridge: Cambridge University Press.

– HUFFORD, L. D., 1989: The diversity of stamen structures and dehiscence patterns among *Magnoliidae*. – Bot. J. Linn. Soc. **100**: 45–85.

FERGUSON, I. K., 1975: Pollen morphology of the tribe *Triclisieae* of the *Menispermaceae* in relation of its taxonomy. – Kew Bull **30**: 49–75.

FORMAN, L. I., 1986: *Menispermaceae*. – In VAN STEENIS, C. G. G. J., (Ed.): Flora Malesiana, Ser. I. **10, 2**, pp. 157–253. – Dordrecht: Nijhoff.

GUÉDÈS, M., 1977: Le gynécée de *Podophyllum* (Berbéridacées): monomérie vraie et placentation suturale de la portion congénitalement close du carpelle. – Compt. Rend: Acad. Sci. Paris **285**: 755–758.

HEEL, W. A. VAN, 1983: The ascidiform early development of free carpels, a S. E. M. – investigation. – Blumea **28**: 231–270.

HIEPKO, P., 1965: Vergleichend-morphologische und entwicklungsgeschichtliche Untersuchungen über das Perianth bei den *Polycarpicae*. – Bot. Jahrb. Syst. **84**: 359–508.

HOOT, S. B., 1994: Phylogenetic relationships within the *Ranunculaceae* based on rbcL and atpB gene sequences. – Amer. J. Bot. **81**: (6, Abstr.): 161.

– CULHAM, A., CRANE, P. R., 1993: Phylogenetic relationships of the *Lardizabalaceae* based on rbcL, atpB, and 18S gene sequences. – Amer. J. Bot. **80**: (6, Abstr.): 119–120.

JENSEN, U., 1968: Serologische Beiträge zur Systematik der *Ranunculaceae*. – Bot. Jahrb. Syst. **88**: 204–268.

JOHANSSON, J. T., JANSEN, R. K., 1993: Chloroplast DNA variation and phylogeny of the *Ranunculaceae*. – Pl. Syst. Evol. **187**: 29–49.

KADEREIT, J. W., 1993: *Papaveraceae*. – In KUBITZKI, K., ROHWER, J. G., BITTRICH, V., (Eds): The families and genera of vascular plants, **II**, pp. 494–506. – Berlin: Springer.

KARRER, A. B., 1991: Blütenentwicklung und systematische Stellung der *Papaveraceae* und *Capparaceae*. – Doctoral Dissertation, University of Zurich.

KEENER, C. S., 1993: A review of the classification of the genus *Hydrastis* (*Ranunculaceae*). – Aliso **13**: 551–558.

KESSLER, P. J. A., 1993: *Menispermaceae*. – In KUBITZKI, K., ROHWER, J. G., BITTRICH, V., (Eds): The families and genera of vascular plants, pp. 402–418. – Berlin: Springer.

KOSUGE, K., 1994: Petal evolution in *Ranunculaceae*. Pl. Syst. Evol. [Suppl.] **8**: 185–191.

KOSUGE, K., TAMURA, M., 1989: Ontogenetic studies on petals of the *Ranunculaceae*. – J. Japan. Bot. **64**: 65–74.

KUMAZAWA, M., 1938: On the ovular structure in the *Ranunculaceae* and *Berberidaceae*. – J. Japan. Bot. **14**: 10–25.

LEINFELLNER, W., 1956: Zur Morphologie des Gynözeums von *Berberis*. – Österr. Bot. Z. **103**: 600–612.

– 1959: Über die röhrenförmige Nektarschuppe an den Nektarblättern verschiedener *Ranunculus*-und *Batrachium*-Arten. – Österr.-Bot. Z. **106**: 88–103.

LOCONTE, H., ESTES, J. R., 1989: Phylogenetic systematics of *Berberidaceae* and *Ranunculales* (*Magnoliidae*). – Syst. Bot. **14**: 565–579.

– STEVENSON, D. W., 1991: Cladistics of the *Magnoliidae*. – Cladistics **7**: 267–296.

MACIOR, L. W., 1970: The pollination ecology of *Dicentra cucullaria*. – Amer. J. Bot. **57**: 6–11.

MIERS, J., 1864–1871: Monograph of the *Menispermaceae*. – London: Williams & Norgate.

MURBECK, S., 1912: Untersuchungen über den Blütenbau der Papaveraceen. – Kungl. Svenska Vetenskapsakad. Handl. (1) **50**: 1–168.

NOWICKE, J. W., SKVARLA, J. J., 1981: Pollen morphology and phylogenetic relationship of the *Berberidaceae*. – Smithsonian Contrib. Bot. **50**: 1–83.

PAYNE, W. W., SEAGO, J. I., 1968: The open conduplicate carpel of *Akebia quinata* (*Berberidales: Lardizabalaceae*). – Amer. J. Bot. **55**: 575–581.

PELLMYR, O., 1992: The phylogeny of a mutualism: evolution and coadaptation between *Trollius* and its seed-parasitic pollinators. – Biol. J. Linn. Soc. **47**: 337–365.

PETERSON, R. L., SCOTT, M. G., MILLER, S. L., 1979: Some aspects of carpel structure in *Caltha palustris* L. (*Ranunculaceae*). – Amer. J. Bot. **66**: 334–342.

QIN, H.-N., 1989: An investigation on carpels of *Lardizabalaceae* in relation to taxonomy and phylogeny. – Cathaya **1**: 61–82.

ROHWEDER, O., 1967: Karpellbau und Synkarpie bei Ranunculaceen. – Ber. Schweiz. Bot. Ges. **77**: 376–432.

ROHWER, J. G., 1994: A note on the evolution of the stamens in the *Laurales*, with emphasis on the *Lauraceae*. – Bot. Acta **107**: 103–110.

RONSE DECRAENE, L. P., SMETS, E. F., 1992: An updated interpretation of the androecium of the *Fumariaceae*. – Canad. J. Bot. **70**: 1765–1776.

– – 1993: The distribution and systematic relevance of the androecial character polymery. – Bot. J. Linn. Soc. **113**: 285–350.

SCHAEPPI, H., 1976: Über die männlichen Blüten einiger Menispermaceen. – Beitr. Biol. Pfl. **52**: 207–215.

– FRANK, K., 1962: Vergleichend-morphologische Untersuchungen über die Karpellgestaltung, insbesondere die Plazentation bei Anemoneen. – Bot. Jahrb. Syst. **81**: 337–357.

SCHNYDER, N., 1982: Analyse divergenter Fruchtdifferenzierung bei einheitlichem Bauplan am Beispiel von *Berberis*, *Vancouveria* und *Caulophyllum* (*Berberidaceae*). – Diploma Thesis, University of Zurich.

SCHÖFFEL, K., 1932: Untersuchungen über den Blüteanbau der Ranunculaceen. – Planta **17**: 315–371.

SOEJIMA, A., 1990: The histogenesis of integuments in some species of *Menispermaceae*. – Bot. Mag. (Tokyo) **103**: 209–216.

SUZUKI, K., 1984: Pollination system and its significance on isolation and hybridization in Japanese *Epimedium* (*Berberidaceae*). – Bot. Mag. (Tokyo) **97**: 381–396.

TAKHTAJAN, A., 1987: Systema Magnoliophytorum. – Leningrad: Nauka.

TAMURA, M., 1965: Morphology, ecology and phylogeny of the *Ranunculaceae* IV. – Sci. Rep. Coll. Gen. Educ. Osaka Univ. **14**: 53–71.

– 1984: Phylogenetical consideration of the *Ranunculaceae*. – Korean J. Pl. Taxon. **14**: 33–42.

– 1993: *Ranunculaceae*. – In KUBITZKI, K., ROHWER, J. G., BITTRICH, V., (Eds): The families and genera of vascular plants, II, pp. 563–583. – Berlin: Springer.

TERABAYASHI, S., 1985: The comparative floral anatomy and systematics of the *Berberidaceae*. I. Morphology. – Mem. Fac. Sci., Kyoto Univ., Ser. Biol., **10**: 73–90.

THANIKAIMONI, G., 1984: Ménispermacées: Palynologie et systématique. – Travaux de l'Institut Français de Pondichéry, Section Scientifique et Technique, **18**: 1–135.

– 1986: Evolution of *Menispermaceae*. – Canad. J. Bot. **64**: 3130–3133.

THORNE, R., 1992: Classification and geography of the flowering plants. – Bot. Rev. **58**: 225–348.

TROUPIN, G., 1962: Monographie des *Menispermaceae* africaines. – Mém. Acad. Sci. Outre-Mer, Cl. Sci. Nat. Méd., n.s., **13**, 2: 1–313.

VOGEL, S., 1993: Pollination systems [in *Ranunculaceae*]. – In KUBITZKI, K., ROHWER, J. G., BITTRICH, V., (Eds): The families and genera of vascular plants, **II**, pp. 568–569. – Berlin: Springer.

WEBER, A., 1993: Struktur, Antheseverlauf und Bestäubung der Blüte von *Nigella arvensis* (*Ranunculaceae*). – Verh. Zool.-Bot. Ges. Österr. **130**: 99–125.

Address of the author: PETER K. ENDRESS, Institute of Systematic Botany, University of Zurich, Zollikerstrasse 107, CH-8008 Zurich, Switzerland.

Accepted December 14, 1994

Pl. Syst. Evol. [Suppl.] 9: 63–70 (1995)

Evolution of the androecium in the *Ranunculiflorae*

L. P. RONSE DECRAENE and E. F. SMETS

Received September 27, 1994

Key words: *Ranunculiflorae*.-Androecium, merosity, reduction, secondary increase, nectar-leaves, petals.

Abstract: Four different evolutionary lines can be recognised for the androecium of the *Ranunculiflorae*: (1) a progressive reduction in the number of stamens and stamen whorls. (2) The transformation of the outermost stamens into nectar-leaves or petals. (3) Meristic changes of the flower. (4) Secondary increases in the number of stamens. It is shown that the position and inception of the stamens is valuable to understand relationships of taxa. *Ranunculaceae* occupy a basal position with the greatest androecial diversity and a reductive trend, linked with a shift from trimery to pentamery and the transformation of the outer stamen pairs into (nectariferous) petals. *Berberidaceae* and satellite families have androecia with an alternation of trimerous (dimerous) whorls and outer staminodial nectar-leaves. In *Papaverales* the trimerous flower with a polycyclic androecium often becomes dimerous and is progressively reduced. Secondary increases occur occasionally in *Berberidaceae*, *Papaveraceae* and perhaps *Glaucidiaceae*.

Androecial characters, especially the position and initiation of the stamens, have shown to be valuable tools for understanding systematic relationships. RONSE DECRAENE & SMETS (1993) considered the *Magnoliidae* to be the basic group of their polymerous complex with a wide array of androecial configurations. Polymery reflects a semophyletic trend between several character-states. In this scheme spiral polyandry, i.e. a multistaminate and spiral androecium, occupies the basal end of an evolutionary line to a lower stamen number, substantiating the ranalean view introduced since ARBER & PARKIN (1908). From this character-state several reductive steps arose along diverging lines.

For classificatory purposes most authors rely on a high number of selected characters from different disciplines. Structural aspects (e.g., the position of organs related to one another in the flower) are of utmost importance, but are sometimes ignored. Authors often restrict their approach of the androecium to the number of stamens, not referring to their exact position in the flower, or worse, non-homologous stamen configurations are confused and described as similar. The *Ranunculiflorae* are an excellent illustration of the semophyletic processes that we recognised for polymery. Four different evolutionary steps can be recognised; they can be analysed separately but ought to be taken together to better understand the process of evolution of the androecium: (1) reductive trends affecting the number of stamens and their arrangement in the flower; (2) the derivation of sterile structures (i.e. petals and nectar-leaves) from the stamens; (3) meristic changes in the flower (see also Ronse Decraene & Smets 1994); (4) seemingly unidirectional trends can be switched by secondarily operating increases.

Reductive trends of the androecium. Within the *Ranunculiflorae* a global reductive trend can be postulated, starting from a spiral multistaminate androecium and ending in androecia with very few stamens. A high number of spirally arranged stamens are characteristic for several genera of the *Ranunculaceae* (e.g., *Caltha*, *Helleborus*, *Pulsatilla*; Fig. 1a) and occur in the *Menispermaceae* (Kessler 1993, Endress 1995, this volume). The stamens are characteristically inserted in a spiral on a conical apex and arise centripetally. This character-state links the *Ranunculiflorae* with other (primitive) taxa of the *Magnoliidae* (see Kubitzki 1993, Ronse Decraene & Smets 1993).

An important phylogenetic step is the cyclisation of the original spiral into whorls (polycycly). The transition between a spiral and whorls is often not clearly settled, as the flower may show a mixture of ortostichies and parastichies and whorls may be compound, arising in two times (so-called pseudowhorls; Figs. 1b, 1c). Polycycly is most clearly visible in trimerous or dimerous flowers, but becomes altered in pentamerous *Ranunculaceae* (see below). The presence of outer stamen pairs is characteristic for polycyclic flowers and tends to be an essential step in the reduction of the multistaminate androecium, viz. the first stamens arise as three (two) pairs in alternation with the two innermost tepaline whorls (see also, e.g., Erbar & Leins 1983, Leins & Erbar 1991, Ronse Decraene & Smets 1993). Within the *Ranunculiflorae* polycyclic androecia occur in many *Papaveraceae* (e.g., *Platystemon*, *Chelidonium*, *Eschscholzia*; Fig. 1b) and *Ranunculaceae* (*Eranthis*, *Ranunculus*, *Cimicifuga*; Fig. 1e; Schöffel 1932). In the genera *Meconella* and *Canbya* of the *Papaveraceae* and in the *Fumariaceae* (where the outer stamen pairs are apparently the inner whorl: Fig. 5, Ronse Decraene & Smets 1992) or *Ranunculaceae* with few stamens, the outer stamen pairs remain present when inner whorls have become progressively lost.

A different trend leads to the replacement of the outer pairs by single stamens. Such a switch, as has also been explained elsewhere (Ronse Decraene & Smets 1993), leads to an alternation of simple (often trimerous) whorls. It is characteristic for a separate branch of the *Ranunculiflorae* (*Berberidaceae*, *Menispermaceae*, *Lardizabalaceae*) with tetracyclic or tricyclic androecia. The outer stamen whorls are transformed into nectar-leaves (Fig. 1d). Further loss of whorls leads to dicycly (Fig. 1f; e.g., *Decaisnea*, *Akebia* of the *Lardizabalaceae*) or monocycly (e.g., *Beirnaertia* of *Menispermaceae*), ending with a single stamen in *Odontocarya* (Kessler 1993).

Petals and nectar-leaves. The more primitive taxa of the *Ranunculiflorae* possess an undifferentiated perianth of bracteolate origin (e.g., *Anemone*, *Hydrastis*: Hiepko 1965). One should designate these as tepals. In some cases there is a gradual transition of outer bract-like sepals and inner coloured petals (e.g., *Anemone*, *Coptis*). Increasing adaptations to entomophily are responsible for the abundant presence of petals and nectar-leaves in the *Ranunculiflorae*. Apart from the *Papaveraceae* where nectaries are absent (excluding *Fumariaceae*) and where petals have a tepalar origin (see BERSILLON 1955, RONSE DECRAENE & SMETS 1990), petals represent transformed stamens. This is particularly clear in the pentamerous *Ranunculaceae*, where the petals and nectar-leaves are homologous to the outer stamen pairs. Their ontogeny, position and morphology (Figs. 1a, 1c; HIEPKO 1965, KOSUGE & TAMURA 1989) is reminiscent of staminodial structures. When outer staminodes bear nectarial tissue, they are usually described as nectar-leaves; when they are showy and petaloid, they are termed petals. However, the difference between both structures tends to be blurred by transitions between flowers (1) without nectaries and outer stamens (e.g., *Thalictrum*), (2) inconspicuous staminodes without nectar production and resembling stamens (e.g., *Actaea*, *Anemopsis*) or with some nectar production (e.g., *Myosurus*, *Xanthorhiza*) and (3) nectar-leaves appearing as showy coloured "petals" (e.g., *Aquilegia*, *Ranunculus*). This transformation is either linked with a progressive combination of nectar-production (e.g., "honey-cups" of *Helleborus*, *Nigella*) and a pronounced petaloidy (petals of *Ranunculus*, *Aquilegia*), or with loss of nectaries and the retention of the nectar-bearer (some *Ranunculus*). In other cases nectar-leaves are associated with showy spurs or hoods of tepalar origin (e.g., *Delphinium*, *Aconitum*). As the floral attraction is primitively provided by the coloured perianth (as in *Trollius*, *Eranthis*, *Caltha*), an increased petaloidy of the staminodes also goes hand in hand with an increased sepaloidy of the original tepals (see also FROHNE & JENSEN 1973).

In the *Berberidaceae*-related groups a similar trend has to be considered where the lower stamens have become transformed into nectar-leaves (even becoming spurred: e.g., *Epimedium*) and sometimes arising from common primordia (e.g., SCHMIDT 1928). In the *Lardizabalaceae* and *Menispermaceae* the nectar-leaves may occasionally be reduced or lost.

Changes in the merosity of the flower. As we recently emphasized (RONSE DECRAENE & SMETS 1994), changes in the merosity of the flower play an essential rôle in modelling the androecial configuration of flowers. Changes in merosity (or meristic variations: see SALISBURY 1919, LEPPIK 1964) have been understood to reflect important phylogenetic steps in the *Ranunculaceae*, but the consequences for the arrangement of the androecium have mostly been ignored.

A spiral flower with numerous parts forms a plesiomorphic condition for all *Magnoliidae* (see KUBITZKI 1987, RONSE DECRAENE & SMETS 1994). This spiral configuration has often become rearranged as an alternation of trimerous whorls. The shift of the perianth parts in threes has the direct consequence of disrupting the spiral of the stamens into whorls and is characteristically related with an insertion of the outermost stamens in pairs. Strictly trimerous flowers are abundant in the *Berberidaceae*-associated groups (Fig. 1d), but rare in *Ranunculaceae* and *Papaveraceae*.

The shift from trimery to pentamery is important for understanding the androecial arrangement of many *Ranunculaceae*. As illustrated by RONSE DECRAENE & SMETS (1994), it may be assumed that a pentamerous condition resulted from the fusion of a member of a more external perianth whorl with one of the inner perianth whorl (cf. SALISBURY 1919). This fusion extends to the stamens as whorls become conglomerated into pseudowhorls and some stamens disappear in the process. Most illustrative is the case of the pentamerous topflower of *Berberis*, reported by EICHLER (1878) and ENDRESS (1987).

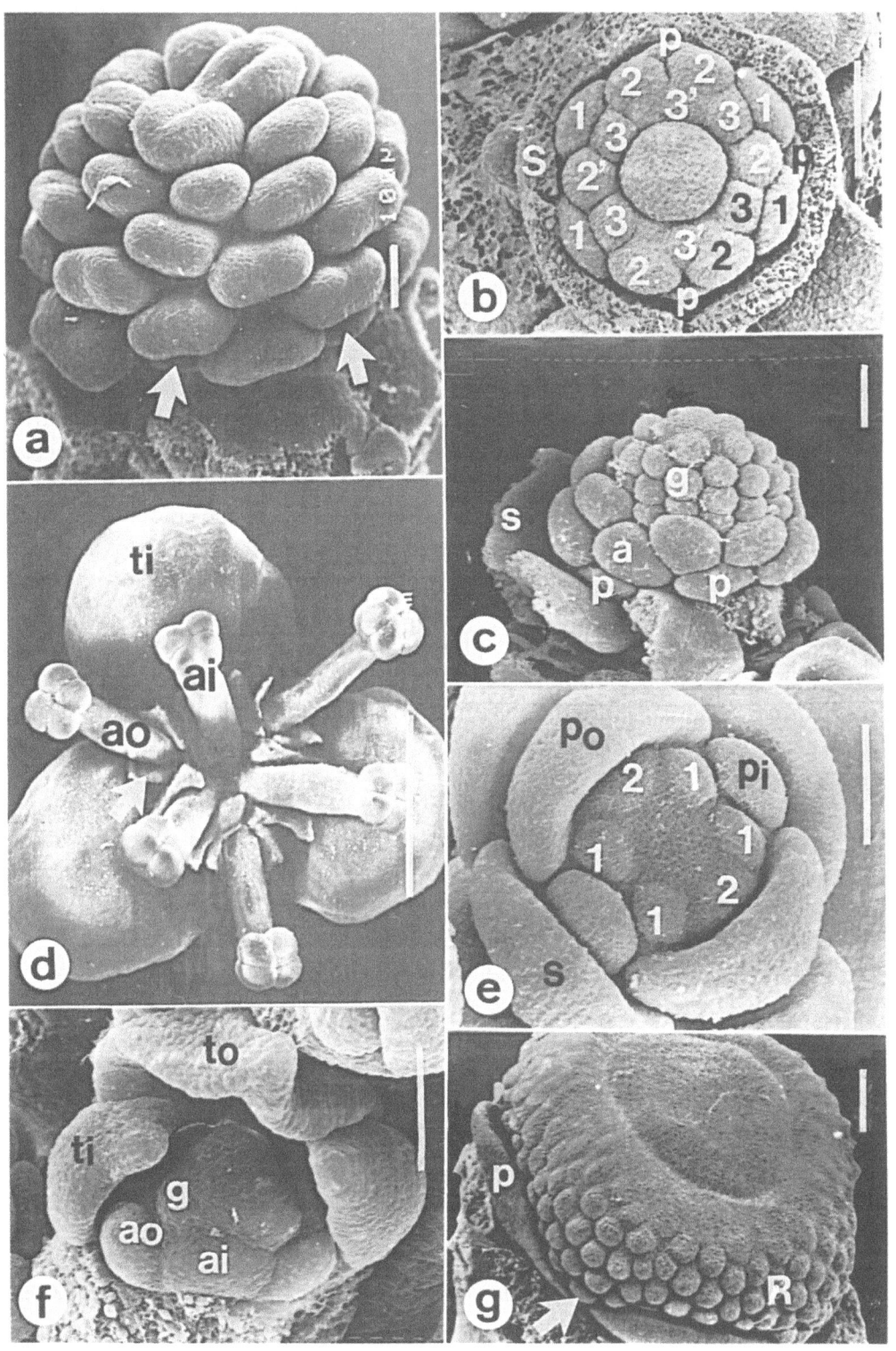

In the *Ranunculaceae* a switch must have occurred relatively early as the androecium is often multistaminate and has a spiral or cyclic inception. Most important is the loss of one stamen of the three outer stamen pairs. This results in five stamens alternating with the five sepals. The outer pairs may easily interfere with a higher trimerous whorl and become conglomerated into a whorl of eight. ENDRESS (1987: 425, Table 1) has illustrated the correlation between successive floral whorls arising along jumping Fibonacci numbers (e.g., *Helleborus, Nigella, Aconitum*). There is an obvious positional link between the remaining five stamens of the pair and the petals or nectar-leaves (Fig. 1a, c).

Another important shift in merosity affects the derivation of dimery. Dimerous flowers are easily obtained from trimery by the loss of a sector as, e.g., in *Clematis* and *Thalictrum* in the *Ranunculaceae, Cissampelos* in the *Menispermaceae, Epimedium* in the *Berberidaceae*, most *Papaveraceae* (Fig. 1b, g) and all *Fumariaceae* (Fig. 1e). Dimerous flowers will have the same stamen arrangement as trimerous flowers, except for the loss of one stamen in each whorl.

It is clear that changes in the merosity of the flower have important consequences for the number of stamens and the appelation of certain stamen configurations. The extensive development of petals has a stabilizing effect on a variable merosity, as is well illustrated by *Aquilegia* having a polycyclic androecium of regularly alternating fives.

Secondary increases. A reduction of the androecium is the common trend in the *Ranunculiflorae*. However, this pattern has been reversed in some cases by a secondary increase. The number of stamens may become multiplied by the division of complex primordia. This is characteristic for some *Berberidaceae* (e.g., *Podophyllum*: DE MAGGIO & WILSON 1986, though their figures are not convincing). In some *Papaveraceae* with a strongly developed circular primordium or ringwall and a very high number of unordered stamens (e.g., *Romneya, Papaver, Meconopsis*: Fig. 1g) there is probably a secondary increase from a lower number of stamens (NEMIROVICH-DANCHENKO 1980, KARRER 1991, RONSE DECRAENE & SMETS 1990). However, to extent this interpretation to all multistaminate *Papaveraceae* is difficult to support. In most cases (e.g., *Eschscholzia, Platystemon*) the development of the androecium runs as in polycyclic flowers (Fig. 1b); the circular primordium is only

---

Fig. 1. *a Helleborus foetidus* (*Ranunculaceae*). Lateral view of young flower bud with parastichies between stamen primordia; arrows point to petal primordia. *b Eschscholzia caespitosa* (*Papaveraceae*). Young flower bud. Numbers indicate the order of inception of successive stamen whorls; note the mixed whorls arising in two times. *c Ranunculus* subg. *Batrachium* (*Ranunculaceae*). Lateral view; note the orthostichies, carpel, stamen and petal primordia. *d Cocculus laurifolius* (*Menispermaceae*). Mature staminate flower; note the tetracyclic arrangement with two whorls of nectar-leaves (arrow) sheathing the base of two inner stamen whorls. *e Corydalis lutea* (*Fumariaceae*). Flower primordium; note the dicyclic androecium with outer pair and inner whorl (numbers). *f Decaisnea fargesii* (*Lardizabalaceae*). Flower with two tepal whorls and dicyclic androecium; note the larger primordia of the inner stamen whorl. *g Papaver somniferum* (*Papaveraceae*). Lateral view of flower bud with large number of stamen primordia arising on a ringwall; the first stamens arise in alternation with the petals. (*ai/ao* inner/outer stamen, *g* carpel, *pi/po* inner/outer petal, *R* ringwall, *ti/to* inner/outer tepal). All bars: 100 μm, except *d*: 1 mm

weakly developed or inexistent (see NEMIROVICH-DANCHENKO 1980, RONSE DECRAENE & SMETS 1990). A higher stamen number tends to be superimposed on a lower number of primordia, linked with the extensive development of a ringwall. Stamens arise centrifugally on a strongly convex apex in *Glaucidium* (*Glaucidiaceae*: TAMURA 1972), though this is questioned by Endress (pers. comm.).

Taxonomic considerations. Because of the easy ordination of some androecial character states as plesiomorphic and apomorphic characters, some considerations about the taxonomic and phylogenetic relationships within the *Ranunculiflorae* are permitted. However, affinities can only be constructed on a wide array of characters and androecial character states are interesting in so far as they are congruent to other.

The *Ranunculaceae* occupy a basal position in the *Ranunculiflorae*. The combination of a reductive trend in the number of stamens, meristic changes and an intergradation between fertile stamens, nectar-leaves and petals gives the *Ranunculaceae* a great diversity of floral and androecial patterns. Occasionally an androecium may be obtained resembling the diplostemony and haplostemony current in other subclasses (e.g., *Xanthorrhiza*). *Circaeaster* (usually placed in *Circaeasteraceae*) represents an extreme in the floral reduction of the *Ranunculaceae*. The flower has only two or three tepals and one to three alternitepalous stamens (obmonocycly). *Kingdonia*, placed in *Circaeasteraceae* by CRONQUIST (1981) conforms to other primitive *Ranunculaceae* by its spiral flower and numerous outer staminodes.

A number of separate evolutionary lines may be distinguished, each one being closely related to *Ranunculaceae*-like ancestors. *Papaveraceae* and *Fumariaceae* have very close links with the *Ranunculaceae*, morphologically as well as phytochemically. They have been derived from a trimerous polycyclic ancestor with three alternating whorls of tepals. A switch to dimery is common and is linked with a pronounced reduction of the stamen numbers, which takes an extreme form in the *Fumariaceae* with the fusion between stamens and the partial reduction of the pollen sacs of the outer stamens (Fig. 1e). In the monocyclic *Hypecoum* the two outer monothecal pairs become fused laterally into two stamens (RONSE DECRAENE & SMETS 1992). In *Pteridophyllum* (sometimes placed in *Pteridophyllaceae*: see LIDÉN 1993) only the outer stamen pairs remain in the obmonocyclic flower. According to LIDÉN several characteristics suggest a closer relation to *Ranunculaceae* than to *Papaverales*.

A second evolutionary line is found in families grouped around the *Berberidaceae*. The main characteristics are the presence of trimerous (occasionally dimerous) flowers with polycyclic to monocyclic androecia and the two outer stamen whorls being transformed into nectar-leaves (Fig. 1d). Stamen pairs have since long been replaced by single stamens. *Lardizabalaceae* are often considered as the most primitive group (KUBITZKI 1993, CARLQUIST 1995, this volume), but on the basis of their floral construction *Menispermaceae* have the greatest diversity. *Menispermaceae*, *Berberidaceae* and *Lardizabalaceae* seem to be strongly linked with each other and based on the association of different floral traits an order *Berberidales* containing the above-mentioned families may be considered. *Glaucidium* has commonly been associated with *Hydrastis*. *Hydrastis* does not differ much from *Ranunculaceae* in having numerous spirally arranged stamens (TOBE & KEATING 1985). The position of *Glaucidium* in the *Ranunculaceae* is supported by its strongly convex apex as a characteristic feature of the latter family. However, the androecium arises centrifugally (TAMURA 1972), contrary to the centripetal inception of the *Ranunculaceae*. The development of the androecium is also different from *Paeoniaceae* where five stamen groups are formed on a sunken apex (e.g., HIEPKO 1964, LEINS & ERBAR 1991). More should be known about the floral development of the genus, as there might be a reversal in stamen inception from the normal acropetal sequence.

Conclusions. The *Ranunculiflorae* remain a fascinating superorder regarding floral and androecial evolution. Especially in the *Ranunculaceae* there is an enormous diversity with numerous "evolutionary trials" and the budding of androecial structures approaching the current configuration of the *Dilleniidae*, *Rosidae* and *Asteridae* (diplostemony and haplostemony). The androecial arrangement strongly suggests the existence of separate lines departing from the *Ranunculaceae*. In fact the creation of three orders *Ranunculales* (including *Ranunculaceae* and *Circaeasteraceae*), *Berberidales* (including *Berberidaceae*, *Lardizabalaceae* and *Menispermaceae*) and *Papaverales* (including *Papaveraceae* and *Fumariaceae*) might be considered.

We wish to thank the Belgian National Fund for Scientific Research (N.F.W.O. – Kredieten aan navorsers) for financial support. This research was supported by grants from the N.F.W.O. projects no. G.0143.95 (general research project) and no. 2.0038.91 (scanning electron microscope). The leading author is a postdoctoral researcher of the N.F.W.O.

References

ARBER, E. A. N., PARKIN, J., 1908: The origin of Angiosperms. – Bot. J. Linn. Soc. **38**: 29–80.

BERSILLON, G., 1955: Recherches sur les Papavéracées. Contribution à l'étude du développement des *Dicotylédones* herbacées. – Ann. Sci. Nat. Bot. **11**: 225–443.

CARLQUIST, A., 1995: Wood anatomy of the *Ranunculiflorae*, a summary. – Pl. Syst. Evol. [Suppl.] **9**: 11–24.

CRONQUIST, A., 1981: An integrated system of classification of flowering plants. – New York: Columbia University Press.

DE MAGGIO, A. E., WILSON, C. L., 1986: Floral structure and organogenesis in *Podophyllum peltatum* (*Berberidaceae*). – Amer. J. Bot. **73**: 21–32.

EICHLER, A. W., 1878: Blüthendiagramme **II**. – Leipzig: Engelmann.

ENDRESS, P. K., 1987: Floral phyllotaxis and floral evolution. – Bot. Jahrb. Syst. **108**: 417–438.

ENDRESS, P. K., 1995: Floral structure and systematics of the *Ranunculaceae*. – Pl. Syst. Evol. [Suppl.] **9**: 47–61.

ERBAR, C., LEINS, P., 1983: Zur Sequenz von Blütenorganen bei einigen Magnoliiden. – Bot. Jahrb. Syst. **103**: 433–449.

FROHNE, D., JENSEN, U., 1973: Systematik des Pflanzenreichs. – Stuttgart: G. Fischer.

HIEPKO, P., 1964: Das zentrifugale Androeceum der *Paeoniaceae*. – Ber. Deutsch. Bot. Ges. **77**: 427–435.

– 1965: Vergleichend-morphologische und entwicklungsgeschichtliche Untersuchungen über das Perianth bei den *Polycarpicae*. – Bot. Jahrb. Syst. **84**: 359–508.

KARRER, A. B., 1991: Blütenentwicklung und systematische Stellung der *Papaveraceae* und *Capparaceae*. – Doctoral dissertation, University of Zürich.

Kessler, P. J. A., 1993: *Menispermaceae.* – In Kubitzki, K., Rohwer, J. R., Bittrich, V., (Eds): The families and genera of vascular plants, **2**, pp. 402–418. – Berlin: Springer.

Kosuge, K., Tamura, M., 1989: Ontogenetic studies on petals of the *Ranunculaceae.* – J. Japan. Bot. **64**: 65–74.

Kubitzki, K., 1987: Origin and significance of trimerous flowers. – Taxon **36**: 21–28.

– 1993: Introduction. – In Kubitzki, K., Rohwer, J. G., Bittrich, V., (Eds): The families and genera of vascular plants, **2**, pp. 1–12. – Berlin: Springer.

Leins, P., Erbar, C., 1991: Fascicled androecia in *Dilleniidae* and some remarks on the *Garcinia* androecium. – Bot. Acta **104**: 336–344.

Leppik, E. E., 1964: Floral evolution in the *Ranunculaceae.* – Iowa State Coll. J. Sci. **39**: 1–101.

Lidén, M., 1993: *Pteridophyllaceae.* – In Kubitzki, K., Rohwer, J. G., Bittrich, V., (Eds): The families and genera of vascular plants, **2**, pp. 556–557. – Berlin: Springer.

Nemirovich-Danchenko, E. N., 1980: The development of the androecium and the nature of polyandry in some *Papaveraceae.* – Bot. Ž. **65**: 1088–1100. (In Russian.)

Ronse Decraene, L. P., Smets, E. F., 1990: The systematic relationship between *Begoniaceae* and *Papaveraceae*: a comparative study of their floral development. – Bull. Jard. Bot. Natl. Belg. **60**: 229–273.

– 1992: An updated interpretation of the androecium of the *Fumariaceae.* – Can. J. Bot. **70**: 1765–1776.

– 1993: The distribution and systematic relevance of the androecial character polymery. – Bot. J. Linn. Soc. **113**: 285–350.

– 1994: Merosity in flowers: definition, origin, and taxonomic significance. – Pl. Syst. Evol. **191**: 83–104.

Salisbury, E. J., 1919: Variation in *Eranthis hyemalis, Ficaria verna*, and other members of the *Ranunculaceae* with special reference to trimery and the origin of the perianth. – Ann. Bot. **33**: 47–79.

Schmidt, E., 1928: Untersuchungen über Berberidaceen. – Beih. Bot. Centralbl. **45**: 329–396.

Schöffel, K., 1932: Untersuchungen über den Blütenbau der Ranunculaceen. – Planta **17**: 315–371.

Tamura, M., 1972: Morphology and phyletic relationship of the *Glaucidiaceae.* – Bot. Mag. (Tokyo) **85**: 29–41.

Tobe, H., Keating, T. C., 1985: The morphology and anatomy of *Hydrastis (Ranunculales)*: systematic reevaluation of the genus. – Bot. Mag. (Tokyo) **98**: 291–316.

Address of the authors: L. P. Ronse Decraene and E. Smets, Laboratory of Systematics, Botanical Institute, Katholieke Universiteit Leuven, Kard. Mercierlaan 92, B-3001 Heverlee, Belgium.

Accepted December 23, 1994

Pl. Syst. Evol. [Suppl.] 9: 71–82 (1995)

Palynology and systematics of *Ranunculiflorae*

STEPHEN BLACKMORE, PETER STAFFORD, and VIVECA PERSSON

Received November 11, 1994

Key words: *Ranunculiflorae, Berberidaceae, Fumariaceae, Lardizabalaceae, Menispermaceae, Nelumbonaceae, Papaveraceae Ranunculaceae.* -Evolution, palynology, pollen morphology, spiralization, successiformy, systematics.

Abstract: The pollen morphological variation in the families of the *Ranunculiflorae* (*Berberidaceae, Fumariaceae, Lardizabalaceae Menispermaceae Nelumbonaceae, Papaveraceae* and *Ranunculaceae*) is summarised and discussed in relation to the systematics of the group. The pollen grains are generally dispersed as monads, but permanent tetrads occur in a few instances. Aperture numbers range from one, in spiraperturate pollen of *Berberis* and some species of *Anemone* and in the rare monosulcate grains that occur in some individuals of *Nelumbo*, to three in the majority of species, to many. These tricolpate-derived forms range from zonocolpate to pantocolpate and pantoporate forms. With the exception of some *Papaver* species, the occurrence of endoapertures appears restricted to *Menispermaceae*, where they are generally present. The ornamentation is generally punctate to microreticulate, often with spinules or striate to striate-reticulate. The exine stratification is generally tectate-columellate, with a granular infratectum in some taxa, and the endexine is well developed. Pollen grains of *Ranunculiflorae* exhibit successiformy and spiralization, two of the evolutionary patterns that were first recognised and defined by VAN CAMPO (1967, 1976). These pollen morphological patterns result in a high level of homoplasy (similarity arising through convergent or parallel evolution). Many of the palynological characters are autapomorphic and thus serve to identify distinctive taxa (such as *Berberis* or *Fumaria*). Fewer characters can clearly be recognised as synapomorphies at the familial or subfamilial level, especially in the absence of a detailed character analysis integrating palynological and other kinds of evidence. One potentially important character that may not be subject to extensive homoplasy is the occurrence of a compound layer of striae (sensu NOWICKE & SKVARLA 1982) in contrast to psilate, punctate or microreticulate ornamentation with scabrae or microechinae.

This paper sets out to review the diversity of pollen morphology in the *Ranunculiflorae* and to consider the contribution of palynology to our knowledge of systematic relationships within the group. The pollen morphology of the *Ranunculiflorae* is well documented as a result of numerous studies at family or genus level. A selective bibliography for each family, extracted from over 300 publications on pollen of the group, is provided here. A few important studies of broader scope, or focused at higher taxonomic levels, have also considered the implications of the pollen morphology of the *Ranunculiflorae*. In addition to the most comprehensive modern studies of NOWICKE & SKVARLA (1980, 1981, 1982, 1983) and FERGUSON (1975, 1978, as well as HARLEY & FERGUSON 1982), there are also the classic works of WODEHOUSE (1935) and ERDTMAN (1952). This review draws heavily on previously published works and, whilst we have investigated selected representatives of all the families, previously unpublished investigations have been undertaken only for the *Nelumbonaceae* and *Lardizabalaceae*.

Following KUBITZKI (1993), the *Ranunculiflorae* is taken to include the families *Berberidaceae* (including *Nandinaceae*), *Fumariaceae*, *Lardizabalaceae* (including *Sargentodoxaceae*), *Menispermaceae*, *Nelumbonaceae*, *Papaveraceae* and *Ranunculaceae* (including *Circaeasteraceae*, *Kingdoniaceae* and *Hydrastidaceae*). For each family a list of key references to previous palynological studies is followed by a summary of the pollen morphological diversity together with discussions of some palynological characters and their patterns of occurrence.

Pollen morphology varies extensively within the *Ranunculiflorae* and often there are similar patterns of variation found within families or genera. The repetitive patterns of variation, especially in aperture number and position, were recognised by VAN CAMPO (1967, 1976) from the *Ranunculidae* and other taxa. In an analysis of the variation of pollen morphology in angiosperms VAN CAMPO noted that some individual species, such as *Claytonia virginica* L. (*Portulacaceae*), exhibit a range of infraspecific variation equal to that found within the entire family. Elsewhere, she observed that the same modes of interspecific variation are found in the related families which constitute an order. The three major patterns of variation she recognised as arising from tricolpate pollen were successiformy, breviaxy and spiralization. According to VAN CAMPO (1976: 129). "Successiformy is characterised by the establishment of a high level of symmetry on subspherical pollen grains by a regular arrangement of the apertures". She attributed the repetitious morphological patterns to the fact that apertural configurations in angiosperms obey a limited number of geometric rules, expressed during development, which produce a limited number of patterns of variation. This explanation she considered to elaborate upon concepts proposed by WODEHOUSE (1935) and D'ARCY THOMPSON (1917). The consequences of these morphological patterns are discussed in relation to the systematics of the *Ranunculiflorae*.

Material and methods

Pollen samples were taken from specimens in the herbarium of the Natural History Museum, London (BM) and prepared by acetolysis (ERDTMAN 1960). Measurements and basic descriptive information were obtained by light microscopy. Scanning electron microscopy was carried out using a Hitachi S800 field emission SEM. Details of the herbarium specimens from which pollen was sampled are only provided for those species that are illustrated and are given in the captions to the figures.

Berberidaceae (Fig. 1 a–d)

Literature. KUMAZAWA (1936), ARCHANGELSKY (1973), ROLAND-HEYDACKER (1974), NOWICKE & SKVARLA (1981), BLACKMORE & HEATH (1984), CHANG & WANG (1983).

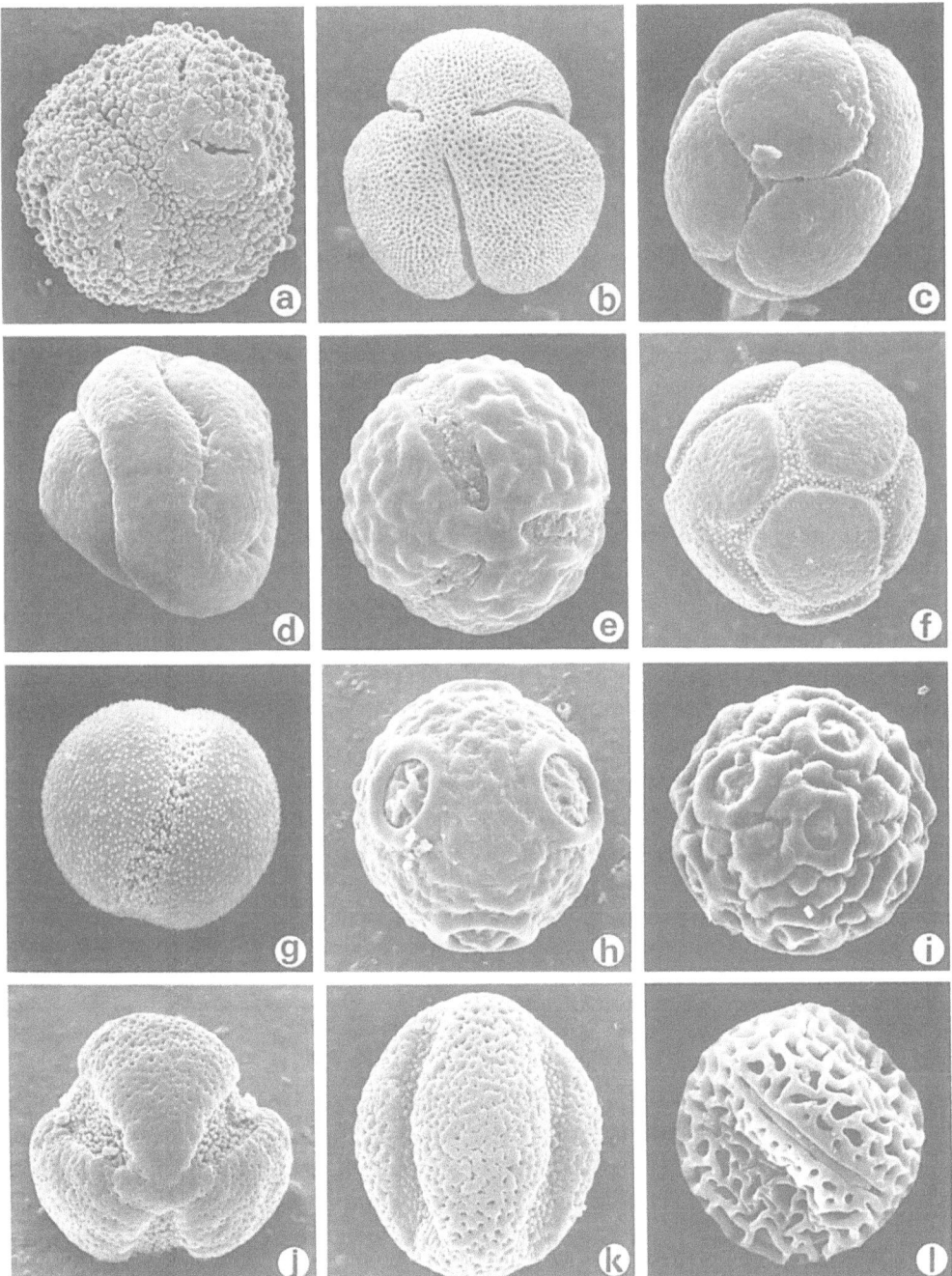

Fig. 1. Scanning electron micrographs of selected *Ranuculiflorae* pollen grains. *Berberidaceae a–d. a Podophyllum hexandrum* ROYALE. Pollen tetrad. STAINTON, SYKES & WILLIAMS *1005* (x 670). *b Bongardia chrysogonum* (L.) GRISEBACH. Polar view. BALLS B. *2174* (x 1100). *c Mahonia aquifolium* (PURSH) NUTALL. Poly-syncolpate grain. HERB REDGROVE s.n. (x 1100). *d Berberis vulgaris* L. Spiraperturate pollen grain. ARMITI s.n. (x 1100). *Fumariaceae e–i. e Corydalis yunnanensis* FRANCH. Polar view. FORREST s.n. (x 1400). *f Platycapnos saxicola* WILLK. Poly-syncolpate Grain R. M. & A. M. HARLEY *630* (x 1100). *g Hypecoum leptocarpum* HOOK. f. & THOMS. Di-colpate grain. *Anon. 10610* (x 1700). *h Rupicapnos africana* POMEL. Pantoporate grain. DAVIS *51489* (x 1100). *i Fumaria capreolata* L. Pantoporate grain. GADECAU s.n. (x 950) *Lardizabalaceae j, k . j Akebia pinnata* DECNE. Polar view, tricolpate grain. BISSET *22* (x 1700). *k Akebia lobata* DECNE. Equatorial view, tricolpate grain. LICENT *4990* (x 1700). *Menispermaceae l. l Cyclea hypoglauca* DIELS. GRESSITT *898* (x 2800)

Dispersal unit. Generally monads but tetrads occur in *Podophyllum* L. and dyads in *Ranzania* Itô.

Apertures. Tricolpate in all taxa except in *Berberis* L. which is spiraperturate, *Mahonia* NUTT. which is irregularly polysyncolpate and *Ranzania* which is 6-pantocolpate.

Ornamentation. Ranging from psilate (in *Berberis* and *Mahonia*), echinate, micro-reticulate or striate (in *Achlys* DC, *Epimedium* L. *Jeffersonia* Barton, *Vancouveria* Morren & Decne.).

Exine. Tectate columellate except in *Berberis* and *Mahonia* which have an amorphous or somewhat granular ectexine structure. A distinctive feature in *Nandina* is the thick endexine, which is thicker than the ectexine.

Size and shape. Spherical to prolate or prolately spheroidal, ranging from 22–26 µm in *Achlys japonica* Maxim. to 50–58 µm in *Bongardia chrysogonum* Boiss.

Comments. NOWICKE & SKVARLA (1981) concluded that pollen morphology did not support the concept of a close relationship between the *Berberidaceae* and the *Ranunculaceae* or *Lardizabalaceae*. However, they identified one character, aperture columellae (columellae-shaped ectexine units that penetrate the endexine at apertures) that is a potential synapomorphy of the *Berberidaceae* and *Ranunculaceae*. Other features were regarded as unusual parallelisms with *Cistaceae* and between *Podophyllum* and *Croton* L. Striate exine ornamentation of the kind referred to as a compound layer of striae (NOWICKE & SKVARLA 1982) also occurs elsewhere among the *Ranunculiflorae* in *Circaeaster Maxim.*, *Hydrastis* L. *Megalantheris* OHWI, *Kingdonia* BALF. F. & W. SM. and *Trollius*.

Fumariaceae (Fig. 1 e–i)

Literature. BELMONTE & al. (1986), CANDAU & SOLER (1981), HEUSSER (1971), IKUSE (1956), KALIS (1979), RACHELE (1974), STERN (1962), VAN CAMPO (1978).

Dispersal unit. Monads.

Apertures. Variable. *Hypecoum* L. has two longitudinal colpi, in subg. *Chiazospermum* the colpi are fused. Tricolpate pollen occurs in many species, frequently in the same anthers as 6-pantocolpate grains. Panto-porate grains of subtribe *Fumariinae*, have distinctive protruding annulli around the ectoapertures. In *Platycapnos* (DC) Bernh. the colpi are fused to form a network.

Ornamentation. Generally psilate or verrucate.

Exine. Tectate-columellate with very small columellae that are not usually visible in the light microscope.

Size and shape. Generally spheroidal to prolate-spheroidal. The outline of unacetolysed *Fumaria* L. pollen grains is distinctive because of the protruding intine at the apertures. Sizes in the range of 15–55 µm have been reported.

Comments. There are palynological differences between the *Fumariaceae* and the *Papaveraceae*. The former are usually pantoaperturate, with six or twelve apertures, verrucate or psilate ornamentation and very small columellae while the latter are typically 3-colpate, punctate or microreticulate with scabrae and have larger columellae that are arranged in curving rows or form an infratectum (KALIS

1979). Palynological characters also appear to define some of the groups within the family. Subtribe *Discocapninae* is characterised by short broad colpi and subtribe *Fumariinae* by the distinctly annulate apertures.

Lardizabalaceae (Fig. 1 j–k)

Literature. AGABABIAN (1973), HEUSSER (1971), KUMAZAWA (1936), (NOWICKE & SKVARLA (1982), WALKER (1976), WALKER & DOYLE (1975), XIA & PENG (1989).
    Dispersal unit. Monads.
    Apertures. Tricolpate with granular aperture membranes.
    Ornamentation. Punctate or foveolate to microreticulate.
    Exine. Tectate-columellate with a thick tectum and small columellae. The endexine has a solid outer zone and a plate-like inner zone (NOWICKE & SKVARLA 1982).
    Size and shape. Prolate to prolate spheroidal. Sizes in the range from 20–40 µm.
    Comments. Pollen grains of the family are all tricolpate. The colpi are broad with granular aperture membranes. The ornamentation varies from punctate to microreticulate. *Sargentodoxa* Rehder & E. Wilson conforms to the typical pollen morphology of the family (NOWICKE & SKVARLA 1982).

Menispermaceae (Fig. 1 l; Fig. 2 a,b)

Literature. FERGUSON (1975, 1978), HARLEY & FERGUSON (1982), IKUSE (1956), THANIKAIMONI (1968, 1984).
    Dispersal unit. Monads.
    Apertures. Inaperturate in *Tiliacora* COLEBR. and *Disciphania* EICHLER. Tricolpate (in *Rhaptonema* MIERS, *Spirospermum* THOUARS, and *Strychnopsis* BAILLON), tricolporate in the vast majority of species (with 4–6 aperturate grains in some species) and triporate (in *Stephania* LOUR.). The colpi may be free or fused (as in *Tinospora* MIERS). An operculum is present in some genera.
    Ornamentation. Microreticulate to coarsely reticulate. The muri in some cases are themselves ornamented.
    Exine. Tectate-columellate.
    Size and shape. Variable in shape from spheroidal to prolate. Generally rather small, in the range of (10–20 µm, with grains up to 45 µm reported in *Legnephora* MIERS.
    Comments. Despite the diversity of pollen encountered in the family, there appears to be relatively little correlation between pollen morphology and gross morphological variation or the subtribal classification FERGUSON (1975) found palynological characters of the *Triclisieae* helpful in grouping some genera and in providing support for the recent classifications and pointed out that "there is a complex reticulation of 'advanced' and 'primitive' pollen characters in the tribe and pollen morphology alone cannot really be considered to contribute any single linear trend in the systematic relationships of the group" (FERGUSON 1975: 73).

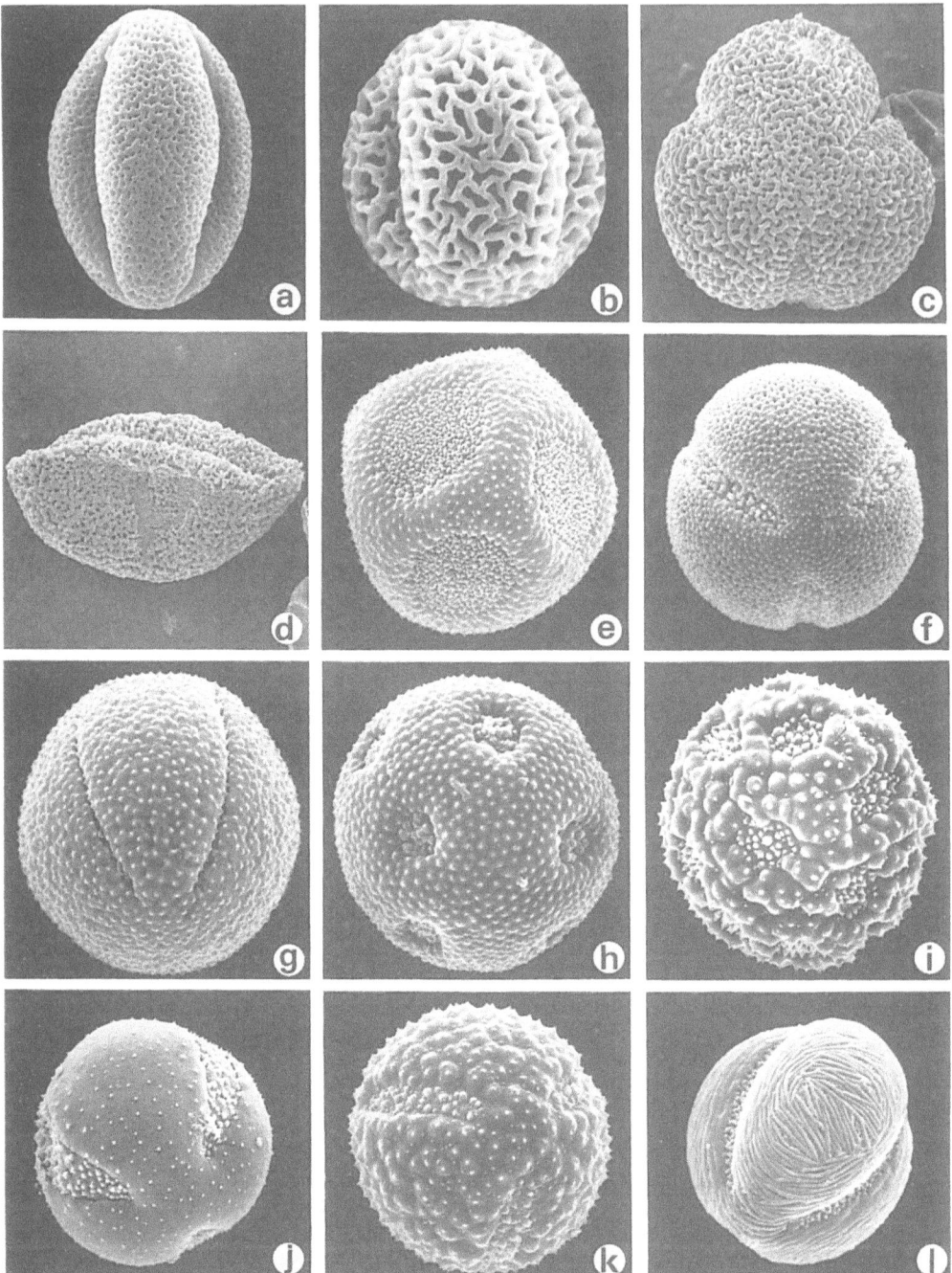

Fig. 2. Scanning electron micrographs of selected *Ranuculiflorae* pollen grains. *Menisper-maceae a,b. a Menispermum dauricum* DC. Equatorial view, tricolpate grain. NECZAEVA s.n. (x 2200). *b Strychnopsis thouarsii* BAILL. Equatorial view, coarsely reticulate tricolpate grain. *Hildebrandt 3160* (x 3300). *Nelumbonaceae c,d. c Nelumbo nucifera* GAERTN. Polar view, tricolpate grain. LUDLOW & SHERRIFF *7797* (x 850). *d Nelumbo nucifera* GAERTN. Equatorial view, monosulcate grain. LUDLOW & SHERRIFF *7797* (x 670). *Papaveraceae e–g. e Papaver argemone* L. Pantoporate grain. SELL *67/310* (x 1400). *f Chelidonium majus* L. Polar view. SHORT *175* (x 1700). *g Eschscholzia californica* CHAMISSO. Poly-zonocolpate grain. KELLOG & HARFORD *32* (x 1700). *Ranunculaceae h–l. h Thalictrum flavum* L. Pantoporate grain. GERRANS *428* (x 2200). *i Ranunculus parviflorus* L. Pantoporate grain. MARSHALL s.n. (x 2200). *j Nigella damascena* L. Oblique polar view, tricolpate grain. SIMPSON *36543* (x 950). *k Ranunculus acris* L. Polar view FRIDTZ *28908 753* (x 1700). *l Trollius europaeus* L. Tricolpate, striate pollen. LAUNERT s.n. (x 2200).

Nelumbonaceae (Fig. 2 c,d)

Literature. Huang (1972), Ikuse (1956), Kuprianova (1979), Walker (1976).

Dispersal unit. Monads.

Apertures. Tricolpate, mixed with monosulcate grains in some individual plants (Kuprianova 1979).

Ornamentation. Microreticulate/rugulate.

Exine. Tectate-columellate.

Size and shape. 50–60 µm. Pollen grains globose-spheroidal or, in the mono-sulcate grains, boat-shaped.

Comments. Various workers, from Erdtman (1952) onwards, have pointed out that *Nelumbo* pollen is morphologically isolated from the *Nymphaeaceae* and from the *Ranunculaceae* or *Berberidaceae*. The mixed occurrence of monosulcate and tricolpate grains in at least some samples of *Nelumbo* Adans., reported by Kuprianova (1979) and confirmed by our observations, has profound implications for the evolutionary relationships between these two aperture types.

Papaveraceae (Fig. 2 e–g)

Literature. Belmont & al. (1986). Candau & Fernandez-Paniagua (1985), Henderson (1965), Ikuse (1956), Kalis (1979), Layka (1976, 1977), Rachele (1974), Van Campo (1978).

Dispersal unit. Monads, with tetrads in a few species (for example, *Meconopsis sherriffi* and reported in cultivated plants of *Eschscholzia* Cham.).

Apertures. Inaperturate (in *Meconopsis sherriffii* G. Taylor), tricolpate, poly-colpate or polyforate. Endoapertures are absent except in *Papaver argemone* L.

Ornamentation. Punctate or microperforate with minute scabrae.

Exine. Tectate-columellate with distinct columellae (in light microscopy).

Size and shape. Prolate to prolate spheroidal. Size ranges from 14–69 µm have been reported.

Comments. As mentioned above, the distinctions between *Fumariaceae*, and *Papaveraceae* are reasonably clear. Within the *Papaveraceae* the subfamilies show some correlations with aperture type. The *Platystemonoideae* are tricolpate through-out whilst in the *Chelidonioideae*, *Chelidonium* L., *Dicranostigma* Hook. f. & Thomson, *Glaucium* Miller and *Hylomecon* Maxim., are tricolpate but *Bocconia* L., *Eomecon* Hance, *Macleaya* R.Br. and *Sanguinaria* L. are polyforate. The *Eschscholzioideae* are polycolpate with 4–11 colpi. Most *Papaveroideae* are tricolpate, exceptions being *Roemeria* Medikus and *Papaver* sect. *Argemonidium* with 4–5 rounded apertures, and some species of *Meconopsis* Viguier, which can be inaperturate, have more than 3 colpi, or 6–8 rounded apertures. Tetrads occur in *Meconopsis sherriffii* (Henderson 1965) and cultivated plants of *Eschscholzia* (Clark 1978).

Ranunculaceae (Fig. 2 h–l)

Literature. Al Eisawi (1986), Fernandez (1985, 1986), Heusser (1971), Kumazawa (1936), Lee & Blackmore (1992), Nowicke & Skvarla (1980, 1981, 1982, 1983), Clarke & al. (1991), Santisuk (1979).

Dispersal unit. Monads.

Apertures. Simple, ranging from the most widespread tricolpate arrangement in *Adonis* L., *Callianthemum* C. MEYER (pro parte), *Hamadryas* COMM. JUSS. EX, *Myosurus* L., *Paroxygraphis* W. W. SM., *Ranunculus* L. (pro parte) and *Trautvetteria* FISCHER & C. MEYER to pantocolpate, pantoporate in *Clematis* L. (pro parte) or spiraperturate in *Anemone* L. (pro parte). Irregular apertures frequently occur, colpi are often partially or completely syncolpate, especially in *Ranunculus* species.

Exine. Tectate-columellate, most with a single thin nexine layer and 3 layers of sexine, but up to 5 sexine layers may be present in *Nigella* L. Dimorphic columellae are widespread in the family with monomorphic columellae present in *Adonis*, *Aquilegia* L., *Caltha* L., *Helleborus* L. and *Trollius* L., and granular columellae in *Aconitum* L. (SANTISUK 1979; NOWICKE pers. comm.). In pollen with dimorphic columellae, the larger columellae project through the tectum as the spinules (NOWICKE & SKVARLA 1981, Figs. 115, 120, 121). Aperture columellae are present in *Anemone*, *Batrachium* (DC) GRAY, *Clematis* and *Ranunculus*.

Ornamentation. Variable, but most commonly involving a perforate tectum with scabrate or microechinate ornamentation. Striate ornamentation occurs in *Trollius* and *Megalantheris*, striate reticulate in *Circaeaster, Kingdonia* and *Hydrastis*, whilst *Helleborus* is reticulate.

Shape and size. Prolate or prolate spheroidal to spheroidal. Sizes ranging from 16–60 μm have been reported.

Comments. The palynological diversity of the *Ranunculaceae* is almost as great as that of the *Ranunculiflorae*. Apertures range from tricolpate to multiaperturate and, rarely, spiraperturate. Both of the major exine types recognised by NOWICKE & SKVARLA (1982) are present, the compound layer of striae in *Trollius* and *Megalantheris* (LEE & BLACKMORE 1992), and the perforate scabrate type elsewhere in the family. This great diversity makes the identification of dispersed pollen grains possible to generic and sometimes even to species level (CLARKE & al. 1991). Such identification is complicated by the fact that many *Ranunculaceae* pollen grains are similar in form to those of other families. This diversity of form is less helpful from a systematic standpoint in that many characters vary within a genus, hence pollen characters do not readily serve to group together related genera.

Discussion

There is wide variation in pollen morphology in the *Ranunculiflorae* and hence palynological characters have the potential to contribute to systematic analyses. However, as with all morphological characters, the ability of this variation to contribute to the resolution of relationships between taxa depends on the context of the analysis both in terms of the taxa and the characters that are compared. The tendency in much of the palynological literature on the *Ranunculiflorae* has been to draw conclusions independently from the other characters of the plants. It is not possible to make universal interpretations of the taxonomic significance of individual characters since synapomorphies at one level of analysis may be plesiomorphies or autapomorphies in the context of other levels of analyses (SCOTLAND 1993). We have resisted this temptation to generalise because, although the distribution and systematic significance of palynological characters is often unequivocal, few such examples are apparent at the higher level within the *Ranunculiflorae*. We discuss some of the palynological characters below.

Apertures. Within the *Ranunculiflorae* there are examples of virtually all the tricolpate-derived aperture configurations found in angiosperms and occasionally distally monosulcate grains occur in *Nelumbo*. The repeating patterns of tricolpate to pantocolpate and pantoporate apertures occur in several families of *Ranunculiflorae*, conforming to the successiform series of VAN CAMPO (1967, 1976). VAN CAMPO proposed that similar aperture configurations arise in independent evolutionary lineages because there is a limited repertoire of developmental processes that define the aperture configurations. Even now, relatively little is known about the precise mechanisms that determine aperture configuration but there are established links to the behaviour and symmetry of the meiotic spindle and to the mode and tempo of cytokinesis (HESLOP-HARRISON 1966, HUYHN 1967, BLACKMORE & BARNES 1995). These developmental mechanisms exert a profound influence over pollen morphology and clearly there are high levels of parallelism and convergence. As a result, although aperture configurations are valuable in the identification of dispersed pollen of *Ranunculiflorae*, they are not appropriate indicators of relationships, at family or tribal level.

Nevertheless, some apertural features have more systematic potential. NOWICKE & SKVARLA (1981) considered aperture columellae (columellae-shaped ectexine units that penetrate the apertural endexine) to be a potential synapomorphy of the *Berberidaceae* and *Ranunculaceae*. Endoapertures are absent from the majority of *Ranunculiflorae*, but present in *Menispermaceae* and some species of *Papaver*. However, ERDTMAN (1952) sometimes described other *Ranunculiflorae* grains as colporoidate, indicating that ill-defined endoapertures are present. The pantoporate grains of subtribe *Fumariinae* have distinctive protruding annuli around the apertures that provide an example of a clear synapomorphy.

Exine stratification and ornamentation. The detailed ultrastructural studies of NOWICKE & SKVARLA (1981, 1982, 1983) and FERGUSON (1975, 1978) show the significant systematic potential of variations in exine stratification, especially at the level of generic relationships. At higher taxonomic levels, the most important character is the distinction between the striate ornamentation found in members of several families of the *Ranunculiflorae* and the more widespread simple, often spinulose, tectum (NOWICKE & SKVARLA 1982). Striate exine is generally interpreted as derived and occurs in *Circaeaster*, *Hydrastis*, *Megalantheris*, *Kingdonia* and *Trollius*. Although the simplest interpretation is that striate pollen has arisen several times within the *Ranunculiflorae*, it is possible that some of these genera represent close relatives. *Megalantheris* and *Trollius* may even be congeneric (LEE & BLACKMORE 1992).

Conclusion

The high levels of convergence in palynological characters, especially apertures, means that caution is needed in interpreting the systematic significance of pollen characters. Ideally, they should be incorporated in a larger analysis, together with other sources of characters. However, the degree of convergence is such that at many levels of analysis the inclusion of palynological characters may actually increase the number of trees obtained. This should not discourage such analyses, it is clear that there are also situations where pollen characters will contribute to improved resolution!

We are grateful to Dr UWE JENSEN and Dr JOACHIM KADEREIT for the invitation to contribute to this stimulating symposium and to Dr JOAN NOWICKE for helpful suggestions made in the process of reviewing the manuscript.

References

AGABAVIAN, V. C., 1973: Pollen of primitive angiosperms. – Erevan Bot. Inst. Akad. Nauk. Armenia SSR.

AL– EISAWI, D., 1986: Pollen morphology of *Ranunculaceae* in Jordan. – Pollen & Spores **28**: 311–328.

ARCHANGELSKY, D. B., 1973: Palynological taxonomy of *Berberidaceae*. – In KUPRIANOVA, L. A. (Ed.): Pollen and spore morphology of recent plants, pp. 18–21 – Leningrad: Nauka.

BELMONTE, J., PEREZ–OBIOL, R., ROURE, J. M., 1986: Clavas para la determinación de los pólenes de las principales especies melíferas de la Península Iberica. – Orsis **2**: 27–54.

BLACKMORE, S., HEATH, G. L. A., 1984: *Berberidaceae*. Northwest European pollen flora 30. – Rev. Palaeobot. Palynol. **42**: 7–21.

– BARNES, S. H., 1995: Garside's rule and the microspore tetrads of *Grevillea rosmarinifolia* A. CUNN. and *Dryandra polycephala* BENTH. (*Proteaceae*). *Rev. Palaeobot. Palynol.* **85**: 111–121.

CANDAU, P., SOLER, A., 1981: Contribucion à la palinologia de la familia *Fumariaceae* en la Peninsula Iberica. – Bot. Macaronesica **8–9**: 147–162.

– FERNANDEZ-PANIAGUA, I., 1985: Polen en *Papaveraceae* de Andalucia occidental. – Anal. Asoc. Palinol. Leng. Espan. **2**: 25–34.

CHANG, K. T., WANG, P. L., 1983: Study on the pollen morphology of the family *Berberidaceae*. – Acta Phytotax. Sinica **21**: 130–142.

CLARK, C., 1978: Pollen shed as tetrads by plants of *Eschscholzia californica* (*Papaveraceae*). – Madroño **25**: 59–60.

CLARKE, G. C. S., PUNT, W., HOEN, P. P., 1991: *Ranunculaceae*. Northwest European pollen flora 51. – Rev. Palaeobot. Palynol. **69**: 117–271.

D'ARCY THOMPSON, W., 1917: Growth and form. – Cambridge: Cambridge University Press.

ERDTMAN, G., 1952: Pollen morphology and plant taxonomy – Angiosperms. Stockholm: Almqvist Wiksell.

– 1960: The acetolysis method, a revised description. – Svensk Bot. Tidskr. **54**: 561–564.

FERGUSON, I. K., 1975: Pollen morphology of the trible *Triclisieae* of the – *Menispermaceae* in relation to its taxonomy. – Kew Bull. **30**: 49–75.

– 1978: Pollen morphology of the tribe *Coscinieae* of the *Menispermaceae* in relation to its taxonomy. – Kew Bull. **32**: 339–346.

FERNANDEZ, I., 1985: Morfologia polinica de *Ranunculus arvensis* L. – Anal. Asoc. – Palinol. Leng. Espan. **2**: 19–23.

– 1986: Contribucion al conocimiento palinologico de la familia *Ranunculaceae* en Andalucia: l. Subf. *Helleboroideae*. – Lagascalia **14**: 13–23.

HARLEY, M. M., FERGUSON, I. K., 1982: Pollen morphology and taxonomy of the tribe *Menispermeae* (*Menispermaceae*). – Kew Bull. **37**: 353–366.

HENDERSON, D. M., 1965: The pollen morphology of *Meconopsis*. – Grana Palynol. **6**: 191–209.

HESLOP–HARRISON, J., 1966: Cytoplasmic continuities during spore formation in flowering plants. – Endeavour **25**: 65–72.

HEUSSER, C. J., 1971: Pollen and spores of Chile. – Tucson: University of Arizona Press.

HUANG, T. C., 1972: Pollen flora of Taiwan. – Taipei National Taiwan University of Botany Press.

HUYNH, K.-L., 1970: Le pollen de genre *Anemone* et du genre *Hepatica* (*Ranunculaceae*) et leur taxonomi. – Pollen Spores **12**: 329–364.

– 1976: Arrangement of some monosulcate, disulcate, dicolpate and tricolpate pollen types in the tetrads, and some aspects of evolution in angiosperms. – In FERGUSON, I. K., MULLER, J., (Eds): The evolutionary significance of the exine. – Linn. Soc. Symp. Ser. **1**: 101–124.

IKUSE, M., 1956: Pollen grains of Japan. – Tokyo: Hirokawa.

KALIS, A. J., 1979: *Papaveraceae*. Northwest European pollen flora. – Rev. Palaeobot. Palynol. **28**: 209–260.

KUBITZKI, K., 1993: Some aspects of the classification and evolution of higher taxa. In KUBITZKI, K., ROHWER, J. G., BITTRICH, V., (Eds): The families and genera of vascular plants. II: Flowering plants – dicotyledons: magnoliid, hamamelid and caryophilid families. – Berlin: Springer.

KUMAZAWA, M., 1936: Pollen grain morphology in *Ranunculaceae*, *Lardizabalaceae*, and *Berberidaceae*. – Japan. J. Bot. **8**: 19–46.

KUPRIANOVA, L. A., 1979: On the possibility of the development of tricolpate pollen from monosulcate. – Grana **18**: 1–4.

LAYKA, S., 1976: Le polymorphisme pollinique dans le genre *Argemone* (*Papaveraceae*). – Pollen & Spores **18**: 351–375.

– 1977: Les caractères de l'endexine chez les Papaveracees. – Bull. Soc. Bot. Franco **122**: 103–107.

LEE, S., BLACKMORE, S., 1992: Pollen morphology in the genus *Trollius* (*Ranunculaceae*). – Grana **31**: 81–100.

NOWICKE, J. W., SKARVLA, J. J., 1980: Pollen morphology: the potential influence in higher order systematics. – Ann. Missouri Bot. Gard. **66**: 633–700.

– 1981: Pollen morphology and phylogenetic relationships of the *Berberidaceae*. – Smithsonian Contrib. Bot. **50**: 1–83.

– 1982: Pollen morphology and the relationships of *Circaeaster, of Kingdonia* and of *Sargentodoxa* to the *Ranunculales*. – Amer. J. Bot. **69**: 990–998.

– 1983: A palynological study of the genus *Helleborus* (*Ranunculaceae*). – Grana **22**: 129–140.

RACHELE, L. D., 1974: Pollen morphology of the *Papaveraceae* of the northeastern United States and Canda. – Bull. Torrey Bot. Club **101**: 152–159.

ROLAND-HEYDACKER, F., 1974: Caractères ultrastructuraux et cytochimiques particuliers du sporoderme des polen de *Berberis vulgaris* L. et de *Mahonia aquifolium* NUTT. – Compt. Rend. Acad. Sci. Paris **278**: 1475–1477.

SANTISUK, T., 1979: A palynological study of the tribe *Ranunculeae*. – Opera Bot. **48**: 1–74.

SCOTLAND, R., 1993: Pollen morphology of *Contortae* (*Acanthaceae*). – Bot. J. Linn. Soc. **111**: 471–504.

STERN, K. R., 1962: The use of pollen morphology in the taxonomy of *Dicentra*. – Amer. J. Bot. **49**: 362–368.

THANIKAIMONI, G., 1968: Morphologie des pollens des Menispermacees. – Inst. Fr. Pondichery, Trav. Sect. Sci. Tech. **5**: 1–57.

– 1984: Menispermacees: palynologie et systematique. – Inst. Fr. Pondichery, Trav. Sect. Sci. Tech. **18**: 1–135.

VAN CAMPO, M., 1967: Pollen et classification. – Rev. Palaeobot. Palynol. **3**: 65–71.

– 1976: Patterns of pollen morphological variation within taxa. – In FERGUSON, I. K., MULLER, J., (Eds): The evolutionary significance of the exine. – Linn. Soc. Symp. Ser. **1**: 125–137.

– 1978: Phylogenie des angiospermes. Approche palynologique. – Mem. Trav. Inst. Ecole Practique des Hautes Etudes Montpellier **4**: 73–90.

WALKER, J. W., 1976: Evolutionary significance of the exine. – In FERGUSON, I. K., MULLER, J., (Eds): The evolutionary significance of the exine. – Linn. Soc. Symp. Ser. **1**: 251–308.

– DOYLE, J. A., 1975: The bases of angiosperm phylogeny: palynology. – Ann. Missour Bot. Garden **62**: 644–723.

WODEHOUSE, R., 1935: Pollen grains: their structure, identification and significance in science and medicine. – New York: McGraw-Hill.

XIA, Q., PENG, Z. X., 1989: A study on the pollen morphology of *Lardizabalaceae, Sargentodoxaceae* and its significance in taxonomy. – Bull. Bot. Res. North-East. Forest. Inst. **9**: 99–114.

Addresses of the authors: STEPHEN BLACKMORE, PETER STAFFORD and VIVECA PERSSON, Department of Botany, The Natural History Museum, Cromwell Road, London SW7 5BD, England.

Accepted January 27, 1995

Pl. Syst. Evol. [Suppl.] 9: 83–84 (1995)

Comparative seed structure in the *Ranunculiflorae*

CLAUDIA BRÜCKNER

Received September 11, 1994

Most *Ranunculiflorae* seeds develop from anatropous, bitegmic, and crassinucellate ovules (CORNER 1976, TAKHTAJAN 1988). In regard to seed structure, four alliances can be distinguished.

Perhaps the most primitive group is formed by *Lardizabalaceae, Sargentodoxaceae*, and *Berberidaceae*. Their rather large seeds commonly have a smooth and shiny surface, raphal arils are frequent. The seed-coat is exotestal to exo-mesotestal in dehiscent as well as in indehiscent fruit types. The seed epidermis with thick outer periclines serves as protective layer. This function is often shared by some subtending parenchyma layers with thickened cell walls. The unspecialized tegmen becomes compressed during seed maturation. The copious endosperm is oily. Its development is cellular in *Lardizabalaceae* which indicates a primitive state of the family. The monotypic, berry-fruited *Nandinaceae* that are traditionally included in *Berberidaceae* deviate in seed-coat structure. The poorly differentiated sarcotesta collapses at maturity, and the thick-walled endotegmen becomes the principal mechanical layer. A thickened endotegmen as well as the occurrence of calcium oxalate crystals in the endotesta point to relations to the papaveraceous stock.

Menispermaceae are classified usually with the families mentioned. However, their position cannot be ascertained by the structure of their curved or amphitropous seeds. Included in drupelets, they lack an aril and show a strongly reduced seed-coat that may be of unitegmic origin in some genera. The endotegmen is its only persisting layer in the majority of taxa. The endosperm is often ruminate but may also be absent when the nutrients are being stored in the cotyledons of the large embryo. These characters isolate *Menispermaceae* in the first group.

The second alliance is made up of *Ranunculaceae, Hydrastidaceae*, and, perhaps, *Circaeasteraceae*. In *Ranunculaceae*, the seeds of the numerous taxa display a considerable amount of structural variety, partly depending on dehiscence versus indehiscence of fruits. Some genera have unitegmic ovules, a condition probably achieved phylogenetically by integumentary shifting. This special process is evident in the ontogeny of transitional stages with two integuments being distinct only around the micropyle (BOUMAN & CALIS 1977). Such a single but bifid integument characterizes the ovules of several taxa. Individually dispersed seeds have a protective exotesta. Their surface is often sculptured by radially elongate epidermal cells projecting in ridges or clusters, which recalls comparable patterns of papaveraceous seeds. The exotestal seed of the monotypic *Hydrastidaceae* fits well the ranunculaceous character set. Both families show close relations to group 1 but have no arils. – The monotypic *Circaeasteraceae* have a very peculiar unitegmic seed tightly enclosed by the carpel. As integument and nucellus degenerate early, the cellular endosperm is in contact with the endocarp. The outer endosperm layer has protective suberized cell walls. The chalazal part of the endosperm is haustorial and, at germination, breaks off together with the pericarp. The position of *Circaeasteraceae* is still an open question. The seed structure of *Kingdoniaceae*, another monotypic family split off from *Ranunculaceae* and probably related to *Circaeasteraceae*, has not been studied yet. The endosperm development follows a specialized cellular mode (MU 1984).

Papaveraceae and *Fumariaceae* form the third group. Here the best differentiated seed-coat layers are found. *Papaveraceae* are characterized by the following layer types: exotesta with thick, stratified outer periclines; occasionally parenchymatous mesotesta; crystal-bearing endotesta filled with cellulosic fibrils; fibrous exotegmen; unspecialized mesotegmen; endotegmen with finely striate walls (BRÜCKNER 1983). In *Fumariaceae*, the seed-coat is simplified but has a prominent exotesta, its prismatic cells being completely filled by cellulosic material. The endotesta may lack crystals, the fibrillar network is mostly inconspicuous. The thin-walled tegmen layers become crushed. Arils are frequent in both families and are to be regarded as primitive. – The seed-coats of two much discussed genera, *Hypecoum* L. and *Pteridophyllum* SIEB. & ZUCC., agree well with *Papaveraceae*. The seeds of the indehiscent *Hypecoum* taxa display the typical layers but have a deciduous exotesta. *Pteridophyllum* is also essentially papaveraceous showing a coarse fibrillar network within the endotesta.

The fourth group consists of the isolated monotypic *Nelumbonaceae*. Within the nutlet, the seed-coat obliterates early, and the endosperm forms a thin film around the massive green embryo. The latter has storage cotyledons, several leaf primordia, an abortive radicle, and adventitious roots. This type of embryo is unique among *Ranunculiflorae*.

References

BOUMAN, F., CALIS, J. I. M., 1977: Integumentary shifting – a third way to unitegmy. – Ber. Deutsch. Bot. Ges. **90**: 15–28.

BRÜCKNER, C., 1983: Zur Morphologie der Samenschale in den *Papaveraceae* JUSS. s. str. und *Hypecoaceae* (PRANTL et KÜNDIG) NAK. – Feddes Repert. **94**: 361–405.

CORNER, E. J. H., 1976: The seeds of dicotyledons I, II. – Cambridge, London, New York, Melbourne: Cambridge University Press.

MU, XI-JIN, 1984: Early development of the endosperm in *Kingdonia uniflora*. – Acta Bot. Sinica **26**: 668–671. (In Chinese.)

TAKHTAJAN, A. (Ed.), 1988: Anatomia seminum comparativa II. Dicotyledones *Magnoliidae, Ranunculidae*. – Leningrad: Nauka. (In Russian.)

Address of the author: Dr CLAUDIA BRÜCKNER, Humboldt-Universität, Institut für Biologie, Spezielle Botanik/Arboretum, Späthstrasse 80/81, D-12437 Berlin, Federal Republic of Germany.

Accepted December 15, 1994

Pl. Syst. Evol. [Suppl.] 9: 85–97 (1995)

Secondary compounds of the *Ranunculiflorae*

UWE JENSEN

Received December 5, 1994

Key words: *Ranunculiflorae*, *Berberidaceae Fumariaceae*, *Menispermaceae*, *Papaveraceae*, *Ranunculaceae*. Benzylisoquinoline alkaloids, natural products, repellents, secondary substances.

Abstract: Natural products can provide additional arguments for the relationships within families of the *Ranunculiflorae*, although parallel evolution and loss of chemical characters probably occurred frequently. Especially the accumulation of benzylisoquinoline alkaloids is a characteristic feature, however, in *Ranunculaceae* restricted to *Coptideae* and *Isopyreae*; they are common in the *Berberidaceae*, but missing in *Podophyllum* and *Diphylleia* and probably other *Berberideae*. Quinolizidine alkaloids are typical products of the *Cimicifugeae* (*Ranunculaceae*) and *Leonticeae* (*Berberidaceae*). The *Delphiniinae* (*Ranunculaceae*) are characterized by diterpene alkaloids and the accumulation of mannitol. Cyanogenic compounds occur in *Isopyreae* (*Ranunculaceae*), but they are observed elsewhere in the *Ranunculiflorae*. Ranunculins are present only in the *Ranunculeae* and *Anemoneae* and additionally in *Helleborus* (all *Ranunculaceae*). Lignans are accumulated in the closely related *Podophyllum* and *Diphylleia*, but also in *Epimedium* (all *Berberidaceae*). For the *Fumariaceae* the free amino acid acetylornithine is suspected to be the main nitrogen transport molecule.

Angiosperms produce an immense number of natural compounds. We interprete the production (and especially the accumulation) of such chemicals in plants as a defence against herbivores and phytopathogens. Polyphenols are considered unspecific repellents which are able to inactivate the digestion enzymes of predators, while specifically reacting chemicals like alkaloids are able to deter specific herbivores. It is a matter of fact that the resources of a plant are not sufficient to allow the production of a great diversity of such substances. In most cases only one or two toxic substance groups are accumulated in a plant species. Their distribution in the plant kingdom is ruled by – largely unknown – selection factors, and can be successfully used to additionally characterize plant taxa, and to deliver further arguments for natural relationships (see, e.g., FROHNE & JENSEN 1992). At least since DE CANDOLLE (1804), petal colours, essential oil constituents, or poisonous metabolites have been included in the description of taxa and were used to improve the natural classification. Only modern phytochemistry and the intensified chemical analysis of plants, however, have accumulated adequate data for relevant systematic conclusions. HEGNAUER's (1962 ff) gigantic work summarized this mass of data and prepared them for use in systematics.

In some cases natural products characterize particular plant categories and indicate their presumed monophyletic origin. Well known examples are: Betalains are only produced in *Caryophyllales* as far as flowering plants are concerned, iridoids might characterize subclasses *Cornidae* and *Lamiidae*, and indicate their derivation from some primitive hamamelids (Frohne & Jensen 1992), or polyins are interpreted to characterize *Apiales* and *Asterales* as closely related plant groups. In most cases, however, such a monophyletic origin is not obvious, and the distribution of a certain substance group might be interpreted in terms of parallel evolution. Since such independent evolutionary origin of a given biogenetic pathway is very unlikely, it is conceivable that molecular processes of deactivation and reactivation of genes responsible for pathway enzymes are involved. This consideration concerns also the chemical compounds of the taxa of the *Ranunculiflorae*.

This contribution does not aim to enumerate the hitherto identified secondary substances of the *Ranunculiflorae*. This information can best be obtained by consulting Hegnauer's "Chemotaxonomie der Pflanzen" (1962 ff), or Ruijgrok (1966, 1967) for ranunculins, and Phillipson & al. (1985) for benzylisoquinoline alkaloids.

Instead, distribution patterns of natural compounds have been worked out, and such information was selected which might be useful as additional taxonomic markers, either characterizing plant taxa, or supporting arguments for relationships between taxa. The value of natural compounds as taxonomic markers is not reduced by the great significance of protein and gene data in modern chemotaxonomy.

Chemical characterization of the *Ranunculiflorae*

The *Ranunculiflorae* are chemically well characterized; in other angiosperm categories such substances are of much less importance, and other chemical markers are known (Table 1). Especially characteristic is the production of benzylisoquino- line alkaloids. They share this characteristic with the woody *Magnoliiflorae* (*Magnoliidae* s. str.), particularly by the occurrence of benzyltetrahydroisoquinoline alkaloids (A, see Table 2), their dimers (B), pro-aporphines (F), aporphines (G) and oxoaporphines (H). Magnoflorine (Fig. 1) can be considered to be the major alkaloid structure. However, in *Ranunculiflorae* (*Ranunculidae*) many additional benzylisoquinoline alkaloids are produced, especially the biochemically derived berberines and morphines (Table 2). These alkaloids, although not present throughout all *Ranunculiflorae*, represent the most valid chemical characters of all its families. Special alkaloid types, or compounds, or even their stereochemical properties can be characteristic for individual taxa (Tolkachev & Schroeter 1979). Ferreira & al. (1980) interpreted the production and accumulation of isoquinoline alkaloids in the "*Polycarpicae*" as a primitive character which in derived groups has been replaced by the accumulation of, e.g., lignanes and other cinnamomic acid metabolites.

The accumulation of benzylisoquinoline alkaloids could be interpreted in terms of protection against herbivores. For humans, besides the toxic character, also the medical use of several drug plants is important. "Hydrastis rhizoma" (*Hydrastis canadensis* L., *Hydrastidaceae*) has been used instead of ergot. Because of similar compounds and colours, the drug has been adulterated with *Xanthorhiza*- and *Coptis* roots and rhizoms (both *Ranunculaceae*). "Aquilegiae herba" (*Aquilegia vulgaris* L., *Ranunculaceae*) as well as "Berberidis radicis cortex" (*Berberis vulgaris*

Table 1. Phytochemical characterization of the *Ranunculiflorae* including *Ranunculaceae, Berberidaceae, Menispermaceae, Papaveraceae* and *Fumariaceae.* – not present, x in some species, xx in several genera, xxx widespread, xxxx in all genera/species, ? questionable

	Ran	Ber	Men	Pap + Fum
Major compounds of the *Ranunculiflorae*				
Benzylisochinoline alkaloids	xxx	xx	xxxx	xxxx
Tyrosine derived cyanogenic glycosides	xx	x	x	xx
Polyphenols: kaempferol, quercetin; frequently additional oxygenation at carbons 6 or 8 of ring A (in contrast to *Magnoliiflorae*)	xxx	xx	?xxx	xxx
Caffeic acid, in *Ranunculaceae* often esterified with sugars, not with quinic acid	xx	xxx	x	?
Major compounds of other angiosperm groups, only rarely produced in *Ranunculiflorae*				
Etheric oils	–	–	x	–
Tannins and flavonoids with trihydroxylated B-ring, e.g., ellagic and gallic acid derivates; prodelphinidins; myricetin	–	?–	–	–
Triterpenoids in cuticular waxes	–	?–	?–	?–
Iridoids	–	–	–	–

L., *Berberidaceae*) have been used as cholagoga. Berberine sulfate is a component of some eye-lotions. Diluted alkaloid extracts from cortex tissue of the closely related *Mahonia aquifolium* (PURSH) NUTT. are applied against psoriasis. "Chelidonii herba" (*Chelidonium majus* L., *Papaveraceae*) is a frequently used component of liver/gall teas, where the alkaloids sanguinarine and chelerythrine seem to be the main active principles. Sanguinarine from *Sanguinaria canadensis* L. (*Papaveraceae*) is of anti-plaque usage of the mouth epithel. The special importance of *Papaver somniferum* L., as the source of opium alkaloids should at least be mentioned.

Besides benzylisoquinoline alkaloids, many other natural compounds characterize individual families, tribes or genera of the *Ranunculiflorae*. Most information is available for the *Ranunculaceae*. Also *Berberidaceae, Menispermaceae* and *Papaveraceae* will be considered in detail. For *Lardizabalaceae* and the segregate families, e.g., *Circaeasteraceae, Pteridophyllaceae, Hydrastidaceae, Glaucidiaceae,* sufficient chemical data are not available. For *Hydrastis canadensis*, however, at least the abundance of benzylisoquinoline alkaloids is known with berberine (Fig. 1) as the major component. The rare alkaloid berberastine has also been found in *Hydrastis*; it is absent from most taxa of the *Ranunculiflorae*, but found in *Coptis*

Table 2. The occurrence of different benzylisoquinoline alkaloid types in the *Ranunculiflorae* (Gottlieb & al. 1993) including *Berberidaceae*, *Ranunculaceae*, *Menispermaceae*, *Papaveraceae* and *Fumariaceae*. + present. *relatively abundant

	Ber	Ran	Men	Pap	Fum
Aporphines and other simple derivatives					
A benzyltetrahydroisoquinolines	+	+	+	+	+
B A. A-dimers	+	+	*		
C A. G-dimers	+				
D A. H-dimers	+	*			
E H. O-dimers		+			
F proaporphines			+	+	
G aporphines	+	+	+	+	+
H oxoaporphines	+	+	+	+	+
I aristolactams, aristolochic acids			+		
J phenanthrenes		+			
K taspines	+				
L benzylisoquinolines	+	+	+	+	
M oxoisoaporphines			+		
N pavines	+	+		+	
O isopavines		+		+	
P cularines					+
Q quettamines	+				
R tropolone-isoquinolines		+			
S azafluoranthenes		+			
T N-benzyltetrahydroisoquinolines					+
Berberines					
A tetrahydroprotoberberines	+	+	+	+	+
B protoberberines	+	+	+	*	+
C isoindolobenzazepines	+				
D corydamines					+
E B. B-dimers			+		
F benzophenanthridines				*	*
G F. F-dimers				+	
H protopines	+	+	+	+	+
I rhoeadines				*	
J spirobenzylisoquinolines					*
K indenobenzazepines					*
L phtalidoisoquinolines	+	+	+	+	*
M narceines					*
N retroprotoberberines				*	
O hypercorines					*
P peshawarines					*
Morphines and other complex derivatives					
A protostephanines			+		
B erythrinans			+		
C cocculolidines			+		
D morphines	+	+	+	+	+
E quettamines-morphine-dimers					+
F D. D-dimers			+	+	
G hasubanonines			*		
H G. G-dimers			+		
I acutumines			+		

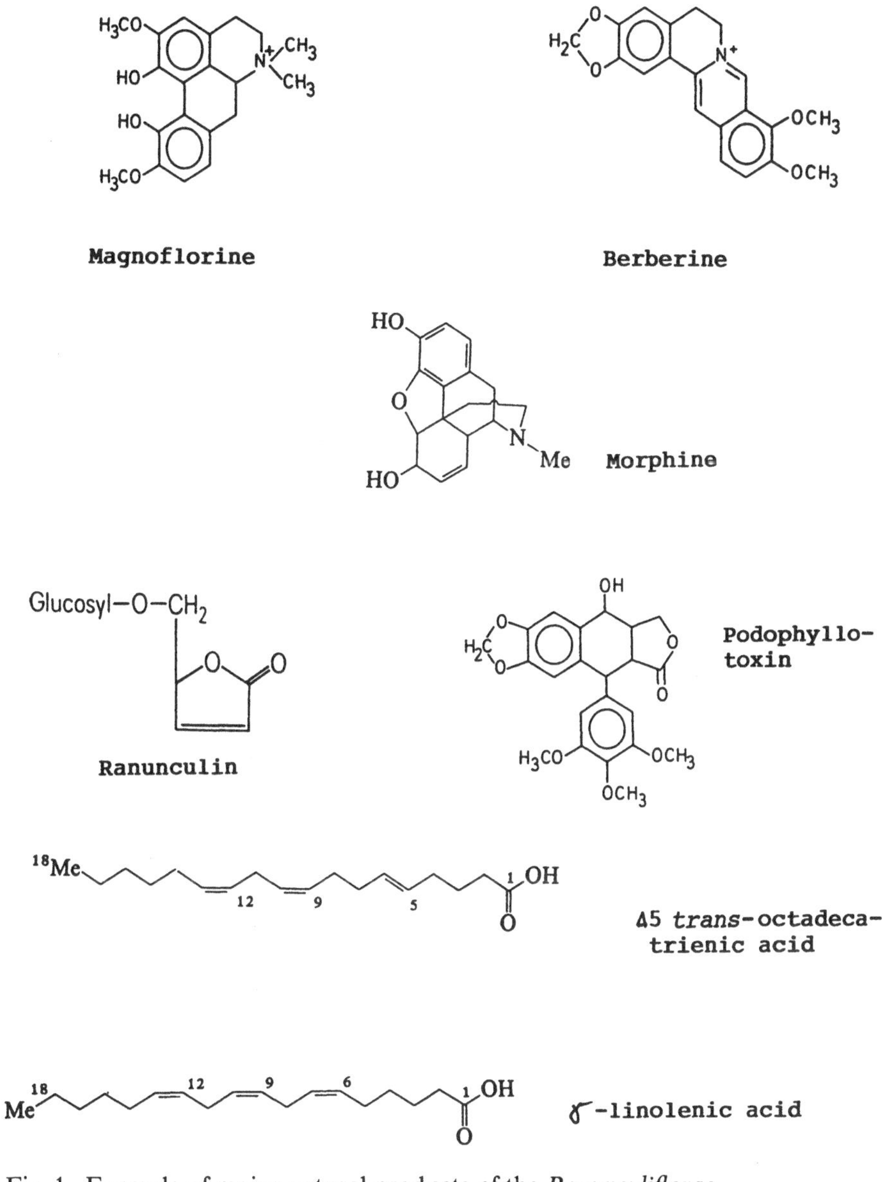

Magnoflorine

Berberine

Morphine

Ranunculin

Podophyllo-toxin

Δ5 *trans*-octadeca-trienic acid

γ-linolenic acid

Fig. 1. Formula of major natural products of the *Ranunculiflorae*

japonica MAKINO and *Xanthorhiza simplicissima* MARSH. (both *Ranunculaceae-Coptideae*) and *Berberis laurina* THUNB. (*Berberidaceae*).

Ranunculaceae

No chemical compounds occur universally in all tribes, subtribes or genera of the family (Table 3). Benzylisoquinoline alkaloids are at least present in almost all genera, and the quaternary magnoflorine (Fig. 1) and its precursor corytuberin are known to be widely distributed within the family. An accumulation of these alkaloids is observed only in *Coptideae* and *Isopyreae*, which are also characterized by T-type chromosomes.

Table 3. Chemical markers of the *Ranunculaceae*, based on the classification of JENSEN & al. (1995) (blank: missing or not yet detected occurrence). *a* Benzylisoquinoline alkaloids; *r* small amounts in some species, mostly quartarnary aporphines. *b* Cyanogenic compounds. *c* Δ 5-*trans* desaturated C18 fatty acids (columbinic acid). *d* Quinolizidine alkaloids. *e* Diterpene alkaloids. *f* Accumulation of sugar alcohols: adonitol (*A*), mannitol (*M*), sorbitol (*S*). *g* Ferulic + sinapic acid. *h* Glycoflavones replacing flavonole glycosides. *i* Ranunculins. *j* γ – Linolenic acid. *k* Saponins. *l* Cardiac glycosides (*C* cardenolides, *B* bufadienolides). *m* Chelidonic acid

		a	b	c	d	e	f	g	h	i	j	k	l	m
Coptideae	*Coptis*	+												
	Xanthorhiza	+												
Isopyreae	*Isopyrum*	+	+	+										
	Aquilegia	+	+	+					+					
	Anemonella	+		?										
	Thalictrum	+	+	+				r	+			+		
Cimicifugeae	*Actaea*	?		?										
	Cimicifuga	−		+								r		
	Anemonopsis	−		?										
	Eranthis	r												
Delphinieae	*Aconitum*	r				+	M							
	Delphinium	r				+	M							
	Consolida	r				+	M							
	Nigella	r					S					+		
Caltheae	*Caltha*	r										+		
Helleboreae	*Helleborus*	r	?							+		?	B	
Adonideae	*Trollius*	r					S	+	+			+		
	Adonis	r					S, A	+	+			+	C	
Ranunculeae	*Trautvetteria*	−								+				
	Myosurus	−	?							+				
	Ranunculus	−	r						?	+		+		
Anemoneae	*Clematis*	r	?							+	+	+		
	Anemone	−	?							+	+	+		r
	Hepatica	−								+		+		
	Pulsatilla	−								+		+		r
	Knowltonia	−								+		+	?	

The natural compounds of the **Coptideae** are not well known. The production and accumulation of benzylisoquinoline alkaloids, however, is established. Berberine is the major alkaloid and responsible for the intensive yellow colour of roots and rhizomes of *Xanthorhiza* MARSH. and *Coptis* SALISB., and also for the bark of the monotypic woody genus *Xanthorhiza* ("yellow root"); these plants were used by the American natives to stain baskets and textiles.

Also the **Isopyreae** are characterized by the accumulation of benzylisoquinoline alkaloids. It is interesting to know that *Thalictrum* L. produces not only genus-specific *Thalictrum* alkaloids such as benzyltetrahydroisosquinoline-aporphine-dimers (cryptopine, pavine- and isopavine alkaloids), but also (as well as *Hydrastis* ELLIS) such benzylisoquinoline alkaloids which otherwise are almost exclusively known from the *Papaveraceae*. Moreover, the alkaloid spectra of the *Isopyreae* (and *Coptideae*) in general are similar to those of the *Menispermaceae* and *Berberidaceae* (HEGNAUER 1973).

In contrast to the preceeding *Coptideae*, cyanogenic compounds derived from tyrosine are present and widely accumulated within the *Isopyreae*. They are responsible for intoxications with these plants. Cyanogenesis, without accumulation of these compounds, however, is not exclusively restricted to this tribe but is also reported for *Ranunculus* L.; the evidence for *Helleborus* L., *Myosurus* L., *Clematis* L., and *Anemone* L. is questionable (HEGNAUER 1990).

A third important chemical marker group for the *Isopyreae* are the fatty acids of seeds. In *Ranunculaceae* quantities of fatty acids between 10 and 50% are reported for seed oils, with olic-, linolic- or linolenic acid being the major compounds. Many species also produce unusual fatty acids. Among them the production of γ-linolenic acid (Fig. 1) in the *Anemoneae* is of special interest (see *Anemoneae*). In the *Isopyreae*, plants belonging to *Isopyrum* L., *Aquilegia* L. and *Thalictrum* produce Δ 5-*trans* desaturated C18 fatty acids (Fig. 1; HEGNAUER 1973, 1990; AITZETMÜLLER & TSEVEGSÜREN 1994; Aitzetmüller 1995, this volume). Although they have not yet been analyzed for *Anemonella* SPACH., they are suspected to occur in this taxon, too. AITZETMÜLLER (1995, this volume) has interpreted this type of fatty acid ("columbinic acid") to be of a biosynthetically derived type and therefore to represent a synapomorphic character of the *Isopyreae*.

Many other substances have been detected in the tribe (HEGNAUER 1973, 1990), such as glycosides of quercetin and kaempferol. Their significance for systematic purposes, however, seems to be lower.

Thus the *Coptideae* (subfam. *Coptidoideae*) and *Isopyreae* (subfam. *Isopyroideae*) with a T-type karyotype are chemically characterized by the accumulation of benzylisoquinoline alkaloids. The following *Ranunculoideae* with an R-type karyotype may produce benzylisoquinoline alkaloids in small amounts, but they never accumulate these compounds.

The **Cimicifugeae** are chemically rather negatively characterized by the lack of benzylisoquinoline alkaloids, cyanogenic compounds, or ranunculin and they do not accumulate toxic substances at all; unknown toxic substances, however, are present in at least *Actaea rubra* (AIT.) WILLD. berries. Of special interest is the presence of quinolizidine alkaloids (e.g., cytisin) in *Cimicifuga* L. (the presence in *Actaea* L. and *Anemonopsis* SIEB. & ZUCC. has not yet been shown). Cytisin is also known from the *Leonticeae* (*Berberidaceae*). It is doubtful whether this can be used as an argument for the common origin of these two groups as done by McFadden (1950).

Saponins are common in the *Ranunculaceae*; in most cases oleanolic acid and hederagenin are the most abundant sapogenins. In *Actaea, Cimicifuga, Beesia* Balf. and *Souliea* Franch. (all *Cimicifugeae*) and also some species of *Thalictrum* (*Isopyreae*), different compounds, i.e., tetracyclic triterpenic saponins with lanostan sapogenins were detected.

The chemistry of *Eranthis* Salisb. (subtribe *Eranthinae*) is not well known; substances which could contribute to a better understanding of the relationships of this genus are almost lacking. Only the detection of acetogenic chromones of the khellol, eranthin, cimifugin structure in *Eranthis* and *Cimicifuga* might provide an interesting argument for relationship between these two genera.

Jensen & al. (1995, this volume) divide the tribe **Delphinieae** into the two subtribes *Delphiniinae* and *Nigellinae* which are chemically different. The *Delphiniinae* are toxic by producing characteristic, diterpene alkaloids; aconitine is a well-known physiologically active substance of the "Aconiti tuber" drug. *Aconitum* species were also used as medical plants in East Asia; in China they might have served to prepare arrow poisons. *Delphinium brownii* Rydb. and its toxic methyllycaconitin with curare effect has caused heavy intoxication to cattle in Canada.

Benzylisoquinoline alkaloids (magnoflorine) are present, but only in small amounts. The *Delphiniinae* also accumulate mannitol in considerable quantities.

Nigella damascena L. seeds contain the alkaloid damascenin, other species produce saponins; magnoflorine is common. Since diterpene alkaloids are completely lacking, no chemical similarity between *Nigella* and the *Delphiniinae* can be observed.

Caltha L. (tribe **Caltheae**) is not characterized by the accumulation of any physiologically effective compounds except the suspected occurrence of pyrrolizidine alkaloids in American species. The presence of magnoflorine and some other general compounds does not contribute to systematics. Several Δ 5-cis desaturated C18 and C20 seed fatty acids remarkably also occur in *Actaea* and *Cimicifuga* (Aitzetmüller 1995).

Helleborus (**Helleboreae**) is toxic by producing the irritating ranunculins (Fig. 1). The production of ranunculins is another answer of the plants to phytophagism and phytopathogenic microorganisms besides benzylisoquinoline alkaloids (magnoflorine and ranunculins are always alternatively produced, Hegnauer 1973), cyanogenic compounds, and diterpene or quinolizidine alkaloids. The phylogenetic significance of the character "ranunculin"is judged controversially in systematics. The presence of ranunculin also in the *Ranunculeae* and *Anemoneae* is probably due to parallel evolution at least in *Helleborus*, but has also been interpreted in terms of putative relationship of all three tribes by Ruijgrok (1967). Another interesting toxic substance group, the cardioactive bufadienolides, is, as far as we know today, exclusively restricted to the genus *Helleborus* within the *Ranunculiflorae*. It occurs in the subterranean plant tissue and is an alternative repellent to the ranunculins in the green tissue. The steroid bufadienolides are accompanied by the equally steroid saponins and phytoecdysones. Relationships of *Helleborus* and *Adonis* based on the presence of cardenolides are rather speculative.

The ***Adonideae*** *Trollius* L. and *Adonis* L. are not characterized by "spectacular" natural compounds, although the production of *Digitalis* cardenolides within *Adonis* (which is unique among the *Ranunculiflorae*) is of considerable pharmaceutical interest and makes "Adonidis herba" a very intensively investigated plant drug.

Two chemical aspects, however, might be important when the close relationship of *Trollius* and *Adonis* is discussed. (1) Kaempferol and quercetin are flavonole-aglyca of general occurrence within the *Ranuculaceae*, but have not yet been detected in *Trollius* and *Adonis*. In these two genera the flavonoles are largely replaced by C-glycoflavones and flavones. This argument in favor of a relationship between the two genera has recently been weakened by similar findings for *Aquilegia vulgaris* and *Thalictrum minus* L. (in *Ranunculus*, C-glycoflavones, flavones and flavonole glyco-sides are abundant). (2) Although ferulic- and sinapic acids are common products in *Ranunculaceae*, a co-occurrence of considerable amounts of these substances has been observed only in *Trollius* and *Adonis*.

The ***Ranunculeae*** and ***Anemoneae*** share significant chemical properties by not producing benzylisoquinoline alkaloids (the only exception is a minor production of magnoflorine in *Clematis*) or cyanogenic compounds except in some species of *Ranunculus* and questionable occurrences in some species of *Myosurus*, *Clematis* and *Anemone*. Instead, the two tribes are characterized by the production of the repellent ranunculins (to which *Ranunculus acris* L. owes its name). At least in *Anemone* and *Clematis* the unusual γ-linolenic acid (Fig. 1) is a constituent of the fatty oils in seeds (AITZETMÜLLER 1995, this volume). Additional substances are included in Table 3.

On the whole, the *Ranunculeae* and *Anemoneae* are chemically very similar. This fact might suggest their common (monophyletic) origin. It should be kept in mind, however, that the suppression of the production of benzylisoquinolines and cyanogenic compounds can also be found in other tribes of the *Ranunculoideae*, and that the alternative production of repellent ranunculins might have originated in parallel. Alternatively, a common root can be suggested for all *Ranunculoideae*, i.e. the *Helleboreae*, *Ranunculeae*, *Anemoneae*, and the other tribes which in that case lost the capability of ranunculin production.

Berberidaceae

Also the *Berberidaceae* produce benzylisoquinoline alkaloids. Lignans and quinolizidine alkaloids occur, and are interesting for the discussion of relationships (Table 4).

The monotypic ***Nandineae*** are characterized by a rich spectrum of ben-zylisoquinoline alkaloids including protopine which is abundant in *Papaveraceae*. Also the content of cyanogenic compounds is relatively high. The biosynthesis of amentoflavon is interpreted as plesiomorphic (HEGNAUER 1989).

The ***Leonticeae*** are rich in secondary compounds with a corresponding pattern in *Leontice* L., *Caulophyllum* MICHX. and *Gymnospermium* SPACH. Magnoflorine and other benzylisoquinoline alkaloids have been detected. The dominating alkaloids, however, are the quinolizidine alkaloids which were found in all genera; they belong to the cytisine-, sparteine- and matrine-type, and within *Berberidaceae* are restricted to this tribe. These alkaloids frequently occur in *Fabaceae* (*Genisteae, Thermop-sideae, Podalyrieae, Sophoreae*). A convergent occurrence is known from the

Table 4. Chemical markers of the *Berberidaceae*, based on the classification of Loconte (1993). *a* Lignans, *b* Isoquinoline alkaloids, *c* Quinolizidine alkaloids, *d* Saponins, *e* Cyanogenic compounds

		a	b	c	d	e
Nandinoideae						
Nandineae	*Nandina*		+			+
Berberidoideae						
Leonticeae	*Leontice*		+	+	+	
	Caulophyllum		+	+	+	
	Gymnospermium		+	+	?	
Berberideae-	*Ranzania*					
Berberidinae	*Berberis*		+			
	Mahonia		+			
Berberideae-	*Jeffersonia*		?		+	
Epimediinae	*Achlys*					
	Vancouveria	?				
	Epimedium	+	+		+	
	Bongardia				+	+
	Podophyllum	+	−			
	Diphylleia	+	−			

Cimicifugeae (*Ranunculaceae*). The presence of quinolizidine alkaloids in the *Leonticeae* can be seen as an important chemical argument for the monophyly of this tribe.

The presence of saponins with hederagenin as aglycon should at least be mentioned, although such substances also occur in *Berberideae*.

The **Berberideae-Berberidinae** according to Loconte (1993) include the very rare and monotypic Japanese *Ranzania* T. Ito, *Berberis* L. and *Mahonia* Nutt. The latter two genera are chemically very similar, and are characterized by a rich production of benzylisoquinoline alkaloids. Berberine (Fig. 1) is the major alkaloid.

Within the **Berberideae-Epimediinae** chemically two groups of genera exist. The group *Jeffersonia* Barton, *Achlys* DC., *Vancouveria* Morr. & Decne., *Epimedium* L. and *Bongardia* C. A. Mey. is chemically heterogeneous. At least in *Epimedium* the benzylisoquinoline alkaloid magnoflorine has been detected. Its occurrence in *Jeffersonia* is questionable. Some genera contain saponins (Table 4). Cyanogenic compounds have been detected in *Bongardia chrysogonum* (L.) Griseb.

Podophyllum L. and *Diphylleia* Michx. are deviating from the former group of genera, and are characterized by the accumulation of lignan-β-glycosides, also called podophyllotoxin-glycosides (Fig. 1), and related compounds ("peltatins") replacing the toxic benzylisoquinoline alkaloids. Because of the lignans the resin-like extract of

Podophyllum peltatum L. rhizoms has been used as a drastic laxans. Recently their antimitotic effects have been investigated for medical purposes.

Menispermaceae

Benzylisoquinoline alkaloids are a universal chemical component also of the *Menispermaceae*, especially those of the bis-benzylisoquinoline- and berberine type. The South American Indians and African natives have used many plants rich in alkaloids as arrow or dart poisons to paralyze food animals. Especially "tubo curare" prepared from *Chondrodendron tomentosum* RUIZ & PAV. and containing d-tubocurarine became the prototype for the synthesis of structurally analogous curarizing pharmaca.

Many *Menispermaceae*, particularly those included in tribes *Fibraureae* and *Menispermeae*, produce N-free sesqui- or di-terpenoid toxic and bitter principles exemplified by picrotoxin and columbin.

Saponins are a minor chemical component of *Menispermaceae* similar to *Ranunculaceae* and *Berberidaceae*. Cyanogenic compounds were detected at least in *Stephania* LOUR. Essential oils are found in roots of *Cissampelos* L. and *Jatrorrhiza* MIERS species. The detection of typical oil cells in *Abuta rufescens* AUBL. (MAHEU 1905, 1906) is of interest because such idioblasts are characteristic for the *Magnoliiflorae* but otherwise absent from the *Ranunculiflorae*.

Papaveraceae and Fumariaceae

A chemical characterization of the tribes is not possible but arguments for the separation of the *Papaveraceae* and *Fumariaceae* are available.

All members of the *Papaveraceae* and *Fumariaceae* produce benzylisoquinoline alkaloids. They are found in laticifers (*Papaveraceae*) or idioblasts (*Fumariaceae*). The major alkaloids belong to the berberine and morphine types (GOTTLIEB & al. 1993, see Table 1), and in many cases are biosynthetic end products (HEGNAUER 1969). Besides benzylisoquinoline alkaloids, the scattered occurrence of some other compounds should be mentioned. The alkaloids in the laticifers are often bound to chelidonic, fumaric, or meconic acid. Chelidonic acid is known from *Chelidonium* L. (*Chelidonioideae*) and *Stylophorum* NUTT., meconic acid from *Meconopsis* VIG., *Papaver* L. and *Roemeria* MEDIK. (all *Papaveroideae*), fumaric acid from *Glaucium* MILL. and *Papaver* (*Papaveroideae*). Cyanogenic compounds are known from *Eschscholzia* CHAM., *Meconopsis* and *Papaver*. Fumaric acid is common in the *Fumariaceae*.

An interesting argument for the independent status of the *Fumariaceae* is the detection of the free amino acid acetylornithine as the main transport molecule for nitrogen. In *Papaveraceae* this kind of nitrogen transport at least is much less pronounced.

This description discovers unequivocal chemical similarities between the families of *Ranunculiflorae* and demonstrates the consistency of the *Ranunculiflorae* as a systematic unit. Chemical arguments for the natural classification within the families are presented but no such arguments are available for the phylogenetic relations between them.

Table 5. Chemical markers of the *Papaveraceae* and *Fumariaceae*, based on the classification of KADEREIT & al. (1994) for the *Papaveraceae*, and of LIDÉN (1993) for the *Fumariaceae*. *a* Benzylisoquinoline alkaloids, *b* Acetylornithine, *c* Cyanogenic glycosides, *d* Chelidonic acid, *e* Meconic acid, *f* Fumaric acid; + present, *r* rare

		a	b	c	d	e	f
PAPAVERACEAE							
Chelidonioideae	*Bocconia*	+					
	Chelidonium	+			+		
	Dicranostigma	+					
	Sanguinaria	+					
	Stylophorum	+			+		
	Glaucium	+					+
Eschscholzioideae	*Dendromecon*	+					
	Eschscholzia	+		+			
	Hunnemannia	+		+			
Papveroideae	*Platystemon*	+					
	Argemone	+					
	Romneya	+					
	Meconopsis	+		+		+	
	Papaver	+		+		+	+
	Roemeria	+				r	
FUMARIACEAE							
Hypecoideae	*Hypecoum*	+					
Fumarioideae	*Adlumia*	+	+				
	Corydalis	+	+				+
	Dicentra	+	+	+			+
	Fumaria	+	+				+

References

AITZETMÜLLER, K., 1995: Fatty acid patterns of *Ranunculaceae* seed oils: phylogenetic relationships. – Pl. Syst. Evol. [Suppl.] **9**: 229–240.

– TSEVEGSÜREN, N., 1994: Seed fatty acids, "Front-end" – desaturases and chemotaxonomy – A case study in the *Ranunculaceae*. – J. Pl. Physiol. **143**: 538–543.

DE CANDOLLE, A. P., 1804: Essai sur les propriétes medicales des plantes, comparées avec leurs formes extérieures et leur classification naturelle. – Paris: I. Méquignon.

FERREIRA, Z. S., GOTTLIEB, O. R., ROQUE, N. F., 1980: Chemosystematic implications of benzyltetrahydroisoquinolines in *Aniba*. – Biochem. Syst. Ecol. **88**: 51–54.

FROHNE, D., JENSEN, U., 1992: Systematik des Pflanzenreichs, 4th edn. – Stuttgart, Jena, New York: G. Fischer.

GOTTLIEB, O. R., KAPLAN, M. A. C., ZOCHER, D. H. T., 1993: A chemosystematic overview of *Magnoliidae*, *Ranunculidae*, *Caryophyllidae* and *Hamamelidae*. – In KUBITZKI, K.,

ROHWER, J. G., BITTRICH, V., (Eds): The families and genera of vascular plants, **II**, pp. 20–31. – Berlin: Springer.

HEGNAUER, R., 1962 ff: Chemotaxonomie der Pflanzen. – Basel: Birkhäuser; especially for *Ranunculaceae* **6** (1973) and **9** (1990), for *Berberidaceae* **3** (1964) and **8** (1989), for *Menispermaceae* **5** (1969) and **9** (1990), for *Papaveraceae* **5** (1969) and **9** (1990).

JENSEN, U., HOOT, S. B., JOHANSSON, J. T., KOSUGE, K., 1995: Systematics and phylogeny of the *Ranunculaceae* – a revised family concept on the basis of molecular data. – Pl. Syst. Evol. [Suppl.] **9**: 273–280.

KADEREIT, J. W., BLATTNER, F. R., JORK, K., SCHWARZBACH, A., 1994: Phylogenetic analysis of the *Papaveraceae* s.l. (incl. *Fumariaceae*, *Hypecoaceae*, and *Pteridophyllum*) based on morphological characters. – Bot. Jahrb. Syst. **116**: 361–390.

LIDÉN, M., 1993: *Fumariaceae*. – In KUBITZKI, K., ROHWER, J. G., BITTRICH, V., (Eds): The families and genera of vascular plants, **II**, pp. 310–318. – Berlin: Springer.

LOCONTE, H., 1993: *Berberidaceae*. In KUBITZKI, K, ROHWER, J. G., BITTRICH, V., (Eds): The families and genera of vascular plants, **II**, pp. 147–152. – Berlin: Springer.

MCFADDEN, S. E., 1950: A series of related cytological and biochemical studies of the *Berberidaceae* and its alliance with *Ranunculaceae*. – Thesis, University of Virginia, Charlottesville, VA.

MAHEU, J., 1905: Sur l'existence de lactifères à caoutchouc dans un genre de Ménispermacées: *Tinomiscium*. – Compt. Rend. **141**: 958.

– 1906: Les organes sécréteures des Ménispermacées. – Bull. Soc. Bot. France **53**: 651.

PHILLIPSON, J. D., ROBERTS, M. F., ZENK, M.H., (Eds), 1985: The chemistry and biology of isoquinoline alkaloids. – Berlin: Springer.

RUIJGROK, H. W. L., 1966: The distribution of ranunculin and cyanogenic compounds in the *Ranunculaceae*. – In SWAIN, T., (Ed.): Comparative phytochemistry, pp. 175–186. – London, New York: Academic Press.

– 1967: Over de verspreiding van ranunculine en cyanogene verbindingen bij de *Ranunculaceae*. Een bijdrage tot de chemotaxonomie van de familie. – Dissertation, University of Leiden.

TOLKACHEV, O. N., SCHROETER, A. I., 1979: The value of stereochemical characters in the chemotaxonomical investigation of alkaloid-bearing plants. – Rast. Resur. **15**: 321–343.

Address of the author: Dr UWE JENSEN, Lehrstuhl Pflanzensystematik, Universität Bayreuth, Universitätsstrasse 30, D-95440 Bayreuth, Federal Republic of Germany.

Accepted January 19, 1995

Pl. Syst. Evol. [Suppl.] 9: 99–118 (1995)

Ordinal and familial relationships of Ranunculid genera

Henry Loconte, Lisa M. Campbell, and Dennis Wm. Stevenson

Received September 30, 1994

Key words: *Decaisnea*, Lardizabalaceae, *Paeonia*, Berberidaceae, Hydrastidaceae, *Glaucidium*, Papaveraceae, Ranunculaceae

Abstract: A morphological-based cladistic analysis was conducted on 116 ingroup taxa and 5 outgroups coded for 109 characters and 192 apomorphic character states. Ranunculid genera are better organized as the two orders Lardizabalales and Ranunculales, rather than Ranunculales and Papaverales of contemporary systems. Within Lardizabalales, the genus *Decaisnea* is segregated as a new family, in contrast to *Sargentodoxa*, which is maintained within Lardizabalalceae. The Berberidaceae and Menispermaceae are early branches of Ranunculales, which is primitively woody. Hydrastidaceae (including *Glaucidium*), Papaveraceae, and Ranunculaceae form a herbaceous clade of derived Ranunculales. The Papaveraceae subfamilies Platystemonoideae, Papaveroideae, and Eschscholzioideae are early branches, whereas Pteridophylloideae, Hypoecoideae, and Fumarioideae are derived. The Chelidonioideae are transitional and paraphyletic and therefore, a new subfamily is recognized for *Glaucium* and *Dicranostigma*. The Ranunculaceae subfamily lsopyroideae is a first branch, and Helleboroideae and Ranunculoideae are sister graups. Within lsopyroideae, the Aquilegieae is a first branch, and lsopyreae and Coptidae (including *Dichocarpum* and *Thalictrum*) are sister groups. Within Helleboroideae, the Caltheae is a first branch, and Helleboreae (including *Nigella*) and Cimicifugeae (including *Aconitum*, *Beesia*, and *Delphinium*) are sister groups. Within Ranunculoideae, the Adonideae is a first branch, and Anemoneae and Ranunculeae are sister groups. *Circaeaster* and *Kingdonia* are sister groups and positioned as a first branch of the Anemoneae, and it is at least 7 steps less parsimonious to remove both from Ranunculaceae, and 14 steps longer to remove only *Circaeaster*. It is 10 steps less parsimonious to consider a Ranunculaceae ancestry for *Paeonia*, which is positioned in Dilleniales. The "ranunculidae" are paraphyletic because Ranunculales are closer to Dilleniales than to Lardizabalales.

The *Ranunculidae* of Takhtajan (1987) represents a segregation from the *Magnoliidae* sensu Cronquist (1981) of those taxa with tricolpate rather than monosulcate pollen. The group includes five major families: *Berberidaceae*, *Lardizabalaceae*, *Menispermaceae*, *Papaveraceae*, and *Ranunculaceae*. Several issues complicate a phylogenetic system for ranunculids, most notably, the segregation of additional families. The major segregation is *Fumariaceae* from *Papaveraceae* (Cronquist 1981, Takhtajan 1987, Dahlgren 1989, Kubitzki 1993). All other segregations are monogeneric and include: *Nandina* from *Berberidaceae* (Takhtajan 1987); *Sargentodoxa* from *Lardizabalaceae* (Cronquist 1981, Takhtajan 1987, Dahlgren 1989,

THORNE 1992); *Hypecoum* (TAKHTAJAN 1987) and *Pteridophyllum* (KUBITZKI 1993) from *Papaveraceae*; and from *Ranunculaceae*: *Circaeaster* (CRONQUIST 1981, TAKHTAJAN 1987, DAHLGREN 1989, THORNE 1992, KUBITZKI 1993), *Glaucidium* (TAKHTAJAN 1987, DAHLGREN 1989, THORNE 1992, KUBITZKI 1993), *Hydrastis* (TAKHTAJAN 1987, DAHLGREN 1989, THORNE 1992), *Kingdonia* (CRONQUIST 1981, TAKHTAJAN 1987), and *Paeonia* (CRONQUIST 1981, TAKHTAJAN 1987, DAHLGREN 1989, THORNE 1992, KUBITZKI 1993). Familial segregation is appropriately addressed by a generic-level cladistic analysis. If a putative segregate renders a major family paraphyletic, the segregation would not be permissible in a phylogenetic classification (AX 1987, JUDD & al. 1994). However, if a putative segregate is isolated or in a phylogenetically primary position, segregation would be supported.

Another unresolved issue is the circumscription of orders. The majority of ranunculids are included in the order *Ranunculales* except for *Papaveraceae*, which is usually placed in *Papaverales* (CRONQUIST 1981, TAKHTAJAN 1987, DAHLGREN 1989). CRONQUIST (1981) transferred *Paeoniaceae* to *Dilleniales* of the *Dillenidae*. THORNE (1992) recognized an order *Paeoniales*, including *Glaucidiaceae*, whereas TAKHTAJAN (1987) isolated *Glaucidiaceae* in *Glaucidiales*.

Traditional hypotheses of interfamilial relationships associate *Berberidaceae* with *Ranunculaceae*, and *Lardizabalaceae* with *Menispermaceae*. LOCONTE & ESTES (1989) proposed a common ancestry between *Berberidaceae* and *Lardizabalaceae* based primarily on ranunculid ingroup comparison, which omitted comparisons with *Papaveraceae*. In a more comprehensive study of magnoliid families, LOCONTE & STEVENSON (1991) proposed a sister-group relationship between *Papaveraceae* and *Ranunculaceae*. CRONQUIST (1981) considered ranunculids to be primitively herbaceous with *Ranunculaceae* the most archaic family, whereas others have considered ranunculids as originally woody based on *Lardizabalaceae* (TAKHTAJAN 1987, LOCONTE & STEVENSON 1991).

A few non-ranunculid taxa have been proposed to be related to ranunculids, including *Coriaria* and *Sabiaceae* (CRONQUIST 1981), *Nelumbo* (DONOGHUE & DOYLE 1989), *Euptelea* (CHASE & al. 1993). We have not considered *Coriaria* and *Sabiaceae* because alternative sister groups for these taxa were beyond the scope of this analysis. The relationship between *Nelumbo* and ranunculids hypothesized by DONOGHUE & DOYLE (1989) is based on the possession of tricolpate pollen and benzylisoquinoline alkaloids. However, the benzylisoquinolines of *Nelumbo* represent the simplest types of these alkaloids, also found in *Aristolochiales* and *Piperales*, in contrast to the complex berberines and morphines found in ranunculids (GOTTLIEB & al. 1993). LOCONTE & STEVENSON (1991) have asserted a nymphaelean position for *Nelumbo* whereas a ranunculid position was 15 steps longer. Furthermore, LOCONTE (1995) hypothesized a sister-group relationship between *Nelumbo* and *Ceratophyllum*, and found it to be at least 12 steps longer to associate *Nelumbo* with ranunculids. Therefore, we excluded *Nelumbo* from this ranunculid analysis. The genus *Euptelea* was considered in our outgroup comparisons.

Relationships of ranunculids to other angiosperms were considered by LOCONTE (1995) in a simultaneous resolution of magnoliids, ranunculids, hamamelids, basal rosids, and basal dilleniids. The hamamelids were recognized as paraphyletic in reference to *Cunoniales*, supporting the concepts of THORNE (1992) and others that hamamelids should be incorporated into the *Roasidae*. Additionally, ranunculids were paraphyletic in reference to the *Dilleniales* and *Theales*. In other words, the ranunculids represent the earliest lineages of the *Dilleniidae*. Therefore, phylogenetic comparisons between ranunculids that do not consider dilleniids would be incomplete. The association of ranunculids with centrosperms is not considered here, based on the hypothesis that the later are advanced *Dilleniidae* related to the *Violales* (LOCONTE, unpubl. data).

Material and methods

Cladistic analyses were computed on a Macintosh Quadra 650 with the software packages MacClade (MADDISON & MADDISON 1993) and PAUP (SWOFFORD 1993). We incorporated 121 taxa including *Lactoris* (root), Paleoherbs (outgroup), *Euptelea* (outgroup), *Platanus* (outgroup), *Dilleniaceae* (outgroup), *Menispermaceae*, 8 genera of *Lardixabalaceae*, 15 genera of *Berberidaceae*, 41 genera of *Papaveraceae*, and 51 genera of *Ranunculaceae*. Nearly all genera for four of the major ranunculid families were coded; only *Menispermaceae* was not treated at the generic level. We coded *Menispermaceae* primarily by the character states of *Penianthus* and *Sphenocentrum*, which were proposed as primitive by THANIKAIMONI (1986). TAMURA (1993) recognized 59 genera for *Ranunculaceae*, but we have omitted 11 of these: *Garidella* and *Komaroffia* were treated as part of *Nigella*; *Consolida* was treated as part of *Delphinium*; *Barneoudia*, *Metanemone*, *Oreithales*, and *Pulsatilla* were treated as part of *Anemone*; and *Callianthemoides*, *Cyrtorhyncha*, *Halerpestes*, and *Peltocalathos* were treated as part of *Ranunculus*.

The majority of the characters were obtained from a synthesis of the available literature supplemented by empirical observations of herbarium specimens, liquid-preserved, and live plants. Characters were taken from LOCONTE & ESTES (1989) for *Berberidaceae* and *Lardizabalaceae*, from LIDÉN (1986) for fumariaceous *Papaveraceae*, and from KADEREIT & al. (1994) for the remaining *Papaveraceae*. Extensive use was made of the various treatments of ranunculid families and genera in KUBITZKI (1993). A total of 109 unordered characters with 192 apomorphic character states (Appendix) were derived from morphology, anatomy, and embryology, with a few chemical and karyological features (data matrix available in electronic format upon request). A total of 13,189 matrix cells includes 1.4% polymorphic codings, 5.9% no comparisons, 14.4% missing data, and 78.3% real data. Character-state optimization of hypothetical ancestors was performed with delayed transformation.

Results and discussion

Analytical aspects. A heuristic PAUP search resulted in only four trees at 568 steps (CI = 0.34, RI = 0.79; counting polymorphisms: 780 steps, CI = 0.52). The four trees differ only in the position of 2 genera in subfam. *Ranunculoideae* of *Ranunculaceae*; otherwise the topology is fully resolved. Searches based on the various addition-sequences also resulted in the same four trees. A search for topologies that are one step longer revealed 2045 trees.

Outgroup relationships. Of the outgroup taxa, *Dilleniaceae* exhibits relationships within the ingroup as the sister group of *Paeonia* (Fig. 1). The genera *Euptelea* and *Platanus* are sister groups forming a clade that we recognize as *Rosidae*. The *Rosidae* is the sister group of the ingroup, which we designate as *Dillenidae*. Alternative positions of *Euptelea* within the ingroup are less parsimonious by a range of two to nine steps from the base of the ingroup to *Ranunculaceae*,

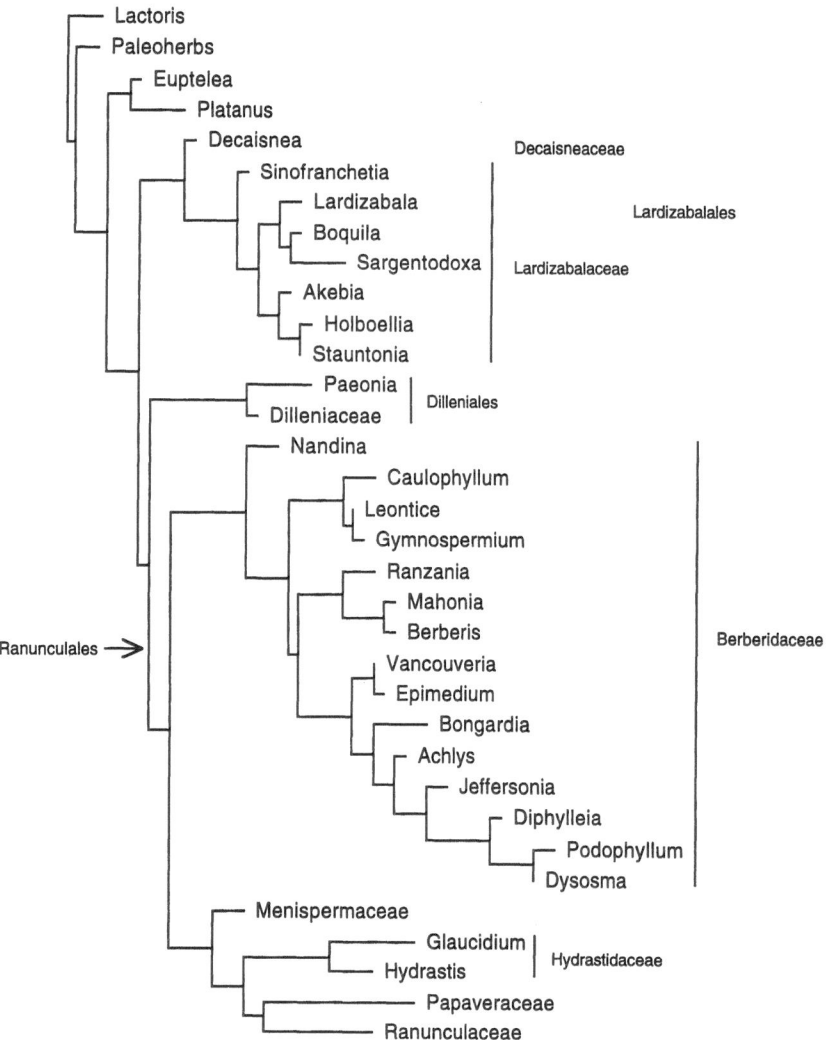

Fig. 1. Phylogenetic relationships for outgroups and basal *Dilleniidae*. *Lactoris*: 2–1, 9–1,11, 61–3, 78–1. Euangiosperms: 11–0. Paleoherbs: 1–2, 19–1. Eudicots (Tricolpates): 26–0, 55–1, 56–1. *Rosidae*: 18–1, 79–6. *Euptelea*: 11. *Platanus*: 10, 46–1, 69–3, 70–2, 104–1. *Dilleniidae*: 51–3, 100. *Lardizabalales*: 12–1, 46–1, 61–3, 76–4. *Decaisneaceae*: 15–1. *Lardizabalaceae*: 1–1, 8, 39–1, 70–3, 79–1. *Sinofranchetioideae*: 65–2. *Lardizabaloideae*: 16, 69–1. *Lardizabaleae*: 46–2, 70–1. *Lardizabala*: 12–2, 104–1. (*Boquila-Sargentodoxa*): 37. *Boquila*: 15–1. *Sargentodoxa*: 16–0, 59, 61–0, 69–3, 100–0. *Stauntonieae*: 12–3, 65–2. *Akebia*: 39–0. (*Holboellia-Stauntonia*): 21, 28. *Holboellia*: 75. *Stauntonia*: zero. *Dillenidae* II: 78–1. *Dilleniales*: 34, 35–2, 38, 50, 51–2, 55–3, 57–1, 73–1, 103. *Paeoniaceae*: 4–1, 7–2, 12–2, 28, 53, 104–3. *Dilleniaceae*: 75. *Ranunculales*: 8, 106. *Berberidaceae*: 7–1, 9–2, 12–2, 59, 61–1, 73–2, 79–1. *Nandinoideae*: 16, 27, 104–6. *Berberidoideae*: 1–2, 2–1, 53, 54. *Leonticeae*: 18–2, 39–1, 69–2, 79–2, 107. *Caulophyllum*: 27, 28, 75. (*Gymnospermium-Leontice*): 2–2. *Gymnospermium*: 103. *Leontice*: zero. *Berberideae*: 34. *Berberidinae*: 12–1, 56–2, 57–4, 104–1. *Ranzania*: 28, 65–2, 70–1. (*Berberis-Mahonia*): 1–0, 2–0, 69–2, 75. *Berberis*: 12–0. *Mahonia*: 16. *Epimediineae*: 14–2, 57–2, 79–7, 97, 104–2. (*Epimedium-Vancouveria*): 15–1, 103. *Epimedium*: 35–1. *Vancouveria*: zero. *Epimediineae* II: 18–2, 65–2. *Bongardia*: 2–2, 12–1, 27, 69–2, 79–2. *Epimediineae* III: 12–0, 19–1. *Achlys*: 100–0. *Epimediineae* IV: 28, 74, 75. *Jeffersonia*: 26, 103. *Epimediineae* V: 12–5, 13–6, 15–1, 18–1, 79–1, 106–0. *Diphylleia*: 14–0. (*Dysosma-Podophyllum*): 14–1, 24, 54–0, 70–1. *Dysosma*: zero. *Podophyllum*: 4–1, 26. *Ranunculales* II: 10, 28, 53, 104–5. *Menispermaceae*: 46–2, 79–5, 100–0. *Ranunculales* III: 1–2, 18–1, 26, 51–1. *Hydrastidaceae*: 2–1, 4–2, 7–2, 13–5, 19–1, 22–0, 75. *Hydrastidoideae*: 15–2, 57–2, 73–2, 79–1. *Glaucidioideae*: 15–1, 35–1, 50, 51–2, 61–2, 74, 104–6, 106–0. *Ranunculales* IV: 9–1, 104–1.

respectively. The sister groups *Dillenidae* and *Rosidae* form the clade Eudicots (= Tricolpates). The sister groups Eudicots and Paleoherbs form the clade Euangiosperms.

Ingroup relationships. After the removal of *Paeonia*, *Hydrastis*, and *Glaucidium* from *Ranunculaceae*, the five major families of ranunculids can be defined as monophyletic. *Paeonia* is isolated (recognized as *Dilleniales*). *Hydrastis* and *Glaucidium* are sister groups; this clade is also isolated and is recognized as *Hydrastidaceae* (Fig. 1). The *Lardizabalaceae* is isolated at the base of the ingroup (recognized as *Lardizabalales*). The ingroup topology can be summarized as: (*Lardizabalales-(Dilleniales-(Berberidaceae-(Menispermaceae-(Hydrastidaceae –* (*Papaveraceae-Ranunculaceae*)))))).

Lardizabalales. In an unpublished dissertation on the comparative anatomy and morphology of *Lardizabalaceae*, Taylor (1967) concluded that *Decaisnea* is the most primitive genus, and that the similarities of *Sargentodoxa* are the result of parallel evolution. Qin (1989) conducted a comparative study of *Lardizabalaceae* carpels and recognized four groups: *Decaisnea*; *Sinofranchetia*; *Boquila* and *Lardizabala*; and *Akebia*, *Holboellia*, and *Stauntonia*. The same four groups were found in a cladistic analysis to Loconte & Estes (1989), except that they included *Sargentodoxa* with *Boguila* and *Lardizabala*. Nowicke & Skvarla (1982) studied the pollen of these genera and demonstrated that the pollen of *Sargentodoxa* is consistent with *Lardizabalaceae* based on SEM and TEM studies. The seed coat anatomy of *Sargentodoxa* is similar to *Sinofranchetia* and *Boquila* (Brückner 1995, this volume; Takhtajan 1988: compare fig. 4 on p. 166 and figs. 8 and 7 on p. 164). The most distinctive feature of *Sargentodoxa* is the gynoecium that is composed of numerous spirally arranged carpels, in contrast to lardizabalaceous gynoecia with three cyclic carpels.

In this analysis, it is most parsimonious to consider *Sargentodoxa* as derived from within *Lardizabalaceae* as the sister group of *Boquila* (Fig. 1). *Sargentodoxa* homologies with *Boquila* and *Lardizabala* include dioecy, hemitropous ovules, and a reduction in ovule number (4–1 in *Boquila*, 1 in *Sargentodoxa*). *Sargentodoxa* is a long branch, and we hypothesize that the autapomorphies of *Sargentodoxa* represent adaptations as a relatively high-climbing vine. Alternative positions for *Sargentodoxa* are less parsimonious: two steps longer as the sister group of *Sinofranchetia* and four steps longer in a position below *Decaisnea*.

Because of the phylogenetically isolated position, we advocate the rank of order for the lardizabalaceous clade, allowing the elevation of *Decaisnea* to a monotypic family (Table 1). The *Lardizabalales* have cellular endosperm development in contrast to the remainder of ingroup taxa which exhibit nuclear endosperm development. Nuclear endosperm development is probably symplesiomorphic throughout the *Dillenidae*. Another plesiomorphy of *Lardizabalales* is their lack of benzylisoquinoline alkaloids, which are characteristic of *Ranunculales*. Instead, the *Lardizabalales* have neolignans (Gottlieb & al. 1993), which may be symplesiomorphic with more primitive taxa such as *Aristolochiales*.

Lardizabalales probably originated in eastern Asia based on the geographic distributions of *Decaisnea* in east Himalaya and China, and *Sinofranchetia* in China. The disjunct distributions of *Lardizabala* and *Boquila* in South America represent a secondary migration via North America. This hypothesis is supported by dis-

Table 1. Phylogenetic classification for ranunculids

Dilleniidae
  Lardizabalales LOCONTE, 1995
    Decaisneaceae[1]
    Lardizabalaceae DECAISNE, 1834
      Sinofranchetioideae[2]
      Lardizabaloideae
        Lardizabaleae
        Stauntonieae H. N. QIN & Y. C. TANG, 1989
  Dillenidae II
    Ranunculales DUMORTIER, 1829
      Berberidaceae A. L. JUSSIEU. 1789
        Nandinoideae (BERNHARDI) HEINTE, 1927
        Berberidoideae
        Leonticeae (SPACH) KOSENKO, 1980
          Berberideae
            Berberidineae
            Epimediineae (DUMORTIER) SPACH, 1839
  Ranunculales II
    Menispermaceae A. L. JUSSIEU, 1789
    Ranunculales III
    Hydrastidaceae MARTINOV, 1820
      Hydrastidoideae
      Glaucidioideae[3]
    Ranunculales IV
      Papaveraceae A. L. JUSSIEU, 1789
        Platystemonoideae ERNST, 1962
        Papaveraceae II
          Papaveroideae
          Papaveraceae III
            Eschscholzioideae ERNST, 1962
            Papveraceae IV
              Glaucioideae[4]
              Papaveraceae V
                Chelidonioideae ERNST, 1962
                Papaveraceae VI
                  Pteridophylloideae MURBECK, 1912
                  Papaveraceae VII
                    Hypoecoideae (DUMORTIER) HUTCHINSON, 1921
                    Fumarioideae (A. P. DE CANDOLLE) ENDLICHER, 1841
      Ranunculaceae A. L. JUSSIEU, 1789
        Isopyroideae SCHRODINGER, 1909
          Aquilegieae (TAMURA) DUNCAN & KEENER, 1991
          Isopyreae II
            Isopyreae
            Coptideae Langlet ex TAMURA & TOSUGE, 1989
        Ranunculaceae II
          Helleboroideae HUTCHINSON, 1923

Table 1. (*Continued*)

   Caltheae J. S. PRESL, 1826
   Helleboroideae II
    Helleboreae
    Cimicifugeae TORREY & A. GRAY, 1838
  Ranunculoideae
   Adonideae KUNTH, 1838
   Ranunculoideae II
    Anemoneae A. P. DE CANDOLLE, 1818
    Ranunculeae
Dillenidea III
 Dilleniales HUTCHINSON, 1924
  Paeoniaceae RUDLOPHI, 1830
  Dilleniaceae SALISBURY, 1807
 Dillenidae IV (remainder of dillenids)

---

¹ *Decaisneaceae* (TAKHTAJAN ex N. H. QIN & Y. C. TANG) LOCONTE, fam. et stat. nov. Basionym: tribe *Decaisneeae* TAKHTAJAN ex N. H. QIN & Y. C. TANG, Cathaya **1**:80. 1989.

² *Sinofranchetioideae* (N. H. QIN & Y. C. TANG) LOCONTE, subfam, et stat. nov. Basionym: tribe *Sinofranchetieae* N. H. QIN & Y. C. TANG, Cathaya **1**:80. 1989.

³ *Glaucidioideae* (TAMURA) LACONTE, subfam. et stat. nov. Basionym: family *Glaucidiaceae* TAMURA, Bot. Mag. (Tokyo) **85**: 40. 1972.

⁴ *Glaucioideae* (MILLER) LOCONTE, subfam. et stat. nov. Basionym: *Glaucium* section *Glaucium* MILLER, Feddes Repert. **89**: 536. 1978.

covery of fossil *Sargentodoxa* fruits from the Eocene of North America (TIFFNEY 1993).

Dilleniales. The position of *Paeonia* as the sister group of *Dilleniaceae* was originally developed by CORNER (1946), and supported by EAMES (1961), KEEFE & MOSELEY (1978), and CRONQUIST (1981). Homologies supported by this analysis include a differentiated five-merous perianth, persistent calyx, complex polyandry with centrifugal development, tricolporate pollen with reticulate exine sculpturing, micropyle formed by the outer integument, and arillate seeds. The alternative position of *Paeonia* at the base of *Ranunculaceae* is ten steps longer.

Ranunculales. Our circumscription of *Ranunculales* includes *Papaveraceae* but excludes *Paeonia* and *Lardizabalaceae*. The traditional recognition of *Papaverales* renders the *Ranunculales* paraphyletic. The most obvious uniquely derived character state for *Ranunculales* is the benzylisoquinoline alkaloids; lost in *Podophyllum* of *Berberidaceae*, *Glaucidium* of *Hydrastidaceae*, and derived *Ranunculaceae*. *Ranunculales* are also diagnosed by vessels with simple perforations, which are mixed with scalariform perforations in *Hydrastidaceae* and several *Ranunculaceae*.

Berberidaceae. A classification for *Berberidaceae* was presented by LOCONTE (1993), who recognized four groups as *Nandinoideae* (monotypic), *Leonticeae* (3 genera), *Berberidineae* (3 genera), and *Epimediineae* (8 genera). This analysis corroborates these relationships (Fig. 1). A striking example of parallel evolution is exhibited by *Bongardia* (*Epimediineae*) and *Leontice* (*Leonticeae*), which are sympatric in arid SW Asia, and possess tubers and yellow flowers. The alternative position of *Bongardia* in the *Leonticeae* rather than the *Epimediineae* is four steps longer. The relationship of *Bongardia* within the *Epimediineae* is also supported by rbcL sequence analysis (KIM 1995, this volume). NICKOL (1995, this volume) supports a sister-group relationship between *Caulophyllum* and *Nandina*, which we find to be six steps less parsimonious. On the basis of our topology the closest relative of *Nandina* is *Caulophyllum*, but we interpret the similarities to be primarily symplesiomorphic.

East Asia is rich in primitive *Berberidaceae* such as *Nandina*, *Caulophyllum*, *Gymnospermium*, and *Ranzania*. Taxa distributed in North America represent independent secondary migrations such as *Caulophyllum*, *Mahonia*, *Berberis*, *Vancouveria*, *Achlys*, *Jeffersonia*, *Diphylleia*, and *Podophyllum*.

Menispermaceae. Infrafamilial classifications for *Menispermaceae* remain relatively subjective because of the high diversity (ca. 65 genera) and mostly tropical distributions. KESSLER (1993) compared previous systems for *Menispermaceae* and recognized five tribes based on fruit and seed characteristics. We suggest that a generic-level cladistic analysis of *Menispermaceae* based on a wide spectrum of characters, including floral and palynological, is necessary and will be facilitated by this analysis.

Hydrastidaceae. It has become popular to disclaim a relationship between *Hydrastis* and *Glaucidium* (TOBE & KEATING 1985, KEENER 1993), and only AIRY-SHAW in WILLIS (1973) appears to have accepted it. A sister-group relationship between the two enigmatic genera is supported based primarily on leaf homologies. Both have a similar type of palmate venation, leaf margin, and an undifferentiated mesophyll. Additionally, the ovules of both genera have a nucellar cap, which is a seres of periclinal divisions in the nucellar epidermis (TOBE, 1981: fig. 1; TOBE & KEATING, 1985: fig. 17). Furthermore, BARTHLOTT & THEISEN (1995, this volume) have observed a unique type of epicuticular wax formation in both genera, and HOOT & CRANE (1995, this volume) find their molecular sequences to be homologous. Both taxa are long branches, with *Hydrastis* exhibiting parallelisms with *Berberidaceae*, and *Glaucidium* exhibiting parallelisms with *Paeonia* and *Papaveraceae*. It is 13 steps longer to position *Glaucidium* as the sister group of *Paeonia*, eight steps longer to position *Glaucidium* as the sister group to *Dilleniales*, and six steps longer to consider *Hydrastidaceae* as the sister group of *Dilleniales*. It is at least 10 steps longer to consider a sister-group relationship between *Hydrastis* and *Berberidaceae*, and 14 steps longer to nest *Hydrastis* within *Berberidaceae* as the sister group of the *Epimediineae*. The hypothesis of KUMAZAWA (1930) that *Glaucidium*, *Hydrastis*, *Podophyllum*, and *Diphylleia* represent an isolated family is at least 10 steps longer. We find it only one step longer to position *Hydrastidaceae* as the sister group of *Ranunculaceae* and this alternative relationship should probably be reconsidered in more detail. RONSE DECRAENE & SMETS (1992) list *Hydrastis* in a table (p. 638) of taxa exhibiting complex polyandry; however, *Hydrastis* exhibits true polyandry that is spirally centripetal (TOBE & KEATING 1985).

(*Papaveraceae-Ranunculaceae*). Of the five major families of ranunculids, we propose *Papaveraceae* and *Ranunculaceae* as the bast candidates for a sister-group relationship based on a chromosome base number of x = 7 and paratracheal wood parenchyma. Both families appear to be primitively herbaceous with solitary flowers rather than inflorescences; each of these characters undergoes reversal within both families. CORNER (1976) proposed that the seeds of both families are homologous, and JENSEN & PENNER (1980) demonstrated similarity in the seed proteins based on serological results. KADEREIT & al. (1994) argue against *Ranunculaceae* as the sister group of *Papaveraceae*. We find the putative sister-group relationship of *Berberidaceae* and *Papaveraceae* (KADEREIT & al. 1994) to be at least three steps longer.

Papaveraceae. Systematists continue to be divided over the treatment of *Papaveraceae* in the broad sense (THORNE 1992), or the restricted sense by segregating *Fumariaceae* (CRONQUIST 1981, TAKHTAJAN 1987, DAHLGREN 1989, KADEREIT 1993). *Papaveraceae* sensu lato has been divided into seven subfamilies: *Platystemonoideae* (3 genera), *Papaveroideae* (8 genera), *Eschscholzioldeae* (3 genera), *Chelidonioideae* (8 genera), *Pteridophlloideae* (1 genus), *Hypecoideae* (1 genus), and *Fumarioideae* (16 genera). Alternatively, all seven of these subfamilies have been recognized as families. CORNER (1976) considered the genus *Bocconia* (of the *Chelidonioideae*) as archetypic based on its relatively large (7 mm) arillate seeds. The *Platystemonoideae* have been considered primitive (GONNERMANN 1980, 1982; BRÜCKNER 1982, 1983; THORNE 1992). However, KADEREIT (1993) has hypothesized a sister-group relationship between *Platystemonoideae* and *Papaveroideae* based on the shared possession of multicellular-multiseriate hairs and polycarpellate gynoecia. The *Chelidonioideae* have also been considered primitive (TAKHTAJAN, 1987, KADEREIT 1993), presumably based on outgroup comparison with the fumarioids.

We have found ample support for the treatment of *Papaveraceae* in the broad sense, including *Pteridophyllum*, *Hypecoum*, and *Fumarioideae* (Fig. 2, discussed below). The seven subfamilies of *Papaveraceae* are maintained with only two changes: (1) *Papaveroideae* sensu ERNST (1962) is polyphyletic and, therefore, the genus *Canbya* is transferred from the *Papaveroideae* to the *Platystemonoideae*; (2) *Chelidonioideae* sensu ERNST (1962) is paraphyletic and, therefore, the genera *Glaucium* and *Dicranostigma* are segregated as a new subfamily (Table 1).

The *Platystemonoideae* are substantiated as a first branch of *Papaveraceae* based on the plesiomorphies of a three-merous perianth, and a gynoecium with free styles. We coded *Platystemon* as paracarpous, rather than apocarpous. On the basis of reciprocal illumination, the paracarpous gynoecium of *Platystemon* could be interpreted as a vestigial plesiomorphy inherited from the apocarpous ancestry of the remaining ranunculids. The inclusion of *Canbya* in the *Platystemonoideae* is based on a common ancestry with *Meconella* (Fig. 2; GREENE 1903). ERNST (1967) compared these taxa and concluded that *Canbya*, with its fused stigmas and valvate fruit dehiscence, is not related to *Platystemonoideae*, which have free stigmas and non-valvate fruit. However, *Canbya* and *Meconella* exhibit the synapomorphies of glabrous vegetative parts, reduced cyclic androecia, and a base chromosome number of x = 8; *Hesperomecon* and *Platystemon* are pubescent, have polyandrous androecia, and base chromosome numbers of x = 7 and 6, respectively. The three genera *Hesperomecon*, *Canbya*, and *Meconella* also share three-carpellate gynoecia, whereas *Platystemon* has several to many carpels. All four genera share the annual habit, leaves with entire margins, nodding flowers, and straight seed axes. Several of these character states are paralleled by the *Papaveroideae*, but the alternative position of *Canbya* in that subfamily is isolated, and four less parsimonious.

The *Papaveroideae* represent a second branch of *Papaveraceae* (Fig. 2). We interpret the similarities between *Papaveroideae* and *Platystemonoideae* as sym-

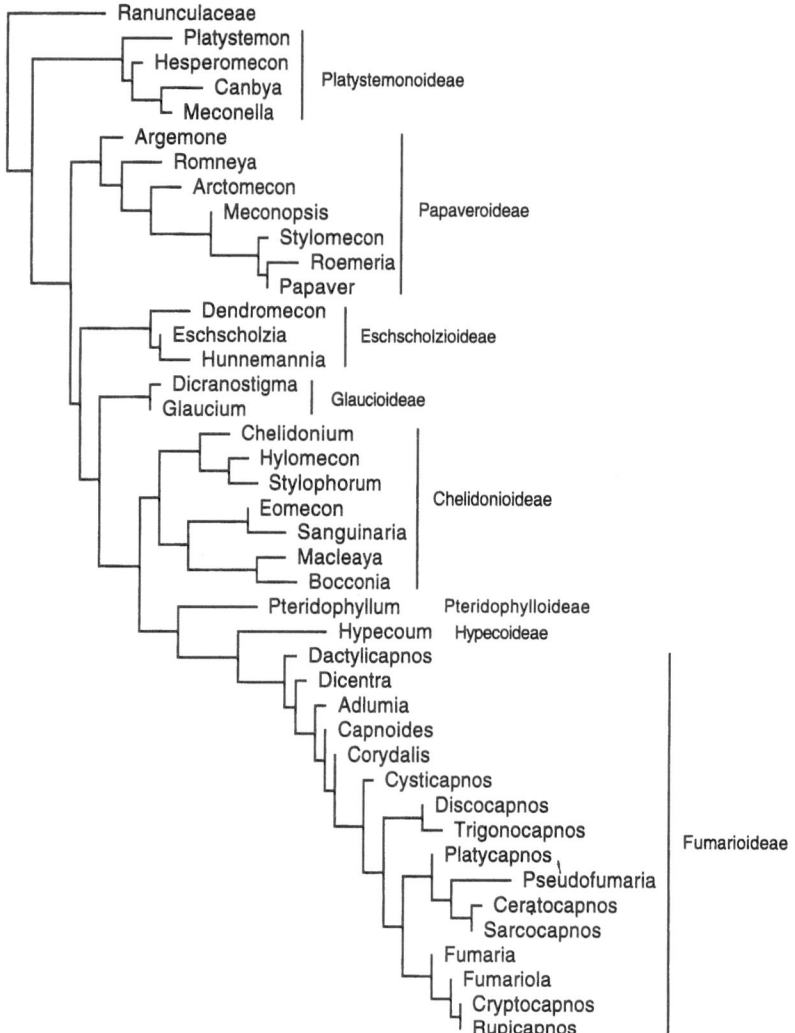

Fig. 2. Phylogenetic relationships for *Papaveraceae*. *Papaveraceae*: 6–1, 15–2, 34, 62–1, 73–2, 76–1, 79–4. *Platystemonoideae*: 1–3, 12–6, 14–1, 18–0, 30, 48–1, 70–3, 102–1. *Platystemon*: 62–2, 88–3, 92–1, 96–1, 104–2. *Platystemonoideae* II: 61–3. *Hesperomecon*: 14–2. (*Canbya-Meconella*): 15–0, 51–0, 104–0. *Canbya*: 48–0, 65–2, 70–0, 89. *Meconella*: 94. *Papaveraceae* II: 65–2. *Papaveroideae*: 89, 93–2, 95–1. *Argemone*: 92–1, 99–1. *Papaveroideae* II. 57–1, 99–2,. *Romneya*: 1–0, 49, 102–1. *Papaveroideae* III: 30, 70–3, 90. *Arctomecon*: 14–2, 92–2, 104–2. *Papaveroideae* IV: 6–2, 57–3, 91–1, 96–1, 101, 102–1. *Meconopsis*: zero. *Papaveroideae* V: 1–3, 35–1, 49, 95–2, 98–1, 109. *Stylomecon*: 65–1. (*Roemeria-Papaver*): 99–0. *Roemeria*: 55–2, 56–2, 90–0. *Papaver*: zero. *Papaverraceae* III: 35–1, 61–2. *Eschscholzioideae*: 5, 15–3, 56–2, 57–1, 88–0, 92–2, 96–3. *Dendromecon*: 1–0, 13–3, 94, 103. (*Eschscholzia-Hunnemannia*): 60. *Eschscholzia*: zero. *Hunnemannia*: 18–0, 57–3, 96–1. *Papaveraceae* IV: 15–1, 65–1, 104–2. *Glaucioideae*: 6–2, 17, 87, 102–1. *Dicranostigma*: 14–2. *Glaucium*: zero. *Papaveraceae* V: 2–1, 26–0, 29–0. *Chelidonioideae*: 6–3, 103. *Chelidonioideae* II: 12–1, 28–0, 96–2. *Chelidonium*: 88–0, 93–1, 102–1. (*Hylomecon-Stylophorum*): 89, 92–1, 94. *Hylomecon*: 88–2, 99–2. *Stylophorum*: 56–2, 90, 104–3. *Chelidonioideae* III: 55–2, 56–2, 92–2. (*Eomecon-Sanguinaria*): 14–2, 15–0, 19–1, 57–5, 88–2, 89, 99–1, 104–4. *Eomecon*: zero. *Sanguinaria*: 13–3, 26, 96–1, 99–1. (*Bocconia-Maclayea*): 2–0, 27, 28–0, 34–0, 47, 94, 104–6. *Bocconia*: 1–0, 57–1, 88–0, 100–0. *Maclayea*: 19–1, 92–1, 93–1. *Papaveraceae* VI: 36, 51–0, 70–0, 102–2.

plesiomorphies inherited from the common ancestry of the family; for example, the polycarpous gynoecia (supported by outgroup comparison). Within *Papaveroideae*, the genus *Papaver*, which has both annual and perennial species, is nested within an annual clade with *Stylomecon* and *Roemeria*. On the basis of a species-level analysis of chloroplast DNA restriction sites, KADEREIT & SYTSMA (1992) concluded that *Stylomecon* and *Roemeria* are nested within the genus *Papaver*. If this is accepted, *Papaver* can be interpreted as primitively perennial, and the sister group of *Papaver* would be *Meconopsis*. Therefore, the generic sequence for *Papaveroideae* can be simplified as *Argemone, Romneya, Arctomecon, Meconopsis, Papaver*.

The remaining *Papaveraceae* are characterized by a two-merous perianth and a two-merous gynoecium. A third branch of *Papaveraceae* is the subfamily *Eschscholzioideae* (Fig. 2), which exhibits the plesiomorphic condition of the gynoecium without a style (shared with *Papaveroideae*). Furthermore, the *Eschscholzioideae* possess a bicarpellate gynoecium with multiple external ridges, which may represent a vestigial plesiomorphy from the polycarpellate ancestry. *Glaucioideae* possess the plesiomorphy of solitary flowers (shared with *Platstemonoideae, Papaveroideae* and *Eschscholzioideae*), whereas *Chelidonioideae* are inflorescent. In an analysis of chloroplast DNA restriction sites, BLATTNER & KADEREIT (1994) found *Glaucium* to be discordant with *Chelidonioideae*.

The genus *Pteridophyllum* is isolated as an independent subfamily (Fig. 2). To consider *Pteridophyllum* as the sister group of *Hypecoum* is only one step less parsimonious. However, to position *Pteridophyllum* at the base of *Papaveraceae* is seven steps longer, and at the base of *Berberidaceae* (as suggested by some authors) is ten steps longer. *Pteridophyllum* exhibits parallelisms with *Sanguinaria* and *Eomecon* of the *Chelidonioideae*, such as basal leaves and a base chromosome number of x = 9.

Our contention that the fumarioids should be considered as an integral part of *Papaveraceae* is based on a series of homologies. Therefore, to consider the fumarioids at the base of *Papaveraceae* sensu stricto is at least four steps less parsimonious, and requires the independent evolution of an inflorescence, two-merous calyx, two-merous corolla, two-merous gynoecium, and fused style. Additionally, LIDÉN (1986) listed the compound acetylornithin as a homology between *Chelidonioideae* and fumarioids. KADEREIT & al. (1994) support the recognition of *Fumariaceae* based on their topology: (*Pteridophyllaceae*-(Fumariaceae-

Fig. 2. (*Continued*)
Pteridophylloideae: 6–0, 14–2, 28–0, 30, 57–4, 73–0, 88–0, 91–2, 104–4. *Papaveraceae* VII: 5, 15–0, 33–2, 52, 76–2, 104–0. *Hypecoideae*: 1–3, 14–2, 48–2, 56–3, 88–3, 94, 96–4, 98–2, 102–1. *Fumarioideae*: 18–2, 48–3, 57–5, 68–3, 103. *Dactylicapnos*: 20–1. *Fumarioideae* II: 12–2. *Dicentra*: 30. *Fumarioideae* III: 1–3, 88–2. *Adlumia*: 20–2. *Fumarioideae* IV: 33–1. *Capnoides*: zero. *Fumarioideae* V: 28–0. *Corydalis*: zero. *Fumarioideae* VI: 56–2, 68–4, 103–0. *Cysticapnos*: 20–1. *Fumarioideae* VII: 79–6, 100–0. (*Discocapnos-Trigonocapnos*): 20–1, 43, 44, 86. *Discocapnos*: zero. *Trigonocapnos*: 64, 102–1. *Fumarioideae* VII: 12–0, 64. *Fumarioideae* IX: 67, 68–2, 102–1. *Platycapnos*: zero. *Fumarioideae* X: 42, 77. *Pseudofumaria*: 1–2, 3, 56–1, 79–4, 100, 103. (*Ceratocapnos-Sarcocapnos*): 15–3, 66. *Ceratocapnos*: 20–1. *Sarcocapnos*: zero. *Fumarioideae* XI: 55–2, 58, 68–1. *Fumaria*: zero. *Fumarioideae* XII: 3, 102–1. *Fumariola*: zero. (*Cryptocapnos-Rupicapnos*): 85. *Cryptocapnos*: zero. *Rupicapnos*: zero.

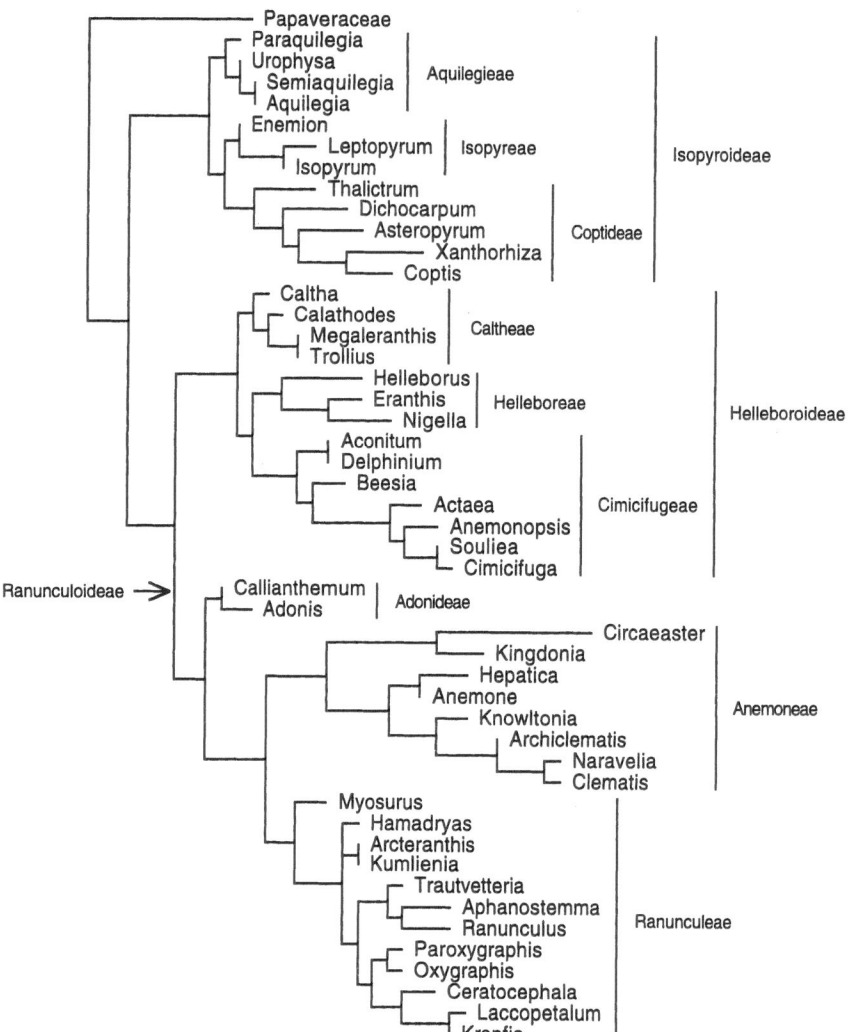

Fig. 3. Phylogenetic relationships for *Ranunculaceae. Ranunculaceae*: 7–1, 35–2, 57–3.
Isopyroideae: 12–2, 18–2, 23. *Aquilegieae*: 39–5. *Paraguilegia*: 40. *Aquilegieae* II: 45.
Urophysa: zero. (*Aquilegia-Semiaquilegia*): 39–2. *Aquilegia*: zero. *Semiaquilegia*: zero.
(*Isopyreae-Coptideae*): 26–0. *Isopyreae*: 63. *Enemion*: zero. (*Isopyrum-Leptopyrum*): 22–0,
23–0, 39–3. *Isopyrum*: zero. *Leptopyrum*: 1–3, 72. *Coptideae*: 55–2, 56–2. *Thalictrum*: 59, 79–3,
100–0, 108. *Coptideae* II: 18–1, 39–4. *Dichocarpum*: 12–4, 61–2, 62–2, 104–2. *Coptideae* III:
80–0. *Asteropyrum*: 12–5, 19–1, 26, 104–0. (*Coptis-Xanthorhiza*): 15–1, 22–2, 104–4. *Coptis*:
13–2, 16, 63. *Xanthorhiza*: 1–0, 12–1, 28–0, 39–1, 40. (*Helleboroideae- Ranunculoideae*):
104–0, 105. *Helleboroideae*: 13–3, 19–1. *Caltheae*: 108. *Caltha*: 22–0. *Caltheae* II: 57–2.
Calthodoides: 63. (*Megaleranthis-Trollius*): 25, 39–1. *Megaleranthis*: zero. *Trollius*: zero.
(*Helleboreae-Cimicifugeae*): 106–0. *Helleboreae*: 39–3, 76–1. *Helleborus*: 12–4, 38, 57–1,
62–2, 72. (*Eranthis-Nigella*): 13–5, 18–2, 40. *Eranthis*: 22–0, 63. *Nigella*: 1–3, 62–1, 104–2,
108. *Cimicifugeae*: 15–1, 26–0, 28–0. (*Aconitum-Delphinium*): 33–1, 39–2. *Aconitum*: zero.
Delphinium: zero. *Cimicifugeae* II: 61–1. *Beesia*: 14–2, 22–0. *Cimicifugeae* III: 12–2, 13–2,
19–0, 23, 39–1. *Actaea*: 65–2, 79–1. *Cimicifugeae* IV: 63. *Anemonopsis*: 15–0, 30. (*Cimicifuga-*
Souliea): 39–5, 40. *Cimicifuga*: 108. *Souliea*: zero. *Ranunculoideae*: 79–3, 100–0. *Adonideae*:
12–1, 18–0, 34. *Adonis*: 13–2, 22–0. *Callianthemum*: zero. (*Anemoneae- Ranunculeae*): 14–2,
72, 74, 106–0. *Anemoneae*: 13–2, 81–0. (*Circaeaster-Kingdonia*): 11, 19–2, 22–0, 51–0, 57–2,

(Chelidonioideae-(Eschscholzioideae-(Glaucioideae-(Platystemonoideae-Papaveroideae)))))), which we find to be six steps less parsimonious. In comparison, the differences between our topology and that of KADEREIT & al. (1994) are the result of outgroup comparison. Our topology for *Papaveraceae* is based on *Ranunculaceae* (first outgroup) and *Hydrastidaceae* (second outgroup). The topology for *Papaveraceae* by KADEREIT & al. (1994) was subjectively rooted at *Pteridophyllum* based on the implicit hypothesis that the lack of latex in *Pteridophyllum* is plesiomorphic (interpreted as a reversal in our topology).

Within *Fumarioideae*, four groups have been recognized: *Corydaleae*, *Discocapninae*, *Sarcocapninae*, and *Fumariinae* (LIDÉN 1986, 1993 a, b); the genus *Cysticapnos* was left unclassified. LIDÉN (1986) presented the relationships of these groups as: (*Corydaleae*-(*Cysticapnos*-(*Discocapninae*-(*Sarcocapninae*-*Fumariinae*)))). Our analysis supports these relationships except that we find the *Corydaleae* to be paraphyletic; it is two steps longer to consider the *Corydaleae* as monophyletic. Therefore, the zygomorphic flowers in *Corydalis* and more derived fumarioids are probably homologous, whereas the bilateral flowers of *Dactylicapnos* and *Dicentra* are symplesiomorphic with *Hypecoum*.

In contrast to other ranunculid families that are basically east Asian, the *Papaveraceae* appear to have originated in western North America. This hypothesis is based on the extant distributions of all *Platystemonoideae*, many *Papaveroideae*, and most *Eschscholzioideae*. KADEREIT & al. (1994) proposed E. Asia as the site of origin for *Papaveraceae* based on the distribution of *Pteridophyllum*, which we interpret as a secondary migration.

Ranunculaceae. The infrafamilial classification of *Ranunculaceae* has been considered most recently by TAMURA (1993), who recognized five subfamilies: *Helleboroideae*, *Ranunculoideae*, *Isopyroideae*, *Thalictroideae*, and *Hydrastidoideae*. We advocate the transfer of *Hydrastis* to the *Hydrastidaceae* (with *Glaucidium*), and *Thalictrum* is subsumed into the subfamily *Isopyroideae* (Fig. 3). Our results are similar to TAMURA's system in that we find support for a relationship between *Helleboroideae* and *Ranunculoideae* (x = 8, r-type chromosomes); however, we propose *Isopyroideae* (x = 7, t-type chromosomes) as ancestral rather than *Helleboroideae*. The chromosome base number of x = 7 occurs in basal lineages of the *Papaveraceae* such as *Hesperomecon* of the *Platystemonoideae*, most of the *Papaveroideae*, and all of the *Eschscholzioideae*. Additionally, the *Isopyroideae* are

---

Fig. 3. (*Continued*)

69–3, 70–2. *Circaeaster*: 1–3, 26–0, 35–1, 59, 61–2, 65–2, 76–3, 78–0, 83–2, 104–7. *Kingdonia*: 39–1, 78–2, 104–4. *Anemoneae* II: 19–1, 55–2, 56–2, 83–1. (*Anemone-Hepatica*): 15–1, 25. *Anemone*: zero. *Hepatica*: 18–2, 76–1, 104–1. *Anemoneae* III: 14–0, 26–0, 108. *Knowltonia*: 12–1, 79–1. *Anemoneae* IV: 1–1, 23, 24, 84. *Archiclematis*: zero. (*Clematis-Naravelia*): 12–1, 14–1, 19–0. *Clematis*: 75. *Naravelia*: 20–1. *Ranunculeae*: 34, 41, 75. *Myosurus*: 1–3, 18–0, 22–2. *Ranunculeae* II: 70–1, 71. *Hamadryas*: 46–2. *Ranunculeae* III: zero. (*Arcteranthis-Kumlienia*): 39–1. *Arcteranthis*: zero. *Kumlienia*: zero. *Ranunculeae* IV: 56–2. *Ranunculeae* V: 14–0, 26–0. *Trautvetteria*: 34–0. (*Aphanostemma- Ranunculus*): 82. *Aphanostemma*: 1–3, 39–1, 81–0. *Ranunculus*: 13–2, 56, 108. *Ranunculeae* VI: 38. (*Oxygraphis-Paroxygraphis*): 16. *Oxygraphis*: 22–2. *Paroxygraphis*: 46–2. *Ranunculeae* VII: 32, 55–2. *Ceratocephala*: 1–3, 104–1. (*Krapfia-Laccopetalum*): 31, 39–5, 81–0. *Krapfia*: zero. *Laccopetalum*: 14–0.

rich in benzylisoquinoline alkaloids, another plesiomorphy, whereas the remaining *Ranunculaceae* have little or no benzylisoquinolines. Furthermore, HOOT (1994) has found support for considering various genera of *Isopyroideae* as primitive based on chloroplast DNA gene sequences.

We divide subfamily *Isopyroideae* into three tribes: *Aquilegieae* (4 genera), *Isopyreae* (3 genera), and *Coptideae* (5 genera). The *Isopyroideae* of TAMURA (1993) are paraphyletic because of the exclusion of *Thalictrum*, which we refer to the tribe *Coptideae* (Fig. 3), based on the homology of polyporate pollen. The genus *Asteropyrum* is tentatively referred to the *Coptideae*. *Asteropyrum* exhibits a base chromosome number of x = 8 as in *Helleboroideae*, and the karyotype has been variously interpreted as t-type (ZHANG, 1982), r-type (YANG & al. 1993), or intermediate (TAMURA 1995, this volume). Our interpretation is t-type and an isopyroid ancestry for *Asteropyrum*; however, a helleboroid position is only one step longer.

The subfamily *Helleboroideae* is divided into three tribes: *Caltheae* (4 genera), *Helleboreae* (3–5 genera), and *Cimicifugeae* (7 or 8 genera). The *Helleboroideae* of TAMURA (1993) are monophyletic, but we find only weak support for the group, and point out that it may be paraphyletic in reference to the *Ranunculoideae*. The group includes several problematic genera, such as *Caltha, Eranthis, Helleborus, Nigella*, and *Trollius* which appear relatively isolated based on various criteria; each was segregated as a separate tribe by DUNCAN & KEENER (1991). The relationship of *Calathodes* and *Megaleranthis* to *Trollius* is supported by their striate pollen exine (LEE 1989, 1992, YING & al. 1993). A sister-group relationship between *Eranthis* and *Nigella* is supported by the homologies of conduplicate-plicate leaf vernation, and apically bilobed staminodia. Apomorphic agreements of *Eranthis, Nigella*, and *Helleborus* include tubular staminodia and solanad embryogeny. The genera *Aconitum, Delphinium*, and *Beesia* are referred to the *Cimicifugeae* based on their indeterminate inflorescences. Additionally, *Beesia* shares a unicarpellate gynoecium with higher *Cimicifugeae*.

The subfamily *Ranunculoideae* is divided into three tribes: *Adonideae* (2 genera), *Anemoneae* (8–12 genera), and *Ranunculeae* (12–16 genera), which is congruent with the classification by TAMURA (1993), except that we include *Circaeaster* in the tribe *Anemoneae* as the sister group of *Kingdonia* (Fig. 3). The putative segregate *Circaeasteraceae* renders the *Ranunculaceae* paraphyletic. To move both *Circaeaster* and *Kingdonia* to a familial position outside *Ranunculaceae* (CRONQUIST 1981) is at least 7 steps longer. To move *Circaeaster* alone (without *Kingdonia*) is 14 steps longer. This is because of the numerous apomorphic agreements between these two genera, the most impressive being the leaves with open dichotomous venation (FOSTER 1961, 1963), pollen exine with striate sculpturing (NOWICKE & SKVARLA 1982), and the orthropous ovules. Within the tribe *Ranunculeae, Myosurus* is substantiated as a first branch (rather than *Trautvetteria*) based on the plesiomorphies of pendulous anatropous ovules, whereas the remainder of *Ranunculeae* exhibit ascending hemitropous ovules. The four equally parsimonious trees of the overall analysis exist in the *Ranunculeae* for the position of the (*Arcteranthis-Kumlienia*) clade; this clade can be considered as: (1) the sister group of *Hamadryas*; (2) after *Hamadryas*; (3) the sister group of (*Trautvetteria-* (*Aphanostemma-Ranunculus*)); (4) the sister group of ((*Paroxygraphis-Oxygraphis*)- (*Ceratocephala-*(*Krapfia-Laccopetalum*))).

Conclusions. The primary relationships found in this analysis can be summarized as: (1) the five major families of ranunculids are monophyletic after the segregation of *Glaucidium, Hydrastis,* and *Paeonia* from *Ranunculaceae*; (2) *Lardizabalaceae* is isolated and recognized as the order *Lardizabalales* including *Decaisnea* as a separate family and *Sargentodoxa* as part of the tribe *Lardizabaleae*; (3) *Paeoniaceae* is the sister group of *Dilleniaceae*; (4) *Berberidaceae, Menispermaceae,* and *Hydrastidaceae* (including *Glaucidium*) are early branches of *Ranunculales*; (5) *Papaveraceae* and *Ranunculaceae* are sister groups; (6) *Papaveraceae* is divisible into eight subfamilies, including the fumarioids; (7) *Ranunculaceae* is divisible into three subfamilies, each with three tribes, including *Circaeaster* and *Kingdonia*. The difference in classification schemes for *Papaveraceae* and *Ranunculaceae* (subfamilies vs. subfamilies and tribes) reflects a difference in their historical treatments as well as their phylogenetic hierarchy (Figs. 2, 3).

Several lines of investigation are possible to recosider the relationships supported by this analysis. First of all, additional characters could be considered such as various types of molecular data. Another approach would be to develop a genericlevel analysis of *Menispermaceae*. Explicit hypotheses of the plesiomorphic conditions for *Menispermaceae* could effect the position of the family as well as basal branching in the surrounding families. Finally, the phylogeny of the ranunculids should be reconsidered in a simultaneous resolution of the families of the *Dillenidae*.

References

Ax, P., 1987: The phylogenetic system, the systematization of organisms on the basis of their phylogenesis. – Chichester: Wiley.

Barthlott, W., Theisen, I., 1995: Epicuticular wax ultrastructure and classification of the *Ranunculiflorae*. – Pl. Syst. Evol. [Suppl.] **9**: 39–45.

Blattner, F. R., Kadereit, J. W., 1994: Phylogenetic analysis of the *Chelidonioideae* (*Papaveraceae*): intercontinental disjunctions. Amer. J. Bot. (6: Suppl.) **81**: 196–197.

Brückner, C., 1982: Zur Kenntnis der Fruchtmorphologie der *Papaveraceae* und der *Hypecoaceae*. – Feddes Repert. Beih. **93**: 153–212.

– 1983: Zur Morphologie der Samenschale in den *Papaveraceae* und *Hypecoaceae*. –Feddes Repert. Beih. **94**: 361–405.

– 1995: Comparative seed structure in the *Ranunculiflorae*. – Pl. Syst. Evol. [Suppl.] **9**: 83–84.

Carlquist, S., 1995: Wood anatomy of the *Ranunculiflorae*. – Pl. Syst. Evol. [Suppl.] **9**: 11–24. – Schneider, E., Miller, R. B., 1994: Wood and bark anatomy of *Argemone* (*Papaveraceae*). – Amer. J. Bot. [Suppl.] **81**: 146.

Chase, M., Soltis, D. E., Olmstead, R. G., Morgan, D., Les, D. H., Mishler, B. D., Duvall, M. R., Price, R. A., Hills, H. G., Qui, Yin-Long, Kron, K. A., Rettig, J. H., Conti, E., Palmer, J. D., Manhart, J. R., Sytsma, K. J., Michaels, H. J., Kress, W. J., Karol, K. G., Clark, W. D., Hedren, M., Gaut, B. S., Jansen, R. K., Kim, Ki-Joong, Wimpee, C. F., Smith, J. F., Furnier, G. R., Strauss, S. H., Xiang, Qiu-Yun., Plunkett, G. M., Soltis, P. S., Swenson, S. M., Williams, S. E., Gadek, P. A., Quinn, C. J., Eguiarte, L. E., Golenberg, E., Learn Jr., G. H., Graham, S. W., Barrett, S. C. H., Dayanandan, S., Albert, V. A., 1993: Phylogenetics of seed plants: an analysis of nucleotide sequences from the plastid gene rbcl. – Ann. Missouri Bot. Gard. **80**: 528–580.

Corner, E. J. H., 1946: Centrifugal stamens. – J. Arnold Arbor. **27**: 423–437.

– 1976: The seeds of dicotyledons. – Cambridge: Cambridge University Press.

CRONQUIST, A., 1981: An integrated system of classification of flowering plants. – New York: Columbia University Press.

CULLEN, J., 1987: A preliminary survey of ptyxis (vernation) in the angiosperms. – Notes Roy. Bot. Gard. Edinburgh **37**: 161–214.

DAHLGREN, G., 1989: The last Dahlgrenogram, system of classification of the dicotyledons. In TAN, K., (Ed.): Plant taxonomy, phytogeography and related subjects, pp. 249–260. – Edinburgh: Edinburgh University Press.

DONOGHUE, M. J., DOYLE, J. A., 1989: Phylogenetic relationships of angiosperms and the relationships of *Hamamelidae*. In CRANE, P. R., BLACKMORE, S., (Eds): Evolution, systematics, and fossil history of the *Hamamelidae*, **1**, pp. 17–45. – Cambridge: Clarendon Press.

DUNCAN, T., KEENER, C. S., 1991: A classification of the *Ranunculaceae* with special reference to the western hemisphere. – Phytologia **70**: 24–27.

EAMES, A. J., 1961: Morphology of the angiosperms. – New York: McGraw–Hill Book Co..

ERDTMAN, G., 1952: Pollen morphology and taxonomy of angiosperms. – Stockholm: Almquist & Wiksell.

ERNST, W. R., 1962: The genera of *Papaveraceae* and *Fumariaceae* in the southeastern United States. – J. Arnold Arbor. **43**: 315–343.

– 1967: Floral morphology and systematics of *Platystemon* and its allies *Hesperomecon* and *Meconella* (*Papaveraceae*: *Platystemonoideae*). – Univ. Kansas Sci. Bull. **47**: 25–70.

FOSTER, A. S., 1961: The floral morphology and relationships of *Kingdonia uniflora*. – J. Arnold Arbor. **42**: 397–416.

– 1963: The morphology and relationships of *Circaeaster*. – J. Arnold Arbor. **44**: 299–321.

GONNERMANN, C., 1980: Beiträge zur Kenntnis der Gynoeceumsstruktur der *Papaveraceae*. – Feddes Repert. Beih. **91**: 593–613.

– 1982: Überblick über die Testa-Morphologie der *Papaveraceae*. – Gleditschia **9**: 17–25.

GOTTLIEB, O. R., KAPLAN, M. A. C., ZOCHER, D. H. T., 1993: A chemosystematic overview of *Magnoliidae*, *Ranunculidae*, *Caryophyllidae* and *Hamamelidae*. – In KUBITZKI, K., (Ed.): Families and genera of vascular plants, **2**, pp. 20–31.– Berlin: Springer.

GREENE, E. L., 1903: *Platystemon* and its allies. – Pittonia **5**: 139–194.

GREY-WILSON, C., 1993: Poppies, Guide to the poppy family in the wild and in cultivation. – Portland, Oregon: Timber Press.

GUNN, C. R., 1980: Seeds and fruits of *Papaveraceae* and *Fumariaceae*. – Seed Sci. Technol. **8**: 3–58.

HOOT, S. B., 1994: Phylogenetic relationships within the *Ranunculaceae* based on rbcL and atpB gene sequences. – Amer. J. Bot. (Suppl.) **81**: 161.

– CRANE, P. R., 1995: Inter-familial relationships in the *Ranunculidae* based on molecular systematics. – Pl. Syst. Evol. [Suppl.] **9**: 119–131.

JENSEN, U., 1995: Secondary compounds of the *Ranunculiflorae*. – Pl. Syst. Evol. [Suppl.] **9**: 85–97.

– PENNER, R., 1980: Investigation of serological determinants from single storage plant proteins. Biochem. Syst. Ecol. **8**: 161–170.

JUDD, W. S., SANDERS, R. W., DONOGHUE, M. J., 1994: Angiosperm family pairs: preliminary phylogenetic analyses. – Harvard Papers Bot. **5**: 1–51.

KADEREIT, J. W., 1993: *Papaveraceae*. – In KUBITZKI, K., (Ed.): Families and genera of vascular plants, **2**, pp. 494–506.– Berlin: Springer.

– SYTSMA, K. J., 1992: Disassembling *Papaver*: a restriction site analysis of chloroplast DNA. – Nordic J. Bot. **12**: 205–218.

– BLATTNER, F. R., JORK, K., SCHWARZBACH, A., 1994: Phylogenetic analysis of the *Papaveraceae* s.l. (incl. *Fumariaceae*, *Hypecoaceae*, and *Pteridophyllum*) based on morphological characters. – Bot. Jahrb. Syst. **116**: 361–390.

KEEFE, J. M., MOSELEY, M. F., 1978: Wood anatomy and phylogeny of *Paeonia* section *Moutan*. – J. Arnold Arbor. **59**: 274–297.

KEENER, C. S., 1993: A review of the classification of the genus *Hydrastis* (*Ranunculaceae*). – Aliso **13**: 551–558.

KESSLER, P. J. A., 1993: *Menispermaceae*. – In KUBITZKI, K., (Ed.): Families and genera of vascular plants, **2**, pp. 402–418. – Berlin: Springer.

KIM, Y.-D., JANSEN, R., 1995: Phylogenetic implications of chloroplast DNA variation in the *Berberidaceae*. – Pl. Syst. Evol. [Suppl.] **9**: 341–349.

KUBITZKI, K., 1993: Families and genera of vascular plants. – Berlin: Springer.

KUMAZAWA, M., 1930: Morphology and biology of *Glaucidium palmatum* with notes on affinities to the allied genera *Hydrastis*, *Podophyllum* and *Diphylleia*. – J. Fac. Sci. Univ. Tokyo, Sect. 3, Bot. **2**: 345–380.

– 1937: Comparative studies on the vernation in the *Ranunculaceae* and *Berberidaceae*. – J. Japan. Bot. **13**: 573–586, 659–669, 713–726.

LEE, SANGTAE, 1989: Palynological evidence for the relationships between, *Megaleranthis saniculifolia* and *Trollius* species. –Pollen & Spores **31**: 173–185.

– 1992: Palynological relationships among *Calathodes* and its relative genera. – Korean J. Pl. Tax. **22**: 23–31.

LIDÉN, M., 1986: Synopsis of *Fumarioideae* (*Papaveraceae*) with a monograph of the tribe *Fumarieae*. – Opera Bot. **88**: 1–133.

– 1993: *Fumariaceae*. – In KUBITZKI, K., (Ed.): Families and genera of vascular plants, **2**, pp. 310–318. – Berlin: Springer.

– 1993: *Pteridophyllaceae*. – In KUBITZKI, K., (Ed.): Families and genera of vascular plants, **2**, pp. 556–557. – Berlin: Springer.

LOCONTE, H., 1993: *Berberidaceae*. – In KUBITZKI, K., (Ed.): Families and genera of vascular plants, **2**, pp. 147–152. – Berlin: Springer.

– 1995: Comparison of alternative hypotheses for the origin of the angiosperms. – In TAYLOR D. W., HICKEY, L., (Eds): Angiosperm origin, evolution and phylogeny. – New York: Chapman & Hall, (In press).

– ESTES, J. R., 1989: Phylogenetic systematics of *Berberidaceae* and *Ranunculales* (*Magnoliidae*). – Syst. Bot. **14**: 565–579.

– STEVENSON, D. WM., 1991: Cladistics of the *Magnoliidae*. – Cladistics **7**: 267–296.

MADDISON, W. P., MADDISON, D. R., 1993: MacClade, version 3.0. – Sunderland, Mass.: Sinauer.

NICKOL, M. G., 1995: Phylogeny and inflorescences of *Berberidaceae*. – a morphological survey. – Pl. Syst. Evol. [Suppl.] **9**: 327–340.

NOWICKE, J. W., SKVARLA, J. J., 1981: Pollen morphology and phylogenetic relationships of the *Berberidaceae*. – Smithsonian Contr. Bot. **50**: 1–83.

– 1982: Pollen morphology and the relationships of *Circaeaster*, of *Kingdonia*, and of *Sargentodoxa* to the *Ranunculales*. – Amer. J. Bot. **69**: 990–998.

– BITTNER, J. L., SKVARLA, J. J., 1986: *Paeonia*, exine substructure and plasma ashing. – In BLACKMORE, S., FERGUSON, I. K., (Eds). Pollen and spores: form and function, pp. 81–95. – London: Academic Press.

QIN, HAI-NING, 1989: An investigation on carpels of *Lardizabalaceae* in relation to taxonomy and phylogeny. – Cathaya **1**: 61–82.

RONSE DECRAENE, L.-P., SMETS, E. F., 1992: Complex polyandry in the *Magnoliatae*: definition, distribution and systematic value. – Nordic. J. Bot. **12**: 621–649.

SWOFFORD, D. L., 1993: PAUP, version 3.1.1. – Champaign, Illinois: Illinois Natural History Survey.

TAKHTAJAN, A., 1987: Systema Magnoliophytorum. – Leningrad: Nauka.

– 1988: Anatomia Seminum Comparativa. 2, *Dicotyledons: Magnoliidae, Ranunculidae.* – Leningrad: Nauka.

– MELIKIAN, A. P., 1972: Comparative anatomical study of seed coat anatomy of *Leontice, Gymnospermium, Caulophyllum* and allied genera in relation to their systematics. – Bot. Z. (Moscow, Leningrad) **57**: 1271–1278.

TAMURA, M., 1962: Petiolar anatomy in the *Ranunculaceae.* – Sci. Rep. Osaka Univ. **11**: 19–47.

– 1993: *Ranunculaceae.* – In KUBITZKI, K., (Ed.): Families and genera of vascular plants, **2**, pp. 563–583. – Berlin: Springer.

– 1995: Phylogeny and classification of the *Ranunculaceae.* – Pl. Syst. Evol. [Suppl.] **9**: 201–206.

TAYLOR, S. A., 1967: Comparative morphology and anatomy of the *Lardizabalaceae.* – Ph.D. dissertation, Indiana University, Bloomington.

THANIKAIMONI, G., 1986: Evolution of *Menispermaceae.* – Canad. J. Bot. **64**: 3130–3133.

THORNE, R. F., 1992: Classification and geography of the flowering plants, – Bot. Rev. **58**: 225–348.

TIFFNEY, B. H., 1993: Fruits and seeds of the Tertiary Brandon lignite: VII. *Sargentodoxa* (*Sargentodoxaceae*). – Amer. J. Bot. **80**: 517–523.

TOBE, H., 1981: Embryological studies in *Glaucidium palmatum* with a discussion of the taxonomy of the genus. – Bot. Mag. (Tokyo) **94**: 207–224.

– KEATING, R. C., 1985: The morphology and anatomy of *Hydrastis* (*Ranunculales*): systematic reevaluation of the genus. – Bot. Mag. (Tokyo) **98**: 291–316.

WILLIS, J. C., 1973: A dictionary of the flowering plants and ferns. 8th edn. – Cambridge: Cambridge University Press.

WU, CHENG-YIH, KUBITZKI, K., 1993: *Circaeasteraceae.* – In KUBITZKI, K., (Ed.): Families and genera of vascular plants, **2**, pp, 288–289. – Berlin: Springer.

– – 1993: *Lardizabalaceae.* – In KUBITZKI, K., (Ed.): Families and genera of vascular plants, **2**, pp. 361–365. – Berlin: Springer.

YANG, QING-ER, GONG XUN, GU AHI-JIAN, WU, QUAN-AN, 1993: A karyomorphological study of five species in the *Ranunculaceae*, with a special consideration on systematic positions of *Asteropyrum* and *Calathodes*. – Acta Bot. Yunnanica **15**: 179–190.

YING, TSUN-SHEN, YU-LONG ZHANG, BOUFFORD, D. E., 1993: The endemic genera of seed plants of China. – Beijing: Science Press.

ZHANG, Z. Y., 1982: Chromosome observation of three ranunculaceous genera in relation to their systematic positions. – Acta Phytotax. Sinica **20**: 402–409.

Address of the authors: HENRY LOCONTE, LISA M. CAMPBELL, and DENNIS WM. STEVENSON, New York Botanical Garden, Bronx, New York 10458, USA.

Accepted February 13, 1995

Appendix. Characters used in the Cladistic analysis

1. Habit: 0 = shrub or tree, 1 = liana, 2 = rhizomatous herb, 3 = annual herb. 2. Rhizome: 0 = absent, 1 = present, 2 = tuberous. 3. Chasmophytic habit: 0 = absent, 1 = present (LIDÉN 1986). 4. Epicuticular waxes: 0 = clustered hollow tubules, 1 = transversely ridged rodlets, 2 = platelets (BARTHLOTT & THEISEN 1995, this volume). 5. Collenchyma: 0 = absent, 1 = present (KADEREIT & al. 1994). 6. Latex: 0 = absent, 1 = clear, 2 = yellow, 3 = orange (CARLQUIST & al. 1994, GREY-WILSON 1993). 7. Vascular bundles: 0 = collateral, 1 = semi-amphivasal (v-shaped xylem), 2 = amphicribal. 8. Vessel perforations: 0 = scalariform, 1 = predominately simple. 9. Axial parenchyma: 0 = apotracheal, 1 = paratracheal, 2 = absent (CARLQUIST 1995, this volume). 10. Wood rays: 0 = uniseriate and multiseriate,

1 = multiseriate only (CARLQUIST 1995, this volume). 11. Nodal anatomy: 0 = multilacunar or trilacunar, 1 = unilacunar. 12. Leaf type: 0 = simple, 1 = pinnately compound (including trifoliate), 2 = ternately decompound, 3 = palmately compound, 4 = pedately compound, 5 = peltate, 6 = entire, linear. 13. Leaf vernation: 0 = conduplicate, 1 = adplicate, 2 = involute, 3 = supervolute, 4 = revolute, 5 = conduplicate plicate, 6 = reclinate (CULLEN 1978, KUMAZAWA 1937). 14. Leaf arrangement: 0 = alternate, 1 = opposite, 2 = basal only. 15. Leaf pubescence: 0 = absent, 1 = multicellular-uniseriate, 2 = multicellular-multiseriate, 3 = unicellular. 16. Leaf duration: 0 = deciduous, 1 = evergreen. 17. Leaf base amplexicaule:
0 = absent, 1 = present (GREY-WILSON 1993). 18. Leaf margin: 0 = entire, 1 = toothed, 2 = mucronate. 19. Leaf venation: 0 = pinnate, 1 = palmate, 2 = dichotomous. 20. Leaf tendril: 0 = absent, 1 = terminal, 2 = petiolulate (LIDÉN 1986). 21. Leaf hypodermis: 0 = absent, 1 = present (TAYLOR 1967). 22. Petiole bundle sheath: 0 = absent, 1 = anemone- type, 2 = ranunculus-type (TAMURA 1962). 23. Petiole interfasicular lignification: 0 = absent, 1 = present (TAMURA 1962). 24. Petiole medullary lignification: 0 = absent, 1 = present (TAMURA 1962). 25. Involucral leaves: 0 = absent, 1 = present. 26. Inflorescence: 0 = present, 1 = solitary flower. 27. Inflorescence axis: 0 = simple, 1 = compound. 28. Inflorescence development: 0 = indeterminate, 1 = determinate. 29. Floral bracts: 0 = bracteose, 1 = frondose (KADEREIT & al. 1994). 30. Flower buds: 0 = erect, 1 = nodding. 31. Receptacular open zone: 0 = absent, 1 = present (TAMURA 1993). 32. Receptacular enlargement: 0 = absent, 1 = present (TAMURA 1993). 33. Floral symmetry: 0 = actinomorphic, 1 = zygomorphic, 2 = disymmetrical. 34. Perianth organization: 0 = undifferentiated, 1 = differentiated. 35. Perianth merousity: 0 = three, 1 = two, 2 = five. 36. Calyx: 0 = enclosing bud, 1 = not enclosing bud (KADEREIT & al. 1994). 37. Sepal traces: 0 = three, 1 = two (TAYLOR 1967). 38. Sepal persistence: 0 = absent, 1 = present. 39. Staminodia type: 0 = absent, 1 = simple, 2 = spurred, 3 = tubular, 4 = cupular, 5 = concave. 40. Staminodia apex: 0 = simple, 1 = bilobed. 41. Petals nectariferous: 0 = absent, 1 = present. 42. Outer petal broadly spathulate: 0 = absent, 1 = present (LIDÉN 1986). 43. Upper petal spurred: 0 = absent, 1 = present (LIDÉN 1986). 44. Inner petal winged: 0 = absent, 1 = present (LIDÉN 1986). 45. Intrastaminal scales: 0 = absent, 1 = present (TAMURA 1993). 46. Sexual expression: 0 = monoclinous, 1 = monoecious, 2 = dioecious. 47. Stamens pendulous: 0 = absent, 1 = present (KADEREIT & al. 1994). 48. Filament shape: 0 = filiform, 1 = dilated, 2 = narrow towards apex, 3 = connate (KADEREIT & al. 1994). 49. Filament color: 0 = pale, 1 = dark (KADEREIT & al. 1994). 50. Stamen development: 0 = centripetal, 1 = centrifugal. 51. Stamen arrangement: 0 = cyclic, 1 = polyandrous, 2 = complex polyandrous, 3 = bicyclic. 52. Staminal nectaries: 0 = absent, 1 = present (KADEREIT 1994). 53. Stamen connective protrusion: 0 = present, 1 = absent. 54. Stamen anther dehiscence: 0 = longitudinal, 1 = valvate. 55. Pollen aperture type: 0 = sulcate, 1 = colpate, 2 = porate, 3 = colporate, 56. Pollen aperture number: 0 = one, 1 = three, 2 = more than three, 3 = two. 57. Pollen exine sculpturing: 0 = scrobiculate (microreticulate), 1 = reticulate, 2 = foveolate, 3 = striate, 4 = spinose, 5 = psilate, 6 = crassimurate (ERDTMAN 1952). 58. Pollen intine: 0 = nonprotruding, 1 = protruding aperture (LIDÉN 1986). 59. Carpel developement: 0 = plicate (conduplicate), 1 = ascidate (peltate). 60. Gynoecium position: 0 = hypogynous, 1 = perigynous (KADEREIT & al. 1994). 61. Carpel number: 0 = more than three, 1 = one, 2 = two, 3 = three. 62. Carpel fusion: 0 = apocarpous, 1 = syncarpous, 2 = paracarpous. 63. Carpel base: 0 = sessile, 1 = stipitate. 64. Style duration: 0 = persistent, 1 = caducous (LIDÉN 1986). 65. Styles: 0 = free, 1 = fused, 2 = absent. 66. Stigma symmetry: 0 = symmetrical, 1 = asymmetrical (LIDÉN 1986). 67. Stigma with apical appendage: 0 = absent, 1 = present (LIDÉN 1986). 68. Stigmatic papillae: 0 = absent, 1 = two large, 2 = diffuse, 3 = four, 4 = two small (LIDÉN 1986). 69. Placentation: 0 = submarginal, 1 = laminar, 2 = basal, 3 = apical. 70. Ovule type:

0 = anatropous, 1 = hemitropous, 2 = orthropous, 3 = camplyotropous. 71. Ovule orientation: 0 = pendulous, 1 = ascending (Tamura 1993). 72. Integuments: 0 = bitegmic, 1 = unitegmic. 73. Micropyle formation: 0 = inner integument, 1 = outer integument, 2 = both integuments. 74. Nucellus type: 0 = crassinucellate, 1 = tenuinucellate. 75. Nucellar epidermis: 0 = without periclinar divisions, 1 = with periclinal divisions. 76. Embryogeny: 0 = onagrad, 1 = solanad, 2 = caryophyllad, 3 = chenopodiad, 4 = asterad. 77. Embryonic suspensor: 0 = absent, 1 = present (Lidén 1986). 78. Endosperm development: 0 = cellular, 1 = nuclear, 2 = helobial. 79. Fruit type: 0 = ventricidal follicle, 1 = berry, 2 = urticle, 3 = achene, 4 = capsule, 5 = drupe, 6 = nut, 7 = circumcissile follicle. 80. Fruit venation: 0 = without transverse veins, 1 = with transverse veins. 81. Achene veins: 0 = absent, 1 = present. 82. Achene sclerenchyma: 0 = absent, 1 = present. 83. Achene body: 0 = glabrous, 1 = pubescent, 2 = with uncinate hairs. 84. Achene tail plumose: 0 = absent, 1 = present. 85. Coarse fruit ornamentation: 0 = absent, 1 = present. 86. Fruit puberulent: 0 = absent, 1 = present. 87. Fruit parenchyma spongy: 0 = absent, 1 = present (Brückner 1982). 88. Fruit dehiscence direction: 0 = acropetal, 1 = basipetal, 2 = lateral, 3 = indehiscent. 89. Fruit valve bundle anastomoses: 0 = rare, 1 = frequent (Brückner 1982). 90. Radial groups of reticulate tracheids in fruit: 0 = absent, 1 = present (Brückner 1982). 91. Seed crystals: 0 = coarse, 1 = sand (Brückner 1993). 92. Seed parenchyma zone: 0 = raphe chalazal region only, 1 = dorsal only, 2 = throughout (Brückner 1993). 93. Fruit secretory cavities: 0 = scattered, 1 = zonal, 2 = phloem (Brückner 1982). 94. Fruit stomata: 0 = adaxial and abaxial, 1 = abaxial only (Brückner 1982). 95. Dorsal nerves of fruit valve: 0 = well developed, 1 = rudimentary to absent, 2 = pseudodorsal nerve (Brückner 1982). 96. Outer wall of seed epidermis: 0 = with dark inner face structure, 1 = unstructured, 2 = granular radial striations, 3 = inner face thickened, 4 = resorbed (Brückner 1983). 97. Seed coat: 0 = thick, 1 = thin (Takhtajan & Melikian 1972). 98. Anticlinal wall of seed epidermis: 0 = straight, 1 = sinuous, 2 = absent (Brückner 1983). 99. Inner epidermis of seed: 0 = very conspicuous, 1 = conspicuous, 2 = inconspicuous (Brückner 1983). 100. Seed number: 0 = one, 1 = two or more. 101. Seeds reniform: 0 = absent, 1 = present. 102. Seed axis: 0 = straight, 1 = curved, 2 = folded (Gunn 1980). 103. Seed aril: 0 = absent, 1 = present. 104. Chromosome base number: 0 = eight, 1 = seven, 2 = six, 3 = five, 4 = nine, 5 = thirteen, 6 = ten, 7 = fifteen. 105. Chromosome size: 0 = t-type, 1 = r-type. 106. Benzylisoquinolines: 0 = absent, 1 = present. 107. Cytisin alkaloids: 0 = absent, 1 = present (Jensen 1995, this volume). 108. Saponines: 0 = absent, 1 = present (Jensen 1995, this volume). 109. Meconic acid: 0 = absent, 1 = present.

Pl. Syst. Evol. [Suppl.] 9: 119–131 (1995)

Inter-familial relationships in the *Ranunculidae* based on molecular systematics

SARA B. HOOT and PETER R. CRANE

Received January 12, 1995

Key words: Eudicots, *Ranunculidae, Papaverales, Berberidaceae, Lardizabalaceae, Menispermaceae, Ranunculaceae, Euptelea, Circaeaster, Kingdonia* – Molecular systematics, *rbc*L, *atp*B, chloroplast DNA, 18S, nuclear ribosomal DNA.

Abstract: Nucleotide sequences from the *rbc*L and *atp*B genes (chloroplast DNA) and 18S nuclear ribosomal DNA provide important new data with which to test previous hypotheses of inter-familial relationships in the angiosperm subclass *Ranunculidae*. Preliminary conclusions based on cladistic analysis of the combined molecular data sets for all three genes are broadly congruent with previous analyses based on *rbc*L data alone. The genus *Euptelea*, which has been placed traditionally among the "lower" *Hamamelididae*, is resolved as more closely related to ranunculids. *Papaverales* (represented by *Pteridophyllaceae* and *Fumariaceae* including *Hypecoum*) are strongly supported as a monophyletic group, and are resolved as basal to all other ranunculids ("core" ranunculids) plus *Euptelea*. The monotypic genera *Kingdonia* and *Circaeaster* are placed as sister taxa and together are sister to *Lardizabalaceae* s.l. (including *Sargentodoxa*). The *Lardizabalaceae-Circaeaster-Kingdonia* clade is the sister taxon to *Menispermaceae-Ranunculaceae-Berberidaceae*. *Menispermaceae* is the sister taxon to *Ranunculaceae* and *Berberidaceae*.

Current models of angiosperm evolution recognize two major clades – monocotyledons and eudicots – both embedded in a systematically depauperate assemblage of "basal" dicotyledons at the magnoliid grade (CRANE 1989, DRINNAN & al. 1994). Morphologically, the eudicots are diagnosed by the possession of triaperturate or triaperturate-derived pollen. Monophyly of the group is further supported by two recent phylogenetic analyses of combined morphological and molecular data (ALBERT & al. 1994, DOYLE & al. 1994). Approximately 70–75% of extant angiosperm species are included in the eudicot group and therefore phylogenetic patterns at the base of this clade are of great interest, for resolving relationships among angiosperms as a whole. However, until recently relationships at the base of the eudicot group have received relatively little attention compared to the number of phylogenetic studies focused at the magnoliid grade.

Recent discussions of eudicot phylogeny have highlighted two assemblages of taxa – "lower" hamamelidids (probably a grade) and ranunculids (perhaps monophyletic) – as potentially "basal" taxa (CRANE 1989, DRINNAN & al. 1994). "Lower" hamamelidids, such as *Tetracentron* and *Trochodendron*, have been thought to retain a variety of unmodified plesiomorphic features from their magnoliid ancestors (cf. ENDRESS 1986, CRANE 1989). *Ranunculidae* (broadly equivalent to *Ranunculiflorae*) have sometimes been allied with the *Magnoliidae* (e.g., CRONQUIST 1981), largely on the basis of their frequently polymerous flowers (DRINNAN & al. 1994), but are more appropriately excluded because of their triaperturate pollen (TAKHTAJAN 1987, CRANE 1989, DRINNAN & al. 1994). In this paper we present a preliminary analysis of inter-familial relationships in ranunculids as part of an ongoing analysis of phylogenetic patterns among 'basal' eudicots (CRANE 1989, HOOT 1995, this volume; HOOT & al. 1995). All 10 families of *Ranunculiflorae* sensu DAHLGREN (1980) are included in the analysis presented here and are represented by a sample of 13 species (Tables 1, 2). These data are a subset of our overall data that include approximately 120 species from families of magnoliids, ranunculids, "lower" hamamelidids and basal rosids.

Table 1. Classifications of the *Ranunculidae* sensu TAKHTAJAN (1987) and *Ranunculiflorae* sensu DAHLGREN (1980). Figures in parentheses indicate the number of genera and species recognized by TAKHTAJAN (1987) in each family, *indicates families included in the molecular systematic studies presented in this paper

TAKHTAJAN (1987)	DAHLGREN (1980)
RANUNCULIDAE	*RANUNCULIFLORAE*
Ranunculanae	*Ranuncuales*
Ranunculales	*Lardizabalaceae*
Lardizabalineae	*Sargentodoxaceae*
Lardizabalaceae (8/35)	*Menispermaceae*
Sargentodoxaceae (1/1)	*Kingdoniaceae*
Menispermineae	*Circaeasteraceae*
Menispermaceae (67/400)	*Ranunculaceae*[3]
Ranunculineae	*Berberidaceae*[4]
Ranunculaceae (66/2000)	*Nandinaceae*
Circaeasteraceae[1] (2/2)	*Papaverales*
Berberidineae	*Papaveraceae*
Hydrastidaceae (1/1)	*Fumariaceae*[5]
Berberidaceae (14/650)	
Nandinaceae (1/1)	
Glaucidiales	
Glaucidiaceae (1/1)	
Paeoniales	
Paeoniaceae (1/35–40)	
Papaverales	
Papaveraceae[2] (24/250)	
Hypecoaceae (1/15)	
Fumariaceae (17/470)	

[1] includes *Kingdonia*; [2] includes *Pteridophyllum*; [3] includes *Hydrastidaceae*; [4] includes *Glaucidiaceae*, *Leonticaceae* and *Podophyllaceae*; [5] includes *Hypecoaceae*

Table 2. Comparison of taxa of ranunculids sampled in the analysis of *rbc*L sequence data by CHASE & al. (1993) with species sequenced for *rbc*L, *atp*B and 18S genes in this analysis. *rbc*L data accessed from GenBank. Outgroups and "lower" hamamelid taxa omitted

Family	*rbc*L (CHASE & al. 1993)	*rbc*L, *atp*B and 18S (this study)
Eupteleaceae	*Euptelea*	*Euptelea polyandra**
Papaveraceae	*Papaver*	not included. – *Dicentra,*
	Sanguinaria	*Hypecoum* and
		Pteridophyllum used as
		placeholders for Papaverales
Fumariaceae	*Dicentra*	*Dicentra eximia*
Hypecoaceae	–	*Hypecoum imberba*
Pteridophyllaceae	–	*Pteridophyllum racemosum*
Circaeasteraceae	–	*Circaeaster agrestis*
Kingdoniaceae	–	*Kingdonia uniflora*
Lardizabalaceae	*Akebia*	*Decaisnea fargesii*
		Sinofranchetia chinensis
Berberidaceae	*Caulophyllum*	*Caulophyllum thalictroides**
	Mahonia	*Nandina domestica*
Menispermaceae	*Cocculus*	*Menispermum canadensis*
		Tinospora caffra
Glaucidiaceae	–	*Glaucidium palmatum*
Hydrastidaceae	–	*Hydrastis canadensis*
Ranunculaceae	*Caltha*	not included. – *Glaucidium*
	Ranunculus	and *Hydrastis* used as
	Xanthorhiza	placeholders

As in our previous phylogenetic study of the *Lardizabalaceae* (HOOT & al. 1995), relationships are assessed by parsimony analysis of molecular sequence data derived from three genes – *rbc*L and *atp*B in the chloroplast genome (cpDNA), and 18S ribosomal DNA in the nuclear genome (nrDNA). Each of the gene sequences are analyzed separately and then together as a single data set (HOOT & al. 1995). Results from the combined data are then examined to determine the extent of character support for specific nodes. Experimental manipulations of the combined data set are used to compare the results with previous hypotheses on the relationships of ranunculid families.

Material and methods

Taxon sampling. In the analyses presented here we include 13 of the total 70 genera in our data set (Table 1) for the 10 families of *Ranunculiflorae* recognized by DAHLGREN (1980). Selection of taxa was designed to maximize systematic coverage within the group and expand the sampling of previous studies based on *rbc*L sequence data (CHASE & al. 1993). Of the 13 families recognized in the *Ranunculidae* by TAKHTAJAN (1987) only the *Paeoniaceae* are not considered here. Previous cladistic results, based on *rbc*L sequence data, suggest that *Paeonia* is distantly related to ranunculids and is nested well within the main group of rosids (Chase & al. 1993). Species included in the analysis are listed in Table 2. Details of sources and vouchers of material are included in a forthcoming paper (Hoot & Crane unpubl.). The genera included here were selected based on previous analyses of relationship within specific families, using both molecular and morphological data. *Platanus* (*Platanaceae*) was included as a placeholder for all non-ranunculid eudicots because of its near basal position in eudicots according to previous phylogenetic analyses using *rbc*L data (see also Discussion).

Pteridophyllum, *Dicentra* and *Hypecoum* were used as placeholders for the *Papaverales*. Morphological analyses (Kadereit & al. 1995, this volume), suggest that *Pteridophyllum* (best recognized as *Pteridophyllaceae*) is basal to both *Fumariaceae* and *Papaveraceae* s. str., while *Hypecoum* (sometimes placed in *Hypecoaceae*) is basal with respect to other *Fumariaceae* (Lidén 1995). *Euptelea* was included both as a second representative of the "lower" hamamelidid grade, and on the basis of previous analyses of chloroplast DNA sequences that ally it with *Ranunculidae* (Chase & al. 1993, Drinnan & al. 1994).

Kingdonia and *Circaeaster* were included as two taxa that are treated frequently as monotypic families. *Sargentodoxa* is also often assigned to its own family (*Sargentodoxaceae*, Table 1) but is sometimes included in the *Lardizabalaceae* s.l. (Loconte & Estes 1989). *Decaisnea* and *Sinofranchetia* were included as placeholders for *Lardizabalaceae* s. str. based on previous analyses of morphology (Loconte & Estes 1989) and combined *atp*B, *rbc*L and 18S DNA sequence data (Hoot & al. 1995).

Hydrastis and *Glaucidium* were incl'ıded as placeholders for the diverse family *Ranunculaceae* based on an extensive molecular phylogenetic study of 25 genera in the family (Hoot 1995, this volume). A close relationship between *Hydrastis* and *Ranunculaceae* is also supported by previous cladistic analyses based on morphological characters (Hoot 1991) as well as chloroplast restriction site data (Johansson & Jansen 1993).

Nandina and *Caulophyllum* were selected as placeholders for *Berberidaceae* based on previous cladistic results, which suggest that one or both genera are probably basal within the family (Kim & Jansen 1995, this volume; Loconte & Estes 1989; Nickol 1995, this volume), and also the frequent separation of *Nandina* as *Nandinaceae* (e.g., Dahlgren 1980, Table 1).

Currently, the poorest sampling for molecular studies in the entire *Ranunculidae* is in the large, predominantly tropical, family, *Menispermaceae*. We are currently acquiring additional material that will expand our sampling in this family. In the analyses presented here, we selected *Tinospora* and *Menispermum* as placeholders for the family based on their putatively divergent systematic positions suggested by fruit and pollen morphology (Thanikaimoni 1984).

Molecular techniques. Total cellular DNA was isolated from fresh, silica-dried, or herbarium material according to the miniprep method of Doyle & Doyle (1987). The amplification primers and PCR protocol for the chloroplast genes, *atp*B and *rbc*L, are described in Hoot & al. (1995). The polymerase chain reaction and sequencing protocols used for the three genes (*rbc*L, *atp*B and 18S nrDNA) are described in Hoot (1995, this volume).

Data collection. Sequence comparisons for the genes *rbc*L, *atp*B, and 18S included 1397, 1468, and 1671 base pairs, respectively. Both strands of DNA were sequenced for both *atp*B and *rbc*L with approximately 80% overlap. Both strands were also sequenced for 18S, but with less overlap between the two directions (30–40%). Nucleotide sequences were recorded in MacClade (Maddison & Maddison 1992). See Hoot (1995, this volume) for details of data collection and alignment. There were several regions in the 18S nrDNA sequences where alignment was impossible because of compressions or base insertion/deletion events. These regions are readily recognizable and were deleted from the data matrix. They are located at the following positions in relation to the soybean 18S sequence (ECKENRODE & al. 1985): 224–231, 666–670, 1180–1181, and the very end of the amplified region, 1711–1761. The possibility of PCR-generated anomalous sequences was checked by comparison of sequences from closely related taxa.

Phylogenetic analyses. Phylogenetic analyses were performed using PAUP 3.1 (SWOFFORD 1993) using the heuristic search option with 100 random additions, and TBR (tree bisection – reconnection branch swapping) and MULPARS (retention of all equally parsimonious trees) in effect. *Asarum* and *Illicium* were designated as outgroups. PAUP was also used to perform bootstrap analysis with 100 replications using the heuristic search option (FELSENSTEIN 1985). Decay indices (the number of steps that must be added to the minimal-length tree before a clade collapses) were computed for all trees also using the heuristic search option (DONOGHUE & al. 1992). Calculations of decay indices were limited by computation time and were undertaken as follows for each analysis; *rbc*L – up to five steps longer; *atp*B – up to five steps; 18S – up to two steps; combination of three data sets – up to 15 steps. Alternative tree topologies and resultant changes in tree length were explored using MacClade 3.0 (MADDISON & MADDISON 1992).

Analyses were performed separately on each of the *rbc*L, *atp*B and 18S sequence data sets, and then the data from all three genes were combined and reanalyzed. In the results presented here one of the most parsimonious trees produced from each analysis is selected to illustrate the support for different nodes. Dotted lines indicate branches which collapse in the consensus of all the most parsimonious trees. Numbers indicated above the branches are the number of base substitutions supporting each node, while numbers below the branches are bootstrap percentages. The number in italics below each branch is the decay index for the corresponding clade. In the results and discussion presented here, clades with bootstrap values of 70% or more are regarded as 'well supported' (HILLIS & BULL 1993). Phylogenetic results presented here should be regarded as preliminary pending more detailed analyses of our complete data set (HOOT & CRANE unpubl.).

Results

Results based on *rbc*L sequence data. Analysis of the *rbc*L sequence data (191 informative characters) produced seven equally parsimonious trees of 708 steps. One of the shortest trees, and those nodes which collapsed in the strict consensus of the equally parsimonious trees, are shown in Fig. 1. *Papaverales* (including *Dicentra, Hypecoum, Pteridophyllum*), *Menispermaceae, Ranunculaceae* (as represented by *Glaucidium* and *Hydrastis*), *Berberidaceae* and *Circaeasteraceae* (*Circaeaster, Kingdonia*) are all well-supported. However, other relationships are generally weakly-supported and the wide separation of the two genera representing *Lardizabalaceae* s. str. is anomalous. *Sinofranchetia* is placed in a relatively distant position (albeit by relatively weakly-supported nodes) with respect to *Decaisnea* and *Sargentodoxa*.

Results based on *atp*B sequence data. Analysis of the *atp*B sequence data (189 informative characters) produced eight equally parsimonious trees of 690 steps (Fig. 1). All of the well-supported clades in the *rbc*L analyses are also present in the *atp*B results. Monophyly of the *Lardizabalaceae* (represented by *Sinofranchetia, Decaisnea*) and a close relationship to *Sargentodoxa* are also well-supported. Above the family level there is good support for a clade of 'core' ranunculids comprised of *Lardizabalaceae, Circaeasteraceae, Menispermaceae, Ranunculaceae* and

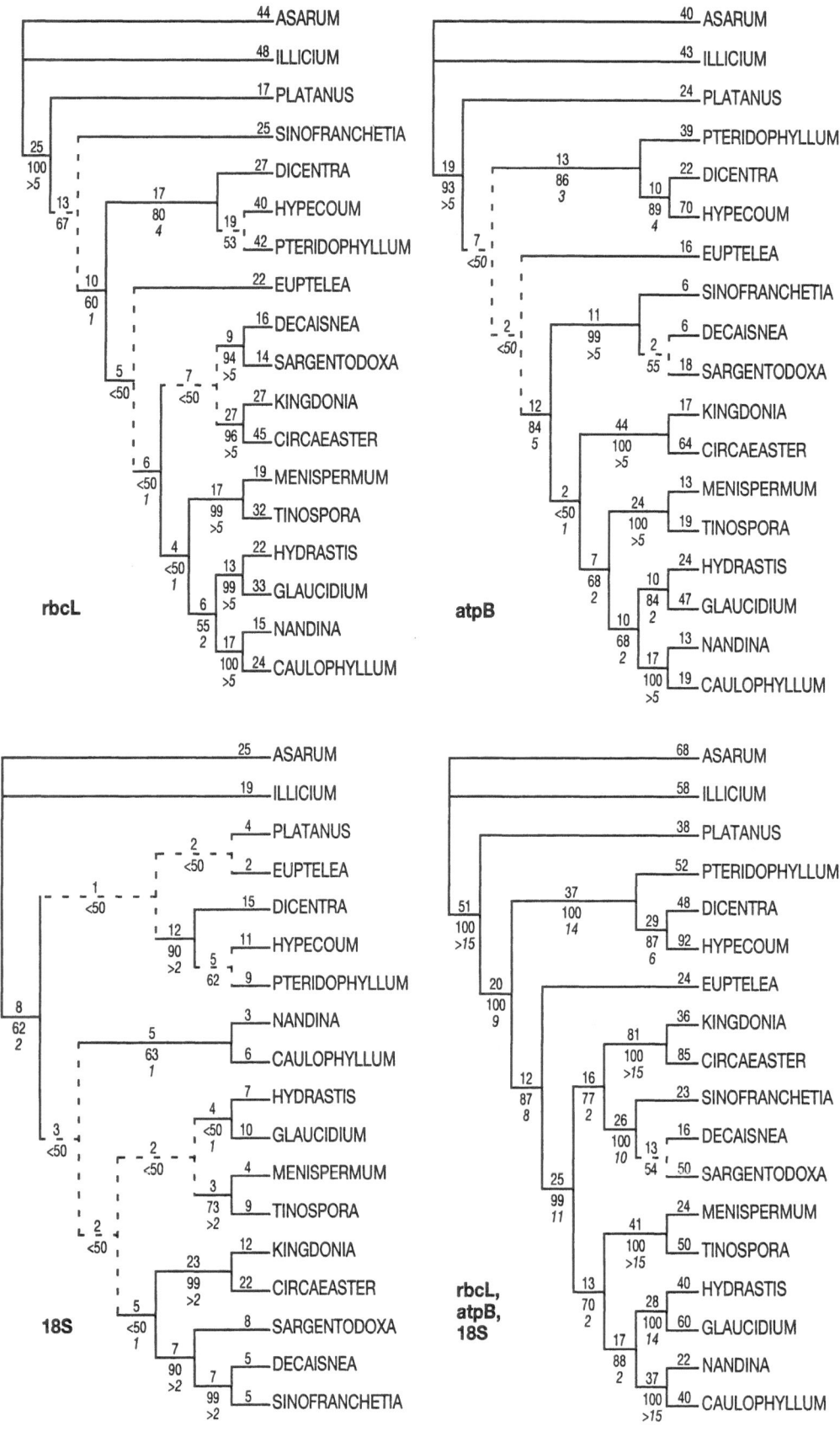

Berberidaceae (84% bootstrap), while below the family level a close relationship between *Dicentra* and *Hypecoum* is well-supported (89% bootstrap).

Results based on 18S sequence data. Analysis of the 18S sequence data (74 informative characters) produced 14 equally parsimonious trees of 265 steps (Fig. 1). The tree based on 18S sequence is considerably less resolved than those based on the two chloroplast genes. *Papaverales, Menispermaceae, Circaeasteraceae, Lardizabalaceae* s. str. and *Lardizabalaceae* s.l. (including *Sargentodoxaceae*) are well-supported. However, support for the *Menispermaceae* is much weaker than in the *rbc*L and *atp*B data (73% versus 99% and 100% bootstrap percentages respectively) and the monophyly of *Ranunculaceae* and *Berberidaceae* is poorly supported. Except for *Papaverales*, none of the clades above the family level are well-supported. The general lack of branch support and resolution in the strict consensus tree reflects the lower number of informative characters in the 18S data set.

Results based on combined *rbc*L, *atp*B and 18S sequence data. Analysis of the combined *rbc*L, *atp*B and 18S sequence data (454 informative characters) produced two equally parsimonious trees of 1272 steps. The two trees differed only in the precise relationships between the three placeholders for *Lardizabalaceae* s.l. (*Sargentodoxa, Decaisnea, Sinofranchetia*). One of the shortest trees and the branches which collapsed in the strict consensus of the two trees, is shown in Fig. 1.

As in the trees based on *rbc*L and *atp*B data, there is strong support in the combined molecular data for the monophyly of traditionally recognized families including *Circaeasteraceae, Menispermaceae, Ranunculaceae* and *Berberidaceae*. As in all three data sets, the *Papaverales* are also well-supported. However, the combined data also provides strong support for clades which were only recognized in one or two of the previous analyses. In particular, the combined data support the monophyly of the *Lardizabalaceae* s.l. (not supported by the *rbc*L data), the monophyly of the "core" ranunculid group comprised of *Lardizabalaceae, Circaeasteraceae, Menispermaceae, Ranunculaceae*, and *Berberidaceae* (not well-supported

◀—————————————————————————————————————

Fig. 1. Results from preliminary cladistic analyses based on *rbc*L, *atp*B, and 18S data, and combined *rbc*L, *atp*B, 18S data, for the families and taxa indicated in Table 2. Dotted lines indicate branches that collapse in the strict consensus tree derived from the shortest trees. Numbers above branches indicate the number of nucleotide changes supporting each branch. Numbers below the branches indicate the percentage of times that the branch was recovered in 100 bootstrap replications. Italicized numbers below the bootstrap values are decay indices. *Asarum* and *Illicium* were designated as outgroups. *rbc*L: One of seven equally most parsimonious trees resulting from a preliminary cladistic analysis based on *rbc*L sequence data; tree length = 708 steps, CI excluding autapomorphies = 0.50, RI = 0.42. *atp*B: One of eight equally most parsimonious trees resulting from a preliminary cladistic analysis based on *atp*B sequence data; tree length = 690 steps, CI excluding autapomorphies = 0.54, RI = 0.50. 18S: One of 14 equally most parsimonious trees resulting from a preliminary cladistic analysis based on 18S sequence data; tree length = 265 steps, CI excluding autapomorphies = 0.54, RI = 0.54. Combined *rbc*L, *atp*B and 18S: One of two equally most parsimonious trees resulting from a preliminary cladistic analysis based on combined *rbc*L, *atp*B and 18S sequence data; tree length = 1272 steps, CI excluding autapomorphies = 0.52, RI = 0.47

Table 3. Comparison of branch support (bootstrap values, decay indices) for various ranunculid families and 'higher–level' systematic groups based on parsimony analyses of the three different gene sequences analysed separately and combined. – indicates that the group is not supported or that the strict consensus tree is unresolved for that clade

Family or "higher–level" systematic group	Bootstrap values				Decay indices			
	rbcL	atpB	18S	rbcL atpB and 18S	rbcL	atpB	18S	rbcL atpB and 18S
Ranunculids (*Papaverales* + *Euptelea* + "core" ranunculids)	67	< 50	–	100	–	–	–	9
Papaverales (*Pteridophyllum, Dicentra, Hypecoum*)	80	86	90	100	4	3	> 2	14
Fumariaceae s.l. (*Dicentra, Hypecoum*)	–	89	–	87	–	4	–	6
Euptelea + "core" ranunculids	< 50	< 50	–	87	–	–	–	8
"Core" ranunculids *Circaeasteraceae, Lardizabalaceae* s.l. *Ranunculaceae, Berberidaceae*)	< 50	84	< 50	99	1	5	–	11
Circaeasteraceae + *Lardizabalaceae* s.l.	< 50	–	< 50	77	–	–	1	2
Circaeasteraceae (*Circaeaster, Kingdonia*)	96	100	99	100	> 5	> 5	> 2	> 15
Lardizabalaceae s.l. (*Sinofranchetia, Decaisnea, Sargentodoxa*)	–	99	90	100	–	> 5	> 2	10
Menispermaceae + *Ranunculaceae* + *Berberidaceae*	< 50	68	–	70	1	2	–	2
Menispermaceae (*Menispermum, Tinospora*)	99	100	73	100	> 5	> 5	> 2	> 15
Ranunculaceae + *Berberidaceae*	55	68	–	88	2	2	–	2
Ranunculaceae (*Hydrastis, Glaucidium*)	99	84	< 50	100	> 5	2	1	14
Berberidaceae *Nandina, Caulophyllum*)	100	100	63	100	> 5	> 5	1	> 15

by *rbc*L and 18S analyses) and the monophyly of the *Fumariaceae* s.l. represented by *Dicentra* and *Hypecoum* (not well-supported by *rbc*L and 18S analyses).

Surprisingly, the combined analysis also provides significant support for several clades that were not recognized or only poorly supported in the individual analyses of *rbc*L, *atp*B and 18S sequences. The most significant of these is the 100% bootstrap support for the monophyly of the *Ranunculidae* (plus *Euptelea*). Other relationships in this category, which are less strongly supported, are: i) inclusion of *Euptelea* with the "core" ranunculids (87% bootstrap value), ii) the sister group relationship of *Circaeasteraceae* and *Lardizabalaceae* (77% bootstrap value), iii) the *Menispermaceae-Ranunculaceae-Berberidaceae* clade (70% bootstrap value), and iv) the sister group relationship between the *Ranunculaceae* and *Berberidaceae* (88% bootstrap value). Table 3 lists the 13 well-supported clades recognized in the combined data set and compares their support in the four separate analyses.

Discussion

Comparison with the *rbc*L analysis of CHASE & al. (1993). In many respects the results from phylogenetic analysis of the combined *rbc*L, *atp*B and 18S sequence data are very similar to the results obtained from the preliminary analyses of the large seed plant *rbc*L data set by CHASE & al. (1993). Despite the more in depth systematic coverage of the *Ranunculidae* in this study, both analyses place *Ranunculaceae* and *Berberidaceae* as sister taxa, and together they comprise the sister-group to *Menispermaceae*. The 'core' ranunculids are also recognized by CHASE & al. (1993) (although *Circaeasteraceae* were not considered), and *Lardizabalaceae* are resolved as basal within the group. The CHASE & al. (1993) analysis also allies *Euptelea* with the 'core' ranunculids, rather than with other 'lower' hamamelidids, and places *Papaverales* as the basal clade among ranunculids sensu lato.

The only two results from the analysis of CHASE & al. (1993) that are surprising in the context of our study are the supposed paraphyly of *Berberidaceae* (represented by *Caulophyllum* and *Mahonia*) with respect to *Ranunculaceae*, and the paraphyly of *Papaveraceae* (represented by *Papaver* and *Sanguinaria*) with respect to *Fumariaceae*. The four relevant genera were not included in the analysis presented here but have been studied as part of more inclusive preliminary analyses of our data. In these trees with more extensive sampling, both *Berberidaceae* and *Papaveraceae* are clearly resolved as monophyletic.

One highly anomalous result in the initial CHASE & al. (1993) study was the placement of *Sargentodoxa* within *Fabaceae*. This result is now known to have been based on misidentified sterile material of *Sargentodoxa* (CHASE, pers. comm.).

Euptelea and the monophyly of the ranunculids. The most surprising result of the CHASE & al. (1993) analysis, as well as our own work based on two additional genes and more extensive sampling, is the incorporation of *Euptelea* within the *Ranunculidae* (above the level of *Papaverales*) as sister taxon to the "core" ranunculids. To test the possibility that *Euptelea* may be more closely related to certain "lower" hamamelidids we conducted a further analysis of our combined *rbc*L, *atp*B and 18S data using species from additional non-ranunculid taxa: *Platanus* (*Platanaceae*) *Nelumbo* (*Nelumbonaceae*), *Sabia* (*Sabiaceae*), *Pachysandra*, *Buxus* (*Buxaceae*),

Trochodendron, Tetracentron (Trochodendraceae), Myrothamnus (Myrotham-naceae), and *Gunnera (Gunneraceae)*. Parsimony analysis of this more inclusive data set (using the same outgroups) provided two most parsimonious trees that differed only in the resolution of relationships within the *Lardizabalaceae* s.l. (trees not presented here). The results provided strong support (100% bootstrap value) for the monophyly of the *Ranunculidae* s.l. (*Papaverales, Euptelea*, "core" ranunculids). All other relationships inferred from analysis of the less inclusive data set (Fig. 1) were also supported. Experimental manipulations showed that placing *Euptelea* as sister taxon to *Papaverales* plus "core" ranunculids added only two steps, while an additional 15 steps were incurred by placing *Euptelea* at the base of the group of "lower" hamamelidids.

Circaeaster and Kingdonia. Based on analysis of the combined sequence data from three genes, *Circaeaster* and *Kingdonia* are strongly supported (100% boot-strap value) as a monophyletic group (Fig. 1). This is consistent with traditional interpretations (FOSTER 1961, 1963), but conflicts with the suggestion (TAMURA 1962, 1993; WU & KUBITZKI 1993) that *Kingdonia* should be placed within the *Ranun-culaceae*. Based on our combined data set it adds 68 steps to move *Kingdonia* to the base of the *Ranunculaceae* s.l. (represented here by *Glaucidium* and *Hydrastis*). Experiments that included a larger sample of *Ranunculaceae* were similarly unpar-simonious.

Relationships of Papaverales. While there have been relatively few explicit phylogenetic analyses of the *Ranunculidae*, most considerations of the group have emphasized the distinctiveness of the *Papaverales* (e.g., DAHLGREN 1980, TAKHTAJAN 1987, Table 1). Results of the analyses presented here, which place *Papaverales* in a relatively basal position with respect to other ranunculids, broadly support these interpretations. However, in two explicit cladistic analyses of the *Ranunculidae* that have been published (LOCONTE & STEVENSON 1991; LOCONTE & al. 1995, this volume), the *Papaverales* are resolved as sister group to the *Ranunculaceae* and are placed in a relatively derived position with respect to other ranunculid families. Based on our combined sequence data this conclusion is highly unparsimonious. With our data set, moving the *Papaverales* into a derived position as sister group to the *Ranunculaceae* adds an additional 22 steps. Moving the *Ranunculaceae* into a relatively basal position as sister group to the *Papaverales* incurs an additional 19 steps.

Based on our results it is also unlikely that the *Papaverales* and *Berberidaceae* are closely related as is suggested by some aspects of their floral morphology (ENDRESS 1995, this volume). Placing the *Papaverales* with the *Berberidaceae* in a sister group relationship with the *Ranunculaceae* requires an additional 21 steps in our combined sequence data.

Relationships among "core" Ranunculidae. Excluding the position of the *Papaverales*, the pattern of the relationships based on the combined *rbc*L, *atp*B and 18S data is similar to that hypothesized by LOCONTE & STEVENSON (1991 and LOCONTE & al. 1995, this volume) based on cladistic analyses of morphological data, especially in the relatively distant position of *Lardizabalaceae* with respect to other "core" ranunculids. The close relationship between *Berberidaceae* and *Ranun-culaceae* suggested by features of floral form is also supported (ENDRESS 1995, this volume). However, it is important to note that in analyses of our combined data set none of the relationships among *Lardizabalaceae*, *Menispermaceae*, *Ranunuculaceae*, and *Berberidaceae* are as strongly supported as the monophyly of the families themselves, the monophyly of the *Ranunuculidae* s.l., or the monophyly of "core" ranunculids (Fig. 1). In this context, it remains an open question whether the similarities in floral architecture between *Menispermaceae* and *Lardizabalaceae* (ENDRESS 1995, this volume) reflect close phylogenetic relationships or the basic floral condition for the "core" ranunculids as a whole.

Conclusions

Parsimony analysis of combined sequence data for the *rbc*L, *atp*B and 18S genes provides strong support for the monophyly of the *Ranunuculidae* s.l., and also a variety of relationships within the group. Although not adequately tested by the analyses presented here, results from previous phylogenetic analyses of *rbc*L sequence data indicate that the *Ranunculidae* are a basal or near basal clade of eudicots that must have diverged at an early stage from other eudicot taxa at the "lower" hamamelidid grade. These results suggest that the ranunculid lineage is ancient and perhaps diverged soon after the first appearance of triaperturate pollen in the fossil record in the late Barremian (c. 126 myr B.P.). Fruits morphologically similar to those of *Circaeaster*, but associated with monosulcate pollen, are known from the Early Cretaceous (CRANE & al. 1994, FRIIS & al. 1995), but otherwise ranunculids are almost absent from the Cretaceous fossil record. This contrasts with the diversity of several ranunculid groups (especially *Menispermaceae*) during the Early Tertiary (CHANDLER 1964, COLLINSON 1983, COLLINSON & al. 1993, MANCHESTER 1994) and suggests that more intensive paleobotanical investigations may ultimately reveal much earlier evidence of ranunculid lineages. Based on previous reviews (DRINNAN & al. 1994), as well as the results presented here, the earliest ranunculids may have had small, few-parted, perhaps fly-pollinated (ENDRESS 1995) flowers, with cyclic, dimerous or trimerous, floral phyllotaxy.

We are grateful to Professors UWE JENSEN and JOACHIM KADEREIT for the opportunity to participate in the international symposium on 'The Systematics and Evolution of the *Ranunculiflorae*.' We thank D. NICKRENT and E. A. ZIMMER for ribosomal DNA amplification and sequencing primers, and G. ZURAWSKI for *rbc*L sequencing primers. We also thank J. HALL for technical advice and for laboratory assistance. This work was supported in part by National Science Foundation Grants DEB-9020237 and INT-9015123 to PETER R. CRANE and DEB-9306533 to SARA B. HOOT.

Note added in proof

Because of the anomalous placement of *Sinofranchetia* in the tree based on *rbc*L sequence data reported here, we recently obtained fresh material from the same specimen of *Sinofranchetia* (Royal Botanic Garden, Edinburgh). DNA from this new leaf material amplified readily (unlike the original amplifications) and generated a different *rbc*L sequence. Reanalysis of the *rbc*L data alone (including this new sequence for *Sinofranchetia*), now places *Sinofranchetia* within the *Lardizabalaceae* (see HOOT et al. 1995, this volume).

References

Albert, V., Backlund, A. A., Bremer, K., Chase, M. W., Manhart, J. R., Mishler, B. D., Nixon, K. C., 1994: Functional constraints and *rbc*L evidence for land plant phylogeny. – Ann. Missouri Bot. Gard. **81**: 534–567.

Chandler, M. E. J., 1964: The Lower Tertiary floras of Southern England. IV. A Summary and survey of findings in the light of recent botanical observations. – London: British Museum (Natural History).

Chase, M. W., Soltis, D. E., Olmstead, R. G., Morgan, D., Les, D. H., Mishler, B. D., Duvall, M. R., Price, R. A., Hills, H. G., Qiu, Y.-L., Kron, K. A., Rettig, J. H., Conti, E., Palmer, J. D., Manhart, J. R., Sytsma, K. J., Michaels, H. J., Kress, W. J., Karol, K. G., Clark, W. D., Hedrén, M., Gaut, B. S., Jansen, R. K., Kim, K.-J., Wimpee, C. F., Smith, J. F., Furnier, G. R., Strauss, S. H., Xiang, Q. -Y., Plunkett, G. M., Soltis, P. S., Swensen, S. M., Williams, S. E., Gadek, P. A., Quinn, C. J., Eguiarte, L. E., Golenberg, E., Learn, G. H., Graham, S. W., Barrett, S. C. H., Dayanandan, S., Albert, V. A., 1993: Phylogenetics of seed plants: an analysis of nucleotide sequences from the plastid gene *rbc*L. – Ann. Missouri Bot. Gard. **80**: 528–580.

Collinson, M. E., 1983: Fossil plants of the London Clay. – London: The Palaeontological Association.

– Boulter, M. C., Holmes, P. L., 1993: *Magnoliophyta* ("*Angiospermae*"). – In Benton, M. J., (Ed.): The fossil record **2**, pp. 809–841. – London: Chapman & Hall.

Crane, P. R., 1989: Paleobotanical evidence on the radiation of non-magnoliid dicotyledons. – Pl. Syst. Evol. **162**: 165–191.

– Friis, E. M., Pedersen, K. R., 1994: Paleobotanical evidence on the early radiation of magnoliid angiosperms. – Pl. Syst. Evol. [Suppl.] **8**: 51–72.

Cronquist, A., 1981: An integrated system of classification of flowering plants. – New York: Columbia University Press.

Dahlgren, R. M. T., 1980: A revised system of classification of angiosperms. – Bot. J. Linn. Soc. **80**: 91–124.

Donoghue, M. J., Olmstead, R. G., Smith, J. F., Palmer, J. D., 1992: Phylogenetic relationships of *Dipsacales* based on *rbc*L sequences. – Ann. Missouri Bot. Gard. **79**: 333–345.

Doyle, J. A., Donoghue, M. J., Zimmer, E. A., 1994: Integration of morphological and ribosomal RNA data on the origin of angiosperms. – Ann. Missouri Bot. Gard. **81**: 419–450.

Doyle, J. J., Doyle, J. L., 1987: A rapid DNA isolation procedure for small quantities of fresh tissue. – Phytochem. Bull. **19**: 11–15.

Drinnan, A. N., Crane, P. R., Hoot, S. B., 1994: Patterns of floral evolution in the early diversification of non-magnoliid dicotyledons (eudicots). – Pl. Syst. Evol. [Suppl.] **8**: 93–102.

Eckenrode, V. K., Arnold, J., Meagher, R. B., 1985: Comparison of the nucleotide sequence of soybean 18S rRNA with the sequences of other small-subunit rRNA's. – J. Mol. Evol. **21**: 259–269.

Endress, P. K., 1986: Floral structure, systematics and phylogeny in *Trochodendrales*. – Ann. Missouri Bot. Garden **73**: 297–324.

– 1995: Floral structure and evolution in *Ranunculanae*. – Pl. Syst. Evol. [Suppl.] **9**: 47–61.

Felsenstein, J., 1985: Confidence limits on phylogenies: an approach using the bootstrap. – Evolution **39**: 783–791.

Foster, A. S., 1961: The floral morphology and relationships of *Kingdonia uniflora*. – J. Arnold Arbor. **42**: 397–416.

– 1963: The morphology and relationships of *Circaeaster*. – J. Arnold Arb. **44**: 299–321.

Friis, E. M., Pedersen, K. R., Crane, P. R., 1995: *Appomattoxia ancistrophora* gen. et sp. nov., a new Early Cretaceous plant with similarities to *Circaeaster* and extant *Magnoliidae*. – Amer. J. Bot. **82**(7): in press.

HILLIS, D. M., BULL, J. J., 1993: An empirical test of bootstrapping as a method for assessing confidence in phylogenetic analyses. – Syst. Biol. **42**: 182–192.

HOOT, S. B., 1991: Phylogeny of the *Ranunculaceae* based on epidermal microcharacters and macromorphology. – Syst. Bot. **16**: 741–755.

– 1995: Phylogeny of the *Ranunculaceae* based on preliminary *atp*B, *rbc*L and 18S nuclear ribosomal DNA sequence data. – Pl. Syst. Evol. [Suppl.] **9**: 241–251.

– CULHAM, A., CRANE, P. R., 1995: The utility of *atp*B gene sequences in resolving phylogenetic relationships: comparison with *rbc*L and 18S ribosomal DNA sequences in the *Lardizabalaceae*. – Ann. Missouri Bot. Gard. **82**: 194–207.

JOHANSSON, J. T., JANSEN, R. K., 1993: Chloroplast DNA variation and phylogeny of the *Ranunculaceae*. – Pl. Syst. Evol. **187**: 29–49.

KADEREIT, J. W., BLATTNER, F. R., JORK, K. B., SCHWARZBACH, A., 1995: The phylogeny of the *Papaveraceae* s.l.: morphological, geographical and ecological implications. – Pl. Syst. Evol. [Suppl.] **9**: 133–145.

KIM, Y.-D., JANSEN, R., 1995: Phylogenetic implications of chloroplast DNA variation in the *Berberidaceae*. – Pl. Syst. Evol. [Suppl.] **9**: 341–349.

LIDÉN, M., 1995: *Fumariaceae*. – In KUBITZKI, K., ROHWER, J. G., BITTRICH, V., (Eds): The families and genera of vascular plants, **VII**, pp. 310–318. – Berlin, Heidelberg, New York: Springer.

LOCONTE, H., ESTES, J. R., 1989: Phylogenetic systematics of *Berberidaceae* and *Ranunculales* (*Magnoliidae*). – Syst. Bot. **14**: 565–579.

– STEVENSON, D. W., 1991: Cladistics of the *Magnoliidae*. – Cladistics **7**: 267–296.

– CAMPBELL, L. M., STEVENSON, D. W., 1995: Ordinal and familial relationships of ranunculid genera. – Pl. Syst. Evol. [Suppl.] **9**: 99–118.

MADDISON, W. P., MADDISON, D. R., 1992: MacClade: interactive analysis of phylogeny and character evolution, version 3.0. – Sunderland, Massachusetts: Sinauer.

MANCHESTER, S. R., 1994: The Flora of the Clarno Chert. – Palaeontographica Americana **58**: 1–205.

NICKOL, M. G., 1995: Phylogeny and inflorescences of *Berberidaceae* – a morphological survey. – Pl. Syst. Evol. [Suppl.] **9**: 327–340.

SWOFFORD, D. L., 1993: PAUP: Phylogenetic analysis using parsimony, version 3.1. – Champaign, IL: Illinois Natural History Survey.

TAKHTAJAN, A., 1987: Systema Magnoliophytorum. – Leningrad: Nauka, (in Russian).

TAMURA, M., 1962: Morphology, ecology and phylogeny of the *Ranunculaceae* I. – Sci. Rept. Osaka Univ. **11**: 115–126.

– 1993: *Ranunculaceae*. – In KUBITZKI, K., ROHWER, J. G., BITTRICH, V., (Eds): The families and genera of vascular plants, **VII**, pp. 563–583. – Berlin, Heidelberg, New York: Springer.

THANIKAIMONI, G., 1984: Menispermacées: Palynologie et systematique. – Institute Français de Pondichery, Travaux Sec. Sci. Tech. **18**: 1–135.

WU, C.-Y., KUBITZKI, K., 1993: *Circaeasteraceae*. – In KUBITZKI, K., ROHWER, J. G., BITTRICH, V., (Eds): The families and genera of vascular plants, **VII**, pp. 288–289.– Berlin, Heidelberg, New York: Springer.

Addresses of the authors: SARA B. HOOT, Department of Biological Sciences, University of Wisconsin – Milwaukee, Lapham Hall, P. O. Box 413, Milwaukee, WI 53201, USA – PETER R. CRANE, The Field Museum, Roosevelt Road at Lake Shore Drive, Chicago, IL 60605, USA.

Accepted March 13, 1995

Pl. Syst. Evol. [Suppl.] 9: 133–145 (1995)

The phylogeny of the *Papaveraceae* sensu lato: morphological, geographical and ecological implications

J. W. KADEREIT, F. R. BLATTNER, K. B. JORK, and A. SCHWARZBACH

Received October 27, 1994

Key words: *Fumariaceae, Papaveraceae, Pteridophyllum, Hypecoum.* – Androecium evolution, gynoecium evolution, geographical distribution, intercontinental disjunction, aridification.

Abstract: On the background of the phylogeny of the *Papaveraceae* s.l., the evolution of the androecium and gynoecium, the geographical distribution, and the ecological differentiation of the family is discussed. An effort is made to homologize the diverse androecia of *Pteridophyllum, Hypecoum, Fumariaceae* and *Papaveraceae* s.str., and it is concluded that both the polyandrous androecium as found in the *Papaveraceae* s.str. and the polycarpellate gynoecium as found in *Papaveraceae* s. str. subfam. *Papaveroideae* incl. *Platystemonoideae* are secondary and derived from whorled oligomerous androecia and bicarpellate gynoecia, respectively. The comparative consideration of the geographical distribution, ecology and diversity of various monophyletic units within the family suggests (1) that forest floor habitats might be ancestral in the family, (2) that in parts of the family generic diversification preceded the break-up of the northern continents, and (3) that aridification of the environment was a major stimulus for both diversification and morphological divergence in the family.

The Papaveraceae s.l. discussed here contain *Pteridophyllum* SIEB. & ZUCC., *Hypecoum* L., *Fumariaceae* and *Papaveraceae* s.str. Within the *Ranunculiflorae* the family in this circumscription is unambiguously characterized by its paracarpous gynoecium of two or more carpels. Other families of the group have an either apocarpous (*Lardizabalaceae* incl. *Sargentodoxaceae, Menispermaceae, Ranunculaceae* incl. *Circaeasteraceae* and *Hydrastidaceae*) or monomerous (*Berberidaceae* incl. *Nandinaceae*) gynoecium. Comprehensive summaries of the morphology, distribution, systematics, ecology etc. of *Pteridophyllum, Hypecoum, Fumariaceae* and *Papaveraceae* s.str. were recently presented by LIDÉN (1993a,b) and KADEREIT (1993). A phylogenetic analysis of generic affinities within *Fumariaceae* was conducted by LIDÉN (1986), and generic affinities with the *Papaveraceae* s.str. and affinities among *Pteridophyllum, Hypecoum, Fumariaceae* and *Papaveraceae* s.str. were analyzed by KADEREIT & al. (1994). Both these analyses were based on morphological characters.

In the present paper these phylogenetic hypotheses by LIDÉN (1986) and KADEREIT & al. (1994) will be used for a discussion of certain aspects of the morphology, distribution and ecology of the *Papaveraceae* s.l.

The phylogeny of the *Papaveraceae* s.l.

The phylogenetic affinities among *Pteridophyllum, Hypecoum, Fumariaceae* and *Papaveraceae* s.str. as found by KADEREIT & al. (1994) are illustrated in Fig. 1. This analysis was based on 39 morphological characters. Within *Papaveraceae* s.l., *Hypecoum, Fumariaceae*, and *Papaveraceae* s.str. share the possession of idioblasts or laticifers (FRIEDEL 1938, LIDÉN 1993a), and all have calcium oxalate crystals in the inner epidermis of the outer integument of the seeds (BRÜCKNER 1985). These two characters are absent from *Pteridophyllum*. On the assumption that *Berberidaceae* is the sister group of *Papaveraceae* s.l. (KADEREIT & al. 1994), which according to ENDRESS (1995, this volume) is strongly supported by several characters of mainly the gynoecium, this character distribution identifies *Pteridophyllum* as basal in the *Papaveraceae* s.l. because both these characters are absent also in the *Berberidaceae* (for exceptions see KADEREIT & al. 1994). The sequence data of HOOT & CRANE (1995, this volume) do not confirm the sistergroup relationship of *Berberidaceae* to *Papaveraceae* s.l. These authors, however, in agreement with the hypothesis of KADEREIT & al. (1994), also identify *Pteridophyllum* as the basal clade of the *Papaveraceae* s.l., which according to their evidence as a whole occupy a basal position within the *Ranunculiflorae*.

The following discussion summarizes the results of KADEREIT & al. (1994). Within *Papaveraceae* s.l., the sistergroup relationship of *Fumariaceae/Hypecoum* to *Papaveraceae* s.str. is supported by their possession of idioblasts or laticifers with watery or milky sap or latex, and calcium oxalate crystals in the inner epidermis of the outer integument of the seeds. *Fumariaceae* and *Hypecoum* are united as a monophyletic group by the shared possession of a subepidermal collenchyma (LÉGER

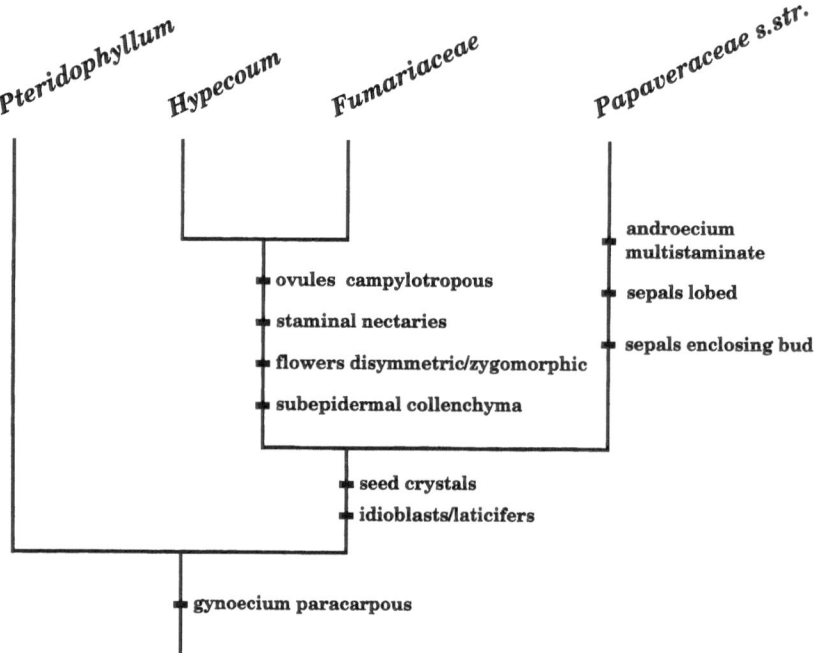

Fig. 1. Phylogenetic affinities within the *Papaveraceae* s.l. (*Pteridophyllum, Hypecoum, Fumariaceae, Papaveraceae* s.str.). For detailed discussion see text and KADEREIT & al. (1994)

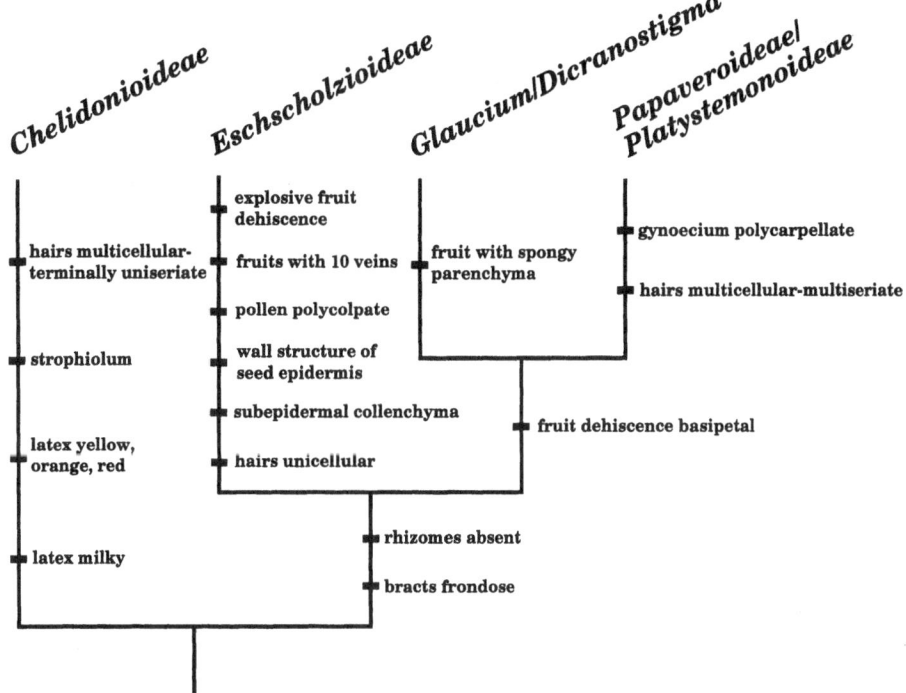

Fig. 2. Phylogenetic affinities within the *Papaveraceae* s.str. For detailed discussion see text and KADEREIT & al. (1994)

1894), disymmetric or zygomorphic flowers, staminal nectaries and campylotropous ovules (BRÜCKNER 1983). The monophyly of the *Papaveraceae* s.str. is supported by the full enclosure of the flower buds by the sepals and the presence of marginal lobes on the sepals (ERNST 1962), and the possession of a multistiminate androecium.

Within *Papaveraceae* s.str. (Fig. 2) the basal *Chelidoniodeae* (*Bocconia* L., *Chelidonium* L., *Eomecon* HANCE, *Hylomecon* MAXIM., *Macleaya* R. BR., *Sanguinaria* L., *Stylophorum* NUTT.) are characterized by milky latex of yellow, orange or red colour, the possession of a strophiolum on the seeds (BRÜCKNER 1983) and multicellular, terminally uniseriate hairs. All of these characters occasionally also occur in other taxa of the *Papaveraceae* s.str. and accordingly are not unique synapomorphies. The remainder of *Papaveraceae* s.str. share frondose flower bracts and lack rhizomes.

Eschscholzioideae (*Dendromecon* BENTH., *Eschscholzia* CHAM. in NEES, *Hunnemannia* SWEET) are united by unicellular hairs (ERNST 1962), the possession of a subepidermal collenchyma (LÉGER 1894), the structure of the seed epidermis (outer wall unstructured, inner and anticlinal walls with thickenings; BRÜCKNER 1983) and by polycolpate pollen grains (LAYKA 1976). The bicarpellate fruits of the *Eschscholzioideae* also have 10 conspicuous longitudinal veins and open explosively in an acropetal direction (ERNST 1962). *Glaucium* MILLER and *Dicranostigma* J. D. HOOK. & THOMS. and *Papaveroideae* incl. *Platystemonoideae* share only the mode of fruit dehiscence. Capsules in this group mostly open basipetally.

Glaucium and *Dicranostigma*, which have been regarded as part of *Chelidoniodeae* by most authors in the past but differ from this group by having frondose instead of bracteose flower bracts, and by lacking a seed appendage and certain alkaloids characteristic of the *Chelidoniodeae* (VENT & MORY 1973), are united by the presence of spongy septa in their fruits (BRÜCKNER 1982).

Papaveraceae (*Arctomecon* TORR. & FRÉM. in FRÉM., *Argemone* L., *Canbya* PARRY ex A. GRAY, *Meconopsis* VIG., *Papaver* L., *Roemeria* MEDIK., *Romneya* HARV., *Stylomecon* TAYLOR) incl. *Platystemonoideae* (*Hesperomecon* GREENE, *Meconella* NUTT. in TORR. & A. GRAY, *Platystemon* BENTH.) share the possession of multicellular-multiseriate hairs (ERNST 1962) and polycarpellate gynoecia. *Platystemonoideae* were recognized to be sistergroup to the North American *Canbya* (KADEREIT & al. 1994) and needs to be abandoned as a subfamily. The above results have been discussed in detail by KADEREIT & al. (1994).

In *Fumariaceae*, the cladistic analysis by LIDÉN (1986) resulted in the recognition of two tribes. Tribe *Corydaleae*, containing five genera (*Dicentra* BERNHARDI, *Capnoides* MILLER, *Adlumia* DC., *Dactylicapnos* WALL., *Corydalis* DC.), is character-ized by a peristent style with a usually large and flattened stigma, and tribe *Fumarieae* with 11 genera (*Pseudofumaria* MEDIK., *Sarcocapnos* DC., *Ceratocapnos* DURIEU, *Platycapnos* (DC.) BERNHARDI, *Fumariola* KORSH., *Fumaria* L., *Cryptocap-nos* RECH. fil., *Rupicapnos* POMEL, *Discocapnos* CHAM. & SCHLECHT., *Trigonocapnos* SCHLECHTER, *Cysticapnos* MILLER) in three subtribes by a caducous style without conspicuously flattened stigma.

The evolution of the androecium and gynoecium in the *Papaveraceae* s.l.

The configuration of the androecium in the *Papaveraceae* s.l. is diverse and has been debated extensively.

In *Pteridophyllum* one whorl of four dithecal stamens can be found. These have been considered either to be arranged in two pairs opposite the outer petals by MURBECK (1912), or individually on radii more or less diagonal to those of the two pairs of petals (Fig. 3) by LEHMINGER & LEINS (1985).

The early development of the androecium appears to be identical in *Hypecoum* and *Fumariaceae*. Whereas some authors (PAYER 1857, EICHLER 1865, BUCHENAU 1866, LEHMINGER & LEINS 1985) observed two large transversal primordia differenti-ating into one central dithecal and two lateral monothecal stamens each (Fig. 3), six independent primordia were described by others (MURBECK 1912, MAIR 1977, RONSE DECRAENE & SMETS 1992a). Two of these six primordia are found in a transversal position opposite the outer pair of petals and develop into dithecal stamens, and the remaining four primordia, which develop into monothecal stamens, are arranged on radii more or less diagonal to those of the two pairs of petals. In *Fumariaceae* (Fig. 3) two transversal stamen complexes consisting of one central dithecal and two lateral monothecal stamens can be found in the adult flowers. In *Hypecoum* (Fig. 3), the monothecal stamens do not fuse with the transversal dithecal stamens, but adjacent monothecal stamens fuse with each other in the flower median (LEHMINGER & LEINS 1985, RONSE DECRAENE & SMETS 1992a). Four primordia in transversal and median position instead of two or six primordia were observed by KARRER (1991) in *Hypecoum*. The stamens in median position according to this author usually do not appear in pairs but as single organs. They may, however, be broader than the transversal primordia. If they appear as pairs, which is the case only very rarely, they

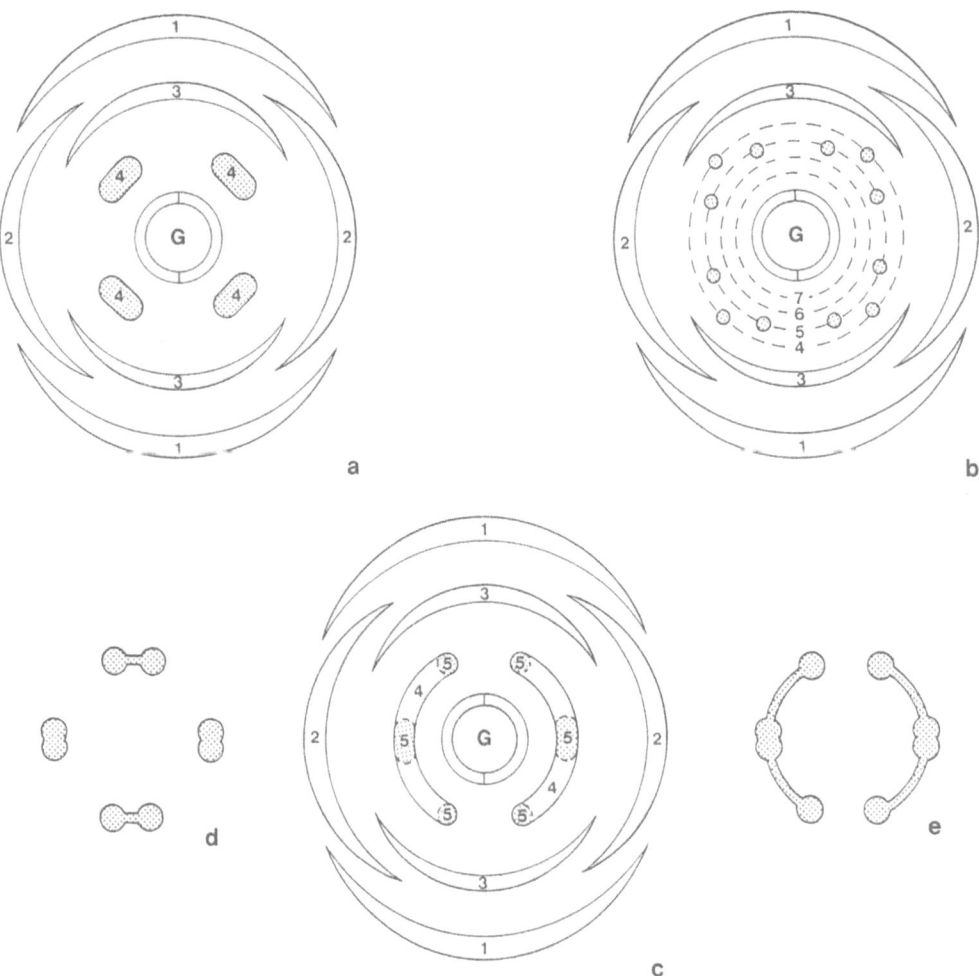

Fig. 3. Ontogenetic floral diagrams of *a Pteridophyllum, b Chelidonium, c Hypecoum* and *Fumariaceae (Dicentra) d* adult androecium of *Hypecoum, e* adult androecium of *Fumariaceae (Dicentra; a, c–e* after LEHMINGER 1985, *b* after KARRER 1991). Figures indicate sequence of organ initiation; *G* gynoecium. For detailed explanation see text

differentiate as dithecal and not as monothecal stamens (ENDRESS, pers. comm.). The various interpretations of the androeceum of *Hypecoum* and *Fumariaceae* have been discussed in detail by RONSE DECRAENE & SMETS (1992a).

In the *Papaveraceae* s.str. the multistaminate androecium consists of whorls arising in a centripetal sequence. In all dimerous genera except the apetalous *Bocconia* and *Macleaya* the outermost (first) whorl comprises four stamens (Fig. 3) on radii more or less diagonal to those of the two pairs of petals (MURBECK 1912, ENDRESS 1987, RONSE DECRAENE & SMETS 1990, KARRER 1991). Formally, they can be described as two stamen pairs alternating with the inner pair of petals (ENDRESS 1987). Subsequent whorls normally consist of a larger number of stamens. These whorls can arise on a ring primordium particularly in flowers with a large number of stamens. In *Bocconia* and *Macleaya* the first two whorls of the androecium consist of two stamens each. These dimerous whorls alternate regularly with the calyx and with each other (RONSE DECRAENE & SMETS 1990, KARRER 1991).

The androecia found in *Pteridophyllum* and *Papaveraceae* s.str. clearly are homologous except for the fact that only one whorl of stamens is found in the former, but mostly several in the latter. The homology of these two groups with *Hypecoum* and *Fumariaceae* depends on the interpretation of their androecium. If interpreted as consisting of an outer whorl of four monothecal stamens and an inner whorl of dithecal stamens, as done by, e.g., ARBER (1932) and RONSE DECRAENE & SMETS (1992a), homology would be established for the outer whorl, and the inner whorl would have to be regarded as reduced in number in *Hypecoum* and *Fumariaceae*. This interpretation, however, leads to difficulties in explaining the position of the two transversal carpels in relation to the stamens. If, alternatively, interpreted as consisting of an outer whorl of two dithecal stamens and an inner whorl of four monothecal stamens, as done, e.g., by CELAKOVSKY (1894) and MURBECK (1912), homology could be established when the four stamens of *Pteridophyllum* and the four stamens of the outermost whorl of the *Papaveraceae* s.str. are interpreted as a doubled pair of stamens opposite the outer pair of petals, as suggested for the *Papaveraceae* s.str. by ENDRESS (1987). *Hypecoum* and *Fumariaceae* then would lack this doubling in the outer whorl of stamens. This interpretation postulates a basically dimerous rather than tetramerous (RONSE DECRAENE & SMETS 1992a) androecium in the *Papaveraceae* s.l. Assuming an origin of the *Papaveraceae* s.l. from an ancestor with spiral androecium, and following the argumentation of ERBAR & LEINS (1981, 1983; for *Magnoliaceae*) and LEINS & ERBAR (1985; for *Aristolochiaceae*), it is also possible to interpret the tetramerous androecium as found in *Pteridophyllum* as plesiomorphic and derived from a spiral androecium through cyclisation of the perianth. The androecium of *Fumariaceae* (and *Berberidaceae*, should this be sistergroup to *Papaveraceae* s.l.) then would have resulted from the transition to regularly alternating whorls through adjustment (harmonization, LEINS & ERBAR 1991) of relative primordium size.

Irrespective of these various interpretations of the androecium the phylogeny of the *Papaveraceae* s.l. suggested here implies that the multistaminate androecium of *Papaveraceae* s.str., in accordance with the opinion expressed by, e.g., MURBECK (1912) and KARRER (1991), is derived rather than primitive as suggested by, e.g., MERXMÜLLER & LEINS (1967) and RONSE DECRAENE & SMETS (1992b). In connection with the absence of nectariferous stamens the origin of the multistaminate androecium of the *Papaveraceae* s.str. can be understood as an adaptation to pollination by pollen collecting insects.

One structural argument supporting the derived rather than primitive status of the multistaminate androecium of the *Papaveraceae* s.str. might be the position of the gynoecium. Irrespective of the number of whorls in the androecium the carpels in all bicarpellate *Papaveraceae* s.str. (and *Pteridophyllum*, *Hypecoum* and *Fumariaceae*) occupy a transversal position, as would be expected in flowers with a biseriate androeceum and regularly alternating whorls.

Not only the multistaminate androecium, but also the polycarpellate gynoecium of the *Papaveroideae* incl. *Platystemonoideae* must be regarded as derived in the family. From the figures given by KARRER (1991) it seems, in a rough approximation, that among the *Papaveraceae* s.str. *Papaveroideae* incl. *Platystemonoideae* have, on average, the highest number of stamens. Although no detailed knowledge of variation in stamen number, breeding system or pollination efficiency is available, it appears possible that the increase in carpel number could reflect an adjustment in pollen/ovule ratio in the polycarpellate members of the *Papaveraceae* s.str. Interestingly, those genera among the bicarpellate *Papaveraceae* s.str. with a high number of stamens, as, e.g., *Glaucium* and *Hunnemannia* (KARRER 1991), also have very long gynoecia with comparatively many ovules. This might represent an alternative way of pollen/ovule ratio adjustment.

The geographical distribution and ecology of the family

The main purpose of this short review of the geographical distribution and ecology of the *Papaveraceae* s.l. is to place the phylogenetic diversification of the family into a geographical and historical context, and to identify ecological factors of possible importance in the process of diversification and morphological evolution.

Comprehensive reviews of the geographical distribution of the entire family, which is found mostly in north temperate regions, have been given by FEDDE (1905, 1909, 1936). ZHUANG (1993) recently tried to interpret the phylogeny of the family on the basis of its geographical distribution. The distribution of the *Fumariaceae* has been discussed by HUTCHINSON (1921), RYBERG (1960) and LIDÉN (1986, 1993b), and short summaries of the distribution of the genera of the *Papaveraceae* s.str. are available in GÜNTHER (1975a, b) and KADEREIT (1993).

The monotypic genus *Pteridophyllum* is found only on Honshu, Japan (LIDÉN 1993a), and the genus *Hypecoum* is distributed from the Mediterranean area into C Asia. An Asian origin of this genus has been postulated by DAHL (1990).

In *Fumariaceae*, tribe *Corydaleae* is distributed in both the Old and New World, whereas tribe *Fumarieae* is found in the Old World only. Intercontinental disjunctions in tribe *Corydaleae* can be observed on different taxonomic levels (LIDÉN 1986). Whereas *Dactylicapnos* is entirely Asian in distribution, and *Capnoides* is limited to northern N America, *Adlumia* is distributed with one species in E Asia and one species in eastern N America. *Dactylicapnos* (STERN 1961), *Capnoides* and *Adlumia* (LIDÉN 1986) may all from part of a paraphyletic *Dicentra*. *Dicentra* as perceived by LIDÉN (1986) can be found in Asia (subgenera *Macranthos* (STERN) STERN and *Hedycapnos* (PLANCH). STERN), in western N America (subg. *Chrysocapnos* (ENGELM.) STERN), and in E Asia, eastern and western N America (subg. *Dicentra*). Because the phylogeny of this group is poorly understood (LIDÉN, pers. comm.) we refrain from further interpretation of its distributional range. Of the 19 sections recognized by LIDÉN (1986) in the large genus *Corydalis*, 16 are limited to the Old World with a conspicuous concentration in the Sino-Himalayan region (LIDÉN 1986). Of these, sect. *Ramoso-sibiricae* FEDDE ex WENDELBO extends with one species, *Corydalis mildbraedii* FEDDE, into E Africa. Sect. *Archaecapnos* POPOV ex MICHALKOVA shows an E Asian/western N American disjunction. The representatives of sect. *Sophorocapnos* (TURCZ.) POPOV are distributed in E Asia and across much of N America. Sect. *Dactylotuber* (RUPRECHT) POPOV is distributed in East Siberia and Alaska, but also in much of C Asia. Similar intercontinental disjunctions have been discussed repeatedly by, e.g., MEUSEL (1969), RAVEN (1972), RAVEN &

AXELROD (1974), WU (1983) and HONG (1993). An historical account of the discussion of this topic was presented by BOUFFORD & SPONGBERG (1983).

Tribe *Fumarieae* with three subtribes (*Sarcocapninae* LIDÉN, *Fumariinae*, *Discocapninae* LIDÉN) and 11 genera is distributed from the Atlantic Islands to Afghanistan with a conspicuous concentration of taxa in the Mediterranean area (LIDÉN 1986). *Fumaria abyssinica* HAMMAR of subtribe *Fumariinae* can be found, like the above *Corydalis mildbraedii*, in E Africa. Subtribe *Discocapninae* with *Discocapnos* and *Trigonocapnos* as well as the genus *Cysticapnos* of uncertain affinities are S African in distribution. Mediterranean area to C Asia/E and S Africa disjunctions have been discussed in some detail by, e.g., MEUSEL (1969) and BURTT (1971) and are believed to reflect the existence of a once widespread Tertiary arid flora in western Eurasia and Africa.

The same patterns of distribution as just described for the *Fumariaceae* can also be found in the *Papaveraceae* s.str. In the *Chelidonioideae*, *Hylomecon* and *Chelidonium* are of Asian and Eurasian distribution, and *Eomecon* (E Asia) and *Sanguinaria* (eastern N America) and the species of *Stylophorum* (E Asia/eastern N America) should be regarded as disjuncts comparable to the above *Adlumia*. The disjunction of the E Asian *Macleaya* and the C to S American *Bocconia* should be regarded as derived from the above pattern. The disjunctions found in the *Chelidonioideae* are discussed in detail by BLATTNER & KADEREIT (1995, this volume).

Eschscholzioideae are exclusively western N American in distribution, and *Glaucium* and *Dicranostigma* are distributed in the Old World with a range similar to that of the above *Fumarieae* excl. subtribe *Discocapninae* and the genus *Cysticapnos*.

Papaveroideae exhibit Old World/New World disjunctions on different levels. The basal differentiation of the group is into a western North American clade consisting of *Arctomecon*, *Argemone*, *Romneya*, *Canbya* and the three genera of the former *Platystemonoideae*, *Hesperomecon*, *Meconella* and *Platystemon*, and an Eurasian clade comprising *Meconopsis*, *Papaver* s.str. (sects. *Pilosa* PRANTL, *Pseudopilosa* M. POP. ex GÜNTHER, *Meconidium* SPACH, *Macrantha* ELKAN, *Rhoeadium* SPACH, *Papaver*, *Carinatae* FEDDE), *Papaver* sect. *Horrida* ELKAN, *Papaver* sect. *Argemonidium* SPACH, the larger part of *Papaver* sect. *Meconella* SPACH and *Roemeria*. This Eurasian group, however, contains two representatives in western North America, namely *Stylomecon* and *P. californicum* A. GRAY. In addition to those, species of *Papaver* sect. *Meconella* are found in arctic N America and the Rocky Mountains. A detailed discussion of the phylogeny of these two major groups is given by SCHWARZBACH & KADEREIT (1995, this volume) and JORK & KADEREIT (1995, this volume).

The distribution of the western N American taxa of the *Papaveroideae* thus resembles that of the N American representatives of *Corydalis* sect. *Archaecapnos*. Within the Eurasian clade the distribution of *Papaver* s.str. and *Papaver* sect. *Horrida*, which are sister taxa (JORK & KADEREIT 1995, this volume), strongly resembles the distribution of *Fumariaceae* tribe *Fumarieae*. Here the S African *P. aculeatum* THUNB. of sect. *Horrida* corresponds to *Fumarieae* subtribe *Discocapninae* and the genus *Cysticapnos*. The distribution of the arctic-alpine *Papaver* sect. *Meconella*, which has been discussed by KADEREIT (1988), in its arctic and C Asian range perhaps is somewhat comparable to the above *Corydalis* sect. *Dactylotuber*.

A summary of the major disjunctions within monophyletic groups found in the *Papaveraceae* s.l. is given below.

E Asia/eastern N America:
Adlumia
Eomecon/Sanguinaria
Stylophorum
Macleaya/Bocconia (the distribution of *Bocconia* in C and S America probably is derived from an eastern N American distribution)

E Asia/N America:
Corydalis sect. *Sophorocapnos* (widespread in N America)
Dactylicapnos/Capnoides (northern N America)

E Asia/western N America:
Corydalis sect. *Archaecapnos*

Mediterranean area to C Asia/western N America:
Meconopsis, Papaver s.l. (excl. *P. californicum* and sect. *Meconella* p.p.), *Roemeria/ Arctomecon, Argemone, Romneya, Canbya, Hesperomecon, Meconella, Platystemon*

Papaver s.str., *Papaver* sect. *Horrida, Papaver* sect. *Argemonidium, Papaver* sect. *Meconella* p.p., *Roemeria, Meconopsis/Stylomecon, P. californicum*

Mediterranean area to C Asia/E and S Africa:
Fumarieae subtribes *Sarcocapninae, Fumariineae/Discocapninae, Cysticapnos*
Fumaria
Corydalis
Papaver s.str./*Papaver* sect. *Horrida*

C to arctic Eurasia/N America:
Corydalis sect. *Dactylotuber*
Papaver sect. *Meconella*

From the above review of the geographical distribution of the *Papaveraceae* s.l. it is obvious that the family must have occupied a wide and continuous range covering much of Laurasia in the geological past. The north hemispherical disjunctions seen, except for those of taxa of arctic distribution, are very likely to have resulted from the fragmentation of Laurasia into the continents as we know them today.

In order to use the above observations for an estimate of the age or relative age of the extant taxa, it is necessary to consider the ecology of the family. The distribution of *Pteridophyllum*, parts of *Fumariaceae* tribe *Corydaleae* (*Adlumia, Corydalis* p.p., *Dicentra* p.p.) and most members of *Papaveraceae* s.str. subfam. *Chelidonioideae* in forest floor and ecologically related habitats may indicate that this type of habit is basal for the family. The remainder of the family, namely *Hypecoum, Fumariaceae* tribe *Corydaleae* p.p., tribe *Fumarieae, Papaveraceae* subfam. *Eschscholzioideae* and *Papaveroideae* incl. *Platystemonoideae* and *Glaucium/Dicranostigma* all are distributed in more or less arid and always open and variously disturbed habitats. On the background of the phylogeny of the family it is possible that this transition from forest floor into arid and open habitats has taken place only a few times (*Hypecoum*,

Fumarieae, Papaveraceae s.str. excl. *Chelidonioideae*, probably more than once in *Corydaleae*) and was a major stimulus for the diversification and morphological evolution in the family. This is particularly obvious in *Papaveraceae* s.str., in which 16 of 23 genera and c. 220 of c. 240 species grow in more or less arid and open habitats. The forest floor genera of the *Chelidonioideae* contain only very few species or are monotypic. In principle the same consideration applies to the *Fumariaceae*, in which only *Corydalis* with its 250 to 300 species (LIDÉN 1986), of which many seem to grow in more or less mesic and partly forest floor habitats, perhaps must be regarded as an exception. Aridification thus has acted not only as a stimulus for diversification, but also for morphological divergence. This is recognizable from the comparison of taxonomic rank of the taxa involved in the different patterns of disjunctions. Whereas the forest floor taxa of E Asian/eastern N American distribution are disjuncts at infrageneric (e.g., *Adlumia, Stylophorum*) rank, most groups with a Mediterranean area to C Asia/western N America distribution are recognized as morphologically divergent distinct genera (e.g., *Papaveroideae*).

As a consequence, it seems that in the forest floor taxa phylogenetic differentiation partly down to generic rank had taken place before the land connection between Eurasia and N America surmountable by members of the *Papaveraceae* s.l. (except perhaps *Corydalis* sect. *Dactylotuber* and *Papaver* sect. *Meconella*) broke up. It is generally believed that such a land connection existed until the Eocene, although the possibility of a Miocene Beringia passage should not be entirely excluded. Those taxa extant in arid and open habitats, however, may well be of much younger age (RAVEN & AXELROD 1974), as has also been suggested for the western N American clade of the *Papaveroideae* by SCHWARZBACH & KADEREIT (1995, this volume). The geographical distribution of *Papaver* s.str. and *Papaver* sect. *Horrida* and *Fumariaceae* tribe *Fumarieae* in both the Mediterranean to C Asian region and in E and S Africa suggests that these taxa must have existed in the middle to late Tertiary when drastic aridification allowed an expansion of the distributional range across the equator.

We gratefully acknowledge helpful comments on earlier versions of this manuscript by Drs PETER ENDRESS (Zürich), ECKEHART JÄGER (Halle), PETER LEINS (Heidelberg), MAGNUS LIDÉN (Gothenburg), and AARON LISTON (Corvallis/Oregon). ANKE BERG (Mainz) kindly drew the figures. The work on the *Papaveraceae* is supported by grants of the Deutsche Forschungsgemeinschaft (DFG) to J. W. K.

References

ARBER, A., 1932: Studies in floral morphology. IV. On the *Hypecoideae*, with special reference to the androecium. – New Phytol. **31**: 145–173.

BLATTNER, F. R., KADEREIT, J. W., 1995: Three intercontinental disjunctions in *Papaveraceae* subfam. *Chelidonioideae*: evidence from chloroplast DNA. – Pl. Syst. Evol. [Suppl.] **9**: 147–157.

BOUFFORD, D. E., SPONGBERG, S. A., 1983: Eastern Asian – eastern North American phytogeographical relationships – a history from the time of LINNAEUS to the twentieth century. – Ann. Missouri Bot. Gard. **70**: 423–439.

BRÜCKNER, C., 1982: Zur Kenntnis der Fruchtmorphologie der *Papaveraceae* JUSS. s.str. und der *Hypecoaceae* (PRANTL & KÜNDIG) NAK – Feddes Repert. **93**: 153–212.

– 1983: Zur Morphologie der Samenschale in den *Papaveraceae* Juss. s.str. und *Hypecoaceae* (Prantl & Kündig) Nak. – Feddes Repert. **94**: 361–405.

– 1985: Frucht- und Samenanatomie von *Pteridophyllum racemosum* Sieb. et Zucc. und die Position der monotypischen Gattung in den *Papaverales*. – Feddes Repert. **96**: 199–213.

Buchenau, F., 1866: Bemerkungen über den Blüthenbau der Fumariaceen und Cruciferen.– Flora **49**: 39–46.

Burtt, B. L., 1971: From the South: an African view of the floras of Western Asia. – In Davis, P. H., Harper, P. C., Hedge, I. C., (Eds): Plant life of South-West Asia, pp. 135–149. – Edinburgh: The Botanical Society of Edinburgh.

Celakovsky, L. J., 1894: Das Reductionsgesetz der Blüten. – Sitzungsber. Böhm. Ges. Math. Naturwiss. **150**: 1–142.

Dahl, A. E., 1990: Infrageneric division of the genus *Hypecoum* L. (*Papaveraceae*).– Nordic J. Bot. **10**: 129–140.

Eichler, A. W., 1865: Über den Blüthenbau der Fumariaceen, Cruciferen und einiger Capparideen. – Flora **48**: 433–444, 449–460.

Endress, P. K., 1987: Floral phyllotaxis and floral evolution. – Bot. Jahrb. Syst. **108**: 417–438.

– 1995: Floral structure and evolution in *Ranunculanae*. – Pl. Syst. Evol. [Suppl.] **9**: 47–61.

Erbar, C., Leins, P., 1981: Zur Spirale in Magnolien-Blüten. – Beitr. Biol. Pfl. **56**: 225–241.

– 1983: Zur Sequenz von Blütenorganen bei einigen Magnoliiden. – Bot. Jahrb. Syst. **105**: 379–400.

Ernst, W. R., 1962: A comparative morphology of the *Papaveraceae*. – Unpublished Ph.D. Dissertation, Stanford University.

Fedde, F., 1905: Die geographische Verbreitung der *Papaveraceae*. – Englers Bot. Jahrb. **36**, Beiblatt 81: 28–43.

– 1909: *Papaveraceae-Hypecoideae* et *Papaveraceae-Papaveroideae*.– In Engler, A. (Ed.): Das Pflanzenreich IV, 104, pp. 1–430. – Leipzig: Engelmann.

– 1936: *Papaveraceae*. – In Engler, A., Harms, H. (Eds): Die natürlichen Pflanzenfamilien **17b**, 2nd edn., pp. 5–145. – Leipzig: Engelmann.

Friedel, J., 1938: Note sur la Structure anatomique du *Pteridophyllum racemosum* Sieb. et Zucc. – Bull. Soc. Bot. France **85**: 406–408.

Günther, K.-F., 1975a: Beiträge zur Morphologie und Verbreitung der *Papaveraceae*. 1. Teil: Infloreszenzmorphologie der *Papaveraceae*; Wuchsformen der *Chelidonieae*.– Flora **164**: 185–234.

– 1975b: Beiträge zur Morphologie und Verbreitung der *Papaveraceae*. 2. Teil: Die Wuchsformen der *Papaveraceae*, *Eschscholzieae* und *Platystemonoideae*.– Flora **164**: 393–436.

Hong, De-Yuan, 1993: Eastern Asian North American disjunctions and their biological significance. – Cathaya **5**: 1–39.

Hoot, S., Crane, P., 1995: Inter-familial relationships in the *Ranunculidae* based on molecular systematics. – Pl. Syst. Evol. [Suppl.] **9**: 119–131.

Hutchinson, J., 1921: The genera of *Fumariaceae* and their distribution. – Bull. Misc. Inf., Royal Bot. Gardens, Kew **3**: 97–115.

Jork, K. B., Kadereit, J. W., 1995: Molecular phylogeny of the Old World representatives of *Papaveraceae* subfam. *Papaveroideae* with special emphasis on the genus *Meconopsis* – Pl. Syst. Evol. [Suppl.] **9**: 171–180.

Kadereit, J. W., 1988: Sectional affinities and geographical distribution in the genus *Papaver* L. (*Papaveraceae*). – Beitr. Biol. Pfl. **63**: 139–156.

– 1993: *Papaveraceae*. – In Kubitzki, K., (Ed): The families and genera of vascular plants **2**, pp. 494–506. – Heidelberg: Springer.

– BLATTNER, F. R., JORK, K., SCHWARZBACH, A., 1994: Phylogenetic analysis of the
    Papaveraceae s.l. (incl. Fumariaceae, Hypecoaceae, and Pteridophyllum) based on mor-
    phological characters. – Bot. Jahrb. Syst. **116**: 361–390.

KARRER, A. B., 1991: Blütenentwicklung und systematische Stellung der Papaveraceae und
    Capparaceae. – Unpublished Ph.D. Dissertation, University of Zürich.

LAYKA, S., 1976: Les methodes modernes de la palynologie appliquées a l'étude des
    Papaverales. – Unpublished Dissertation, University of Montpellier.

LÉGER, L. J., 1894: Recherches sur l'appareil végétatif des Papavéracées (Papavéracées et
    Fumariacées D. C.). – Mém. Soc. Linn. Normandie **18**: 195–623.

LEHMINGER, R., 1985: Entwicklungsgeschichtliche Untersuchungen an Fumariaceen-Blüten.
    – Unpublished Diplomarbeit, University of Bonn.

– LEINS, P., 1985: Besonderheiten in der Entwicklungsgeschichte einiger Fumariaceen-
    Blüten. – In EHRENDORFER, F., (Ed.): 8. Symposium Morphologie, Anatomie und Sys-
    tematik. Zusammenfassungen; 58.

LEINS, P., ERBAR, C., 1985: Ein Beitrag zur Blütenentwicklung der Aristolochiaceae, einer
    Vermittlergruppe zu den Monokotylen. – Bot. Jahrb. Syst. **107**: 343–368.

– 1991: Entwicklungsmuster in Blüten und ihre mutmaßlichen phylogenetischen Zusam-
    menhänge. – Biol. Unserer Zeit **21**: 197–204.

LIDÉN, M., 1986: Synopsis of Fumarioideae (Papaveraceae) with a monograph of the tribe
    Fumarieae. – Opera Bot. **88**: 1–133.

– 1993a: Pteridophyllaceae. – In KUBITZKI, K., (Ed.): The families and genera of vascular
    plants **2**, pp. 556–557. – Heidelberg: Springer.

– 1993b: Fumariaceae. – In KUBITZKI, K., (Ed): The families and genera of vascular plants **2**,
    pp. 310–318. – Heidelberg: Springer.

MAIR, O., 1977: Zur Entwicklungsgeschichte monosymmetrischer Dicotylen-Blüten. –
    Dissertationes Botanicae **38**: 1–88.

MERXMÜLLER, H., LEINS, P., 1967: Die Verwandtschaftsbeziehungen der Kreuzblütler und
    Mohngewächse. – Bot. Jahrb. Syst. **86**: 113–129.

MEUSEL, H., 1969: Beziehungen in der Florendifferenzierung von Eurasien und Nord-
    amerika. – Flora **158**: 537–564.

MURBECK, S., 1912: Untersuchungen über den Blütenbau der Papaveraceen. – Kongl.
    Svenska Vetensk. Acad. Handl. **50**: 1–168.

PAYER, J. B., 1857: Traité d'organogénie comparée de la fleur. – Paris: V. Masson.

RAVEN, P. H., 1972: Plant species disjunctions: a summary. – Ann. Missouri Bot. Gard. **59**:
    234–246.

– AXELROD, D. I., 1974: Angiosperm biogeography and past continental movements. – Ann.
    Missouri Bot. Gard. **61**: 539–673.

RONSE DECRAENE, L. P., SMETS, E. F., 1990: The systematic relationship between Begoniaceae
    and Papaveraceae: a comparative study of floral development. – Bull. Jard. Bot. Belg. **60**:
    229–273.

– 1992a: An updated interpretation of the androecium of the Fumariaceae. – Canad. J. Bot.
    70: 1765–1776.

– 1992b: Complex polyandry in the Magnoliatae: definition, distribution and systematic
    value. – Nordic J. Bot. **12**: 621–649.

RYBERG, M., 1960: A morphological study of Fumariaceae and the taxonomical significance
    of the characters examined. – Acta Horti Bergiani **19**: 121–248.

SCHWARZBACH, A., KADEREIT, J. W., 1995: Rapid radiation of North American desert genera
    of the Papaveraceae: evidence from restriction site mapping of PCR-amplified
    chloroplast DNA fragments. – Pl. Syst. Evol. [Suppl.] **9**: 159–170.

STERN, K. R., 1961: Revision of Dicentra. – Brittonia **13**: 1–52.

VENT, W., MORY, B., 1973: Beiträge zur Kenntnis der Sippenstruktur der Gattungen *Glaucium* ADANS. und *Dicranostigma* HOOKER. & THOMSON (*Papaveraceae*). – Gleditschia **1**: 33–41.

WU, ZHENYI, 1983: On the significance of Pacific intercontinental discontinuity. – Ann. Missouri Bot. Gard. **70**: 577–590.

ZHUANG, XUAN, 1993: The taxonomic and evolution and distribution of *Papaveraceae*. – Acta Bot. Yunnanica **15**: 137–148.

Address of the authors: J. W. KADEREIT, F. R. BLATTNER, K. B. JORK, A. SCHWARZBACH, Institut für Spezielle Botanik und Botanischer Garten, Johannes Gutenberg-Universität, D-55099 Mainz, Federal Republic of Germany

Accepted November 21, 1994

Pl. Syst. Evol. [Suppl.] 9: 147–157 (1995)

Three intercontinental disjunctions in *Papaveraceae* subfamily *Chelidonioideae*: evidence from chloroplast DNA

FRANK R. BLATTNER and JOACHIM W. KADEREIT

Received October 25, 1994

Key words: *Papaveraceae*, *Chelidonioideae*, *Bocconia*, *Chelidonium*, *Eomecon*, *Hylomecon*, *Macleaya*, *Sanguinaria*, *Stylophorum*. – Biogeography, intercontinental disjunctions, cpDNA.

Abstract: An RFLP analysis of the chloroplast genome of *Papaveraceae* subfam. *Chelidonioideae* resulted in one most parsimonious tree consisting of three monophyletic groups. Each group contains intercontinental disjunct taxa. Whereas *Eomecon/Sanguinaria* and *Stylophorum* are examples for the well known E Asian – eastern N American disjunction, the E Asian – C and S American disjunction in *Macleaya/Bocconia* is rare. The genus *Stylophorum* in this analysis is paraphyletic. The N American *Stylophorum diphyllum* is sister group to the Old World members of *Stylophorum* and *Chelidonium*. All American representatives of the subfamily possess distinctive morphological features. This might be the result of Tertiary and Quaternary climatic oscillations leading to severe bottleneck situations in ancient N American populations.

The *Papaveraceae* s. str. (excl. *Pteridophyllum* SIEB. & ZUCC., *Hypecoum* L. and *Fumariaceae*) traditionally have been divided into four subfamilies: *Chelidonioideae*, *Eschscholzioideae*, *Papaveroideae* and *Platystemonoideae* (e.g., BRÜCKNER 1982, CRONQUIST 1981, ERNST 1962, TAKHTAJAN 1987). Certain changes to this have been suggested in a recent cladistic analysis of the *Papaveraceae* based on morphological characters (KADEREIT & al. 1994). This analysis resulted in a basal position of the *Chelidonioideae* within *Papaveraceae* s. str., and excluded *Glaucium* MILLER and *Dicranostigma* HOOK f. & THOMS. from the subfamily. The remaining seven genera of the *Chelidonioideae* s. str. are *Bocconia* L., *Chelidonium* L., *Eomecon* HANCE, *Hylomecon* MAXIM. (incl. *Coreanomecon* NAK.), *Macleaya* R.BR., *Sanguinaria* L., and *Stylophorum* NUTT. The *Chelidonioideae* in this circumscription share bicarpellate gynoecia [*Stylophorum diphyllum* (MICH.) NUTT. has three to five carpels], arillate seeds [not in *Macleaya microcarpa* (MAXIM) FEDDE], multicellular, terminally uniseriate hairs (*Eomecon* and *Sanguinaria* are glabrous), orange to red latex in all organs, and small floral bracts. Within the subfamily three monophyletic groups could be recognized by KADEREIT & al. (1994): (1) *Eomecon* and *Sanguinaria* are glabrous, white flowering herbs with palmately veined leaves, (2) *Macleaya* and *Bocconia* as wind pollinated shrubs and trees without petals, and (3) *Hylomecon*,

Chelidonium and *Stylophorum* as yellow flowering herbs with pinnate leaves and umbellate inflorescences. Each of these three groups contains a pair of intercontinental disjuncts. Whereas *Eomecon/Sanguinaria* and the species of *Stylophorum* show an E Asian – eastern N American disjunction, *Macleaya/Bocconia* are E Asian–C and S American disjuncts.

The E Asian – temperate N American floristic relationship has long been known (reviewed in BOUFFORD & SPONGBERG 1983). Since the time of GRAY (1859) many authors have studied this intercontinental distribution pattern on the basis of floristic comparisons (e.g., HARA 1972, HONG 1993, HU 1935, LI 1952, LITTLE 1983, SARGENT 1894, TIFFNEY 1985a, WU 1983), and up to 120 genera (WU 1983) have been listed as disjuncts. Phylogenetic studies of such disjuncts based on molecular evidence recently were published by, e.g., HOEY & PARKS (1991, 1994, *Liquidambar* L.), PARKS & WENDEL (1990, *Liriodendron* L.), and SHINWARI & al. (1994, *Disporum* SALISB. ex D. DON.). In contrast to the E Asian – N American disjunction a temperate E Asian–C and S American distribution is far less common in higher plants (SHARP 1972).

In search for further support for the morphology-based phylogenetic hypothesis by KADEREIT & al. (1994) we conducted an RFLP-analysis of the chloroplast genome. Particularly in *Bocconia* and *Macleaya*, which are quite dissimilar vegetatively, it seems possible that the shared reproductive characters related to wind-pollination are the result of convergent evolution rather than true homologies (LI 1952).

Material and methods

Species of all genera of *Papaveraceae* subfam. *Chelidonioideae* were cultivated in the Botanical Garden of the Johannes Gutenberg-Universität, Mainz, Germany. The taxa used in this study are: *Chelidonium majus* L., *Bocconia frutescens* L., *Eomecon chionantha* HANCE, *Glaucium flavum* CRANTZ, *Hylomecon vernalis* MAXIM., *Macleaya cordata* R.BR., *Sanguinaria canadensis* L., *Stylophorum diphyllum*, and *Stylophorum lasiocarpum* (OLIV.) FEDDE. The wild origin of all seeds is unknown except for *Chelidonium majus*, of which material was collected from wild populations in Mainz (BLATTNER 7–92). Voucher specimens of all investigated plant material are deposited in JGM.

Total DNAs were isolated from leaf tissue using the hot CTAB procedure of DOYLE & DOYLE (1987) as modified by SMITH & al. (1991). The DNAs were digested with 7 restriction enzymes (ApaI, AvaI, BamHI, CfoI, SacII, SmaI, XhoI) and electrophoresed in 0.6% to 1.2% agarose gels. The DNA fragments were transferred to Hybond-N nylon filters (Amersham) with a vacuum blotting system (Pharmacia), and fixation was carried out via UV cross linking on a transilluminator at 312 nm wavelength. The filters were sequentially hybridized with 40 cloned tobacco cpDNA probes (T1 to T40, kindly provided by K. J. SYTSMA, originally developed by J. D. PALMER) covering the complete chloroplast genome. The probes were ^{32}P labelled with a random primed DNA labelling kit (USB). The protocols of prehybridization, hybridization and washes follow SYTSMA & SCHAAL (1985). After filter probing the filters were exposed for 24 to 48 hours to a Kodak XAR film. Restriction site differences among the species were coded for phylogenetic analysis. A list of characters is available upon request from the authors.

The analysis of cpDNA restriction site mutations was performed with the computer program PAUP 3.1.1 (SWOFFORD 1993). Wagner parsimony (KLUGE & FARRIS 1969, FARRIS

1970) and the "exhaustive" option were used to find the shortest possible Wagner tree. Bootstrap values (FELSENSTEIN 1985) with 1000 replicates (branch-and-bound search algorithm) and decay indices (BREMER 1988, DONOGHUE & al. 1992) were calculated. Because the variation of restriction sites in *Pteridophyllum*, the basal taxon in *Papaveraceae* s.l. (KADEREIT & al. 1994; HOOT 1995, this volume), was generally too high for direct comparison with the *Chelidonioideae*, *Glaucium* was used as outgroup.

Results

The size of the chloroplast genome in the taxa examined ranges between c. 152 to 155 kb. Approximate values for the four cpDNA regions are: large single copy 78 kb, small single copy 22 kb, and inverted repeat 27 kb. The genome is mostly colinear with that of *Nicotiana* L. except for *Eomecon* and *Sanguinaria* which share a 2 kb inversion in the inverted repeat (tobacco probes T29a to T30).

Approximately 250 restriction sites were detected in the cpDNA of each of the species examined, representing 1.5 kb or c. 1% of the chloroplast genome. 67 mutational differences could be mapped, of which 35 were informative. The Wagner parsimony analysis resulted in one most parsimonious tree of 58 steps length (Fig. 1) with a consistency index of 0.72 (excluding uninformative characters). Dollo parsimony generated the same tree topology. The cpDNA tree contains three major clades identical to those obtained in the analysis of morphological characters (Fig. 2). In this tree *Sanguinaria* and *Eomecon* occupy a basal position and are sister

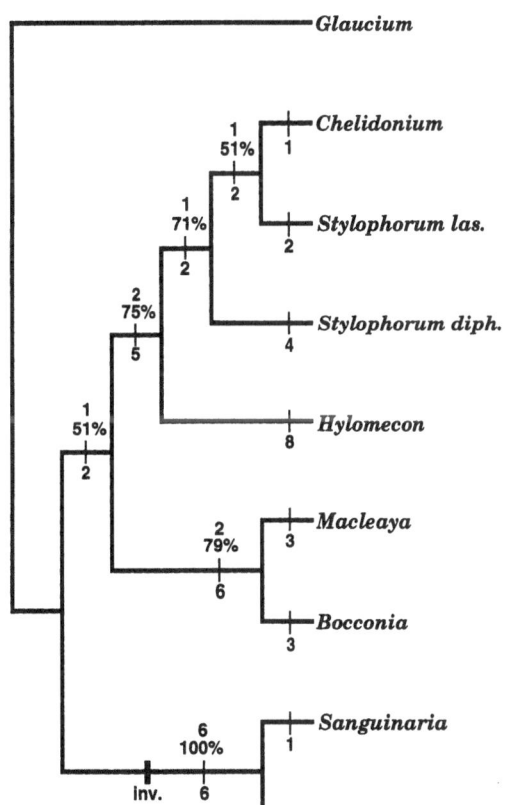

Fig. 1. Most parsimonious tree obtained from cpDNA RFLP-data (90 steps, CI 0.72, excluding uninformative characters). The number of unambiguous character changes is indicated below the branches, bootstrap values (%) and decay indices above. *Inv.* indicates a 2 kb inversion shared by *Eomecon* and *Sanguinaria*

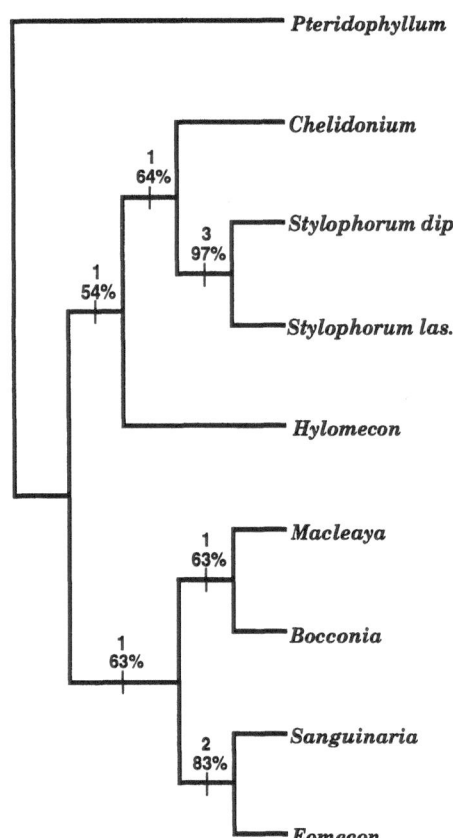

Fig. 2. Topology of the most parsimonious tree based on morphological characters of subfam. *Chelidonioideae* as published in KADEREIT & al. (1994). Bootstrap values (%) and decay indices are indicated

group to the remaining taxa. *Sanguinaria* and *Eomecon* differ by only one mutation. *Bocconia* and *Macleaya* are sister group to the yellow flowering herbs *Hylomecon*, *Stylophorum*, and *Chelidonium*. Six mutational differences between *Macleaya* and *Bocconia* were found. *Stylophorum* is paraphyletic. *Stylophorum diphyllum* is sister group to the Chinese *Stylophorum lasiocarpum* plus *Chelidonium*. The two species of *Stylophorum* are separated by eight, and *Stylophorum lasiocarpum* from *Chelidonium* by three mutations. The bootstrap values for the three main groups (*Hylomecon/ Stylophorum/Chelidonium* 75%, *Macleaya/Bocconia* 79%, and *Eomecon/Sanguinaria* 100%) are higher than those of the branches defining the position of *Macleaya/Bocconia* in relation to the other genera (51%), and those within the paraphyletic *Stylophorum* (51%).

Discussion

The cpDNA tree (Fig. 1) supports the three major groups found in the morphological analysis of KADEREIT & al. (1994). In view of this observation it seems likely that the morphologically obvious groups are natural entities and not the result of convergent evolution.

The first difference between the cpDNA tree and the tree based on morphological characters is the position of the *Macleaya/Bocconia* clade. In this study these two genera are sister group to the yellow flowering herbs, whereas morphological characters support a sister group relationship to *Eomecon/Sanguinaria* (Fig. 2). The weak

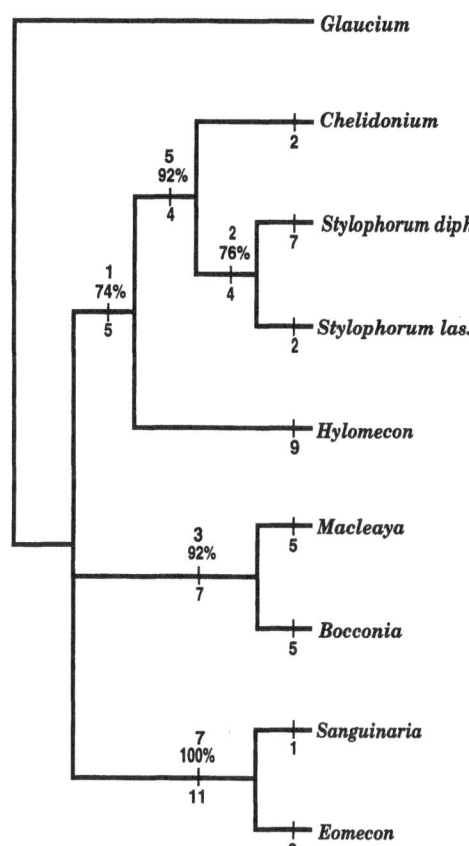

Fig. 3. Strict consensus tree of three most parsimonious Wagner trees (89 steps, CI 0.71) resulting from the combination of morphological and cpDNA characters in a single data matrix. For numbers see Fig. 1

support for the position of *Macleaya/Bocconia* as indicated by the bootstrap and decay values (Fig. 1) does not allow us to decide between the two alternative topologies. The analysis of the combined data sets resulted in three most parsimonious trees. The consensus cladogram from these three trees failed to resolve the relationships among the three major clades and shows a basal trichotomy (Fig. 3). The inconsistency between the two data sets might be the result of the use of different outgroups in the two analyses. In the context of this study the poor resolution in the basal part of the tree is irrelevant because the pairs of intercontinental disjuncts are supported as monophyletic groups in both studies and in the consensus tree of the combined data.

The second inconsistency between the two data sets concerns the position of *Chelidonium*. In the morphological analysis *Chelidonium* is sister group to *Stylophorum* (Fig. 2). In this study *Stylophorum diphyllum* is sister group to *Chelidonium* plus the Old World *Stylophorum lasiocarpum*. This would make *Stylophorum* paraphyletic (Fig. 1). In the analysis of the combined data sets *Stylophorum* is a monophyletic genus (Fig. 3). Morphologically *Stylophorum* and *Chelidonium* are very similar (GÜNTHER 1975), and can be distinguished mainly by the absence of nearly opposite leaves below the umbellate inflorescences and glabrous instead of hairy fruits in *Chelidonium*. The two genera are further united by their alkaloid contents (SLAVÍK & al. 1991) and the occurrence of chelidonic acid, which as far as tested has not yet been found in any other species of the subfamily or the family (RAMSTADT 1953; BLATTNER, unpubl.). The similarity in morphology, secondary chemistry and cpDNA structure does not contradict the position of *Chelidonium* within *Stylophorum*. Based on the results shown in Figs. 2 and 3, however, we prefer to regard *Stylophorum* as a monophyletic genus separate from the closely related *Chelidonium*. This hypothesis is supported by the results of Brückner (1982, 1983), who found that characters of fruit and seed morphology and anatomy clearly separate *Chelidonium* from *Stylophorum*.

In summary we can regard the monophyly of the intercontinental disjuncts in subfam. *Chelidonioideae* as well supported by the results of the RFLP analysis of the chloroplast genome. *Macleaya/Bocconia* are distributed in E Asia and C and S America (Fig. 4). *Eomecon/Sanguinaria* show an E Asian–eastern N American distribution pattern which is very similar to the distribution of the genus *Stylophorum* (Fig. 5). If *Stylophorum* should be paraphyletic, the pattern of geographical disjunction depends on the sister group relationship of the second Asian species, *Stylophorum sutchuense* (Franch.) Fedde. In case this species (not included in this study) should be sistergroup to *S. diphyllum*, an E Asian-eastern N American disjunction would exist also within a paraphyletic *Stylophorum*. If, as suggested by morphological similarity, *S. sutchuense* should be sister group to *S. lasiocarpum/Chelidonium*, however, the disjunction would be eastern N American–Eurasian (*S. sutchuense/S. lasiocarpum/Chelidonium*). Such a phylogeny would require a different phytogeographical interpretation (see below) than the assessment of *Stylophorum* as monophyletic.

Eomecon/Sanguinaria as well as *Stylophorum* fit into the commonly found distribution pattern of plants growing in mesophytic forests of Asia and eastern N America (Li 1952, Hong 1993). The omission of *Eomecon/Sanguinaria* from all lists of E Asian–eastern N American disjuncts clearly illustrates that phytogeographical comparison should rely on phylogenetic analysis rather than on the comparison of generic names contained in floristic lists. Most herbaceous species involved in this disjunction are forest dwellers and grow in damp and shady woods.

Fig. 4. Geographical distribution of the two intercontinental disjunct genera *Macleaya* and *Bocconia*

Fig. 5. Geographical distribution of the E Asian–eastern N American disjunct taxa *Eomecon/Sanguinaria* and *Stylophorum*

Members of this vegetational unit are best interpreted as relics of a once widely distributed geoflora which covered the northern continents at appropriate latitudes during the Miocene. This temperate flora might have arisen from a more tropical Eocenean precursor when the climate deteriorated in the late Oligocene (McCartan & al. 1990, Smiley & Rember 1981). During the Eocene a connection between Europe and North America still existed. This North Atlantic land bridge was interrupted in the late Eocene by climatic cooling and the opening of the North Atlantic as the result of plate tectonical events. In the Oligocene the C Asian Turgai Strait closed and connected Europe with E Asia. The existence of a truly continuous geoflora depended on the negotiability of the Eocenean Bering land bridge, connecting Alaska with NE Asia (Tiffney 1985b). The Turgai Strait and Bering land bridge make it unlikely that the Eocene geoflora ever formed a continuous vegetation belt on the northern continents. This applies even more to the heavily fragmented Miocene flora which was more similar to the extant vegetation of eastern N America and E Asia than to the tropical Eocene plant communities. The Middle Miocene (15 mybp) might have been the last opportunity for members of a deciduous forest flora to migrate between Eurasia and N America via Beringia (Parks & Wendel 1990) before the climate there became too cool for migration. For the last 10 million years long distance dispersal between the two continents would have to be postulated as the only possibility for gene flow. The cooling cycles in the Late Tertiary and the Quaternary influenced the temperate floras particularly in Europe, where many taxa became extinct, and led to the disjunct distribution we observe today. In case *Chelidonium* should be nested within a paraphyletic *Stylophorum* this group would, much in contrast to *Eomecon/Sanguinaria* and *Macleaya/Bocconia*, have been spared from such extinction.

This short summary of geological and climatic changes during the Tertiary shows that the disjunctions observed might be of different age. Common ancestors might have been living in the tropical Eocene or the temperate Miocene, and might have migrated along land bridges or island chains, or via long distance dispersal. In the *Papaveraceae*, where except for pollen grains of *Dendromecon* BENTH. (AXELROD 1950) and a capsule similar to those found in subfam. *Papaveroideae* (SMITH & HICKEY 1994; SMITH, pers. comm.) no fossil record exists, reconstruction of migration routes is difficult. The difference in numbers of mutations found between *Eomecon* and *Sanguinaria* (one) and between the species of *Stylophorum* (eight) may indicate a different age of these disjunctions if one assumes equal mutation rates in both lineages.

The Asian and Latin American distribution of *Macleaya* and *Bocconia* is unusual. On the basis of the extant distribution of the family its northern hemispherical origin can be postulated (RAVEN & AXELROD 1974). The distribution of *Bocconia* should be regarded as derived from the disjunction pattern found in *Eomecon/Sanguinaria* and *Stylophorum*. We assume a "normal" E Asian–N American distribution of the ancestors of *Macleaya* and *Bocconia* and subsequent immigration of *Bocconia* or the *Bocconia* ancestor into the West Indies and S America. A direct land connection between the two Americas exists since the Pliocene (5 mybp). Before that volcanic islands were present in the position of today's Costa Rica and Panama. In the Early Eocene a land bridge formed between S Mexico and Haiti, but this again was disrupted in the Middle Eocene. The complete island chain of the West Indies rose above sea level in the Miocene. The upheaval of the Andes started in the Eocene in northern Colombia, and the main uplift took place in the Miocene and Pliocene (WINKLER 1990). The distribution of *Bocconia* in tropical highlands suggests that this taxon may be a descendant of Eocenean ancestors living in tropical climates. The comparatively large number of species (c. 10) in *Bocconia* might be the result of its immigration into the Caribbean Islands followed by speciation in island habitats. The red and fleshy seed aril of *Bocconia*, an adaptation to bird dispersal, may have supported movement along the islands of the Caribbean arc and into the Andean tropical highlands. After the immigration into Central and South America the North American ancestor of *Bocconia* became extinct.

In *Magnolia* L. and *Nyssa* L. other examples for this distribution pattern occur. *Magnolia* sect. *Rhytidospermum* is distributed in E Asia (3 species), eastern N America (8 species), and SE Mexico (1 species). *Magnolia* sect. *Theorhodon* occurs with 16 species in C and S America and with one species (*M. grandiflora* L.) in Florida (LITTLE 1983, VÁZQUEZ-G. 1994). The disjunctions found in *Magnolia* thus are tropical N and northern S American (sect. *Theorhodon*) and temperate N American–E Asian (sect. *Rhytidospermum*).

In *Nyssa* L. the closest relative of the C American representative is an E Asian species rather than the geographically closer N American representatives of the genus (WEN & STUESSY 1993). In both genera fossils from N America and Europe reveal the once much wider distribution. Already ENGLER (1879) recognized this eastern N American–C American disjunction represented by temperate elements in the C American flora.

The interpretation of *Macleaya/Bocconia* as old (Eocene) disjuncts might explain the high morphological divergence of these two genera. *Bocconia* and *Macleaya* differ in leaf shape, fruit anatomy and, very conspicuously, the woodiness of

Bocconia in contrast to the herbaceous habit of *Macleaya*. On the other hand the difference between the two genera on the molecular level (six mutations) is smaller than in *Stylophorum* (eight mutations). Accordingly the possibility of a more recent immigration (5 mybp) into tropical C and S America of a temperate ancestor should not be excluded until fossil evidence may decide this question. The discovery of fossil *Bocconia* seems not unlikely because this genus contains trees with large seeds (up to 5 mm) and ample pollen production (anemophily).

It is remarkable that in all three cases of disjunction described above the New World representatives are morphologically divergent. *Sanguinaria* possesses eight to 12 petals, which is unique in the family, *Stylophorum diphyllum* has fruits with three to five instead of two carpels as found in all other members of the subfamily, and *Bocconia* is one of the very few woody representatives in this mainly herbaceous family.

The woodiness of *Bocconia* can be interpreted as an adaptation to its climatically constant tropical habitats where trees have ecological advantages. It is more difficult to find an explanation for the aberrant morphology of *Sanguinaria* and *Stylophorum diphyllum*. Evolutionary theory postulates that the fixation of rare characters most easily happens in small populations. Small populations could result either from founder events or from bottleneck situations. The founder event hypothesis in our opinion is less likely (at least if it assumes long distance immigration from E Asia) because *Sanguinaria* and *Stylophorum diphyllum* both are not isolated elements but grow in plant communities which are very similar to their E Asian counterparts. It seems more likely that the oscillating climatic changes during the Late Tertiary and Quaternary affected eastern N America more than E Asia, leading to severe bottleneck situations. These climatic/geographic events might have been the same which caused the extinction of many other taxa in Europe and N America which today exist only in E Asia [e.g., *Cercidiphyllum* SIEB. & ZUCC. (WOLFE 1972), *Pterocarya* KUNTH., or *Zelkova* SPACH. (MCCARTAN & al. 1990)], or show a SE Asian–C and S American distribution [*Symplocos* JACQ. (KRUTZSCH 1989)], but are well known from the European or N American fossil record.

We thank Dr AARON LISTON, Corvallis, Oregon State University, for his critical review of the manuscript. The molecular work of the authors is supported by the German Research Foundation (DFG).

References

AXELROD, D. I., 1950: Evolution of desert vegetation in western North America. – Publ. Carnegie Inst. Washington **590**: 215–306.

BOUFFORD, D. E., SPONGBERG, S. A., 1983: Eastern Asian – eastern North American phytogeographical relationships–a history from the time of LINNAEUS to the twentieth century. – Ann. Missouri Bot. Gard. **70**: 423–439.

BREMER, K., 1988: The limits of amino acid sequence data in angiosperm phylogenetic reconstructions. – Evolution **42**: 795–803.

BRÜCKNER, C., 1982: Zur Kenntnis der Fruchtmorphologie der *Papaveraceae* JUSS. s.str. und der *Hypecoaceae* (PRANTL & KÜNDIG) NAK. – Feddes Rep. **93**: 153–212.

– 1983: Zur Morphologie der Samenschale in den *Papaveraceae* JUSS s.str. und *Hypecoaceae* (PRANTL & KÜNDIG) NAK. – Feddes Rep. **94**: 361–405.

CRONQUIST, A., 1981: An integrated system of flowering plants. – New York: Columbia University Press.

DONOGHUE, M. J., OLMSTEAD, R. G., SMITH, J. F., PALMER, J. D., 1992: Phylogenetic relationships of *Dipsacales* based on *rbc*L sequences. – Ann. Missouri Bot. Gard. **79**: 333–345.

DOYLE, J. J., DOYLE, J. L., 1987: A rapid DNA isolation procedure for small amounts of fresh leaf tissue. – Phytochem. Bull. **19**: 11–15.

ENGLER, A., 1879: Versuch einer Entwicklungsgeschichte der Pflanzenwelt, insbesondere der Florengebiete seit der Tertiärperiode, 1. die extratropischen Gebiete der nördlichen Hemisphäre. – Leipzig: Engelmann.

ERNST, W. R., 1962: A comparative morphology of the *Papaveraceae*. – Ph.D. Thesis Stanford University.

FARRIS, J. S., 1970: Methods for computing Wagner trees. – Syst. Zool. **19**: 83–92.

FELSENSTEIN, J., 1985: Confidence limits on phylogenies: an approach using the bootstrap. – Evolution **39**: 783–791.

GRAY, A., 1859: Diagnostic characters of phanerogamous plants, collected in Japan by Charles Wright, botanist of the U.S. North Pacific Exploring Expedition, with observations upon the relations of Japanese flora to that of North America, and of other parts of the northern temperate zone. – Mem. Amer. Acad. Arts sci. n.s. **6**: 377–453.

GÜNTHER, K.-F., 1975: Beiträge zur Morphologie und Verbreitung der *Papaveraceae*. 1. Teil: Infloreszenzmorphologie der *Papaveraceae*; Wuchsformen der *Chelidonieae*. – Flora **164**: 185–234.

HARA, H., 1972: Corresponding taxa in North America, Japan and the Himalayas. – In VALENTINE, D. H., (Ed.): Taxonomy, Phytogeography, and Evolution, pp. 61–72. – London: Academic Press.

HOEY, M. T., PARKS, C. R., 1991: Isozyme divergence between Eastern Asian, North American, and Turkish species of *Liquidambar* (*Hamamelidaceae*). – Amer. J. Bot. **78**: 938–947.

– – 1994: Genetic divergence in *Liquidambar styraciflua*, *L. formosana*, and *L. acalycina* (*Hamamelidaceae*). – Syst. Bot. **19**: 308–316.

HONG, DE-YUAN, 1993: Eastern Asian–North American disjunctions and their biological significance. – Cathaya **5**: 1–39.

HOOT, S. B., CRANE, P. R., 1995: Inter-familial relationships in the *Ranunculidae* based on molecular systematics. – Pl. Syst. Evol. [Suppl.] **9**: 119–131.

HU, H. H., 1935: A comparison of the ligneous flora of China and eastern North America. – Bull. Chinese Bot. Soc. **1**: 79–97.

KADEREIT, J. W., BLATTNER, F. R., JORK, K. B., SCHWARZBACH, A. 1994: Phylogenetic analysis of the *Papaveraceae* s.l. (incl. *Fumariaceae*, *Hypecoaceae*, and *Pteridophyllum*) based on morphological characters. – Bot. Jahrb. Syst. **116**: 361–390.

KLUGE, A. G., FARRIS, J. S., 1969: Quantitative phyletics and the evolution of anurans. – Syst. Zool. **18**: 1–32.

KRUTZSCH, W., 1989: Paleogeography and historical phytogeography (paleochorology) in the Neophyticum. – Pl. Syst. Evol. **162**: 5–61.

LI, HUI-LIN, 1952: Floristic relationships between Eastern Asia and eastern North America. – Trans. Amer. Phil. Soc. **42**: 371–429.

LITTLE, E. L. Jr., 1983: North American trees with relationships in Eastern Asia. – Ann. Missouri Bot. Gard. **70**: 605–615.

McCARTAN, L., TIFFNEY, B. H., WOLFE, J. A., AGAR, T. A., WING, S. L., SIRKIN, L. A., WARD, L. W., BROOKS, J., 1990: Late Tertiary floral assemblage from upland gravel deposits of the southern Maryland Coastal Plain. – Geology **18**: 311–314.

PARKS, C. R., WENDEL, J. F., 1990: Molecular divergence between Asian and North American species of *Liriodendron* (*Magnoliaceae*) with implications for interpretation of fossil floras. – Amer. J. Bot. **77**: 1243–1256.

RAMSTADT, E., 1953: Über das Vorkommen und die Verbreitung von Chelidonsäure in einigen Pflanzenfamilien. – Pharm. Act. Helvet. **28**: 45–57.

RAVEN, P. H., AXELROD, D. I., 1974: Angiosperm biogeography and past continental movements. – Ann. Missouri Bot. Gard. **61**: 539–673.

SARGENT, C. S., 1894: Forest flora of Japan: notes on the forest flora of Japan. – Boston: Houghton & Mifflin.

SHARP, A. J., 1972: The possible significance of some exotic distributions of plants occuring in Japan and/or North America. – In GRAHAM, A., (Ed.): Floristic and palaeofloristic of Asia and eastern North America, pp. 61–64. – Amsterdam: Elsevier.

SHINWARI, Z. K., TERAUCHI, R., UTECH, F. H., KAWANO, S., 1994: Recognition of the New World *Disporum* section *Prosartes* as *Prosartes* (*Liliaceae*) based on sequence data of the *rbc*L gene. – Taxon **43**: 353–366.

SLAVÍK, J., HANUŠ, V., SLAVÍKOVÁ, L., 1991: Alkaloids from *Stylophorum lasiocarpum* (OLIV.) FEDDE. – Collect. Czech. Chem. Commun. **56**: 1116–1122.

SMILEY, C. J., REMBER, W. C., 1981: Paleoecology of the Miocene Clarkia Lake (northern Idaho) and its environs. – In BORECOT, A. J., GRAY, J., BERRRY, W. B. N., (Eds): Communities of the past, pp. 551–590. – Stroudsburg: Dowden, Hutchinson & Ross.

SMITH, J. F., SYTSMA, K. J., SHOEMAKER, J. S., SMITH, R. L., 1991: A qualitative comparison of total cellular DNA extraction protocols. – Phytochem. Bull. **23**: 2–9.

SMITH, U. R., HICKEY, L. J., 1994: A capsular fruit from the latest Cretaceous of western interior United States. – Abstr. Amer. J. Bot. **81**: 102–103.

SWOFFORD, D. L., 1993: PAUP: phylogenetic analysis using parsimony, vers. 3.1.1. – Champaign, Ill.: Natural History Survey.

SYTSMA, K. J., SCHAAL, B. A., 1985: Genetic variation, differentiation, and evolution in a species complex of tropical shrubs bases on isozymic data. – Evolution **39**: 582–593.

TAKHTAJAN, A. L., 1987: Systema Magnoliophytorum. – Leningrad: Nauka.

TIFFNEY, B. H., 1985a: Perspectives on the origin of the floristic similarity between Eastern Asia and eastern North America. – J. Arnold Arbor. **66**: 73–94.

– 1985b: The Eocene North Atlantic land bridge: its importance in Tertiary and modern phytogeography of the northern hemisphere. – J. Arnold Arbor. **66**: 243–273.

VÁZQUEZ-G., J. A., 1994: *Magnolia* in Mexico and C-America: a synopsis. – Brittonia **46**: 1–23.

WEN, JUN, STUESSY, T. F., 1993: The phylogeny and biogeography of *Nyssa* (*Cornaceae*). – Syst. Bot. **18**: 68–79.

WINKLER, S., 1990: Zur Evolution der Gattung *Tillandsia* L. [The evolution of the genus *Tillandsia* L.] – Bot. Jahrb. Syst. **112**: 43–77.

WOLFE, J. A., 1972: An interpretation of Alaskan Tertiary floras. – In GRAHAM, A., (Ed.): Floristic and palaeofloristic of Asia and eastern North America, pp. 201–233. – Amsterdam: Elsevier.

WU, ZHENGYI, 1983: On the significance of the pacific intercontinental discontinuity. – Ann. Missouri Bot. Gard. **70**: 577–590.

Address of the authors: FRANK R. BLATTNER and JOACHIM W. KADEREIT, Institut für Spezielle Botanik, Johannes Gutenberg-Universität Mainz, D-55099 Mainz, Germany.

Accepted December 14, 1994

Pl. Syst. Evol. [Suppl.] 9: 159–170 (1995)

Rapid radiation of North American desert genera of the *Papaveraceae*: Evidence from restriction site mapping of PCR-amplified chloroplast DNA fragments

Andrea E. Schwarzbach and Joachim W. Kadereit

Received October 31, 1994

Key words: *Papaveraceae, Arctomecon, Argemone, Canbya, Platystemon, Romneya,*–Rapid radiation, molecular evolution, morphological evolution, cpDNA, polymerase chain reaction (PCR).

Abstract: Phylogenetic relationships of a group of North American desert genera of the *Papaveraceae* subfam. *Papaveroideae* and *Platystemonoideae* were investigated with an RFLP analysis of three PCR-amplified chloroplast genome regions. In agreement with earlier results it was found that subfam. *Platystemonoideae* is nested within subfam. *Papaveroideae*. The group under study is characterized by a large number of generic autapomorphies but only few informative synapomorphies. This is interpreted as strong evidence for a rapid radiation event caused by major climatic changes in the past. Poor phylogenetic resolution seems to reflect biological reality and not to be an experimental artifact. There is also evidence for accelerated morphological evolution under arid climatic conditions.

The *Papaveraceae* s. str. currently comprise 23 genera and approximately 240 species of mostly temperate northern hemispherical distributed. The family according to Kadereit & al. (1994) can be divided into four monophyletic groups, the *Chelidonioideae, Eschscholzioideae, Glaucium* Miller/*Dicranostigma* Hook. f. & Thoms. (traditionally regarded as part of the *Chelidonioideae*, Ernst 1962) and *Papaveroideae/Platystemonoideae* (traditionally treated as two subfamilies, Ernst 1962). Whereas *Eschscholzioideae* and *Platystemonoideae* are endemic to western North America, three of seven genera of the *Chelidonioideae* and six of eight genera of the *Papaveroideae* are at least partly distributed in the New World.

The subject of the present study is the evolutionary history of a closely related group of desert genera of the *Papaveroideae* from the American Southwest. This group contains *Arctomecon* Torr. & Frém., *Argemone* L., *Canbya* Parry ex. A. Gray and *Romneya* Harv., and also *Platystemon* Benth. as a member of the former *Platystemonoideae*. The close affinities among these genera had already been shown in a phylogenetic analysis of morphological characters of the family (Kadereit & al. 1994). They are clearly separated from a monophyletic clade of the subfamily including the Old World genera *Papaver* L., *Meconopsis* VIG. and *Roemeria* MEDIK., but also the New World *Papaver californicum* A. GRAY and *Stylomecon heterophylla* (BENTH.) G. TAYLOR. The latter two species are, despite their New World distribution, not closely related to the rest of the New World genera of the *Papaveroideae* (JORK & KADEREIT 1995, this volume). In contrast to the Madro-Tertiary elements *Arctomecon*, *Argemone*, *Canbya* and *Romneya*, *Stylomecon heterophylla* had been suggested to be derived from the Arcto-Tertiary flora already by STEBBINS & MAJOR (1965). *Argemone*, *Arctomecon*, *Canbya*, *Romneya* and the genera of the former *Platystemonoideae* are morphologically highly divergent.

The genus *Argemone* contains 32 species (OWNBEY 1958, 1961; POWELL 1972; JOHNSTON 1976; SORARÚ 1976; MCDONALD 1991) which are distributed mainly in western North America. *Argemone* comprises annual or perennial herbs or rarely shrubs with commonly pinnately incised leaves. All species have three conspicuous sepals terminated by mostly elongated apical sepal horns. The perianth is 3-merous, the number of stamens is large and the paracarpous gynoecium is formed by 3–7 carpels. All parts of the plants are mostly distinctly spiny (OWNBEY 1958, 1961).

Arctomecon is distributed with three species in the Mohave desert (JANISH 1977, NELSON & WELSH 1993). The species share an identical and very unique leaf shape: The leaves are more or less parallel-veined and lobed at the end, each lobe is terminated by a stiff prickle, and the surface is covered with a dense fur of stiff long hairs. The leaves are arranged in a basal rosette. The species can be easily distinguished by their flower colour, the number of petals and the shape of the inflorescences (ERNST 1962, NELSON & WELSH 1993).

The woody *Romneya* contains two species (CLARK 1993). These shrubs have very large white 3-merous flowers with a large number of stamens (up to 700). The genus occurs in coastal sageshrub communities in Southern California and adjacent Baja California.

The genus *Canbya* consists of two ephemeral species (ERNST 1962, 1967). One white-flowering species is endemic to the Mohave desert of California (CLARK 1993), and the other, yellow-flowering one, is distributed in the sagebrush-plains of southeastern Oregon and northwestern Nevada. The annual plants never exceed five cm in height. Vegetatively, they are characterized by linear leaves forming a small rosette. Their 3-merous flowers have a reduced androecium of mostly 12, nine or six stamens. The gynoecium consists of three carpels and the fruit is a capsule opening basipetally with valves (ERNST 1967).

Finally, *Platystemon*, traditionally part of subfam. *Platystemonoideae* (ERNST 1962, 1967), is a monotypic genus, also with linear leaves and 3-merous flowers. The genus is characterized by very unique fruits. They consist of 5–20 carpels which at maturity dissociate and break into one-seeded nutlets. *Platystemon* is widely distributed in the Southwest of North America (ERNST 1967).

Meconella NUTT. with three species and the monotypic *Hesperomecon linearis* (BENTH.) GREENE also belong to the former *Platystemonoideae* (ERNST 1967) but were not available for our molecular study.

The high degree of morphological divergence among the genera mentioned above made it difficult to find phylogenetically informative morphological characters. Accordingly, the hypothesis about phylogenetic affinities in this group of genera presented by KADEREIT & al. (1994) is not very well supported.

In order to circumvent this problem a molecular study was undertaken using restriction site length polymorphisms of PCR amplified chloroplast genome fragments. The PCR method makes possible the detailed restriction site comparison of DNA regions with levels of variability adequate for the taxonomic problem at hand (ARNOLD & al. 1991, LISTON 1992, RIESEBERG & al. 1992). Three different chloroplast regions were used which all have been shown to be variable enough to achieve resolution even at species level in other taxonomic groups (LISTON 1992, LISTON & KADEREIT in press, JOHNSON & SOLTIS 1994, STEELE & VILGALYS 1994, TABERLET & al. 1991, FANGAN & al. 1994). The aim of this analysis was the identification of the phylogenetic relationships within and among the above-mentioned genera.

Material and methods

Plant material. All genera of subfam. *Papaveroideae*, one member of former subfam. *Platystemonoideae* and the genus *Glaucium* were examined. Altogether 42 species were scored and included in the phylogenetic analysis. For the North American desert group which will be discussed here, two (of altogether 32) species of *Argemone*, three (3) of *Arctomecon*, two (2) of *Romneya* and one (2) of *Canbya* were included in the study (Table 1). The remaining 32 taxa investigated are listed in JORK & KADEREIT (1995, this volume). Based on the analysis of morphological characters of the *Papaveraceae* (KADEREIT & al. 1994), *Glaucium* (formerly subfam. *Chelidonioideae*) was used as outgroup because of its sister group relationship to *Papaveroideae/Platystemonoideae*, and *Platystemon* (subfam. *Platystemonoideae*) was included because of its possible close affinity to *Canbya*.

Molecular techniques. Total DNAs were isolated using the method of DOYLE & DOYLE (1987) modified by the addition of sodium metabisulfite (1–2% w/v) to the CTAB buffer. PCR amplification reaction mixtures (100 µl) contained: 50 mM KCl, 10 mM Tris-HCl (pH 9.0), 0.1% Triton X-100, 2.5 mM $MgCl_2$, 0.2 mM each of the dNTPs, 50 pmol of each primer, and 2.5 units of Taq DNA polymerase (Promega Corporation, Madison, WI). Amplification was carried out using a Grant Autogene II (Cambridge, UK) thermocycler programmed for 3 min at 94 °C, followed by 37 cycles of 1 min at 94 °C, 1 min at 52 °C, and 3.5 min at 72 °C, with a final 8 min at 72 °C.

Table 1. Species used in RFLP study. Vouchers: *AS* A. SCHWARZBACH, *FB* F. BLATTNER, *KJ* KIRSTIN JORK & J. W. KADEREIT. All vouchers are deposited at JGM, Germany. Collection permits for endangered plant species are available upon request from A. S.

Species	Voucher	Locality (State: County)
Arctomecon californica TORR. & FRÉM.	AS 93–78	Nevada: Clark
A. humilis COVILLE	AS 93–66	Utah: Washington
A. merriamii COVILLE	AS 93–76	Nevada: Clark
Argemone corymbosa subsp. *arenicola* G. B. OWNBEY.	AS 93–54	Arizona: Coconino
A. polyanthemos (FEDDE) G. B. OWNBEY.	AS 93–25	New Mexico: Dona Ana
Canbya aurea S. WATS.	AS 93–83	Nevada: Humboldt
Glaucium squamigerum KAR. & KIR.	FB 122	Cultivated Bot. Garden Mainz
Platystemon californicus BENTH.	KJ 91/41	Cultivated Bot. Garden Mainz
Romneya coulteri HARV.	AS 93–7	Bot. Garden Berkeley 50.1640
R. trichocalyx EASTW.	AS 93–5	Bot. Garden Berkeley 65.1048

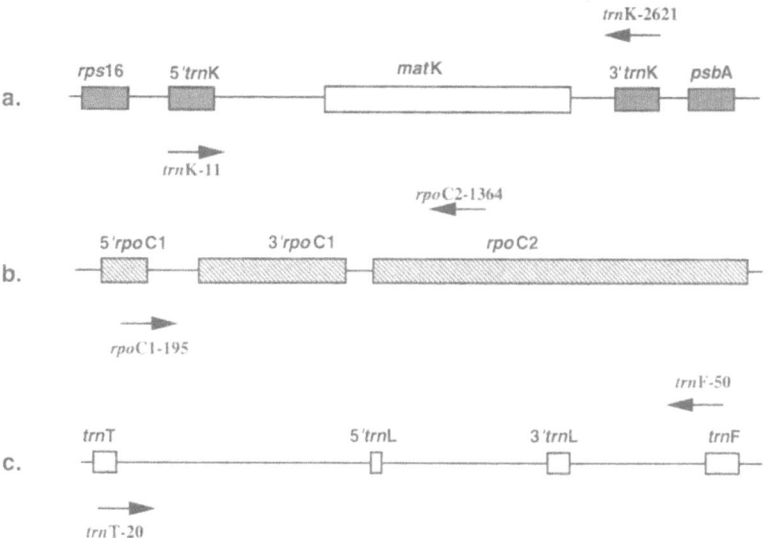

Fig. 1. cpDNA regions used for amplification. *a* Map of the *trn*K-gene including *mat*K (2,500 bp), *b* map of the *rpo*C-region (4,100 bp) and *c* map of the *trn*L-region (1,700 bp). Boxed areas represent coding regions. Arrows indicate the position and direction of the primers used in this study

Molecular characters were obtained from restriction fragment length polymorphisms (RFLPs) of PCR amplified chloroplast genome fragments. Three different regions from within the large single copy region of the chloroplast genome were used (Shinozaki & al. 1986): (1) *trn*K, the tRNA gene for lysine including *mat*K (Neuhaus & Link 1987, Liston & Kadereit 1995) which is 2500 bp long. (2) *rpo*C, a ca. 4100 bp region encompassing the chloroplast genes RNA polymerase Cl and partly RNA polymerase C2 including an intron in *rpo*Cl and the intergenic spacer between the two genes (Liston 1992). (3) *trn*L, a 1700 bp long cpDNA region between *trn*T and *trn*F, including three non-coding regions (Taberlet & al. 1991). The structure of the amplified fragments and the position of the primers is shown in Fig. 1.

The amplification products were digested with mainly frequently cutting restriction enzymes. For *trn*K 21: AluI, BsaJI, BstUI, CfoI, DdeI, HaeIII, RsaI, MspI, Sau96I, ScrFI (4-bp recognition sites); AseI, BamHI, BclI, BglII, BsaBI, BstBI, DraI, EcoRI, EcoRV, SspI, XmnI (6-bp recognition sites); for rpoC eight: CfoI, HaeIII, Sau96I, ScrFI (4-bp recognition sites); AseI, BglII, DraI, EcoRV (6-bp recognition sites), and for *trn*L two: Sau96I (4-bp recognition sites); BclI (6-bp recognition sites) enzymes were used. DNA fragments were electrophoresed in 1.4% agarose gels and stained with ethidium bromide. Restriction sites were mapped by performing double digests.

Cladistic analysis of combined data. All RFLP-characters obtained for the three cp DNA regions were combined in a molecular data matrix comprising 42 species. The characters were treated as unordered. The cladistic analysis was carried out with PAUP 3.1.1. (Swofford 1993). The heuristic search algorithm, with the MULPARS option and TBR branch swapping, and SIMPLE addition was performed. 50 replications of RANDOM addition were also conducted and no additional trees were found. The strict consensus tree was calculated from all most parsimonious trees. The consistency index (CI) and retention index (RI) were calculated. In order to assess statistical support for the different clades, a decay analysis (Bremer 1988, Donoghue & al. 1992) was performed. Due to restricted computer capacity a bootstrap analysis (heuristic search algorithm, TBR branch swapping) could be conducted for only 22 replicates. Although this does not meet statistical standards the results from this analysis will be reported.

Results

The RFLP analysis yielded 190 characters (*trn*K: 133, *rpo*C: 42, *trn*L: 15) of which 121 were potentially informative. There were six most parsimonious trees 240 steps long with a consistency index of 0.5 and a retention index of 0.82 (excluding non-informative characters). The six most parsimonious trees differed only within the Old World clade (JORK & KADEREIT 1995, this volume). The strict consensus tree (242 steps long) of these six most parsimonious trees is shown in Fig. 2. The New

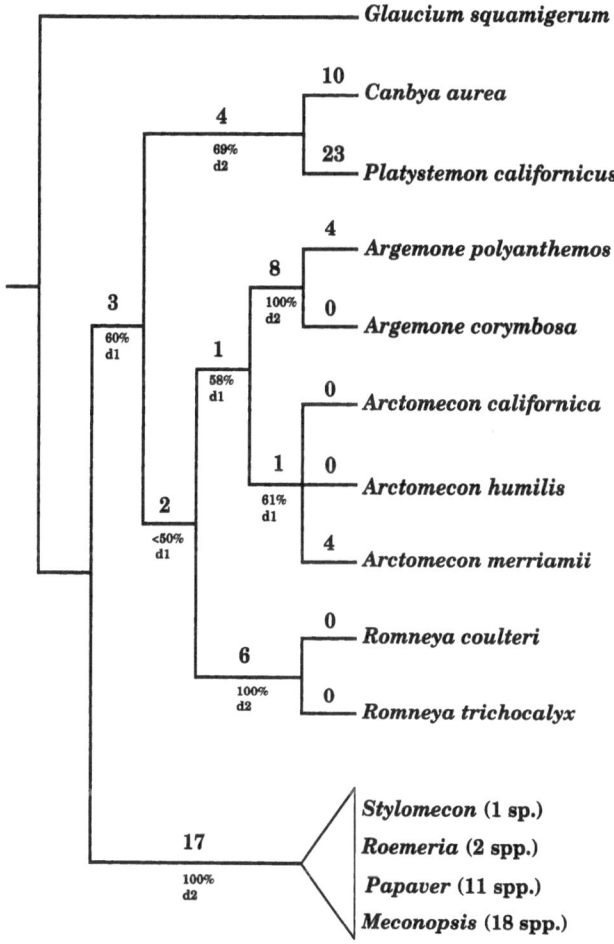

Fig. 2. Strict consensus of six most parsimonious trees obtained from the analysis of 190 (121 informative) restriction site mutations within three chloroplast DNA regions. The length of the consensus tree is 242 steps, the consistency index excluding autapomorphies is 0.5, the retention index is 0.82. The figures above branches indicate numbers of restriction sites supporting each branch (unambiguous characters only), the figures below branches are bootstrap values (%) and decay indices (*d*). *Glaucium squamigerum* was used as outgroup. The Old World group (*Stylomecon, Roemeria, Papaver, Meconopsis*) was represented by altogether 32 species

World clade comprising *Argemone, Arctomecon, Romneya, Canbya* and *Platystemon* forms a monophyletic group. *Platystemon* as a member of former subfam. *Platystemonoideae* is sister group of *Canbya* (*Papaveroideae*). This sister group relationship was also found in the cladistic analysis of morphological characters (Kadereit & al. 1994). *Canbya/Platystemon* are sister group to the rest of the New World clade, in which *Romneya* is sister group to *Arctomecon/Argemone*. The genera *Arctomecon* and *Romneya*, of which all species were investigated, and *Argemone*, as far as examined, all are monophyletic.

Whereas the entire Old World *Papaveroideae* (17 synapomorphies, Fig. 2) and some clades within this group are supported by a large number of characters (Jork & Kadereit 1995, this volume), the North American desert group shows more autapomorphies than cladistically informative synapomorphies. *Canbya* and *Platystemon* have a very large number of autapomorphies (10 and 23, respectively) and the number of autapomorphies is fairly high also in *Romneya* (6) and *Argemone* (8). In contrast to this, only one synapomorphy supports the sister group relationship of *Argemone* and *Arctomecon*, and two that of these two genera to *Romneya*. Accordingly, the basal branches in the New World clade are very poorly supported. Most of the branches collapse in trees one step longer than the most parsimonious trees, as indicated by the decay indices in Fig. 2.

Discussion

The two different data sets, morphological (Kadereit & al. 1994) and molecular, resulted in two at least partly different tree topologies (Fig. 3). Common to both is the sister group relationship of *Canbya* and *Platystemonoideae* and the well-supported monophyly of the Old World clade including the New World *Stylomecon* and *Papaver californicum*. This clade is supported by 17 molecular and four morphological synapomorphies. In the molecular tree the New World clade is

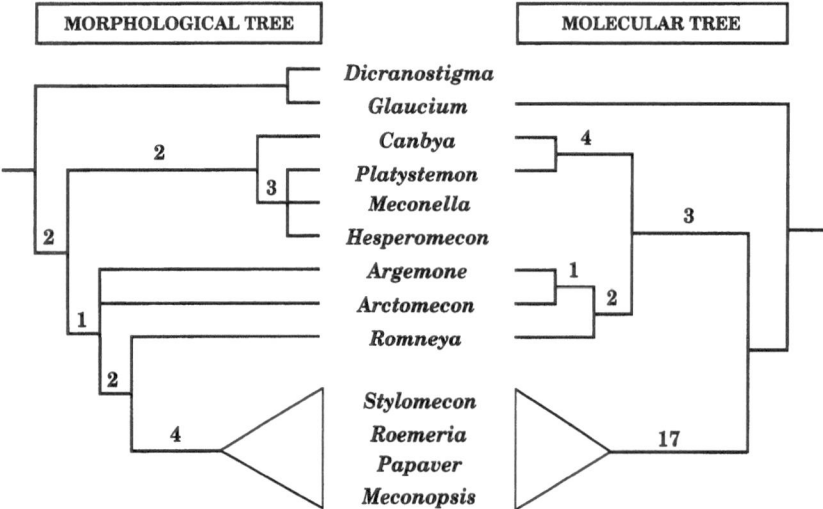

Fig. 3. Comparison of tree topology of strict consensus morphological tree (Kadereit & al. 1994) and strict consensus molecular tree (as in Fig. 2). The figures above branches indicate numbers of unambiguous characters supporting each branch

monophyletic whereas it is paraphyletic in the morphological tree. In the morphological tree *Canbya/Platystemonoideae* is sister group to the rest of the *Papaveroideae*, the relationship between *Argemone* and *Arctomecon* is not resolved, and *Romneya* turned out to be the sister group to the Old World clade (Fig. 3). A morphological synapomorphy which would support the monophyly of the New World group of genera is the presence of 3-merous flowers.

The relationships among *Argemone, Arctomecon, Romneya* and *Canbya/Platystemon* as found in the molecular and morphological analysis are incongruent. As both trees are not very robust it cannot be decided which of the two topologies might be correct. The combination of the morphological and molecular data set resulted in a basal polytomy within the New World clade. On the background of the above results we can ask two major questions: (1). Why is phylogenetic resolution poor in the examined North American clade of genera? and (2). What is a plausible explanation for the high degree of morphological divergence among the genera?

A possible explanation for the poor resolution obtained would be a fast radiation event. MACNAIR (1989) proposed that rapid speciation can occur in response to adaptive challenge when colonizers are faced with vacant adaptive niches and speciate to fill them. Rapid diversification events are widely known from colonizations of islands, where plants are confronted with novel environments. Examples for this are the Hawaiian silversword alliance (WITTER & CARR 1988; BALDWIN & al. 1990, 1991) or the diversification of several plant groups on the Juan Fernandez Islands (STUESSY & al. 1990, CRAWFORD & al. 1992, VALDEBENITO & al. 1992). Other examples of rapid radiation found in spatially non-restricted habitats are the diversification of *Aquilegia* L. (*Ranunculaceae*) which seems to be the result of a key innovation, namely the evolution of multiple nectar spurs which allowed the adaptation to different pollinators (HODGES & ARNOLD 1994), or the rapid evolution and diversification of species of *Helianthus* L. sect. *Helianthus* throughout North America (RIESEBERG & al. 1991).

It can be assumed that the genera discussed here have differentiated within a short period of time when new areas became available for colonization. All species examined are endemics of arid or semiarid habitats of the North American Southwest. These arid regions originated with the progressive aridification of the climate since the Tertiary and early Quaternary. In the early Tertiary tropical seas covered most of coastal and central California. Temperate to subtropical rainforests occurred along the coast, and dry tropical forest and subtropical savanna covered much of the interior (THORNE 1986). Plant communities adapted to drier and cooler conditions like the coastal sageshrub and oak woodland belong to a flora which according to AXELROD (1958) possibly evolved along the drier marginal areas of the North American tropics during the Eocene, and spread across the Southwest in Oligocene and Miocene times and it was largely in the later Pliocene, when environmental diversity became more pronounced across western America, that the flora segregated into a number of derivative communities of more restricted occurrence. According to WOLFE (1987) the terminal Eocene temperature deterioration caused the elimination of many taxa and the relatively depauperate flora of the early oligocene offered opportunities for radiation and diversification. Additionally, major mountain building events during the Quaternary caused a change of habitats. The uplift of the Sierra Nevada and the Transverse and

Peninsular Ranges created a huge rainshadow causing aridification of the regions beyond these new mountain ranges (RAVEN & AXELROD 1978, THORNE 1986, GRAHAM 1993).

Furthermore as a result of Quaternary pluvial-interpluvial cycles vegetation composition has undergone distributional and compositional alterations. The fragmentation of arid and semiarid areas and the establishment of regional endemism might be the result of such climatic cycles (RIDDLE & HONEYCUTT 1990). Most of the plant communities in the desert regions aquired their present composition during the last 8000–11000 years (THORNE 1986, WELLS & HUNZIKER 1976).

Because of the complexity of climatic and geological changes no single pattern seems to explain the origin of the majority of the modern flora. It cannot be decided whether the global climatic changes during the Tertiary or the local aridification due to mountain building events in the Quaternary of the American Southwest caused the rapid diversification of the New World *Papaveroideae*. Although it is thus not known when diversification occurred, where the common ancestor of the group of desert genera examined here came from, or whether diversification took place within or outside its extant distributional range, the small number of synapomorphies and the poor resolution in the basal part of this clade can best be interpreted as the result of a rapid radiation event when the new arid regions became available for colonization.

Alternatively, the poor resolution found may reflect lack of information (MADDISON 1989, HOELZER & MELNICK 1994) resulting from the use of insufficiently variable regions of the genome. This explanation seems not to apply to the taxa examined here because the analysis yielded a large number of characters which, however, are distributed very unevenly. Therefore, the poor resolution of the phylogeny presented here should not be regarded as an experimental artifact but as a reflection of biological reality.

Large differences in morphology which are not paralleled by corresponding divergence of the molecular marker molecules have been described for many taxonomic groups and often resulted in unexpected molecular phylogenies (WEST-EBERHARD 1986; GOTTLIEB 1986; SYTSMA & GOTTLIEB 1986; SYTSMA & SMITH 1988, 1992; KADEREIT 1994). A biological explanation for the high degree of morphological divergence in our desert group might be relatively rapid morphological evolution in arid to semiarid regions (STEBBINS 1952, 1972, 1974; AXELROD 1967, 1972). STEBBINS (1952) pointed out that arid surroundings might have three major effects on evolution: (1). When moisture is limited, local diversity in soil and other environmental factors has a greater effect on the vegetation than in regions where moisture is adequate. (2) The division of larger sized populations into smaller units which are isolated from each other is favoured. (3) Selection for specialized structures which help plants to withstand severe droughts is strong. On the basis of these hypotheses we can explain the high morphological divergence of, e.g., *Arctomecon*. All three species occur on gypsum soil in the Mohave desert and thus occupy a very specialized habitat (MEYER 1986, NELSON & HARPER 1991). They grow in small populations, and the three species are isolated from each other. They are perennials with a persisting leaf rosette and show many adaptations to the dry climate. Vegetatively, the compact shape of the leaf rosettes and the extremely glaucous and hairy leaves can be interpreted as adaptations for the reduction of transpiration. The distinct morphological differences among the species of *Arctomecon* together with their geographical isolation recently led NELSON & WELSH (1993) to suggest that the separation of these taxa is not a recent event. *Arctomecon californica* and *A. humilis* show many morphological differences. Whereas *A. californica* has yellow 3-merous flowers with a caducous corolla and a persistent capsule, *A. humilis* has white 2-merous flowers with a persisting corolla. In this species the entire fruit with dried petals detaches from the plant for dispersal. In contrast to the high level of morphological divergence these two species are identical in their restriction site pattern. This contrast between molecular and morphological divergence within *Arctomecon* may provide indirect support for the rapid radiation hypothesis put forward above for generic diversification.

Because the prickliness of *Argemone* also can be understood as an adaptation to arid regions, where defense against herbivores is very important, and ephemerals like *Canbya* and *Platystemon* by-pass the severe drought of their habitat by a short life cycle, it seems indeed likely that aridity as a major stimulus for evolutionary divergence as suggested by STEBBINS (1952) was highly relevant for the diversification of the New World *Papaveroideae*.

We thank Dr AARON LISTON, Corvallis, Oregon University, for teaching the molecular techniques to A. S. and for kindly providing the primer sequences for PCR amplification, and for critically reviewing the manuscript. We also thank the various people who helped us to obtain collection permits and to collect the plant material. The molecular work is supported by the German Research Foundation (DFG). A grant for the collection of plant material and for a visit to Dr AARON LISTON's laboratory at Oregon State University was given to A. S. by the German Academic Exchange Service (DAAD).

References

ARNOLD, M. L., BUCKNER, C. M., ROBINSON, J. J., 1991: Pollen mediated introgression and hybrid speciation in Louisiana irises. – Proc. Natl. Acad. Sci. USA **88**: 1398–1402.

AXELROD, D. I., 1958: Evolution of the Madro–Tertiary geoflora. – Bot. Rev. **24**: 433–509.

– 1967: Drought, diastrophism, and quantum evolution. – Evolution **21**: 201–209.

– 1972: Edaphic aridity as a factor in angiosperm evolution. – Amer. Naturalist **106**: 311–320.

BALDWIN, B. G., KYHOS, D. W., DVORAK, J., 1990: Chloroplast DNA evolution and adaptive radiation in the Hawaiian silversword alliance (*Asteraceae-Madiinae*). – Ann. Missouri Bot. Gard. **77**: 96–109.

– – – CARR, G. D., 1991: Chloroplast DNA evidence for a North American origin of the Hawaiian silversword alliance (*Asteraceae*). – Proc. Natl. Acad. Sci. USA **88**: 1840–1843.

BREMER, K., 1988: The limits of amino acid sequence data in angiosperm phylogenetic reconstruction. – Evolution **42**: 795–803.

CLARK, C., 1993: *Papaveraceae.* – In HICKMAN, J. C., (Ed.): The Jepson Manual. Higher plants of California, pp. 810–816. – Berkeley: University of California Press.

CRAWFORD, D. J., STUESSY, T. F., COSNER, M. B., HAINES, D. W., SILVA, M. O., BAEZA, M., 1992: Evolution of the genus *Dendroseris* (*Asteraceae: Lactuceae*) on the Juan Fernandez Islands: Evidence from chloroplast and ribosomal DNA. – Syst. Bot. **17**: 676–682.

DONOGHUE, M. J., OLMSTEAD, R. G., SMITH, J. F., PALMER, J. D., 1992: Phylogenetic relationships of *Dipsacales* based on *rbc*L sequences. – Ann. Missouri Bot. Gard. **79**: 333–345.

Doyle, J. J., Doyle, J. L., 1987: A rapid DNA isolation procedure for small quantities of fresh leaf tissue. – Phytochem. Bull. **19**: 11–15.

Ernst, W. R., 1962: A comparative morphology of the *Papaveraceae*. – Ph.D. Thesis, Stanford University.

– 1967: Floral Morphology and systematics of *Platystemon* and its allies *Hesperomecon* and *Meconella* (*Papaveraceae: Platystemonoideae*). – Univ. Kansas Sci. Bull. **47**: 25–70.

Fangan, B. M., Stedje, B., Stabbetorp, O. E., Jensen, E. S., Jacobsen, K. S., 1994: A general approach for PCR-amplification and sequencing of chloroplast DNA from crude vascular plant and algal tissue. – BioTechniques **16**: 484–494.

Gottlieb, L. D., 1986: Genetic differentiation, speciation and phylogeny in *Clarkia* (*Onagraceae*). – In Iwatsuki, K., Raven, P. H., Bock, W. J., (Eds.): Modern aspects of species, pp. 145–160. – Tokyo: University of Tokyo Press.

Graham, A., 1993: History of the vegetation: Cretaceous (Maastrichtian)–Tertiary. – Flora of North America 1, pp. 57–70. – New York: Oxford University Press.

Hodges, S. A., Arnold, M. L., 1994: Columbines: a geographically widespread species flock. – Proc. Natl. Acad. Sci. USA **91**: 5129–5132.

Hoelzer, G. A., Melnick, D. J., 1994: Patterns of speciation and limits to phylogenetic resolution. – Trends Ecol. Evol. **9**: 104–107.

Janish, J. R., 1977: Nevada's vanishing bear-poppies. – Mentzelia **3**: 2–10.

Johnson, L. A., Soltis, D. E., 1994: *mat*K DNA sequences and phylogenetic reconstruction in *Saxifragaceae* s. str. – Syst. Bot. **19**: 143–156.

Johnston, M. C., 1976: A new species of prickly poppy from Mexico. – Wrightia **5**: 259–260.

Jork, K. B., Kadereit, J. W., 1995: Molecular phylogeny of the Old World representatives of *Papaveraceae* subfamily *Papaveroideae* with special emphasis on the genus *Meconopsis* Vig. – Pl. Syst. Evol. [Suppl.] **9**: 171–180.

Kadereit, J. W., 1994: Molecules and morphology, phylogenetics and genetics. – Bot. Acta **107**: 369–373.

– Blattner, F. R., Jork, K. B., Schwarzbach, A., 1994: Phylogenetic analysis of the *Papaveraceae* s.l. (incl. *Fumariaceae*, *Hypecoaceae*, and *Pteridophyllum*) based on morphological characters. – Bot. Jahrb. Syst. **116**: 361–390.

Liston, A., 1992: Variation in the chloroplast genes *rpo*C1 and *rpo*C2 of the genus *Astragalus* (*Fabaceae*): evidence from restriction site mapping of a PCR-amplified fragment. – Amer. J. Bot. **79**: 953–961.

– Kadereit, J. W., 1995: Chloroplast DNA evidence for introgression and long distance dispersal in the desert annual *Senecio flavus* (*Asteraceae*). – Pl. Syst. Evol. **197**: 33–41.

Maddison, W., 1989: Reconstructing character evolution on polytomous cladograms. – Cladistics **5**: 365–377.

Macnair, M. R., 1989: The potential for rapid speciation in plants. – Genome **31**: 203–210.

McDonald, A., 1991: Plantae alpinae Novae Mexicanae: *Argemone subalpina* (*Papaveraceae*). – Brittonia **43**: 120–122.

Meyer, S. E., 1986: The ecology of gypsophile endemism in the eastern Mohave desert. – Ecology **67**: 1303–1313.

Nelson, D. R., Harper, K. T., 1991: Site characteristics and habitat requirements of the endangered dwarf bear-claw poppy (*Artomecon humillis* Coville, *Papaveraceae*). – Great Basin Naturalist **51**: 167–175.

– Welsh, S. L., 1993: Taxonomic revision of *Arctomecon* Torr. & Frém. – Rhodora **95**: 197–213.

Neuhaus, H., Link, G., 1987: The chloroplast tRNA-Lys (UUU) gene from mustard (*Sinapis alba*) contains a class II intron potentially coding for a maturase-related polypeptide. – Curr. Genet. **11**: 251–257.

OWNBEY, G. B., 1958: Monograph of the genus *Argemone* for North America and the West Indies. – Mem. Torrey Bot. Club **21**: 1–159.

– 1961: The genus *Argemone* in South America and Hawaii. – Brittonia **13**: 91–109.

POWELL, A. M., 1972: A new species of *Argemone* (*Papaveraceae*) from Mexico. – Southw. Naturalist **17**: 106–107.

RAVEN, P. H., AXELROD, D. I., 1978: Origin and relationships of the California flora. – Univ. California Publ. Bot. **72**: 1–134.

RIDDLE, B. R., HONEYCUTT, R. L., 1990: Historical biogeography in North American arid regions: an approach using mitochondrial-DNA phylogeny in grasshopper mice (genus *Onychomys*). – Evolution **44**: 1–15.

RIESEBERG, L. H., BECKSTROM-STERNBERG, S. M., LISTON, A., ARIAS, D. M., 1991: Phylogenetic and systematic inferences from chloroplast DNA and isozyme variation in *Helianthus* sect. *Helianthus* (*Asteraceae*). – Syst. Bot. **16**: 50–76.

HANSON, M. A., PHILBRICK, T. C., 1992: Androdioccy is derived in *Datiscaceae:* Evidence from restriction site mapping of PCR-amplified chloroplast DNA fragments. – Syst. Bot. **17**: 324–336.

SHINOZAKI, K., OHME, M., TANAKA, M., WAKASUGI, T., HAYASHIDA, N., MATSUBAYASHI, T., ZAITA, N., CHUNWONGSE, J., OBOKATA, J., YAMAGUCHI-SHINOZAKI, K., OHTO, C., TORAZAWA, K., MENG, B.-Y., SUGITA, M., DENO, H., KAMAGOSHIRA, T., YAMADA, K., KUSUDA, J., TAKAIWA, F., KATO, A., TOHDOH, N., SHIMADA, H., SUGUIRA, M., 1986: The complete nucleotide sequence of the tobacco chloroplast genome: its organisation and expression. – E.M.B.O.J. **5**: 2043–2049.

SORARÚ, S. B., 1976: Nota sobre el género *Argemone* (*Papaveraceae*) en la República Argentina. – Darwiniana **20**: 445–457.

STEBBINS, G. L., 1952: Aridity as a stimulus to plant evolution. – Amer. Naturalist **86**: 33–44.

– 1972: Ecological distribution of centers of major adaptive radiation in angiosperms. – In VALENTINE, D. H., (Ed.): Taxonomy, phytogeography and evolution, pp. 7–34. – London, New York: Academic Press.

– 1974: Flowering plants: evolution above the species level. – Cambridge: Harvard University Press.

– MAJOR, J., 1965: Endemism and speciation in the California flora. – Ecol. Monogr. **35**: 1–35.

STEELE, K. P., VILGALYS, R., 1994: Phylogenetic analyses of *Polemoniaceae* using nucleotide sequences of the plastid gene *mat*K. – Syst. Bot. **19**: 126–142.

STUESSY, T. F., CRAWFORD, D. J., MARTICORENA, C., 1990: Patterns of phylogeny in the endemic vascular flora of the Juan Fernandez Islands, Chile. – Syst. Bot. **15**: 338–346.

SWOFFORD, D. L., 1993: PAUP: Phylogenetic analysis using parsimony, vers. 3.1.1. – Computer program distributed by the Illinois Natural History Survey, Champaign, Ill.

SYTSMA, K. J., GOTTLIEB, L. D., 1986: Chloroplast DNA evidence for the origin of the genus *Heterogaura* from a species of *Clarkia* (*Onagraceae*). – Proc. Natl. Acad. Sci. USA **83**: 5554–5557.

– SMITH, J. F., 1988: DNA and morphology: comparison in the *Onagraceae*. – Ann. Missouri Bot. Gard. **75**: 1217–1237.

– SMITH, J. F., 1992: Molecular systematics of *Onagraceae*: examples from *Clarkia* and *Fuchsia*. – In SOLTIS, P. S., SOLTIS, D. E., DOYLE, J. J., (Eds): Molecular systematics of plants, pp. 295–323. – New York, London: Chapman & Hall.

TABERLET, P., GIELLY, L., PAUTOU, G., BOUVET, J., 1991: Universal primers for amplification of three non-coding regions of chloroplast DNA. – Pl. Molec. Biol. **17**: 1105–1109.

THORNE, R. F., 1986: A historical sketch of the vegetation of the Mohave and Colorado deserts of the American Southwest. – Ann. Missouri Bot. Gard. **73**: 642–651.

Valdebenito, H., Stuessy, T. F., Crawford, D. J., Silva, M. O., 1992: Evolution of *Erigeron* (*Compositae*) in the Juan Fernandez Islands, Chile. – Syst. Bot. **17**: 470–480.

Wells, P. V., Hunziker, J. H., 1976: Origin of the creosote bush (*Larrea*) deserts of southwestern North America. – Ann. Missouri Bot. Gard. **63**: 843–861.

West-Eberhard, M. J., 1986: Alternative adaptations, speciation, and phylogeny (a review). – Proc. Natl. Acad. Sci. USA **83**: 1388–1392.

Witter, M. S., Carr, G. D., 1988: Adaptive radiation and genetic differentiation in the Hawaiian silversword alliance (*Compositae: Madiinae*). – Evolution **42**: 1278–1287.

Wolfe, J. A., 1987: An overview of the origins of the modern vegetation and flora of the northern Rocky Mountains. – Ann. Missouri Bot. Gard. **74**: 785–803.

Address of the authors: Andrea E. Schwarzbach and Joachim W. Kadereit, Institut für Spezielle Botanik, Johannes Gutenberg-Universität, D-55099 Mainz, Federal Republic of Germany.

Accepted January 9, 1995

Pl. Syst. Evol. [Suppl.] 9: 171–180 (1995)

Molecular phylogeny of the Old World representatives of *Papaveraceae* subfamily *Papaveroideae* with special emphasis on the genus *Meconopsis*

KIRSTIN B. JORK and JOACHIM W. KADEREIT

Received October 10, 1994

Key words: *Papaveraceae*, *Papaveroideae*, *Cathcartia*, *Meconopsis*, *Papaver*, *Roemeria*, *Stylomecon*. – Molecular phylogeny, geographical disjunction.

Abstract: The RFLP-analysis of PCR amplified cpDNA fragments of 42 representatives of *Papaveraceae* subfam. *Papaveroideae* resulted in six most parsimonious cladograms. The subfamily can be divided into a New World group (*Arctomecon*, *Argemone*, *Canbya*, *Romneya* and *Platystemon*) and an Old World group (*Meconopsis*, *Papaver* s.l. and *Roemeria*) containing *Stylomecon heterophylla* and *Papaver californicum* as New World taxa. In the Old World group neither *Meconopsis* nor *Papaver* are monophyletic. Whereas *Meconopsis* consists of three clades, *Papaver* comprises five clades, with *Roemeria* as sister group to *P.* sect. *Argemonidium* and *Stylomecon* as sister group to *P. californicum*. Various lines of evidence suggest that *Meconopsis* should be interpreted as a paraphyletic base group of Old World *Papaveroideae* which gave rise to a polyphyletic *Papaver*.

The *Papaveroideae*, one of four subfamilies of the *Papaveraceae* s. str. (23 genera, ca. 240 species, KADEREIT 1993) comprise eight genera with about 176 species (KADEREIT & al. 1994) and have a mainly north-temperate distribution in both the Old and the New World. In the New World, the subfamily is represented by *Argemone* L., *Arctomecon* TORR. & FRÉM., *Canbya* PARRY ex A. GRAY, *Romneya* HARV., *Stylomecon* G. TAYLOR and some species of *Papaver* L. This New World group (excluding *Stylomecon* and *Papaver*, see below) has been discussed in detail by SCHWARZBACH & KADEREIT (1995, this volume). *Meconopsis* VIG. with ca. 50 species, *Papaver* with the majority of its ca. 80 species and *Roemeria* MEDIK. with three species are the Old World representatives of the subfamily. This paper will concentrate on the phylogenetic relationships among the Old World taxa. Since the phylogeny of *Papaver* and *Roemeria* has been investigated by KADEREIT & SYTSMA (1992), here special emphasis will be placed on the phylogeny of the genus *Meconopsis* which after *Papaver* is the second largest genus of the *Papaveraceae*.

Meconopsis was founded by VIGUIER (1814) based on a single species, the West European *Papaver cambricum* L. It was differentiated from *Papaver* by having a distinct style instead of a sessile stigma in form of a disc on top of the ovary. Ten years later, from about 6000 km away, the first Asian representative of *Meconopsis* was recognized on the basis of capsule morphology (DE CANDOLLE 1824). The first detailed study and classification of the genus was published by PRAIN (1895) who recognized 23 species. In a later revision of the genus PRAIN (1906) established two sections based on pubescence characters with a total of 27 species. In the most recent revision by TAYLOR (1934) *Cathcartia* HOOK. f. ex HOOK. was included in *Meconopsis* and 41 species were recognized in two subgenera, *Eumeconopsis* (PRAIN) FEDDE and *Discogyne* G. TAYLOR. Whereas subg. *Discogyne* contained only two species, *Eumeconopsis* was further divided into three sections, *Cambricae* (PRAIN) FEDDE, *Eucathcartia* (PRAIN) G. TAYLOR and *Polychaetia* PRAIN, and the latter section into two subsections with two and six series, respectively. Since then nine additional species have been described (WILLIAMS 1972; FENG 1979; ZHOU 1979, 1980; ASWAL 1985; WU & CHUANG 1985; DEBNATH & NAYAR 1989).

Meconopsis contains perennial monocarpic or polycarpic herbs. The leaves are sessile or petiolate, arranged in a basal rosette or cauline, and entire to pinnately or palmately incised or divided. The flowers are borne on simple scapes or are aggregated in leafy racemes or axillary cymules. The perianth consists of two sepals and four (up to 10) petals. The androecium is polymerous. A distinct style, the morphological character which led VIGUIER (1814) to establish the genus, can be present or absent. In subg. *Discogyne* the ovary is topped by a cartilaginous disc which in contrast to *Papaver* does not bear the stigmatic rays but a style. The majority of the species are native to the Sino-Himalayan region, ranging from Chitral in Pakistan to the southwestern provinces of China. They inhabit alpine woods and scrub, meadows and scree slopes between 1800 and 5500 m altitude.

The type-species of the genus, *Meconopsis cambrica* (L.) VIG., the Welsh Poppy, is found in Western Europe where it has isolated areas of occurrence in Ireland, Wales and Southwestern England, and on the European continent can be found from Western France to Northern Spain (CASTROVIEJO & al. 1986: 417, JALAS & SUOMINEN 1991). Its natural habitats are humid and shady deciduous forest floors and shaded ravines at altitudes between ca. 200 m in Great Britain and up to ca. 1800 m on the European continent. Because of its disjunct distribution and certain morphological characters to be discussed below the close affinity of *M. cambrica* to the remainder of the genus has been doubted repeatedly in the past (e.g., TAYLOR 1934, ERNST 1962, BRÜCKNER 1982).

Material and methods

Plant material. Altogether 42 taxa were examined in the molecular analysis, including all genera of subfam. *Papaveroideae*, one member each of former subfam. *Platystemonoideae* and *Glaucium* MILLER, which was chosen as outgroup on the basis of a cladistic analysis of morphological characters (KADEREIT & al. 1994). Eighteen species of *Meconopsis* representing most of the series recognized by TAYLOR (1934) were included in this study. *Papaver* was represented with 11 spp. of nine of 11 sections recognized (KADEREIT 1988a), *Roemeria* by two of its three species, and *Stylomecon* by its only species, *S. heterophylla* (BENTH.) G. TAYLOR. The New World taxa are listed in SCHWARZBACH & KADEREIT (1995, this volume). The origin of leaf material is listed below in alphabetical order. Voucher specimens are deposited at the place of cultivation of the species.

Meconopsis aculeata ROYLE: Kashmir, cultivated J. COBB (no voucher); *M. betonicifolia* FRANCH.: wild origin unknown, cultivated J. COBB (no voucher); *M. cambrica* VIG.: wild origin unknown, cultivated Botanischer Garten (BG) University of Mainz (KJ 91/46); *M. chelidonifolia* BUR. & FRANCH.: wild origin unknown, cultivated RBG Edinburgh no. 19510280; *M. delavayi* (FRANCH.) FRANCH. ex PRAIN: Sino-British Lijiang Expedit. 573, Yunnan, China, cultivated RBG Edinburgh no. 19871597; *M. discigera* PRAIN: MCBEATH, R 2524, Nepal, cultivated RBG Edinburgh no. 19912814; *M. dhwojii* G. TAYLOR ex HAY: wild origin unknown, cultivated RBG Edinburgh no. 19695307; *M. grandis* PRAIN: MCBEATH, KEKE 490, Kambachen, Nepal, cultivated RBG Edinburgh no. 19892241; *M. henrici* BUR. & FRANCH. aff.: HIRST s.n., cultivated RBG Edinburgh no. 19920705; *M. horridula* HOOK. *f.* & THOMS.: EFHM, SICH 161, Sichuan, China, cultivated RBG Kew no. 1988–4713; *M. integrifolia* (MAXIM.) FRANCH.: Chengdu Edinburgh Exp. 478, China, cultivated RBG Edinburgh no. 19913350; *M. latifolia* (PRAIN) PRAIN: wild origin unknown, cultivated RBG Edinburgh no. 19695312; *M. napaulensis* DC.: CHUDZIAL s.n., Nepal, cultivated J. COBB (no voucher); *M. paniculata* (D. DON) PRAIN: wild origin unknown, cultivated J. COBB (no voucher); *M. punicea* MAXIM.: COX, HUTCHISON & MACDONALD 2586 (1987), Jiuzhaigou-Songpan, Kangali Pass, Sichuan, China, cultivated J. COBB (no voucher); *M. simplicifolia* (D. DON) WALP.: wild origin unknown, cultivated RBG Edinburgh no. 841866; *M. superba* KING ex PRAIN: wild origin unknown, cultivated RBG Edinburgh no. 19820948; *M. villosa* (HOOK. f.) G. TAYLOR: SINCLAIR & LONG 5574, E of Dochong La, Bhutan, cultivated RBG Edinburgh no. 19841951; *Papaver aculeatum* THUNB.: South Africa, cultivated BG University of Mainz (no voucher); *P. alpinum* L.: wild origin unknown, cultivated Oregon State University Corvallis (AS 93/91); *P. argemone* L.: wild origin unknown, cultivated BG University of Mainz (AS 94/117); *P. atlanticum* (BALL) COSS.: wild origin unknown, cultivated BG University of Mainz (KJ 93/7); *P. californicum* A. GRAY: wild origin unknown, cultivated BG University of Mainz (KJ 92/1); *P. croceum* LEDEB.: wild origin unknown, cultivated BG University of Mainz (no voucher); *P. dubium* L.: KJ 93/10, ruderal BG University of Mainz; *P. orientale* LEDEB.: wild origin unknown, cultivated BG University of Mainz (KJ 91/39); *P. rhoeas* L.: KJ 93/7, wayside Gonsenheim/Mainz, Germany; *P. somniferum subsp. setigerum* (DC.) CORB.: wild origin unknown, cultivated BG University of Mainz (AS 94/89); *P. tauricolum* BOISS.: wild origin unknown, cultivated BG University of Mainz (no voucher); *Roemeria hybrida* (L.) DC: wild origin unknown, cultivated BG University of Mainz (AS 94/87); *R. refracta* DC.: wild origin unkonwn, cultivated BG University of Mainz (AS 94/88); *Stylomecon heterophylla*: wild origin unknown, cultivated BG University of Mainz (no voucher).

Molecular techniques and phylogenetic analysis. Molecular characters were obtained from restriction site fragment polymorphisms (RFLP) of three PCR amplified chloroplast genome fragments. These were *trn*K (2500 bp), *rpo*C (ca. 4100 bp) and *trn*L (1700 bp) of the large single copy region. The amplification products were digested with 21 (*trn*K), eight (*rpo*C) and two (*trn*L) mainly frequently cutting enzymes. The consistency index (CI) and retention index (RI) were calculated. In order to assess statistical support for the different clades, a decay and bootstrap analysis was performed. Due to restricted computer capacity only 22 replicates could be conducted for the bootstrap analysis. Although this does not meet statistical standards the results from this analysis are reported in Fig. 1. For further details of molecular techniques and phylogenetic analysis see SCHWARZBACH & KADEREIT (1995, this volume). Experimental manipulations of tree topology were performed with the computer program MacClade version 3.0 (MADDISON & MADDISON1992).

Results

The RFLP analysis yielded 190 characters of which 121 were informative. The cladistic analysis with PAUP 3.1.1. (SWOFFORD 1993) resulted in six most parsimonious trees 240 steps long with a consistency index of 0.5 and a retention index of 0.82 (excl. uninformative characters). Variability among these trees is limited mainly to the relative position of *Meconopsis cambrica*. In three of the six trees *M. cambrica* is sister group to *Papaver* sect. *Argemonidium* SPACH/*Roemeria*. In the other three trees it is sister group to the second clade of *Meconopsis* (see below for details) and *Papaver* sect. *Meconella* SPACH is sister group to *P.* sect. *Argemonidium*/*Roemeria*. Differences within these two groups of three trees concern details of topology within the second clade of *Meconopsis*.

In the strict consensus tree shown in Fig. 1 subfam. *Papaveroideae* consists of two distinct clades. The New World clade, containing *Argemone, Arctomecon, Canbya*,

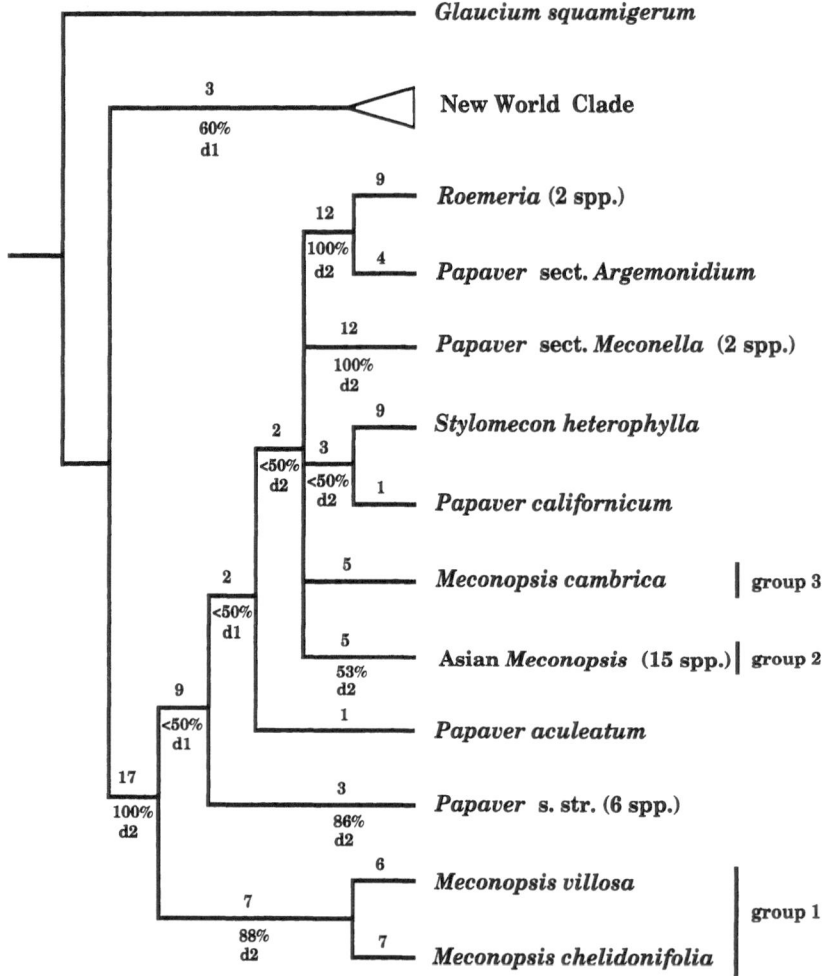

Fig. 1. Strict consensus of six most parsimonious trees obtained from a heuristic search, analysing 190 (121 informative) restriction site mutations within three cpDNA regions. The consensus tree is 242 steps long and has a consistency index (excl. uninformative characters) of 0.5, the retention index is 0.82. The figures above branches indicate the number of restriction sites supporting the branch (unambiguous characters only,) the figures below branches are bootstrap values (%) of 22 bootstrap replications and decay indices (*d*). *Glaucium squamigerum* KAR. & KIR. was used as outgroup. When more than one representative of a genus, section or species group was scored, the number of taxa studied are indicated in brackets. On the right the three groups of *Meconopsis* found are marked with bars

Romneya and *Platystemon* BENTH. is discussed in detail by SCHWARZBACH & KADEREIT (1995, this volume). The Old World clade (incl. *Stylomecon* and *Papaver californicum*) is well supported by 17 synapomorphies (Fig. 1). *Meconopsis* consists of three groups. The first clade of this genus, which is basal to all other *Papaveroideae* of the Old World, contains *M. chelidonifolia* and *M. villosa*. The second clade, containing 15 more species of the genus, is separated from the basal clade by *Papaver* s. str. (sects. *Pilosa* PRANTL, *Macrantha* ELKAN, *Meconidium* SPACH and *Rhoeadium* SPACH) and the South African *P. aculeatum* (sect. *Horrida* ELKAN). This second clade of *Meconopsis* is part of a polytomy comprising four more groups. These are *M. cambrica* as third clade of the genus, *Papaver californicum/Stylomecon*, *P.* sect. *Meconella*, and *P.* sect. *Argemonidium/Roemeria*. As evident from the bootstrap values indicated in Fig. 1, the sister group relationship of *Meconopsis villosa* and *M. chelidonifolia* and *Papaver* s. str. are reasonably well-supported, as is the sister group relationship of *P.* sect. *Argemonidium* to *Roemeria*.

Discussion

The basal clade of *Meconopsis*. Morphologically *M. villosa* and *M. chelidonifolia* as basalmost clade are characterized by their polycarpic habit and their similarity in flower colour. Both have uniformly yellow petals and stamens, whereas other Asian species of the genus have different colours in different parts of the flower. The two species are further united by having four petals, whereas more than four petals can frequently be found in other species, and by having watery instead of milky latex. Both taxa are found in forest habitats or shrub thickets. They are dissimilar in capsule morphology. *Meconopsis villosa* has a sessile stigma on top of a long (up to 7 cm) and narrow capsule, and capsules open with basipetal valves for up to half of their length. In contrast, *M. chelidonifolia* has small ellipsoid capsules with a short but distinct style. The capsules open with interplacental valves for only a short distance from the apex. *M. villosa* was first described as the separate genus *Cathcartia* by J. D. HOOKER (1851) which prior to TAYLOR's (1934) monograph contained up to eight species. *Cathcartia* is maintained by some authors with a single species, *C. villosa* HOOK f. (BISWAS 1966: 125, GRIERSON & LONG 1984), and GREY-WILSON (1993: 82) further adds the closely allied *C.* (*Meconopsis*) *smithiana* HAND.-MAZZ. which was not included in this study. TAYLOR (1934) placed *M. villosa* and *M. chelidonifolia* in two different series of sect. *Eucathcartia* which he regarded, after *M. cambrica*, as the second basal group in the genus. Our data might support either the inclusion of *Cathcartia* into *Meconopsis* or the reinstatement of an enlarged *Cathcartia* probably including *M. villosa*, *M. smithiana*, *M. chelidonifolia* and the closely related *M. oliverana* FRANCH. & PRAIN ex PRAIN.

The basal position among Old World *Papaveroideae* of at least part of Asian *Meconopsis* is congruent with the results of KADEREIT & al. (1994). All Asian *Meconopsis* (group 1 and 2 of the molecular tree) differ from the rest of the Old World taxa of subfam. *Papaveroideae* in one morphological and one chemical character. *Papaver*, *Roemeria*, *Stylomecon heterophylla* and *Meconopsis cambrica* are clearly characterized by the possession of pseudodorsal veins in the gynoecium (ERNST 1962). In this specific vascularisation pattern two adjacent placental traces send out lateral branches which fuse in the carpel median and form a compact basipetal trace.

These pseudodorsal veins occupy the position where other genera have dorsal traces (for example *M. chelidonifolia* and *M. villosa*, ERNST 1962) representing the median vein of the carpel, or no median traces at all (several Asian *Meconopsis*, ERNST 1962). These pseudodorsal traces are absent from all Asian species of *Meconopsis*. The second character shared by all *Papaveroideae* with an Old World distribution (incl. *Stylomecon* and *Papaver californicum*) except Asian *Meconopsis* is the presence of meconic acid in the latex (FAIRBAIRN & WILLIAMSON 1978, JORK unpubl). The absence of pseudodorsal veins and meconic acid identified Asian *Meconopsis* as basal among the Old World representatives of the subfamily. In the analysis of KADEREIT & al. (1994) *Meconopsis* was divided into only two terminal taxa, *M. cambrica* and Asian *Meconopsis*. The molecular data thus support the topology obtained from morphological characters but further divide Asian *Meconopsis* into two subgroups.

The second clade of *Meconopsis*. The second group of *Meconopsis*, which is separated from the basal clade by *Papaver* s. str. and *P. aculeatum*, consists of the remaining species distributed in the Himalayas and China. In our analysis, these are *Meconopsis punicea*, *M. discigera*, *M. integrifolia*, *M. simplicifolia*, *M. betonicifolia*, *M. grandis*, *M. horridula*, *M. delavayi*, *M. henrici*, *M. latifolia*, *M. aculeata*, *M. napaulensis*, *M. dhwojii*, *M. paniculata* and *M. superba*. Details of topology within this clade in relation to TAYLOR's (1934) classification and the morphology of the species involved will be discussed elsewhere. This clade is also characterized by the lack of pseudodorsal traces and meconic acid. In view of the fact that these two characters are absent from all members of subfam. *Papaveroideae* from the New World (except *Papaver* and *Stylomecon*), the absence of these characters in Asian *Meconopsis* must be regarded as plesiomorphic and cannot be used as arguments for the monophyly of all Asian species of the genus. Also, when the branch consisting of Asian *Meconopsis* is experimentally transferred to the basalmost *Meconopsis* clade, a tree 11 steps longer than the most parsimonious trees results. This clearly indicates that the molecular tree is robust with respect to the division of Asian *Meconopsis* into two groups, and that Asian *Meconopsis* thus is not monophyletic. It may well be a polyphyletic group comprising the basal *Cathcartia* (see above) and the remainder of Asian *Meconopsis* as distinct subgroups. Because, however, the genus in Asia has a continuous distributional range, and also shows continuous morphological variation, we believe that it is unlikely that *Meconopsis* in Asia originated twice. Instead, we prefer to regard it as paraphyletic. Following this argumentation, *Papaver* s. str. and *P. aculeatum* would have originated from within *Meconopsis* (Fig. 1).

A final point of interest as regards Asian *Meconopsis* are the chromosome numbers, distribution and ecology of the two clades found. *Meconopsis villosa* (n = 16, RATTER 1968) and *M. chelidonifolia* (n = 14, RATTER 1968) have a lower ploidy level than all other species counted (n = 28–ca. 60, RATTER 1968) except *M. bella* PRAIN (n = 11, LÖVE 1977) and *M. cambrica* (n = 11, 14, 28 (MAUDE 1940, ERNST 1962, RATTER 1968)). As easternmost representative of the genus *M. chelidonifolia* together with its putative close ally *M. oliverana* FRANCH. & PRAIN ex PRAIN is distributed in Western China. This region is geologically older than the Himalayan region (WISSMANN 1959, HARA 1966) and has not been affected by Quarternary glaciation (HSIEH & SALTER 1973, map I–15: 20; RAO 1974; DERBYSHIRE

1992). Moreover, *M. chelidonifolia* and *M. villosa* are species of shady and dense forest floors. These geographical and ecological characteristics do not apply to the majority of species of the second clade of the genus. The comparatively low ploidy level of *M. chelidonifolia* and *M. villosa*, the distribution of *M. chelidonifolia* and the ecology of both species might be regarded as support for their basal position in the genus as found in our molecular phylogeny.

The phylogenetic affinities of *M. cambrica*. The topology of the six most parsimonious trees suggests two possible sister group relationships for *M. cambrica*:

(1) *Meconopsis cambrica* is sister group to the rest of *Meconopsis*. The relationship of *M. cambrica* to the second clade of Asian *Meconopsis*, which above was postulated to be related to the basal clade, finds support in the topology of three of the six most parsimonious trees. The characters pseudodorsal veins and meconic acid, as discussed above, do not support this hypothesis, and the presence of a style which might provide morphological support for the above hypothesis is variable in the genus. The flowers of *M. cambrica*, like those of *M. villosa* and *M. chelidonifolia*, are homogeneously yellow in colour.

(2) The other half of the six shortest trees support a clade consisting of *Papaver* sect. *Argemonidium/Roemeria* as sister group to *M. cambrica*. The glabrous capsules of *M. cambrica* versus the spiny capsules of *P.* sect. *Argemonidium/Roemeria*, the perennial versus annual habit, and the humid versus arid habitat of the two groups in our opinion makes this an unlikely hypothesis.

Morphological and phytochemical evidence allows us to formulate two additional hypotheses about possible relationships of *M. cambrica*:

(3) *Meconopsis cambrica* might be related to the arctic-alpine *Papaver* sect. *Meconella*. One character linking *M. cambrica* with this group is the possession of nudicaulin, a chemically unidentified yellow floral pigment (PRICE & al. 1939, HARBORNE 1967: 149–152). The molecular data give no support for such a relationship, nor does general and capsule morphology or geographical distribution. The ecology of the C Asian and arctic-alpine *Papaver* sect. *Meconella* partly resembles that of *Meconopsis*.

(4) *Meconopsis cambrica* is sister group to *Papaver californicum/Stylomecon*. The only morphological argument in favour of this hypothesis is that *Stylomecon*, which originally had been described as *Meconopsis heterophylla* by BENTHAM (1835), has a distinct style. Otherwise this relationship seems unlikely with respect of growth habit and ecology. Whereas the European *M. cambrica* is perennial, the two N American species are annual and occur in more arid habitats. Irrespective of this difference in ecology, however, the distribution of *M. cambrica* (W Europe) and *Papaver californicum/Stylomecon* (Western N America) might be regarded as derived from a continuous Arcto-Tertiary range. Arcto-Tertiary elements form part of the Californian flora (RAVEN & AXELROD 1974), and indeed *Stylomecon* has been listed as one example for this group by STEBBINS & MAJOR (1965).

Mainly for geographical and ecological reasons we prefer the first of the above four hypotheses. *Meconopsis cambrica* in Oceanic Europe has always been interpreted as a Tertiary relic (BLUNT 1976, CHUANG 1981). Its ecology as a species of humid forest floors is very similar to that of *M. villosa* and *M. chelidonifolia*, which also are found in humid forest floors and thickets in the Eastern Himalaya and China. Also, other taxa of comparable distribution are known (see, e.g., *Sanicula*

europaea L. (MEUSEL & SCHUBERT 1971), *Asarum* L., *Paris* L. (MEUSEL & al. 1965) or
Impatiens L. (MEUSEL & al. 1978)). Although *Meconopsis cambrica* differs from Asian
Meconopsis in having both capsules with pseudodorsal veins and meconic acid, it
resembles them more than any of its alternative relatives in habit, flower colour,
and capsule morphology. If thus *M. cambrica* is most closely related to the remainder
of *Meconopsis*, not only the two basal clades of this group, but the entire genus is best
interpreted as a paraphyletic base group of the Old World *Papaveroideae* which gave
rise to all other clades found in the molecular tree. The implications of this
hypothesis for *Papaver*, namely that this genus is highly polyphyletic, confirms the
results of KADEREIT & SYTSMA (1992). These authors presumed *Papaver* to consist of
three separate lines, *Papaver* s. str. (containing sects. *Rhoeadium*, *Carinatae* FEDDE,
Papaver, *Macrantha*, *Meconidium* and *Pseudopilosa* M. POP. ex GÜNTHER), *P.* sect.
Meconella and *P.* sect. *Argemonidium*, of which the latter was found to be related to
Roemeria. A fourth line, *P. californicum* with close affinities to *Stylomecon*, can now
be added. The South African *P. aculeatum* appears to be related to *Papaver* s. str. as
suggested by KADEREIT (1988b).

   On the basis of the disjunct distribution of *Meconopsis* the following scenario can
be developed. *Meconopsis cambrica* is the westernmost relic of the genus which once
had a continuous distribution in mild and humid habitats ranging across Laurasia.
Aridification of the climate in the late Tertiary and Quaternary and the upheavel of
the Himalayan Mountain range resulted in the fragmentation of the originally
continuous range of *Meconopsis*. At the same time the newly formed arid habitats
could be colonized by the various phylogenetic lines of *Papaver* and the related
Roemeria and *Stylomecon* in both Central to Western Eurasia and W North
America. All these lines thus originated from within *Meconopsis* as did the arctic-
alpine *P.* sect. *Meconella* which already had been postulated to have originated in
C Asia by, e.g., HANELT (1969). If this scenario should be correct, the polytomy of our
molecular tree might be the result of far-reaching extinction and represents a
biological reality rather than an experimental artefact.

   We would like to thank the Royal Botanic Garden Edinburgh, the Royal Botanic Garden
Kew and Dr JAMES L. S. COBB, St. Andrews, for their cooperation and for supplying fresh leaf
material, and Dr AARON LISTON (Corvallis/Oregon) for critical comments on the manuscript.
The molecular work is supported by the German Research Foundation (DFG).

References

ASWAL, B. S., 1985: *Meconopsis bikramii* ASWAL (*Papaveraceae*) a new species from Lahul
   Valley, Himachal Pradesh, India. – Indian J. Forest. **8**: 84–85.
BENTHAM, G., 1835: *Meconopsis heterophylla.* – Trans. Hort. Soc. London, ser. 2, **1**: 408.
BISWAS, K., 1966: Plants of Darjeeling and the Sikkim Himalayas. – Alipore: West Bengal
   Government Press.
BLUNT, W., 1976: Poppycock: The blue poppies. – Garden (London) **101**: 211–212.
BRÜCKNER, C., 1982: Zur Kenntnis der Fruchtmorphologie der *Papaveraceae* JUSS. s. str. und
   der *Hypecoaceae* (PRANTL ex KÜNDIG) NAK. – Feddes Repert. **93**: 153–212.
CANDOLLE, A. P. DE, 1824: *Papaveraceae.* – Prodromus systematis naturalis regni vegetabilis,
   1, pp. 117–124.
CASTROVIEJO, S., LAÍNZ, M., LÓPEZ GONZÁLEZ, G., MONTSERRAT, P., MUNOS GARMENDIA, F.,

PAIVA, J., VILLAR, L., (Eds): 1986: Flora Iberica 1: *Lycopodiaceae-Papaveraceae*. – Madrid: Real Jardín Botánico, C.S.I.C.

CHUANG, HSUAN, 1981: The systematic evolution and geographical distribution of *Meconopsis*. – Proceedings of Symposium on Qinghai–Xizang (Tibet) Plateau (Beijing, China), 2. Environment and ecology of Qinghai–Xizang Plateau, pp. 1263–1268. – Beijing: Science Press; New York: Gordon and Breach.

DEBNATH, H. S., NAYAR, M. P., 1989: A new species of *Meconopsis* VIG. (*Papaveraceae*) from Nepal. – J. Japan. Bot. **64**: 157–159.

DERBYSHIRE, E., 1992: Origine et remise en cause de la théorie des glaciations pléistocènes de la Chine du Sud-Est: le cas des massifs de Lushan et Huangshan. – Ann. Géogr. **566**: 472–490.

ERNST, W. R., 1962: A comparative morphology of the *Papaveraceae*. – Ph.D. Thesis, Stanford University.

FAIRBAIRN, J. W., WILLIAMSON, E. M., 1978: Meconic acid as a chemotaxonomic marker in the *Papaveraceae*. – Phytochemistry **17**: 2087–2089.

FENG, KUO MEI, 1979: *Meconopsis wumungensis* K. M. FENG sp. nov. – In WU, CHENG YIH, (Ed.): Flora Yunnanica, **2**, p. 33. – Beijing: Academia Sinica Press.

GREY-WILSON, C., 1993: Poppies. – Portland, Oregon: Timber Press.

GRIERSON, A. J. C., LONG, D. G., 1984: Flora of Bhutan, **1**, **2**. – Edinburgh: Royal Botanic Garden.

HANELT, P., 1969: Revision der mongolischen Taxa von *Papaver* L. sect. *Scapiflora* RCHB. sowie Studien zur Systematik und Evolution dieser Sektion. – Habil. Gatersleben.

HARA, H., 1966, (Ed.): The flora of Eastern Himalaya. – Tokyo: The University of Tokyo Press.

HARBORNE, J. B., 1967: Comparative biochemistry of the flavonoids. – London, N. Y: Academic Press.

HOOKER, J. D., 1851: *Cathcartia villosa*. – Bot. Mag. Ser. 3, Vol. VII, (Vol. **77** of all), table 4596.

HSIEH, CHIAO MIN, SALTER, C. L., 1973: Atlas of China. – New York: McGraw-Hill.

JALAS, J., SUOMINEN, J., (Eds), 1991: Atlas Florae Europaeae **9**. *Paeoniaceae to Capparaceae*. – Helsinki: Helsinki University Printing House.

KADEREIT, J. W., 1988a: Sectional affinities and geographical distribution in the genus *Papaver* L. (*Papaveraceae*). – Beitr. Biol. Pfl. **63**: 139–156.

– 1988b: The affinities of the south–hemispherical *Papaver aculeatum* THUNB. (*Papaveraceae*). – Bot. Jahrb. Syst. **109**: 335–341.

– 1993: *Papaveraceae*.– In KUBITZKI, K., ROHWER, J. G., BITTRICH, V., (Eds): The families and genera of vascular plants, **2**, pp. 494–506. – Berlin, Heidelberg: Springer.

– SYTSMA, K. J., 1992: Disassembling *Papaver*: a restriction site analysis of chloroplast DNA. – Nordic J. Bot. **12**: 205–217.

– BLATTNER, F. R., JORK, K. B., SCHWARZBACH, A., 1994: Phylogenetic analysis of the *Papaveraceae* s.l. (incl. *Fumariaceae*, *Hypecoaceae*, and *Pteridophyllum*) based on morphological characters. – Bot. Jahrb. Syst. **116**: 361–390.

LÖVE, A., 1977: IOPB Chromosome number reports L VII. – Taxon **26**: 445.

MADDISON, W. P., MADDISON, D. R., 1992: MacClade. Analysis of phylogeny and character evolution, version 3. – Sunderland, Mass.: Sinauer.

MAUDE, P., 1940: Chromosome numbers in some British plants. – New Phytol. **39**: 17–32.

MEUSEL, H., SCHUBERT, R., 1971: Beiträge zur Pflanzengeographie des Westhimalayas 1. Teil: Die Arealtypen. – Flora **160**: 137–194.

– JÄGER, E., WEINERT, E., 1965: Vergleichende Chorologie der zentraleuropäischen Flora. Karten, **I**. – Jena, Stuttgart, New York: G. Fischer.

–– 1978: Vergleichende Chorologie der zentraleuropäischen Flora, Karten, II. – Jena: G. Fischer.

PRAIN, D., 1895: Noviciae Indicae IX. – J. Asiat. Soc. Bengal. **64**: 303–327.

– 1906: A review of the genera *Meconopsis* and *Cathcartia*. – Ann. Bot. (Oxford) **20**: 323–370.

PRICE, J. R., ROBINSON, R., SCOTT-MONCRIEFF, R., 1939: The yellow pigment of *Papaver nudicaule*. Part 1. – J. Chem. Soc. (London): 1465–1468.

RAO, M. A., 1974: Chapter 10. Vegetation and phytogeography of the Himalaya. – In MANI, M. S., (Ed.): Ecology and biogeography in India. – The Hague: Junk.

RATTER, J. A., 1968: Cytological studies in *Meconopsis*. – Notes Roy. Bot. Gard. Edinburgh **28**: 191–200.

RAVEN, P. H., AXELROD, D. I., 1974: Angiosperm biogeography and past continental movements. – Ann. Missouri Bot. Gard. **61**: 539–673.

SCHWARZBACH, A., KADEREIT, J. W., 1995: Rapid radiation of North American desert genera of the *Papaveraceae*: evidence from restriction site mapping of PCR-amplified chloroplast DNA fragments. – Pl. Syst. Evol. [Suppl.] **9**: 159–170.

STEBBINS, G. L., MAJOR, J., 1965: Endemism and speciation in the California flora. – Ecol. Monogr. **35**: 1–35.

SWOFFORD, D. L., 1993: PAUP: Phylogenetic analysis using parsimony, ver. 3.1.1.– Computer program distributed by the Illinois Natural History Survey, Champaign, Ill.

TAYLOR, G., 1934: An account of the genus *Meconopsis*. – London: New Flora & Silva.

VIGUIER, L. G. A., 1814: Histoire naturelle, médicale et économique, des pavots et des argémones. – Dissertation Montpellier.

WILLIAMS, L. H. J., 1972: *Meconopsis taylorii*, a new species from Nepal. – Trans. Bot. Soc. Edinburgh **41**: 347–349.

WISSMANN, H. VON, 1959: Die heutige Vergletscherung und Schneegrenze in Hochasien mit Hinweisen auf die Vergletscherung der letzten Eiszeit. – Akad. Wiss. Mainz, Abh. Math. – Naturwiss. Kl. **14**: 1101–1431.

WU, CHENG YIH, CHUANG, HSUAN, 1985: *Meconopsis pseudohorridula* C. Y. WU et H. CHUANG sp. nov. – In WU, CHENG YIH, (Ed.): Flora Xizangica, **2**, p. 234. – Kunming: Science Press.

ZHOU, LI HUA, 1979: New taxa of *Meconopsis* from Qinghai-Tibet Plateau. – Acta Phytotax. Sin **17**: 112–114.

– 1980: Study on the *Meconopsis* of Qinghai-Tibet Plateau. – Bull. Bot. Lab. N. E. Forest. Inst., Harbin **8**: 91–101.

Address of the authors: KIRSTIN B. JORK and JOACHIM W. KADEREIT, Institut für Spezielle Botanik, Johannes Gutenberg-Universität Mainz, D-55099 Mainz, Federal Republic of Germany.

Accepted January 11, 1995

Pl. Syst. Evol. [Suppl.] 9: 181–182 (1995)

Serotaxonomical investigation in the *Papaverales*

V. S. Shneyer, N. G. Kutyavina, and N. S. Morosova

Received October 25, 1994

Key words: *Papaverales, Papaveraceae, Hypecoaceae, Fumariaceae,* – Immuno-electro-phoresis, immunodiffusion.

Abstract: Seed storage proteins of 14 genera of *Papaverales* are studied by serological methods to elucidate taxonomical relationships within the order.

The phylogenetic connections within the *Papaverales* and the placement of some genera are unclear. The serological method of protein comparison has repeatedly allowed to elucidate the relationships between taxa even in complex groups of plants (Jensen 1991). We used serological techniques to compare the seed storage proteins of representatives of 14 genera of the *Papaverales* with seven reference systems.

Material and methods

Species investigated. *Papaveraceae: Sanguinaria canadensis* L.*, *Hylomecon vernalis* Maxim, *Chelidonium majus* L.; *Glaucium fimbrilligerum* (Trautv.) Boiss.*; *Eschscholzia californica* Cham.*; *Roemeria refracta* DC., *Meconopsis cambrica* (L) Vig., *Argemone mexicana* L., *Papaver orientale* L.*; *Platystemon californica* Benth.*. *Hypecoaceae: Hypecoum parviflorum* Kar. et Kir.*. *Fumariaceae Dicentra eximia* (Ker) Torr.; *Corydalis bracteata* (Steph) Pers.*; *Fumaria vaillantii* Loisel (according to Takhtadjan 1987). Species marked by asterics are used as reference systems. Voucher specimens are departed in the herbarium of Komarov's Botanical Institute of RAS, Department of Biosystematics.

    Methods. Proteins were extracted from delipified seed meal with a Na–K phosphate buffered 1, 2% NaCl pH 7, 0. Immunoelectrophoresis and immunodiffusion were performed on agarose – starch (0, 75%; 2, 5%) gels in borate-phosphate buffer (pH 7, 6) with rabbit antisera. The formation of spurs, double spurs or identical precipitation lines were recorded and used for the assessment of serological similarities to the reference taxa. (Lester & al. 1983).

Results and conclusions

All genera investigated differ remarkably serologically; this may be evidence for their comparatively old origin.

The representatives of subtribe *Papavereae* (*Papaver, Roemeria, Meconopsis, Argemone*) revealed the highest level of serological similarity and in general reacted identically with other reference systems of other taxa. So this group may be the youngest in the family.

Subtribe *Chelidonieae* is rather heterogeneous serologically: *Sanguinaria* and two other genera – *Chelidonium* and *Hylomecon* – are quite dissimilar with respect to most reference systems. The relationships of *Sanguinaria* are uncertain because it shows serological similarity with taxa of both *Chelidonieae* and *Papavereae*.

Glaucium (*Glaucieae*) and *Platystemon* (*Platystemonoideae*) are more similar to the genera of tribe *Papavereae* than to all others.

Eschscholzia (*Eschscholzioideae*) showed the least serological similarity to other genera of the *Papaveraceae*.

Among the members of the *Papaveraceae* none could be shown to be closer to the *Fumariaceae* than others.

Hypecoum (*Hypecoaceae*) is more similar to *Fumariaceae* than to *Papaveraceae*.

References

JENSEN, U., 1991: Steps toward the natural system of the dicotyledons: serological characters. – Aliso **13**: 183–190.

LESTER, R. N., ROBERTS, P. A., LESTER, C., 1983: Analysis of immunotaxonomic data obtained from spur identification and absorption techniques. – In JENSEN, U., FAIRBROTHERS, D. E., (Eds): Proteins and nucleic acids in plant systematics, pp. 275–300. – Berlin, Hiedelberg: Springer.

TAKHTAJAN, A., 1987: Systema Magnoliophytorum. – Leningrad: Nauka. (In Russian)

Address of the authors: V. S. SHNEYER, N. G. KUTYAVINA, N. S. MOROSOVA, V. L. KOMAROV's Botanical Institute, Russian Academy of Science, Prof. Popov Street 2, 197376, St. Petersburg, Russia.

Accepted January 11, 1995

Pl. Syst. Evol. [Suppl.] 9: 183–188 (1995)

Phylogeny of *Corydalis*, ITS and morphology

MAGNUS LIDÉN, TATSUNDO FUKUHARA, and TORSTEN AXBERG

Received October 6, 1994

Key words: *Papaveraceae, Fumariaceae, Corydalis, Dicentra.* – rDNA, ITS, seedcoat anatomy.

Abstract: The phylogeny of *Corydalis* was deduced from sequences of the transcribed spacers in the nuclear ribosomal DNA. A unique derivation of zygomorphic flowers and a basal position of a paraphyletic *Dicentra* is suggested. The section *Strictae* is sistergroup to the rest of the genus, which consists of two well supported clades: (*Sophorocapnos, Thalictrifoliae, Aulacostigma* and *Cheilanthifoliae*), and the majority of the genus, including all the tuberous groups. *Duplotuber* is far removed from the other tuberous sections.

Corydalis DC. in its usual circumscription harbours more than 430 species, but its circumscription and subdivision has been debated. The removal of *Ceratocapnos claviculata* (L.) LIDÉN and the reerection of *Pseudofumaria* MEDIKUS and *Capnoides* MILLER (LIDÉN 1981, 1986) has been adopted in most recent floras. A segregation of /the tuberous groups as a subgenus or genus (*Pistolochia* BERNHARDI) has been put forward by some authors (MOWAT & CHATER 1964, SOJAK 1972). To check the appropriateness of this suggestion and with the goal to find robust clades on which formal subgeneric taxa could be based, a combined approach was attempted, using a molecular cladogram as a skeleton in the light of which morphological characters were evaluated.

The nuclear rDNA consists of numerous repeats of a transcribed core-sequence, separated by non-transcribed, quickly evolving spacers. The transcribed part is, in eukaryotes, built up of three of the four ribosomal genes (18S, 5.8S, and 28S), separated by internal transcribed spacers (ITS1, ITS2), and with the 18S-segment preceded by an external transcribed spacer. The ITSs show too much variation to be useful at higher levels, but have been utilized within plant families and genera: *Compositae* (BALDWIN 1992, 1993), *Astragalus* (WOJCIECHOWSKI & al. 1993), *Winteraceae* (SUH & al. 1993), and *Epilobium* (BAUM & al. 1994).

Material and methods

Plants were grown in the Göteborg Botanical Garden. Vouchers are in GB: *Papaver rhoeas* L. (no voucher); *Hypecoum imberbe* SM. (DNA 9011); *Dicentra spectabilis* (L.) PERS. (LIDÉN

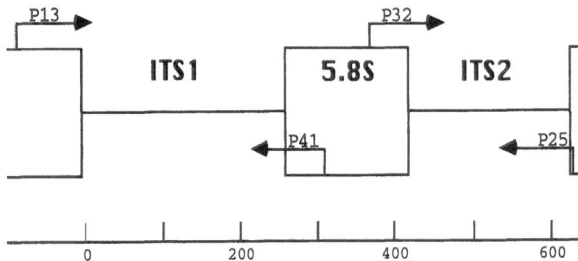

Fig. 1. Amplified region of nuclear rDNA. Primers: P13: 5′-TCCGCCATTGAACC-TTATCATTTAGAGGAA, P25: 5′-GGGTAGTCCCGCCTGACCTG, P32: 5′-GTCT-TTGAACGAAGTTG, P41: 5′-TCGCATTTCGCTACGTTC

DNA 941, garden origin); *D. macrantha* Oliv. (DNA 942, garden origin); *Sarcocapnos enneaphylla* (L.) DC. (S 811, Spain); *Corydalis rupestris* Boiss. (DNA 901, garden origin); *C. ophiocarpa* Hook. f. & Thoms. (DNA 945, Sikkim); *C. pallida* (Thunb.) Pers. (DNA 946, Japan); *C. wilsoni* N. E. Br. (DNA 9010, garden origin); *C. decumbens* (Thunb.) Pers. (DNA 921, Japan); *C. ochotensis* Turcz. (DNA 881, garden weed, Göteborg); *C. cashmeriana* Royle (DNA 943, garden origin); *C. flaccida* Hook. f. & Thoms. (photo, Sikkim); *C. nobilis* (L.) Pers. (DNA 944, garden origin); *C. popovii* Nevski ex Popov (DNA 902, Tadjikistan); *C. darwasica* Prain (DNA 903, Kazakhstan); *C. nariniana* An. Fed. (DNA 904, Armenia); *C. uniflora* (Sieb.) Nyman (DNA 905, Crete); *C. solida* (L.) Clairville subsp. *solida* (DNA 906, garden weed, Göteborg); *C. solida* subsp. *incisa* Lidén (DNA 907, Greece); *C. henrikii* Lidén (holotype, Anatolia); *C. "caucasica alba"* auct. hort. (DNA 908, garden origin, originally from Caucasus); *C. bracteata* (Steph.) Pers. (DNA 909, garden origin); *C. schanginii* (Pall.) Fedtsch. subsp. *schanginii* (DNA 9010, Kazakhstan); *C. ruksansii* Lidén (holotype, Tadjikistan); *C. tauricola* (Cullen & Davis) Lidén (DNA 888, Anatolia); *C. wendelboi* Lidén (DNA 882–887, all Anatolia).

DNA sequencing. DNA were prepared from 50–250 mg fresh tissue according to Rogers & Bendich (1985). A segment including ITS1, 5.8S and ITS2 was amplified (with Perkin Elmer Cetus 480 Thermal Cycler) from total genomic DNA using the polymerase chain reaction and the primer pair P13/P25 (Fig. 1), giving fragments of about 700 bp.

Reaction mixture consisted of 0.1–10 ng plant DNA, 50 mM KCl, 10 mM Tris-HCl pH 8.3, 0.01% gelatin, 200 µM of each dNTP, 1.0 µM of each primer, 1.25 units of a Taq polymerase, and 0.9 or 1.5 mM MgCl$_2$ in a total reaction volume of 50 µl. The reactions were run for 40 cycles (1′ at 94° C; 2′ at 51° C; 3′ at 72° C) followed by 7′ 72° C and 4° C. Products were purified with Qiagen PCR purification kit and fragments were ligated into Smal cleaved pTZ18R or pTZ19R (Pharmacia) plasmids or into the pCR1000 (Invitrogen c.) plasmid using Invitrogen TA Cloning kit. The pTZ constructs were cloned in TG1 cells through electroporation (Bio-Rad Gene-Pulser: 2.5 kV, 200 Ohm, 25 uFD). The pCR1000 constructs were cloned in INV1alphaF' cells included in the kit. Presence of inserts in positive transformants was controlled by cleaving with EcoRI/Pst1 (pTZ) or EcoRI (pCR1000).

Double stranded sequencing with ^{35}S-labelled nucleotides following Sequenase version 2.0 protocol (US Biochemical was made with Boehringer Mannheim "universal sequencing primer" and Pharmacia "reverse sequencing primer", priming in the plasmid or with the amplification primers. Two primers priming in the 5.8s gene were constructed (Fig. 1). Primers were manufactured by Scandinavian Gene Synthesis (Köping, Sweden).

Recent data (from *C. wilsonii*, *C. pallida*, *C. ophiocarpa*, and *Dicentra*) were obtained by direct sequencing of PCR-amplificates as described in Oxelman & Lidén 1995, this volume).

The within-species variation in the ITS sequences was checked with samples from six different populations of *Corydalis wendelboi*. Five of these were identical, while one differed in one position. The closely related *C. tauricola* differed in one position. This also suggests that cloning of "paralogous" repeats of polymorphic rDNA has had little impact on this study, although there is a considerable risk at the lower level of the analysis, as shown by two clones from the same individual of *Vaccaria hispanica* (*Caryophyllaceae*), that differed in three positions in each of the ITSs. A much safer strategy is direct sequencing of PCR-amplificate.

Sequences are deposited in the EMBL database (accession numbers X85445–X85492).

Data analysis. The data were analyzed in two subsets, (Fig. 2A, B) with PAUP 3.1.1 (heuristic method; 10 replicates of random addition sequence; TBR branch swapping, Swofford 1993). Decay indices (Bremer 1988) were approximated from all trees found up to four steps longer than the shortest. One hundred bootstrap replicates were also run with NNI branch swapping, each with 5 replicates of random addition sequence. *Dicentra spectabilis*, *D. macrantha*, *Corydalis pallida*, *C. ophiocarpa*, and *C. nobilis* were represented by partial sequences only.

The sequences were aligned manually. Variable parts that could not be aligned over the whole data set, were excluded from the first subset. Gaps were coded as missing values (or the regions were excluded in case of ambiguous alignment), but with informative gap positions recoded as additional characters. Gap coding is a contraversial issue (e.g., Wojciechowski & al. 1993). It seems more important to recognize presence of indels rather than to risk a false substitution signal, as "indels" often constitute unique synaponorphies for morphologically distinct groups (Baum & al. 1994).

Results and discussion

Molecular trees (Fig. 2). *Papaver*, *Hypecoum*, *Dicentra*, and *Sarcocapnos* form succesively closer outgroups to *Corydalis*, which is in accordance with Lidén (1986), with the exception of the position of *Dicentra spectabilis* as basal within the subfamily. A strongly supported basal dichotomy between *Corydalis rupestris* (clade I) and the rest of the genus was found. A synapomorphy for the latter branch is a large deletion at the 5′ end of ITS1. *C. wilsonii* (sect. *Thalictrifoliae* Fedde). *C. pallida* (sect. *Sophorocapnos* (Turcz.) Popov), and *C. ophiocarpa* (sect. *Cheilanthifoliae* Lidén) together form a strongly supported sister clade (II) to clade III which embraces most of the genus.

Two of the tuberous sections, *Leonticoides DC.* and *Corydalis*, cluster together in a weakly supported clade, as a sistergroup to the non-tuberous *Capnogorium* (Bernh.) Endlicher, whereas sect. *Duplotuber* Ryberg groups with *C. ochotensis* (sect. *Ramoso-Sibiricae* Fedde ex Wendelbo). *Leonticoides* and *Corydalis* are well supported, and most smaller clades agree with morphological evidence. However, the insertion of *C. tauricola* between *C. ruksansii* and *C. schanginii* is surprising, as the latter two species share several morphological synapomorphies. A higher AT-ratio in *C. schanginii* and *C. tauricola* is a potential source of bias.

Morphological evidence. Paraphyly of the tribe *Corydaleae* Reichenb. (Loconte & al. 1995, this volume) and especially of the genus *Dicentra* is evident from seed coat anatomy, where *Dicentra spectabilis* is the only species in the subfamily sharing endotestal seed coat with all potential outgroups (Fukuhara & Lidén, unpubl.). A basal position is in accordance with the ITS1 sequence.

Clade II (including sect. *Aulacostigma* Lidén, not sequenced) is a homogeneous group, with a seed coat anatomy similar to that of *Strictae* (I), but distinguished by keeled, usually ornamented seeds, and different elaiosome, hilum, and stigma structure, (Brückner 1985; 1992; Lidén 1986 and unpubl.) Clade III is characterized

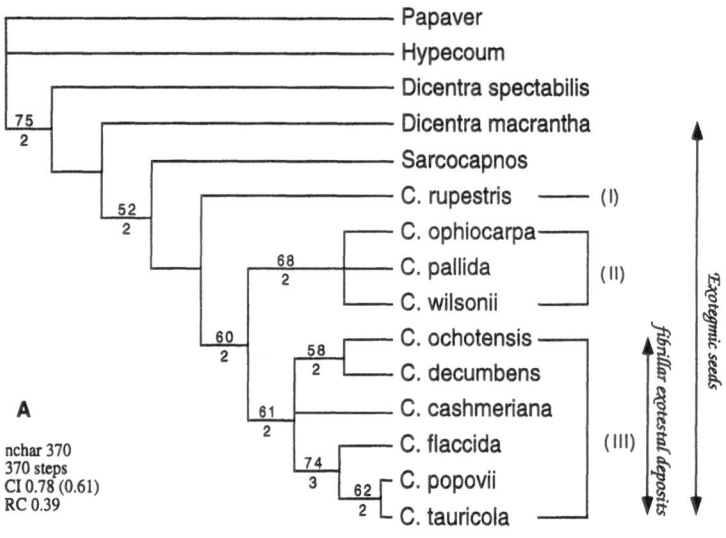

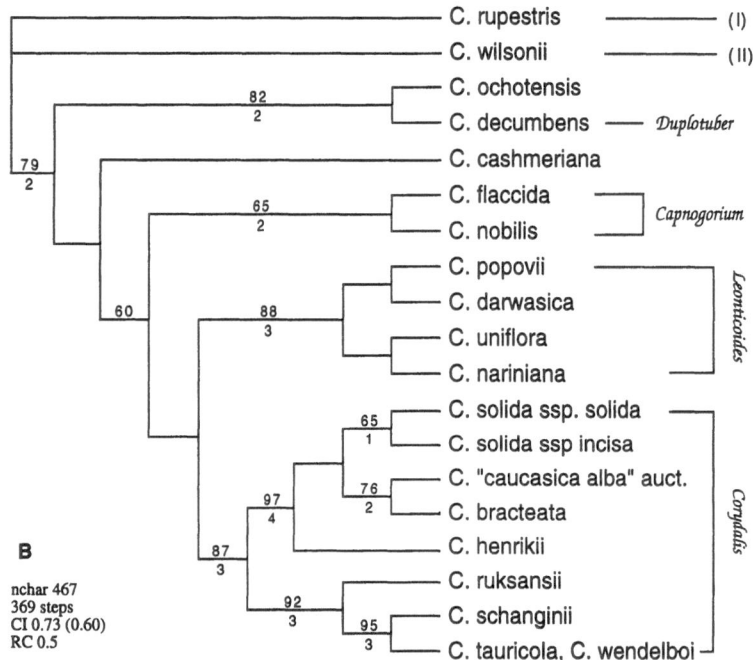

Fig. 2. *A* Basal taxa and conservative regions only: consensus of the 9 shortest trees. *B* Clade III, whole dataset: single most parsimonious tree. Values above branches denote bootstrap values higher than 50%, values below are decay indices

for example by some apomorphic seed characters, like thin-walled endotegmen and fibrillar exotestal deposits (FUKUHARA & LIDÉN, unpubl.).

The placement of sect. *Duplotuber* far away from the other tuberous groups is supported by several morphological characters. Its tubers are not homologous and seed, stigma, and fruit morphology are also widely different (BRÜCKNER 1985, 1993; OH & KIM 1988; FUKUHARA 1992; FUKUHARA & LIDÉN, unpubl. Whether the tubers even of the "core" tuberous group are homologous may be doubted (RYBERG 1960), although monocotyly is a potential synapomorphy for this assemblage. In sect. *Duplotuber* both monocotylous (*C. decumbens*) and dicotylous (*C. buschii* NAKAI) species occur, but the seedling is of very different type. In sect. *Oocapnos* POPOV ex WENDELBO is third type of monocotylous seedling is found, and some species of *Dicentra* also share this oddity.

Duplotuber shares acuminate nectaries with part of sect. *Corydalis* (e.g., *C. solida*). However, forcing the tuberous sections together requires 10 extra steps on the ITS tree, and a position at the base of sect. *Corydalis* is even more expensive. The conclusion that acuminate nectaries evolved twice is inescapable. Papillate inner fruit epidermis is a character common to *Duplotuber*, *Capnogorium*, and most of the other tuberous groups (BRÜCKNER 1993). It is correlated to non-explosive dehiscence, and the polarity is uncertain.

A position of *Capnogorium* close to or in the core tuberous group has been suggested by BRÜCKNER (1985, 1992) on the basis of carpology, and is also supported by seed coat characters. The molecular data may hint in the same direction, although many groups have not been sequenced.

Conclusion

ITS sequences provide a tool for estimation of phylogeny at low and intermediate levels, and "indels" are particularly informative. A division of *Corydalis* in a tuberous and a nontuberous taxon would result in two polyphyletic groups. Three distinct subtaxa can be recognized on a morphological basis, coinciding with the three main clades in the rDNA tree. A unique derivation of zygomorphic flowers (*Capnoides* MILLER excepted) and paraphyly of the tribe *Corydaleae* sensu LIDÉN (1986) seem probable. The paraphyletic *Dicentra* should be split up, but molecular data are wanting.

We are indebted to VIVIAN ALDÉN for her invaluable help with DNA sequencing. HENRIK ZETTERLUND, Göteborg Botanic Garden, contributed fresh plant material. Roses to the organizers, for a marvellous time in Bayreuth.

References

BALDWIN, B. G., 1992: Phylogenetic utility of the internal transcribed spacers of nuclear ribosomal DNA in plants: an example from the *Compositae*. – Mol. Phyl. Evol. **1**: 3–16.

– 1993: Molecular phylogenetics of *Calycadenia* (*Compositae*) based on ITS sequences of nuclear ribosomal DNA: Chromosomal and morphological evolution reexamined. – Amer. J. Bot. **80**: 222–238.

BAUM, D. A., SYTSMA, K. J., HOCH, P. C., 1994: A phylogenetic analysis of *Epilobium* (*Onagraceae*) based on nuclear ribosomal DNA sequences. – Syst. Bot. **19**: 363–388.

BREMER, K., 1988: The limits of amino acid sequence data in angiosperm phylogenetic reconstruction. – Evolution **42**: 795–803.

BRÜCKNER, C., 1985: Zur Samenmorphologie in *Corydalis* VENT. (*Fumariaceae* DC.). – Gleditschia **13**: 53–61.

– 1992: Gynoecium ontogeny and carpology in *Corydalis* DC. sect. *Cheilanthifoliae* LIDÉN. – Flora **187**: 299–316.

– 1993: Comparative carpology in the tuber-bearing sections and sect. *Capnogorium* of *Corydalis* DC. (*Fumariaceae*). – Bot. Jahrb. Syst. **115**: 367–420.

FUKUHARA, T., 1992: Seed-coat anatomy of Japanese species of *Corydalis* and *Dicentra* (*Papaveraceae-Fumarioideae*). – Bot. Mag. (Tokyo) **105**: 303–321.

LIDÉN, M., 1981: On the typification of *Corydalis*. – Taxon **30**: 322–325.

– 1986: Synopsis of *Fumarioideae* (*Papaveraceae*) with a monograph of the tribe *Fumarieae*. – Opera Bot. **88**: 1–133.

LOCONTE, H., CAMPBELL, L., STEVENSON, D. W., 1995: Ordinal and familial relationships of Ranunculid genera. – Pl. Syst. Evol. [Suppl.] **9**: 99–118.

MOWAT, A., CHATER, A. O., 1964: *Corydalis*. – In TUTIN, T. G. & al., (Eds): Flora Europaea **1**: 252–254. – Cambridge: Cambridge University Press.

OH, B. Y., KIM, Y. S., 1988: The taxonomic characters of Korean *Corydalis* (Fumariaceae) and its significance in phylogenetic consideration. – Korean J. Pl. Taxon. **18**: 33–51.

OXELMAN, B., LIDÉN, M., 1995: The position of Circaeaster – evidence from nuclear ribosomal DNA. – Pl. Syst. Evol. [Suppl.] **9**: 189–193.

ROGERS, S. O., BENDICH, A. J., 1985: Extraction of DNA from milligram amounts of fresh, herbarium, and mummified plant tissues. – Pl. Mol. Biol. **5**: 69–76.

RYBERG, M., 1960: A morphological study of *Fumariaceae*. – Acta Horti Bergiani **19**: 121–248.

SOJAK, J., 1972: Nomenklatorische Bemerkungen. – Nationalumus. Praha Naturwiss. **140**: 127–134.

SUH, Y., THIEN, L. B., REEVE, H. E., ZIMMER, E. A., 1993: Molecular evolution and phylogenetic implications of internal transcribed spacer sequences of ribosomal DNA in *Winteraceae*. – Amer. J. Bot. **80**: 1042–1055.

SWOFFORD, D., 1993: PAUP: Phylogenetic analysis using parsimony. – Champaign., Ill.: Illinois Natural History Survey.

WOJCIECHOWSKI, M. F., SANDERSON, M. J., BALDWIN, B. G., DONOGHUE, M. J., 1993: Monophyly of aneuploid *Astragalus* (*Fabaceae*): evidence from nuclear ribosomal DNA internal transcribed spacer sequences. – Amer. J. Bot. **80**: 711–722.

Addresses of the authors: MAGNUS LIDÉN and TORSTEN AXBERG, Systematisk Botanik, Carl Skottsbergs Gata 22, S-413 19 Göteborg, Sverige (Sweden), e-mail: magnus. liden@systbot. gu. se . – TATSUNDO FUKHARA, Department of Botany, Faculty of Science, Kyoto University, Kyoto, 606–01 Japan.

Accepted February 2, 1995

Pl. Syst. Evol. [Suppl.] 9: 189–193 (1995)

The position of *Circaeaster* – evidence from nuclear ribosomal DNA

BENGT OXELMAN and MAGNUS LIDÉN

Received October 1, 1994

Key words: *Ranunculales, Circaeaster, Kingdonia.* – nrDNA, 28S, ITS2.

Abstract: Based on c. 800 bases from the 5′ end of the 28S rRNA gene, *Circaeaster* is shown to be a member of the *Ranunculales*, with *Kingdonia* as most probable sister-taxon. The ITS2 sequences were too variable to reliably align above the family level, but the sister-group relation of *Circaeaster* and *Kingdonia* showed robustness to alterations in alignment parameters and similarity levels comparable to well-established families.

Circaeaster agrestis MAXIM. is usually placed within the *Ranunculales* (e.g., CRONQUIST 1981, WU & KUBITZKI 1993), but alternative positions (*Chloranthaceae*, MAXIMOVICZ 1881; *Saururaceae*, JUNELL 1931) have been proposed.

FOSTER (e.g., 1971) noted the strong similarity of the open dichotomous leaf venation (unique among angiosperms) between *Circaeaster* and *Kingdonia uniflora* BALF. f. & W. W. SMITH. They also have tenuinucellate hemitropous ovules (HU & al. 1990) and similar pollen morphology (NOWICKE & SKVARLA 1982). Despite these similarities, *Circaeasteraceae* is often retained as a monotypic family whereas *Kingdonia* is put in *Ranunculaceae* (e.g., TAKHTAJAN 1969, TAMURA 1993).

Utilizing one of the internal transcribed spacers (ITS2) and about 800 bases from the 5′ end of the 28S gene of the nuclear ribosomal DNA (Fig. 1) we address the following questions: (1) is *Circaeaster* a member of the *Ranunculales*? (2) is *Circaeaster* a member of the *Ranunculaceae*? (3) are *Circaeaster* and *Kingdonia* sister groups?

Material and methods

Plant material. Sequences from 27 taxa were used in this study. Eight were downloaded from the EMBL database. Abbreviations used: GBT = Göteborg Botanical Garden, KGB = Kunming/Göteborg Botanical Expedition to Yunnan 1993, Ox. = BENGT OXELMAN.

Outgroup sequences: *Arabidopsis thaliana* (L.) HEYNH. (EMBL: X52320), *Sinapis alba* L. (EMBL: X57137), *Citrus limon* (L.) BORNM. f. (EMBL: X05910), *Fragaria ananassa* DUCHESNE (EMBL: X58118), *Lycopersicon esculentum* MILL. (EMBL: X52265, X07889 and X13557), *Daucus carota* L. (EMBL: X17534), *Vaccaria hispanica* (MILL.) RAUSCH. (Ox. 1764: Morocco,

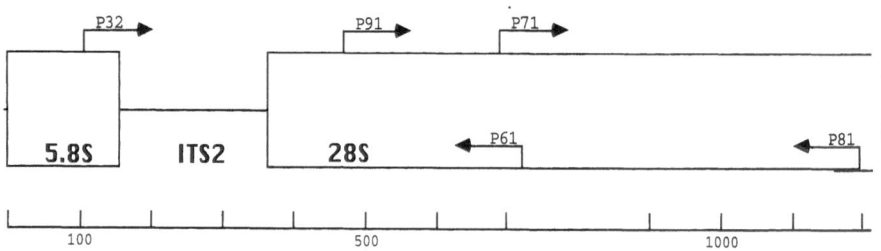

Fig. 1. Nuclear rDNA region sequenced for this study. Primers: P32 = GTCTTTGAAC-GCAAGTTG, P91 = GCGGAGGAAAAGAAACTTACAAGGA, P71 = ACGAGTC-GGGTTGTTTGGGAATG, P61 = CATTCCCAAACAACCCGACT, P81 = CCCGCT-CAGGCATAGTTCACCAT, P32 was designed by Torsten Axberg, the others by Bengt Oxelman. The primers were synthesized by Scandinavian Gene Synthesis, Köping, Sweden

Tselfat), *Silene fruticosa* L. (Ox. 934: Greece, Lakonia, Nea Itilon), *Ceratophyllum demersum* L., (Ox. 2254: Sweden, Lake Tåkern), *Drimys winteri* Forst. (Ox. 2275: GBT), *Nuphar pumila* (Timm) DC. (Ox. 2271: GBT), *Houttuynia cordata* Thunb. (Ox. 2263: GBT).

Ingroup: *Trochodendron aralioides* Sieb. & Zucc. (Ox. 2268: GBT), *Clematis apiifolia* DC. (Ox. 2276: GBT), *Ranunculus enysii* T. Krk. (Ox. 2273: GBT), *Trollius ranunculoides* Hemsl. (KGB 41: Yunnan, Napa Hai), *Coptis trifolia* Salisb. (Ox. 2250: GBT), *Hydrastis canadensis* L. (Ox. 2249: GBT), *Glaucidium palmatum* Sieb. & Zucc. (Ox. 2277: GBT), *Berberis dictyophylla* Franch. (KGB 538: Yunnan, Deqin, Xi Shan), *Podophyllum peltatum* L. (GBT, no voucher), *Bongardia chrysogonum* Boiss. (Ox. 2246: GBT), *Pteridophyllum racemosum* Sieb. & Zucc. (Ox. 2235: GBT), *Papaver rhoeas* L. (GBT, no voucher), *Kingdonia uniflora* (KGB 516: Yunnan, Deqin, Xi Shan), *Circaeaster agrestis*. (KGB 263: Yunnan, Mai Li), *Menispermum canadense* L. (Ox. 2267: GBT). Voucher specimens are deposited at GB.

DNA extraction, amplification and sequencing. The DNA extraction protocol of Yoon & al. (1991) was followed, but for difficult templates (e.g., herbarium material) a phenol/chloroform extraction (Maniatis & al. 1982) preceded the chloroform extraction. ITS2 plus c. 800 bp of the 5′-end of 28S (Fig. 1) was amplified from total genomic DNA using the polymerase chain reaction (PCR) and the primer pair P32/P81. For some difficult templates, it was necessary to amplify shorter fragments. 50 µl reactions were run on a Perkin-Elmer Cetus 480 thermal cycler, each containing 2.5 units Taq-polymerase (Promega or Advanced Biotechnologies), standard reaction buffer, 0.1 mM of each dNTP, 10 pmol of each primer, 0.5–10 ng template DNA and 1 mM $MgCl_2$. Denaturation (5 min at 95 °C) was followed by addition of polymerase at 80 °C, and 40 cycles of 95 °C for 30″, 57 °C for 1′, 72 °C for 2′), ending with 7 min at 72 °C. Fragments were separated on 1% SeaPlaque agarose gels and purified with either the MagicPCRPrep (Promega) kit or the Qiaquick (Qiagen) kit.

For sequencing we used the fmol kit (Promega). An initial labelling program was run involving 5 µl 5x reaction buffer, 5 pmol primer, 50–500 ng template, and 9 pmol each of the two or three dNTPs occurring closest to the 3′-end of the primer (dATP or dTTP labelled with ³⁵S; duPont) in a total volume of 16 µl. The labelling reaction consisted of a denaturation at 98 °C for 3 min, and 50 °C for 1 min, when 2 µl containing 5 units sequencing grade Taq polymerase were added. Then followed 20 cycles of (95 °C 30″, 50 °C 1′). Each reaction was then split into four termination tubes, each containing 2 µl d/ddNTP, and run for 60 cycles of (95 °C for 30″, 58 °C for 1′, 70 °C for 2′). The fragments were separated on polyacrylamide gels (35 × 45 cm) using the salt gradient method of Sheen & Seed (1988). Before loading, the samples were denatured at 90 °C for 2 min and then put on ice. The sequences are deposited in the EMBL database (accession numbers X83829–X83849).

Data analysis. The 28S sequences were aligned manually. Gaps were coded as missing values. Informative gap positions were recoded as additional characters. The ITS2 sequences were aligned with Pileup (GENETICS COMPUTER GROUP 1991). GapWeight/GapLength-Weight were set in the following combinations: 5.0/5.0, 5.0/0.3, 3.0/1.0, 3.0/0.0, 1.0/1.0, and 1.0/0.0.

PAUP version 3.1.1 (SWOFFORD 1993) was used for finding the most parsimonious trees for the aligned sequences. The heuristic search option with 10 random additions and TBR swapping was used. Decay indices (BREMER 1988) were approximated from all trees found up to four steps longer. Three houndred bootstrap replicates were run, each with random addition sequence. Conflicting hypotheses (TAMURA 1993, LOCONTE & al. 1995, this volume) regarding the position of *Circaeaster* were evaluated using the topological constraints option in PAUP.

Results and discussion

For the 28S sequences, a single most parsimonious tree with length 1082, consistency index 0.49 (0.40 with autapomorphies excluded), and retention index 0.40 was found. The ITS2 sequences could not be meaningfully aligned above family level, but the grouping of *Circaeaster* and *Kingdonia* was as robust (or even more robust) to alterations in alignment parameters and resampling of data as were well-established families as *Ranunculaceae* s. str., *Berberidaceae* s.l., and *Papaveraceae*.

Two main conclusions can be drawn from this study. *Kingdonia* and *Circaeaster* appear as sister-groups and they also appear to be well at home in the *Ranunculales*, although the boundaries of the order are uncertain. These conclusions are in agreement with HOOT & CRANE (1995, this volume) and LOCONTE & al. (1995, this

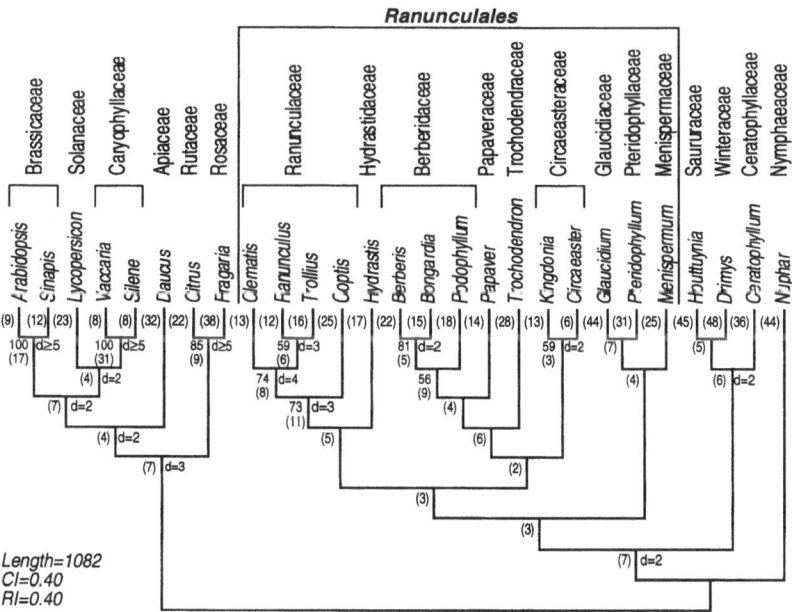

Fig. 2. Single most parsimonious tree found from the 28S sequences. Numbers to the left of branches denote bootstrap support higher than 50%; numbers in parentheses indicate the inferred number of nucleotide substitutions (ACCTRAN optimization); *d* decay index (only shown for d ⩾ 2)

volume). It took eight extra steps to nest *Circaeaster* and *Kingdonia* within *Ranunculaceae* with the constraint [*Coptis*, (*Trollius*, *Clematis*, *Ranunculus*, *Kingdonia*, *Circaeaster*)] (LOCONTE & al. 1995, this volume). The constraint (*Kingdonia*, *Clematis*) (TAMURA 1993) resulted in 16 trees requiring six extra steps, all compatible with keeping *Circaeasteraceae* monotypic. Our data thus supports a ditypic *Circaeasteraceae*, in agreement with CRONQUIST (1981) and HOOT & CRANE (1995, this volume).

Many of the groupings in Fig. 2 are not very robust to resampling of data (as indicated by low bootstrap values), but *Berberidaceae* and *Ranunculaceae* appear well supported. The basal position of *Coptis* in *Ranunculaceae* is in agreement with other recent studies (this volume). Surprising results in our study are the unstable position of *Pteridophyllum* (did not group with *Papaver*) and the position of *Trochodendron* (nested in *Ranunculales*). The long terminal branches (Fig. 2) may indicate that the variable positions in the sequenced part of 28S are in fact too variable to be an efficient tool for phylogenetic estimation.

We express our warm gratitude to VIVIAN ALDÉN for skilful technical assistance.

References

BREMER, K., 1988: The limits of amino acid sequence data in angiosperm phylogenetic reconstruction. – Evolution **42**: 795–803.

CRONOUIST, A., 1981: An integrated system of classification of flowering plants. – New York: Columbia University Press.

FOSTER, A. S., 1971: Additional studies on the morphology of blind vein endings in the leaf of *Circaeaster agrestis*. – Amer. J. Bot. **58**: 263–272.

GENETICS COMPUTER GROUP, 1991: Program manual for the GCG package, ver. 7. – 575 Science Drive, Madison, Wisconsin.

HOOT, S., CRANE, P., 1995: Inter-familial relationships in the *Ranunculidae* based on molecular systematics. – Pl. Syst. Evol. [Suppl.] **9**: 119–131.

HU, Z.-H., YANG, J., JING, R. Q., DONG, Z.-M., 1990: Morphological studies on *Circaeaster agrestis* II. Morphology and anatomy of flower, fruit and seed. – Cathaya **2**: 77–88.

JUNELL, S., 1931: Die Entwicklungsgeschichte von *Circaeaster agrestis*. – Svensk Bot. Tidskr. **25**: 238–270.

LOCONTE, H., CAMPBELL, L., STEVENSON, D., 1995: Ordinal and familial relationships of *Ranunculid* genera – Pl. Syst. Evol. [Suppl.] **9**: 99–118.

MANIATIS, T., FRITSCH, E. F., SAMBROOK, J., 1982: Molecular cloning: a laboratory manual. – Cold Spring Harbor: Cold Spring Harbor Laboratory.

MAXIMOWICZ, C. J., 1881: Diagnoses Plantarum Novarum Asiaticarum, IV. – Bull. Acad. Sci. St. Petersbourg **27**: 425–560.

NOWICKE, J. W., SKVARLA, J. J., 1982: Pollen morphology and the relationships of *Circaeaster*, of *Kingdonia*, and of *Sargentodoxa* to the *Ranunculales*. – Amer. J. Bot. **69**: 990–998.

SHEEN, J.-Y., SEED, B., 1988: Electrolyte gradient gels for DNA sequencing. – BioTechniques **6**: 942–943.

SWOFFORD, D., 1993: PAUP: Phylogenetic analysis using parsimony, ver. 3.1.1. – Champaign, Ill. Illinois Natural History Survey.

TAKHTAJAN, A. L., 1969: Flowering plants: origin and dispersal. – Washington DC.: Smithsonian Institution.

TAMURA, M., 1993: *Ranunculaceae.* – In KUBITZKI, K., ROHWER, J. G., BITTRICH, V., (Eds): Families and genera of vascular plants, **2**, pp. 563–583. – Berlin: Springer.

WU, C.-Y., KUBITZKI, K., 1993: *Circaeasteraceae.* – In KUBITZKI, K., ROHWER, J. G., BITTRICH, V., (Eds): Families and genera of vascular plants, **2**, pp. 288–289. – Berlin: Springer.

YOON, C.-S., GLAWE, D. A., SHAW, P. D., 1991: A method for rapid small-scale preparation of fungal DNA. – Mycologia **83**: 835–838.

Address of the authors: BENGT OXELMAN and MAGNUS LIDÉN, Department of Systematic Botany, Carl Skottsbergs Gata 22, S-413 19 Göteborg, Sweden.

Accepted December 26, 1994

Pl. Syst. Evol. [Suppl.] 9: 195–199 (1995)

Phylogenetic relationships of the Lardizabalaceae and Sargentodoxaceae: Chloroplast and nuclear DNA sequence evidence

Sara B. Hoot, Alastair Culham, and Peter R. Crane

Received February 15, 1995

Key words: *Lardizabalaceae, Sargentodoxaceae, atp*B, *rbc*L, 18S nuclear ribosomal DNA, phylogeny.

Abstract: A combined data set of three genes (*atp*B and *rbc*L from the chloroplast genome and 18S nuclear ribosomal DNA) was used to assess relationships among the genera of *Lardizabalaceae* and to explore the relationships of the family to *Sargentodoxaceae* and other families in the *Ranunculidae* (*Fumariaceae, Eupteleaceae, Circaeasteraceae, Kingdoniaceae, Berberidaceae,* and *Menispermaceae*). *Lardizabalaceae* and *Sargentodoxaceae* together comprise a monophyletic group. *Sargentodoxa, Decaisnea,* and *Sinofranchetia,* in that order, are basal within the *Lardizabalaceae* sensu lato. As in previous studies based on molecular and morphological data, the 'core' *Lardizabalaceae* (*Boquila, Lardizabala, Akebia, Holboellia,* and *Stauntonia*) characterized by laminar placentation, is well-supported as a monophyletic group. A subset of this clade consisting of three Asian genera (*Akebia, Holboellia,* and *Stauntonia*) is also a well-supported monophyletic group.

The *Lardizabalaceae* are a family of twining (rarely erect) shrubs found in temperate areas of Eastern Asia and South America. The family is characterized by alternate, palmate (rarely pinnate) compound leaves; regular unisexual flowers; six overlapping or valvate sepals (three in *Akebia*); staminodia or petals small or absent; three carpels (up to nine in *Akebia*); numerous ovules (four, by abortion, in *Boquila*) with laminar placentation (sub-marginal in *Decaisnea* and *Sinofranchetia*; Taylor 1967, Qin 1989).

Previous phylogenetic analyses based on *rbc*L sequence data for a single species of *Akebia* suggest that the *Lardizabalaceae* occupy a key, and potentially basal position, in the evolution of the *Ranunculidae* (Chase & al. 1993). More recently, phylogenetic analyses using three gene sequences (*rbc*L and *atp*B cpDNA, 18S nrDNA) have clarified intergeneric relationships in the family (Hoot & al. 1995) and have provided a more secure basis for representing the family in our ongoing investigation of the basal eudicot radiation (Hoot & Crane 1995). One issue, however, that remains to be examined in detail is the phylogenetic position of *Sargentodoxa*, which is commonly thought to be close to the *Lardizabalaceae* but has been placed with the legumes by recent *rbc*L sequence data, probably due to misidentification (CHASE & al. 1993). This paper focuses on the placement of *Sargentodoxa* with respect to the *Lardizabalaceae* and the effect of its inclusion on phylogenetic hypotheses within the family. Outgroup sampling was also increased over previous studies (HOOT & al. 1995) to include taxa (*Euptelea*, *Kingdonia*, and *Circaeaster*) which appear to be close to *Larbizabalaceae* based on a broader survey of ranunculids (HOOT & CRANE 1995, unpublished results).

Materials and Methods

The seven genera of *Lardizabalaceae* and seven outgroup genera sampled are listed in Table 1. All 14 taxa were sequenced for each of the following genes: *atp*B and *rbc*L from the chloroplast genome, and 18S nuclear ribosomal DNA (18S nrDNA). DNA extraction, gene. amplification, and sequencing for most genera were perfomed as described in HOOT & al. (1995). The sequencing reactions for *Euptelea*, *Kingdonia*, *Circaeaster*, and *Sargentodoxa* were performed as described in HOOT (1995). Sequencing, accession, and voucher information may be found in HOOT & al. (1995, in preparation).

Data analysis. Phylogenetic analyses of the combined data sets (*atp*B, *rbc*L, 18S nrDNA) were performed using PAUP 3.1 (SWOFFORD 1993) using the branch-and-bound search option to assure recovery of the most parsimonious trees. PAUP was also used to perform bootstrap analysis with 1,000 replications using the heuristic search option (FELSENSTEIN 1985). The decay indices were computed up to ten steps longer than the most parsimonious tree using the heuristic search option (DONOGHUE & al. 1992). Alternative tree topologies and resultant changes in tree length were explored using MacClade 3.0 (MADDISON & MADDISON 1992).

Outgroup taxa selected included those used in a previous analysis of the family (HOOT & al. 1995) together with *Kingdoniaceae*, *Circaeasteraceae*, and *Eupteleaceae*, which recent studies place as relatively basal in the *Ranunculidae* (HOOT & CRANE 1995, work in progress).

Table 1. Families and species sequenced. The *Akebia* gene sequence for *rbc*L is from Qiu (CHASE & al. 1993)

Family	Species
Lardizabalaceae	*Akebia quinata* DECNE.
	Boquila trifoliata DECNE.
	Decaisnea fargesii FRANCHET
	Holboellia latifolia WALLICH
	Lardizabala biternata RUIZ & PAVÓN
	Sinofranchetia chinensis HEMSLEY
	Stauntonia hexaphylla (THUNB.) DECNE.
Berberidaceae	*Nandina domestica* THUNB.
Circaeasteraceae	*Circaeaster agrestis* MAXIM.
Eupteleaceae	*Euptelea polyandra* SIEBOLD & ZUCC.
Fumariaceae	*Dicentra eximia* TORREY
Kingdoniaceae	*Kingdonia uniflora* BALF. f. & W. W. SM.
Sargentodoxaceae	*Sargentodoxa cuneata* REHDER & E. WILSON
Menispermaceae	*Tinospora caffra* MIERS

In previous studies, the *Papaverales* are resolved as basal to the *Lardizabalaceae* and other ranunculids (CHASE & al. 1993; HOOT & CRANE 1995, work in progress) and *Dicentra eximia* (*Fumariaceae, Papaverales*) was therefore used to root the phylogenetic analysis presented here (Table 1).

Results

The analysis performed using a combination of all three data sets (*atp*B, *rbc*L, and 18S nrDNA) resulted in three trees derived from 580 variable sites (228 informative sites) with a treelength = 827 steps, consistency index (CI) excluding autapomorphies = 0.63, and retention index (RI) = 0.68. The clade which collapses in a strict consensus tree of the three most parsimonious trees is shown with dotted lines (Fig. 1). The clades consisting of *Kingdonia/Circaeaster* and *Nandina/Tinospora* form unresolved sister groups to the *Lardizabalaceae* sensu lato (*Lardizabalaceae* plus *Sargentodoxaceae*). The monophyly of the *Lardizabalaceae* sensu lato (including *Sargentodoxa*) is strongly supported with 25 nucleotide changes, a bootstrap value of 100%, and a decay index of > 10. *Sargentodoxa* is resolved as basal to the *Lardizabalaceae* sensu stricto, with a high bootstrap value (100%) and decay index = around 9. *Decaisnea* and *Sinofranchetia* are basal in that order to all other members of the *Lardizabalaceae*, also with strong support (Fig. 1). A clade of 'core'

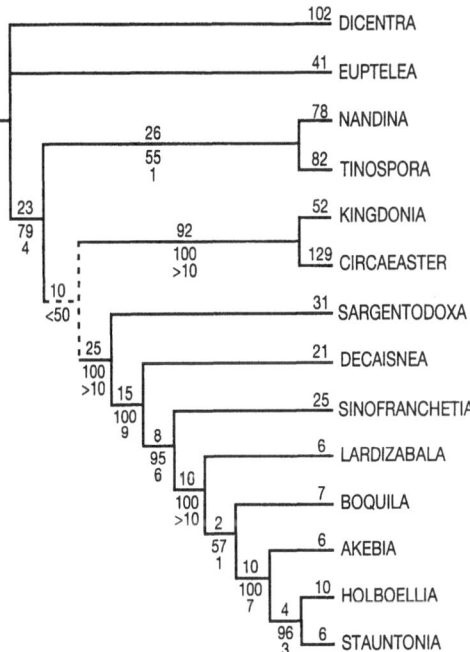

Fig. 1. One of three most parsimonious phylogenetic trees resulting from a combination of *atp*B, 18S nrDNA, and *rbc*L data sets. Numerals above branches indicate the number of nucleotide changes supporting each branch. Numerals below indicate the percentage of times that the branch was recovered in 1000 bootstrap replications. Numerals below the bootstrap values are decay indices, indicating how many additional steps are necessary before the branch collapses. Dotted lines indicate where a branch collapses in the strict consensus tree derived from the three most parsimonious trees. CI excluding uninformative characters = 0.63, RI = 0.68

Lardizabalaceae (*Lardizabala*, *Boquila*, *Akebia*, *Stauntonia*, and *Holboellia*) is strongly supported (100% bootstrap value, decay index > 10). The data resolve the relationship of the two South American taxa, *Boquila* and *Lardizabala*, as paraphyletic with *Lardizabala* basal to all other 'core' *Lardizabalaceae*. There is strong support for the group of three Asian genera (*Akebia*, *Holboellia*, *Stauntonia*; 100% bootstrap, decay index = 7) and for a sister-group relationship between *Holboellia* and *Stauntonia* (96% bootstrap, decay index = 3).

Discussion

In broader analyses of the *Ranunculidae* and 'lower' *Hamamelididae* based on *rbc*L, *atp*B, and 18S nrDNA sequences separately or combined, *Sargentodoxa* is always placed within the *Ranunculidae*, basal to the *Lardizabalaceae* (Hoot & Crane 1995, work in progress). The trees derived from each gene sequence (trees not presented here), or a combination of the three data sets, all support the monophyly of the *Lardizabalaceae* s.l. (Fig. 1). This is in contrast to our earlier report (Hoot et al. 1995) where the tree based on *rbc*L sequence data was anomalous in placing *Sinofranchetia* among the outgroups (unlike *atp*B, 18S nrDNA, and the combined data set). Because of this anomalous position, we recently obtained fresh material from the same specimen of *Sinofranchetia* (Royal Botanic Garden, Edinburgh). This new material amplified readily using several different protocols. This is in contrast to the original sample in which the DNA was degraded and the *rbc*L gene was very difficult to amplify. The new *rbc*L sequence, generated by an automated DNA sequencer (ABI), differs from the original *rbc*L sequence and reanalysis of the *rbc*L data alone places *Sinofranchetia* within the *Lardizabalaceae* in the same position as seen in Fig. 1.

The tree based on the combined data places *Sargentodoxa* as basal to the *Lardizabalaceae* (Fig. 1). This agrees with previous classifications based on morphology, which often place *Sargentodoxa* as a separate family most closely related to the *Lardizabalaceae* (Hutchinson 1973, Cronquist 1981, Takhtajan 1987, but see Loconte & al. 1995). Morphologically, the genus is separated from other *Lardizabalaceae* on the basis of its carpel with a single pendulous ovule. It resembles other members of the *Lardizabalaceae* in its vining habit, alternate trifoliate leaves, dioecy (also found in *Lardizabala* and *Boquila*), flowers in pendulous racemes, and floral parts in sixes.

The analysis presented here (Fig. 1) as well as previous analyses based on the three genes individually (Hoot & al. 1995) strongly support the recognition of a clade of 'core' *Lardizabalaceae* comprising *Akebia*, *Stauntonia*, *Holboellia*, *Lardizabala*, and *Boquila* (Fig. 1). An important morphological feature that is diagnostic of this clade is laminar placentation. The two genera of *Lardizabalaceae* with submarginal placentation (*Decaisnea* and *Sinofranchetia*) are placed external to this group and this is consistent with previous phylogenetic results based on traditional data and classification schemes (Loconte & Estes 1989, Qin 1989).

We thank D. Nickrent and E. A. Zimmer for ribosomal DNA amplification and sequencing primers, and G. Zurawski for *rbc*L sequencing primers. We are grateful to the following for providing us with leaf material: R. McBeath, Royal Botanic Garden, Edinburgh; A. Reznicek, University of Michigan; T. Lammers, The Field Museum; M. Maunder and T.

UPSON, Royal Botanic Gardens, Kew; H. QIN, Chinese Academy of Sciences, Beijing; and E. VAN JAARSVELD, Kirstenbosch Botanical Garden. This work was supported in part by NSF grants DEB–9020237 to PRC and DEB–9596011 to SBH.

References

CHASE, M. W., SOLTIS, D. E., OLMSTEAD, R. G., MORGAN, D., LES, D. H., MISHLER, B. D., DUVALL, M. R., PRICE, R. A., HILLS, H. G., QIU, Y.-L., KRON, K. A., RETTIG, J. H.,CONTI, E., PALMER, J. D., MANHART, J. R., SYTSMA, K. J., MICHAELS, H. J., KRESS, W. J., KAROL, K. G., CLARK, W. D., HEDRÉN, M., GAUT, B. S., JANSEN, R. K., KIM, K.-J., WIMPEE, C. F., SMITH, J. F., FURNIER, G. R., STRAUSS, S. H., XIANG, Q.-Y., PLUNKETT, G. M., SOLTIS, P. S., SWENSEN, S. M.,WILLIAMS, S. E., GADEK, P. A., QUINN, C. J., EGUIARTE, L. E., GOLENBERG, E., LEARN, G. H., GRAHAM, S. W., BARRETT, S. C. H., DAYANANDAN, S., ALBERT, V. A., 1993: Phylogenetics of seed plants: an analysis of nucleotide sequences from the plastid gene rbcL. – Ann. Missouri Bot. Gard. **80**: 528–580.

CRONQUIST, A., 1981: An integrated system of classification of flowering plants. – New York: Columbia University Press.

DONOGHUE, M. J., OLMSTEAD, R. G., SMITH, J. F., PALMER, J. D., 1992: Phylogenetic relationships of Dipsacales based on rbcL sequences. – Ann. Missouri Bot. Gard. **79**: 333–345.

FELSENSTEIN, J., 1985: Confidence limits on phylogenies: an approach using the bootstrap. – Evolution **39**: 783–791.

HOOT, S. B., CULHAM, A., CRANE, P. R., 1995: The utility of atpB gene sequences in resolving relationships in the Lardizabalaceae, including comparisons with rbcL and 18S ribosomal DNA sequenes. –Ann. Missouri Bot. Gard. **82**: 194–207.

– 1995: Phylogeny of the Ranunculaceae based on preliminary atpB, rbcL, and 18S nuclear ribosomal DNA sequence data. – Pl. Syst. Evol. [Suppl.] **9**: 241–251.

– CRANE , P. R., 1995: Interfamilial relationships in the Ranunculidae based on molecular systematics. –Pl. Syst. Evol., [Suppl.] **9**: 119–131.

HUTCHINSON, J., 1973: The families of flowering plants. – Oxford: Clarendon Press.

LOCONTE, H., ESTES, J. R., 1989: Phylogenetic systematics of Berberidaceae and Ranunculales (Magnoliidae). – Syst. Bot. **14**: 565–579.

– CAMPBELL, L. M., STEVENSON, D. W., 1995; Ordinal and familial relationships of the Ranunculiflorae. – Pl. Syst. Evol., [Suppl.] **9**: 99–118.

MADDISON, W. P., MADDISON, D. R., 1992: MacClade: interactive analysis of phylogeny and character evolution, version 3.0. – Sunderland, Massachusetts: Sinauer Assoc.

QIN, H.-N., 1989: An investigation on carpels of Lardizabalaceae in relation to taxonomy and phylogeny. – Cathaya **1**: 61–82.

SWOFFORD, D. L., 1993: PAUP: Phylogenetic analysis using parsimony, version 3.1. Computer program distributed by Illinois Nat. Hist. Surv., Champaign, IL.

TAKHTAJAN, A., 1987: Systema Magnoliophytorum. – Leningrad: Nauka (in Russian).

TAYLOR, B. A. S., 1967: The comparative morphology and phylogeny of the Lardizabalaceae. Ph.D. Thesis, Botany, Indiana University (University Microfilms, Inc., Ann Arbor, Michigan, code 68–7226).

Addresses of authors: Dr SARA B. HOOT, Department of Biological Sciences, Lapham Hall, PO Box 413, University of Wisconsin–Milwaukee, Milwaukee, WI 53201, USA. – Dr ALASTAIR CULHAM, Department of Botany, School of Plant Sciences, University of Reading, Reading, RG6 2AS, UK. – Dr PETER R. CRANE, Departments of Botany and Geology, The Field Museum, Roosevelt Road at Lake Shore Drive, Chicago, Illinois 60605, USA.

Accepted February 24, 1995

Pl. Syst. Evol. [Suppl.] 9: 201–206 (1995)

Phylogeny and classification of the *Ranunculaceae*

MICHIO TAMURA

Received October 11, 1994

Key words: *Ranunculaceae*, *Paeonia*, *Glaucidium*, *Hydrastis*, *Circaeaster*, *Kingdonia*, *Asteropyrum*, – Systematics, chromosomes, carpel.

Abstract: Some problems of the phylogenetic classification of *Ranunculaceae* are discussed. The family is delimited excluding *Paeonia*, *Glaucidium* and *Circaeaster*, and including *Hydrastis* and *Kingdonia*. It becomes difficult to subdivide the family into the R-type and T-type group by chromosome size. These two groups can be maintained by a difference in stainability of the interphase nuclei and the prophase chromosomes with the exception of *Myosurus*. *Asteropyrum* has fairly long chromosomes but contains benzylisoquinoline alkaloids in large quantity as the T-type group. The genus is placed at the top of the T-type group linking the T-type to the R-type group. *Callianthemum* and *Adonis* have achenes but are rather similar to *Helleboreae* with follicles in their mode of vascular supply to the ovule, and these two genera may be directly connected to *Helleboreae* phyletically.

The family *Ranunculaceae* delimited by DE CANDOLLE (1817) has been considered to be one of the typical natural families for a long time. PRANTL (1887) recognized the heterogeneity of the family, and three genera, *Paeonia* L., *Glaucidium* SIEB. & ZUCC., and *Hydrastis* L., which were thought improper as members of *Ranunculaceae*, were classified as tribe *Paeonieae*. PRANTL (1887) wrote that the tribe might be the direct offspring of the ancestral group from which *Ranunculaceae* and *Berberidaceae* were derived, and could be placed in the *Berberidaceae* as its basal-most member. *Circaeaster* MAXIM. was described by MAXIMOWICZ (1881), and classified in the *Ranunculaceae* by HALLIER (1903). *Kingdonia* BALF. f. & W. W. SMITH was described and classified in the *Ranunculaceae* by BALFOUR & SMITH (1914). DIELS (1932) emphasized the relationship between *Circaeaster* and *Kingdonia* mainly based on the similarity in the venation of their leaves, and suggested that they should be included in the *Ranunculaceae*, while HUTCHINSON (1926) excluded *Circaeaster* from the family and created a monotypic family *Circaeasteraceae*. There are still many debatable points on the affinities of these genera and their systematic position is not settled yet.

The family *Ranunculaceae*, excluding *Paeonia*, *Glaucidium* and *Circaeaster*, and including *Kingdonia* and *Hydrastis* can be divided into two groups with karyological characters. Since the time of ADANSON (1763), on the other hand, the family has often been divided into a group with multiovular carpels or follicles and a group with uniovular carpels or achenes.

In this article, the delimitation and subdivision of the *Ranunculaceae* is taken up for discussion.

Paeonia, *Glaucidium* and *Hydrastis*

Among the three genera which were thought improper as members of *Ranunculaceae* by PRANTL (1887), *Paeonia* is decidedly not a member of *Ranunculaceae*. It has no clear relationship not only to *Ranunculaceae* but also to any other family of angiosperms. *Paeonia* has many particular features, and especially its embryogenesis is quite unique and otherwise unknown in angiosperms (YAKOVLEV & YOFFE 1957, CAVE & al. 1961, CARNIEL 1967).

Glaucidium also is not a member of *Ranunculaceae*. I created *Glaucidiaceae* for it based on the following characteristics (TAMURA 1972). (1) Very many stamens initiated centrifugally and supplied with branches of the receptacular bundles. In contrast, in *Ranunculaceae* stamen initiation is centripetal. (2) Thick integuments. The outer integument is 6 to 13, and the inner about 5 cell layers thick, i.e., altogether 11 to 18 cell layers. In *Ranunculaceae* integuments are 6–8 cell layers thick, and either single or double (KUMAZAWA 1938). (3) Dehiscence of carpels along both the ventral and dorsal sutures. (4) Haploid chromosome number of n = 10. (5) Coumarines present in large quantity. Thereafter, I noticed that in *Garidella* L. and sometimes in *Nigella* L., carpels also dehisce along both the ventral and dorsal sutures. Even if this character is omitted, *Glaucidium* is quite different from *Ranunculaceae*.

TOBE & KEATING (1985) emphasized the peculiarity of *Hydrastis* and supported its separation as a monotypic family *Hydrastidaceae*. According to these authors the distinctive features which distinguish the genus from *Ranunculaceae* are the following: (1) unique mode of origin of the vascular supply to stamens and carpels. (2) Micropyle being formed by both integuments. (3) Xylem not V-shaped in cross section. (4) Presence of scalariform vessel perforations. (5) Haploid chromosome number of n = 13. (6) Pollen tectum consisting of a compound layer of striae. (7) Leaf mesophyll not differentiated. (8) Unique course of stem medullary bundles. (9) D-galactose present.

Most of these features, however, are not unusual in *Ranunculaceae*. In *Ranunculus* L., especially in the species with many small carpels, the traces to carpels and sometimes to stamens are more or less fascicular as in *Hydrastis*. The comparative length of the inner and outer integuments was regarded as an important character by PRANTL (1887), but KUMAZAWA (1938) showed that in *Aquilegia* L. the outer integument usually is longer than the inner, and that this character is not stable. Although the xylem of *Ranunculaceae* is usually V-shaped in cross section, this sometimes is obscure as in the petioles of *Helleborus* L., *Calathodes* HOOK f. & THOMS., *Beesia* BALF. f. & W. W. SMITH etc. In *Hydrastis*, scalariform vessel perforations are found, but simple perforations also are common. In *Ranunculaceae*, perforations mostly are simple, but scalariform perforations are not rare as reported by AVITA & INAMDAR (1981). Furthermore, according to CHEN & LI (1990), in *Asteropyrum* DRUMM. & HUTCH., all vessels are typically scalariform with over 30 transverse bars, whereas in *Hydrastis* there are at most three to 10 bars. Accordingly, although *Hydrastis* has some particular features, the genus not necessarily needs to be separated from *Ranunculaceae*.

Circaeaster and *Kingdonia*

Circaeaster and *Kingdonia* attracted the attention of botanists by having open dichotomous leaf-venation and leaf traces in even number. Because these characteristics are unique among angiosperms, these two genera were often included in the same group, e.g., in *Kingdoniinae* or *Kingdonioideae* of *Ranunculaceae* by JANCHEN (1949) and THORNE (1974), respectively, and in *Circaeasteraceae* by CRONQUIST (1968) and THORNE (1983). These two genera, however, have no close relationship to each other.

The embryology of *Circaeaster* was studied by JUNELL (1931) and many particular features were found. (1) The tapetum arises from the innermost layer of the anthers, and not from the sporogenous tissue as in *Ranunculaceae*. (2) Two ovules are attached not to the ventral sutures, as in *Ranunculaceae*, but to the lateral walls of the ovary at the same level. (3) The integument consists of 2–3 cell layers and is much thinner than in *Ranunculaceae*. (4) Endosperm formation of cellular type. (5) Five cells arranged longitudinally in the 5-celled stage of the embryo. (6) In the process of seed formation the nucellus and integument are dissolved and the endosperm lies naked in the ovary. (7) The endosperm is differentiated into a micropylar and a chalazal part, the surface of the former is suberized, and the latter degenerates. These features never can be observed in *Ranunculaceae*.

Considering these characteristics, JUNELL (1931) concluded that *Circaeaster* is related to *Saururaceae*. MAXIMOWICZ (1881) first described the genus and suggested a relationship to *Chloranthaceae*. OLIVER (1895) stated that *Circaeaster* might be close to the *Anemoneae* of *Ranunculaceae*, although ovular attachment is different. HUTCHINSON (1926) created the monotypic *Circaeasteraceae* for it. As stated by many authors *Circaeaster* is a very reduced plant, and its unclear relationships are due to this fact.

In contrast to this, most characters of *Kingdonia* are common in *Ranunculaceae*. Besides the open dichotomous venation, the only other important feature unusual in the family is its helobial endosperm (MU 1983). Even if these characters are unique, *Kingdonia* can be included in the very diverse *Ranunculaceae*.

Subdivision of the family by karyological characters

Chromosome characters were first taken into account for the classification of *Ranunculaceae* by LANGLET (1932). He called long, generally bent or repeatedly curved chromosomes R(anunculus)-type chromosomes, and generally simply curved, smaller chromosomes T(halictrum)-type chromosomes, and divided the family into a group with R-type chromosomes and a group with T-type chromosomes. In *Ranunculus*, the typical genus with R-type chromosomes, however, *R. nipponicus* (MAKINO) NAKAI var. *major* HARA and *R. sceleratus* L. have comparatively small chromosomes, ranging from ca. 3 to 1–1.5 μm long. In *Coptis japonica* (THUNB.) MAKINO with T-type chromosomes, the largest chromosome is ca. 2.5 and the smallest ca. 2 μm long, and in *Aquilegia flabellata* SIEB & ZUCC. and *Dichocarpum stoloniferum* (MAXIM.) W. T. WANG & HSIAO the largest is ca. 2 and the smallest ca. 1.5 μm long (OKADA & TAMURA 1979). These values of chromosome length of R-type and T-type chromosomes are not clearly distinct. It is thus evident that the distinction between R-type and T-type is not always clear as far as the size of the chromosomes is concerned.

Both chromosome types, however, can be distinguished by the stainability of prophase chromosomes and interphase nuclei (KURITA 1960, OKADA & TAMURA 1979). In T-type genera, several deeply stained bodies can clearly be observed in the interphase nuclei in a dilutedly and evenly stained matrix. The prophase chromosomes are stained deeply in the proximal part of the arms, and the boundary between the stained and unstained part is distinct. In the R-type genera, only few dark stained bodies are obscurely observed in the interphase nuclei in a well and unevently stained matrix. The prophase chromosomes are stained throughout their length (OKADA & TAMURA 1979).

An exception to this is *Myosurus* L., which usually is considered to be closely related to *Ranunculus*. According to KURITA (1963), *Myosurus minimus* L. has comparatively small metaphase chromosomes, interphase nuclei of the T-type, and prophase chromosomes of the R-type. It is necessary to confirm these observations in other species.

Asteropyrum has considerably large chromosomes and a diploid chromosome number of $2n = 16$. Its basic number is supposed to be eight. According to YANG & al. (1993), chromosomes of *Asteropyrum* belong to the R-type, but ZHANG (1982) stated that they belong to the T-type. This genus contains benzylisoquinoline alkaloids like berberine and magnoflorine (HSIAO 1980, ZHU & HSIAO 1991). It is a well-known fact that the genera with T-type chromosomes accumulate alkaloids of the benzylisoquinoline type, while genera with R-type chromosomes do not accumulate such alkaloids. Thus, *Asteropyrum* may be said to belong to the T-type group from a phytochemical point of view. The large chromosomes and the basic number of eight, the commonest number in the R-type group, may suggest that *Asteropyrum* links the T-type to the R-type group. The bitegmic ovules (SUN & WANG 1983) and the scalariform vessels (CHEN & LI 1990) of *Asteropyrum* may show the primitiveness of the genus.

The examples of *Myosurus* and *Asteropyrum* lead to doubts about the subdivision of the family into an R- and T-type group. I believe, however, that this karyological character is still of prime importance in the classification of the *Ranunculaceae*. The results from phytochemical studies support the distinction of these two karyological groups.

Subdivision of the family by carpel features

Ranunculaceae have often been divided into a group with uniovular and one with multiovular carpels, or a group with follicles and one with achenes. The uniovular carpels or achenes were usually regarded as being derived from the multiovular carpels or follicles by reduction of number of ovules and fusion of carpellar bundles. LANGLET (1932) suggested that such evolution of the carpels took place independently in the R- and T-type group. PRANTL (1887) suspected two possibilities of carpel evolution in the family. In the first, ovules are produced along both margins of a carpel, and are supplied by branches from both ventral bundles. When the number of ovules is reduced to one, this still is supplied with a branch from a ventral bundle, as in *Callianthemum* C. A. MEY. Alternatively, one fertile ovule is produced at the common base of the free carpel margins, and a few abortive ovules sometimes are observed along both lateral margins. These are regarded as evidence for reduction of ovule number. The former possibility is realized in PRANTL's (1887) tribe *Helleboreae* which usually has follicles, and the latter in his tribe *Anemoneae* which has achenes.

PRANTL (1887) classified *Callianthemum* in his *Helleboreae*, the follicular group, and regarded it as the end member in the evolution of carpel reduction. Except for *Thalictrum* L., the genera of his *Anemoneae*, the achenial group, have unitegmic ovules. In *Callianthemum* the ovules are bitegmic as in the genera of *Helleboreae*. *Adonis* L. has the same carpel features as *Callianthemum*. Both genera are closely related to each other and seem to be considerably isolated from other members of the achenial group, such as *Anemone* L., *Ranunculus*, *Clematis* L., etc. PRANTL's (1887) idea that *Callianthemum* is directly connected to the group with multiovular carpels may be correct at least from a phylogenetic point of view.

References

ADANSON, M., 1763: Familles des plantes. – Paris: Vincent.

AVITA, S., INAMDAR, J. A., 1981: Diversity in the vessel elements of *Ranunculaceae-Paeoniaceae* complex. – Feddes Repert. **92**: 397–411.

BALFOUR, I. B., SMITH, W. W., 1914: *Kingdonia*. – Diagnoses specierum novarum in herbario Horti Regii Botanici Edinburgensis cognitarum 51. – Notes Roy. Bot. Gard. Edinburgh **8**: 191–192.

CARNIEL, K., 1967: Über die Embryologie in der Gattung *Paeonia*. – Österr. Bot. Z. **114**: 4–19.

CAVE, M. S., ARNOTT, H. J., COOK, S. A., 1961: Embryology in the California peonies with reference to their taxonomic position. – Amer. J. Bot. **48**: 397–404.

CHEN, Y. -Z., LI, Z. -L., 1990: Comparative studies of perforation plate structures of vessels in *Ranunculaceae*. – Acta Bot. Sinica **32**: 245–251.

CRONQUIST, A., 1968: The evolution and classification of flowering plants. – London: Nelson.

DE CANDOLLE, A. P., 1817: Regni vegetabilis systema naturale. – Paris: Treuttel & Wurtz.

DIELS, L., 1932: *Circaeaster*, eine hochgradig reduzierte Ranunculacee. – Beih. Bot. Centralbl. **49**, Ergänzungsband: 55–60.

HALLIER, H., 1903: Vorläufiger Entwurf des natürlichen (phylogenetischen) Systems der Blütenpflanzen. – Bull. Herb. Boissier, ser. 2, **3**: 306–317.

HSIAO, P.-K., 1980: A preliminary study of the correlation between phylogeny, chemical constituents and pharmaceutical aspects in the taxa of Chinese *Ranunculaceae*. – Acta Phytotax. Sinica **18**: 142–153.

HUTCHINSON, J., 1926: The families of flowering plants **1**. – London: Macmillan.

JANCHEN, E., 1949: Die systematische Gliederung der Ranunculaceen und Berberidaceen. – Denkschr. Österr. Akad. Wiss., Math.–Naturwiss. KI. **108**: 1–82.

JUNELL, S., 1931: Die Entwicklungsgeschichte von *Circaeaster agrestis*. – Svensk Bot. Tidskr. **25**: 238–270.

KUMAZAWA, M., 1938: On the ovular structure in the *Ranunculaceae* and *Berberidaceae*. – J. Japan. Bot. **14**: 10–25.

KURITA, M., 1960: Chromosome studies in *Ranunculaceae*, 16. Comparison of an aspect of nucleus and chromosome between several genera. – Mem. Ehime Univ., sect. 2, ser. B., **4**: 53–58.

– 1963: Chromosome studies in *Ranunculaceae* 21. Karyotype of *Myosurus* and *Adonis*. – Mem. Ehime Univ., sect. 2, ser. B, **4**: 487–492.

LANGLET, O., 1932: Über Chromosomenverhältnisse und Systematik der *Ranunculaceae*. – Svensk Bot. Tidskr. **26**: 381–400.

MAXIMOWICZ, C. J., 1881: Diagnoses plantarum novarum asiaticarum 4. – Bull. Acad. Sci. St. Petersburg III, **27**: 556–558.

MU, X. -J., 1983: Ovule, female and male gametophyte and fertilization of *Kingdonia uniflora* BALF. f. et W. W. SMITH. – Acta Phytotax. Sinica **25**: 497–504.

OKADA, H., TAMURA, M., 1979: Karyomorphology and relationship on the *Ranunculaceae*. – J. Japan. Bot. **54**: 65–77.

OLIVER, D., 1895: *Circaeaster agrestis* MAXIM. – In OLIVER, D. (Ed.): Icones plantarum **4** (4): pl. 2366.

PRANTL, K., 1887: Beiträge zur Morphologie und Systematik der Ranunculaceen. – Bot. Jahrb. Syst. **9**: 225–273.

SUN, A.-C., WANG, F.-X., 1983: Contribution to the morphology and embryology of *Asteropyrum peltatum*. – Bot. Res. **1**: 85–90.

TAMURA, M., 1972: Morphology and phyletic relationship of the *Glaucidiaceae*. – Bot. Mag. (Tokyo) **85**: 29–41.

THORNE, R. F., 1974: A phylogenetic classification on the *Annoniflorae*. – Aliso **8**: 148–209.

– 1983: Proposed new realignments in the angiosperms. – Nordic J. Bot. **3**: 85–117.

TOBE, H., KEATING, R. C., 1985: The morphology and anatomy of *Hydrastis* (*Ranunculales*): systematic evaluation of the genus. – Bot. Mag. (Tokyo) **98**: 291–316.

YAKOVLEV, M. S., YOFFE, M. D., 1957: On some peculiar features in embryogeny of *Paeonia*. – Phytomorphol. **7**: 74–82.

YANG, Q.-E., GONG, X., GU, Z.-J., WU, Q.-A., 1993: A karyomorphological study of five species in the *Ranunculaceae* from Yunnan, with a special consideration on systematic positions of *Asteropyrum* and *Calathodes*. – Acta. Bot. Yunnan. **15**: 179–190.

ZHANG, Z.-Y., 1982: Chromosome observations of three ranunculaceous genera in relation to their systematic positions. – Acta Phytotax. Sinica **20**: 402–409.

ZHU, M., HSIAO, P.-G., 1991: Distribution of benzylisoquinolines in *Magnoliidae* and other taxa. – Acta Phytotax. Sinica **29**: 142–155.

Address of the author: MICHIO TAMURA, Kinki University, Uchita, Nagagun, Wakayama, 649-64 Japan.

Accepted December 9, 1994

Pl. Syst. Evol. [Suppl.] 9: 207–216 (1995)

Embryo morphology of the *Ranunculaceae*

K. ENGELL

Received October 7, 1994

Key words: *Ranunculaceae.* – Embryo, suspensor, morphology, differentiation, differentiation coefficient.

Abstract: In some species of *Ranunculaceae* the seeds contain embryos with a weak outer as well as inner differentiation; in others the embryos have developed cotyledons, hypocotyl and radicle with a pronounced differentiation in promeristems inside the embryo. Embryos in *Thalictreae* and *Actaeae* are weaker differentiated than embryos from most species belonging to other tribes. The ratios cotyledon length to total embryo length (excl. the suspensor) in these tribes as well as in subtribes *Helleborinae* and *Calthinae* in tribe *Caltheae* are about one third. In other subtribes of *Caltheae*, *Delphiniinae* and *Nigellinae*, the ratios are almost one half as in most of the other tribes. It is therefore obvious that *Caltheae* is a very inhomogeneous tribe.

It is commonly known that some embryos in the *Ranunculaceae* have not completed their growth and morphological differentiation by the time when the seeds or the achenes are shed from the mother plant (EARLE 1938, TAMURA & MIZUMOTO 1972). Another well-known problem is that the variability and irregularity in the embryogenesis of the family are substantial, too (DAVIS 1966; JOHANSEN 1950; LY THY BA 1974, 1980, 1981; STERCKX 1990). Furthermore, it is known that there is variation in the components of the endosperm of the seeds from *Ranunculaceae* (JENSEN, 1968a, b). Therefore, in order to obtain a more complete understanding of the phylogenetic relationships in the family, it was an obvious idea to illustrate whether there exists a relationship between the developmental stages of the embryos in the seeds and the proteins of the endosperm in the same seed. Whereas Prof. UWE JENSEN, Bayreuth, will take care of the endosperm, my part is to investigate the structure of as many embryos as possible from genera of the entire *Ranunculaceae*. The embryos have been investigated by light (LM) and scanning electron microscopy (SEM).

Material and methods

The seeds for these investigations were collected by Prof. UWE JENSEN, Bayreuth, in the Botanical garden in Copenhagen as well as in natural habitats in Denmark and Germany. Some seeds were used for biochemical analyses, others for structural investigations. The smallest seeds were fixed directly; from the bigger seeds the embryos were dissected. For LM the seeds or embryos were placed immediately in the fixative, 2.5% glutaraldehyde + 2% paraformaldehyde in 0.025 M phosphate buffer at pH 7.0. After fixation the material was embedded in glycol methacrylate (according to Feder and O'Brien 1968). 3 μm sections, cut with glass knives, were double-stained with periodic acid – Schiff (PAS) for insoluble polysaccharides and aniline blue black (ABB) for proteins (Fischer 1968). Observations and photomicrographs were made with a Reichert Jung Polyvar microscope and an automatic camera system, Reichert CAM-ES.

Material for SEM was postfixed in 1% OsO_4 in phosphate buffer (0.005 M pH 7.0). After dehydration in a graded acetone series, the material was dried in a critical point dryer (CPD, Polaron E 500) and sputter coated with gold (Polaron E 30). Specimens were viewed with a Jeol JSM-P15 SEM.

According to the classification proposed by Jensen (1968a,b) the *Ranunculaceae* are divided into two subfamilies with the following tribes and subtribes: (1) subfam. *Hydrastidoideae*, tribus *Hydrastideae*, subtribus *Hydrastinae* (*Hydrastis* Ellis, *Glaucidium* Sieb. & Zucc.). (2) subfam. *Ranunculoideae*, tribe *Thalictreae*, subtribes *Isopyrinae* (*Isopyrum* L., *Leptopyrum* Reichb., *Aquilegia* L.) and *Thalictrinae* (*Anemonella* Spach., *Thalictrum* L., *Semiaquilegia* Mak.); tribe *Coptideae*, subtribe *Coptidinae* (*Coptis* Salisb., *Xanthorrhiza* Marsh.); tribe *Actaeae*, subtribe *Cimicifuginae* (*Actaea* L., *Cimicifuga* L., *Anemonopsis* Sieb. & Zucc.); tribe *Caltheae*, subtribe, *Delphiniinae* (*Aconitum* L., *Delphinium* L., *Consolida* (*DC.*) S. F. Gray), *Helleborinae* (*Helleborus* L.), *Eranthidinae* (*Eranthis* Salisb.), *Nigellinae* (*Nigella* L., *Komaroffia* Kuntze, *Garidella* Tourn.) and *Calthinae* (*Caltha* L.), tribe *Ranunculeae*, subtribes *Trollinae* (*Trollius* L.), *Adonidinae* (*Adonis* L., *Callianthemum* C. A. Mey.), *Ranunculinae* (*Myosurus* L., *Ranunculus* L., *Ceratocephalus* Moench., *Batrachium* (*DC.*) S. F. Gray, *Ficaria* Huds.); tribe *Anemoneae*, subtribes: *Clematidinae* (*Clematis* L.), *Anemoninae* (*Anemone* L., *Hepatica* Mill., *Pulsatilla* Mill.).

Results

Embryo morphology. Some genera have species which shed their seeds in early spring, and the embryos in such seeds are morphologically very undifferentiated. These undifferentiated species belong to several of the tribes: *Eranthis* (*Caltheae*: *Eranthidinae*), *Ficaria* (*Ranunculeae*: *Ranunculinae*), *Hepatica* and some *Anemone* (*Anemoneae*: *Anemoninae*) and *Isopyrum* (*Thalictreae*: *Isopyrinae*). The embryos are differentiated in an embryo proper with only the protodermal layer initiated, and an often long suspensor (Fig. 1a,b). In other species of, e.g. *Anemone*, in which seed shedding takes place in the summer period, the embryos are clearly differentiated into cotyledons, hypocotyl and radicle, and the inner structural differentiation is advanced, too. Procambium, a ground meristem and a protoderm are distinguishable, and these primary meristems more or less extend into the cotyledons. On the other hand, species of *Isopyrum*, which also shed their seeds in the summer period (Fig. 1c), have no such well-differentiated embryos in the seeds. Apparently there is no relationship between the length of the developmental period and the developmental stage.

By cultivation of seeds for 2–3 months from the early fruiting *Eranthis hiemalis* (L.) Salisb. *Isopyrum thalictroides* L. and *Anemone nemorosa* L., the embroys were matured. Although at this time they have developed cotyledons, hypocotyl and radicle (Fig. 1d,e), they still show very weak inner differentiation. $1\frac{1}{2}$ months later the embryos of *A. nemorosa* were well-differentiated in the inner part, too, and even the plumule had developed procambial strings (Fig. 1f). In the embryos of *Eranthis*

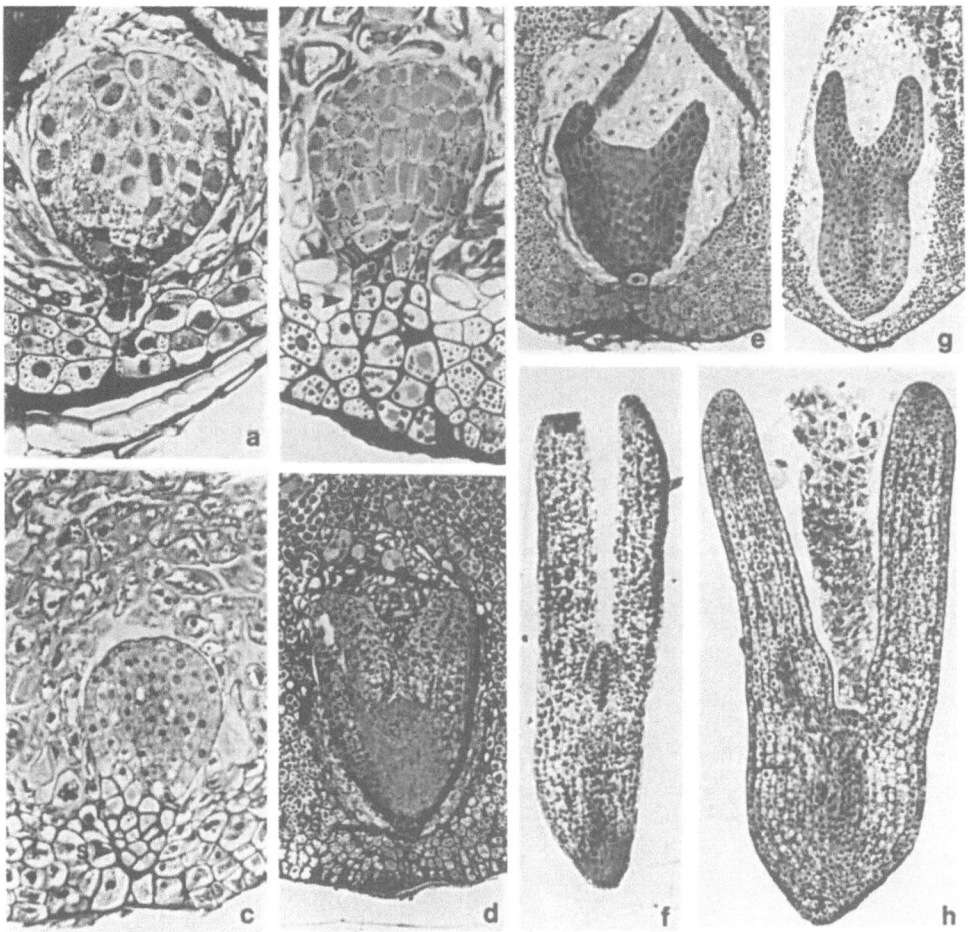

Fig. 1. LM photos of embryos from different *Ranunculaceae*. Most of the embryos are still surrounded by the endosperm and parts of the testa. *a Ficaria verna* HUDS. *subsp. verna*, × 225. *b Hepatica nobilis* SCHREB., × 225. *c Isopyrum hallii* GREY., × 225. *d Isopyrum thalictroides* (cultivated for two months after seed shedding), x 84. *e Anemone nemorosa* (after two months in cultivation), × 84. *f Anemone nemorosa* (cultivated for four months), × 34. *g Caltha palustris* L., × 56. *h Delphinium elatum*, × 56. *a, b, c, f, g, h* Embryos fixed just after seed shedding. *s* suspensor

hiemalis the differentiation of the procambium in the axis was still very weak even after cultivation for 5 months after shedding. Again the developmental stage reached does not only depend on the length of the time for development. In most of the other genera and species in the family the embryos have well-developed procambial strings in the hypocotyl and often also in the cotyledons by the time of seed shedding (Fig. 1 g, h).

Suspensors. It has not yet been possible to obtain information about the suspensors in all the taxa, but the preliminary investigations give information enough to show that there are three possible developmental types. Some embryos have a long suspensor with only one row of cells, as, e.g., *Coptis trifolia* (L.) SALISB., or short or long ones with two juxtaposed rows of cells as in *Ficaria verna* HUDS or

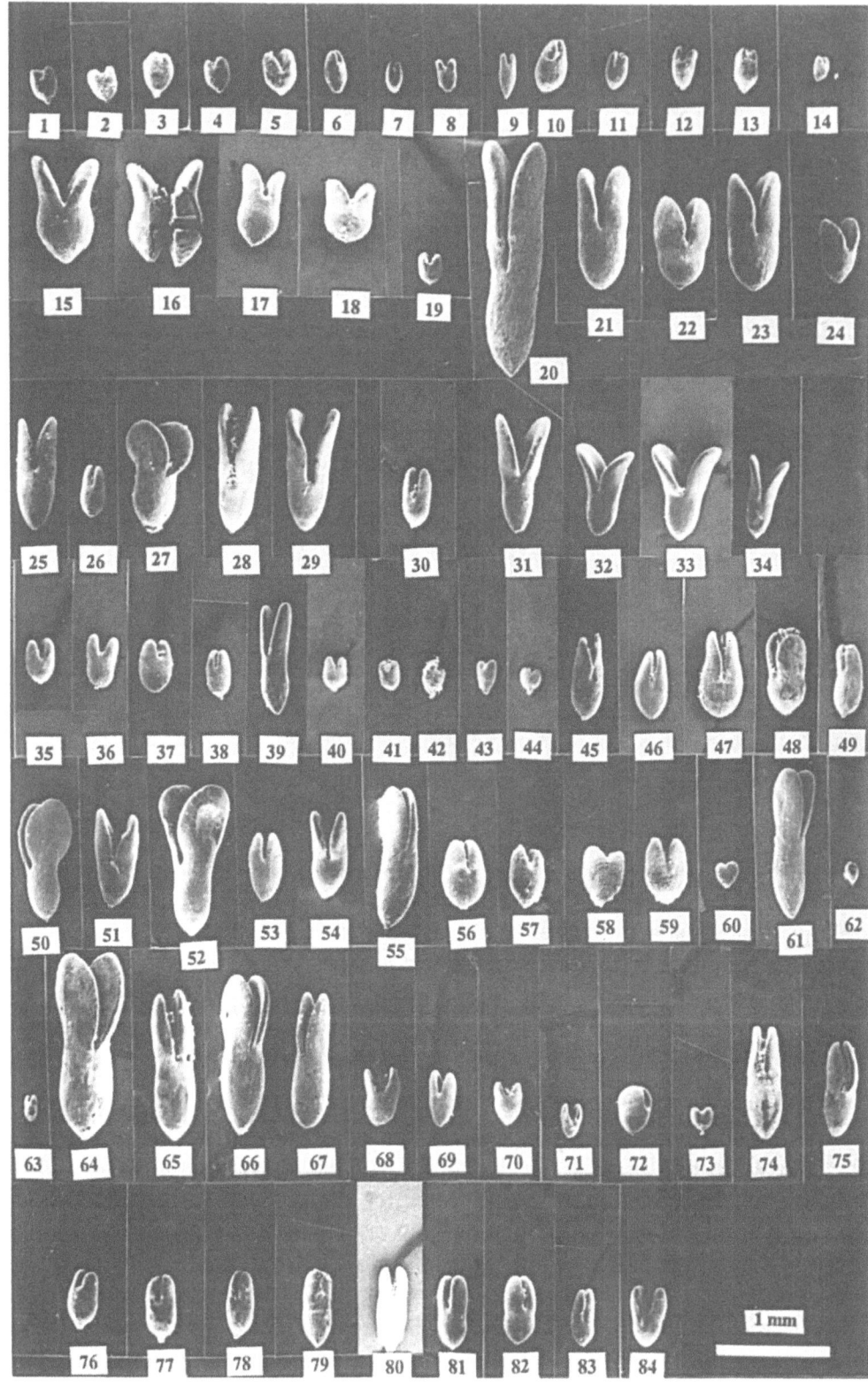

Fig. 2. SEM micrographs of 84 embryos from different species of the *Ranunculaceae*. For species names see Table 1

Anemone nemorosa (Fig. 1a,e). Others have broad suspensors with 6–8 juxtaposed rows or more irregular arrangements of the cells, as e.g., *Thalictrum, Isopyrum* (Fig. 1d) or *Hepatica* (Fig. 1b). A summary of the preliminary results shows that in *Thalictreae* all the investigated suspensors are broad and often short. One-rowed suspensors are found in *Coptideae* and in subtribe *Delphiniinae*, but most of the other investigated genera or species have long or short suspensors with two juxtaposed rows of cells. In subtribe *Anemoninae* there are both broad as well as two-rowed suspensors.

SEM investigations. Another way to look at embryos is to remove them from the seeds just after shedding and to analyse the free embryos in the scanning electron microscope.

84 species belonging to 24 genera and representing nearly all the tribes in the system except for the *Hydrastideae* and *Coptideae* have been studied.

After examinations of all the embryos (Fig. 2) it is evident that they are different. However, it is also possible to demonstrate that the embryos in some tribes are more like each other than in other tribes.

Measurements of the total length (not including the suspensor), the width of the hypocotyl, and the length of the cotyledons were made of all embryos. For each species measurements of 10–15 specimens were made, and the mean values and standard deviations were calculated. Furthermore some differentiation coefficients, the ratio of cotyledon length to embryo length, and the ratio of embryo width to embryo length were calculated (Table 1). The correlations between these two coefficients also were investigated (Fig. 3).

In some tribes these characters are nearly constant. In others it is difficult to show that the embryo structures in a subtribe or even in a genus have anything in common.

Looking at tribe *Thalictreae*, it is obvious that the embryos of subtribe *Isopyrinae* with *Isopyrum* (Fig. 2: 60), *Leptopyrum* (Fig. 2: 62) and *Aquilegia* (Fig. 2: 6–13) are small and only in an early heart-shaped stage with very small cotyledons and a weak inner differentiation. The ratio cotyledon length to total embryo length is close to one third (Fig. 3). The same comments apply to the other subtribe, *Thalictrinae* with *Semiaquilegia* (Fig. 2: 14) and *Thalictrum* (Fig. 2: 76–80). The embryos from these genera are bigger, but the cotyledons still measure about one third of the total embryo length.

In tribe *Actaeae* with *Actaea* (Fig. 2: 1–5) and *Cimicifuga* (Fig. 2: 40–44) the same situation is found. The inner differentiation of the embryos is a little more advanced, but again the cotyledons only make up about one third of the total length (Fig. 3).

The results are different in tribe *Caltheae*. The embryos in some of the subtribes are very well-differentiated, in others their differentiation is intermediate, and finally in some *Eranthidinae* the embryos are very immature when the seeds are shed. In *Delphiniinae* with *Aconitum* (Fig. 2: 19–24), *Delphinium* (Fig. 2: 50–54) and *Consolida* (Fig. 2: 31–34) the embryos are nearly identical. They are well-differentiated in their outer as well as inner morphology; the cotyledons make up about one half of the embryo length (Fig. 3), and they are almost leaf-like with a narrow stalk-like petiole connected to the hypocotyl and an expanded blade or lamina (Fig. 2: 50, 52). Most of the cotyledons in this subtribe have more or less such shape.

Table 1. Embryo measurements of species of the *Ranunculaceae*. Mean values and standard deviations of the measurements and differentiation coefficients are indicated

No.	Species	Embryo-length		Embryo-width		Cotyl.-length		CL/EL	EW/EL
		Mean	STDs	Mean	STDs	Mean	STDs	ratio	ratio
1.	Actaea acuminata	0,33	0,01	0,24	0,01	0,10	0,01	0,31	0,72
2.	A. erythrocarpa	0,30	0,01	0,24	0,02	0,09	0,02	0,31	0,79
3.	A. pachytoda	0,07	0,03	0,27	0,02	0,15	0,04	0,40	0,72
4.	A. rubia	0,29	0,03	0,23	0,02	0,11	0,03	0,38	0,77
5.	A. spicata	0,37	0,01	0,27	0,01	0,14	0,05	0,39	0,74
6.	Aquilegia alpina	0,33	0,03	0,18	0,01	0,09	0,02	0,27	0,55
7.	A. atracta	0,26	0,01	0,15	0,01	0,08	0,00	0,31	0,56
8.	A. canadensis	0,31	0,08	0,16	0,01	0,10	0,02	0,26	0,52
9.	A. canadensis	0,36	0,01	0,15	0,01	0,10	0,01	0,27	0,40
10.	A. chrysantha	0,43	0,05	0,20	0,03	0,15	0,02	0,35	0,47
11.	A. flabellata	0,30	0,02	0,17	0,02	0,10	0,01	0,33	0,58
12.	A. oxysepala	0,36	0,02	0,19	0,02	0,11	0,00	0,31	0,55
13.	A. oxysepala	0,35	0,02	0,18	0,01	0,11	0,01	0,32	0,52
14.	Semiaquilegia ecalcarata	0,22	0,01	0,14	0,01	0,07	0,01	0,27	0,65
15.	Adonis annua	0,87	0,09	0,45	0,05	0,41	0,06	0,47	0,52
16.	A. aestivalis	0,93	0,04	0,46	0,07	0,36	0,05	0,38	0,49
17.	A. flamnea	0,63	0,07	0,34	0,01	0,26	0,04	0,42	0,54
18.	A. vernalis	0,55	0,02	0,40	0,01	0,28	0,02	0,50	0,73
19.	Aconitum anthora	0,27	0,02	0,21	0,02	0,11	0,02	0,41	0,75
20.	A. charmichaelii	1,39	0.28	0,38	0,03	0,76	0,19	0,54	0,28
21.	A. ferox	1,11	0,06	0,43	0,03	0,60	0,04	0,53	0,38
22.	A. lycotonum	0,72	0,06	0,40	0,01	0,34	0,03	0,48	0,55
23.	A. nepellus	0,88	0,09	0,41	0,04	0,43	0,06	0,49	0,47
24.	A. paniculatum	0,56	0,04	0,30	0,02	0,24	0,02	0,43	0,53
25.	Anemone altaica	1,02	0,05	0,35	0,02	0,53	0,05	0,52	0,34
26.	A. baldensis	0,44	0,03	0,22	0,02	0,17	0,02	0,39	0,49
27.	A. narcissiflora	1,05	0,19	0,40	0,05	0,53	0,14	0,50	0,38
28.	A. polyanthes	1,23	0,09	0,37	0,03	0,73	0,05	0,59	0,30
29.	A. rivularis	1,09	0,11	0,33	0,03	0,63	0,06	0,57	0,30
30.	Anemonopsis macrophylla	0,49	0,04	0,24	0,02	0,22	0,03	0,44	0,49
31.	Consolida ajacis	0,84	0,14	0,28	0,02	0,43	0,10	0,50	0,33
32.	C. ambigua	0,78	0,03	0,27	0,03	0,40	0,01	0,51	0,35
33.	C. orientalis	0,85	0,05	0,28	0,01	0,45	0,03	0,53	0,33
34.	C. regalis	0,74	0,04	0,23	0,01	0,39	0,01	0,54	0,32
35.	Caltha biflora	0,38	0,05	0,25	0,01	0,16	0,01	0,43	0,66
36.	C. laeta	0,40	0,05	0,23	0,01	0,15	0,03	0,36	0,59
37.	C. leptosepala	0,43	0,04	0,29	0,02	0,17	0,02	0,40	0,67
38.	C. palustris	0,38	0,02	0,22	0,01	0,13	0,01	0,35	0,58
39.	Ceratocephalus falcatus	0,86	0,06	0,23	0,01	0,43	0,03	0,50	0,27
40.	Cimicifuga americana	0,26	0,01	0,19	0,01	0,11	0,01	0,42	0,73
41.	C. europaea	0,23	0,03	0,17	0,02	0,08	0,01	0,37	0,74
42.	C. cordifolia	0,26	0,01	0,19	0,00	0,05	0,01	0,21	0,73
43.	C. racemosa	0,26	0,02	0,18	0,02	0,07	0,01	0,29	0,69
44.	C. simplex	0,18	0,01	0,17	0,01	0,04	0,00	0,24	0,95
45.	Clematis alpina	0,64	0,05	0,27	0,02	0,33	0,04	0,51	0,42
46.	C. integrifolia	0,60	0,03	0,28	0,02	0,33	0,03	0,54	0,47
47.	C. recta	0,67	0,03	0,32	0,04	0,34	0,04	0,51	0,49
48.	C. viticella	0,70	0,05	0,35	0,02	0,36	0,02	0,51	0,50
49.	C. vitalba	0,55	0,03	0,24	0,01	0,23	0,01	0,42	0,43
50.	Delphinium californium	0,88	0,10	0,28	0,04	0,49	0,09	0,56	0,31
51.	D. cashmerianum	0,82	0,05	0,33	0,02	0,42	0,03	0,52	0,40

Table 1 (*Continued*)

No.	Species	Embryo-length		Embryo-width		Cotyl.-length		CL/EL	EW/EL
		Mean	STDs	Mean	STDs	Mean	STDs	ratio	ratio
52.	D. elatum	1,15	0,10	0,33	0,02	0,59	0,08	0,51	0,28
53.	D. staphisagria	0,59	0,05	0,26	0,01	0,22	0,01	0,37	0,45
54.	D. tatsienense	0,61	0,10	0,27	0,03	0,31	0,07	0,52	0,44
55.	Garidella nigellastrum	1,01	0,10	0,33	0,03	0,47	0,04	0,46	0,33
56.	Helleborus foetidus	0,55	0,02	0,34	0,03	0,25	0,03	0,45	0,62
57.	H. lividus	0,41	0,07	0,28	0,01	0,10	0,03	0,24	0,68
58.	H. niger	0,43	0,03	0,34	0,02	0,13	0,02	0,31	0,80
59.	H. orientalis	0,49	0,05	0,34	0,03	0,17	0,05	0,34	0,69
60.	Isopyrum hallii	0,19	0,02	0,15	0,02	0,05	0,01	0,24	0,82
61.	Komoroffia integrifolia LAMOS	1,08	0,17	0,27	0,02	0,49	0,07	0,46	0,25
62.	Leptopyrum fumarioides	0,16	0,02	0,11	0,01	0,055	0,003	0,35	0,73
63.	Myosurus minimus	0,23	0,01	0,11	0,01	0,08	0,01	0,30	0,50
64.	Nigella damascena	1,47	0,13	0,44	0,03	0,72	0,05	0,49	0,30
65.	N. hispanica	1,07	0,13	0,34	0,06	0,48	0,07	0,45	0,32
66.	N. orientalis	1,24	0,07	0,38	0,02	0,59	0,03	0,47	0,31
67.	Pulsatilla alpina	1,09	0,05	0,31	0,02	0,54	0,04	0,49	0,29
68.	P. styriaca	0,46	0,04	0,25	0,02	0,22	0,01	0,47	0,54
69.	P. vernalis	0,51	0,07	0,26	0,02	0,26	0,05	0,52	0,50
70.	P. vulgaris	0,41	0,07	0,23	0,01	0,19	0,05	0,47	0,56
71.	Ranunculus illyricus	0,28	0,01	0,15	0,01	0,15	0,01	0,52	0,53
72.	R. gramineus	0,40	0,09	0,25	0,04	0,19	0,04	0,49	0,63
73.	R. thora	0,23	0,05	0,19	0,01	0,09	0,03	0,40	0,83
74.	R. arvensis	0,90	0,05	0,32	0,01	0,45	0,03	0,50	0,35
75.	R. auricomus	0,70	0,11	0,27	0,03	0,33	0,06	0,45	0,38
76.	Thalictrum aquilegifolium	0,46	0,02	0,27	0,01	0,16	0,01	0,34	0,58
77.	T. flavum L. subsp. glaucum	0,47	0,04	0,24	0,02	0,18	0,02	0,38	0,50
78.	T. isopyroides	0,47	0,10	0,22	0,03	0,19	0,04	0,42	0,48
79.	T. minus	0,56	0,08	0,26	0,01	0,22	0,02	0,40	0,45
80.	T. petaloideum	0,51	0,13	0,22	0,03	0,22	0,06	0,43	0,42
81.	Trollius chinensis	0,64	0,06	0,26	0,03	0,29	0,11	0,45	0,41
82.	T. europaeus	0,53	0,06	0,26	0,02	0,27	0,04	0,50	0,49
83.	T. pumilus	0,49	0,02	0,21	0,02	0,27	0,02	0,56	0,44
84.	T. yamanensis	0,49	0,03	0,24	0,02	0,26	0,03	0,52	0,50

Subtribe *Nigellinae* with *Nigella* (Fig. 2: 64–66), *Komaroffia* (Fig. 2: 61) and *Garidella* (Fig. 2: 55) are the most uniform of the subtribes with respect to their embryo structures. The embryos are well-differentiated with leaf-like cotyledons, and the total embryo length is about twice that of the cotyledons.

The *Eranthidinae* were not represented in this SEM investigation because the embryos of *Eranthis hiemalis* are too small and undifferentiated for dissection. The results from the LM investigations show, however, that the embryos are very immature and very little if at all differentiated.

The other two subtribes in tribe *Caltheae*, *Helleborinae* (Fig. 2: 56–59) and *Calthinae* (Fig. 2: 35–38), have in common that their embryos are of only weak differentiation, and again the cotyledons form only one third of the total embryo length.

In tribe *Ranunculeae* it is difficult to find similarities in embryo structures. Subtribe *Trollinae* (Fig. 2: 81–84) is homogeneous. The length of the cotyledons is

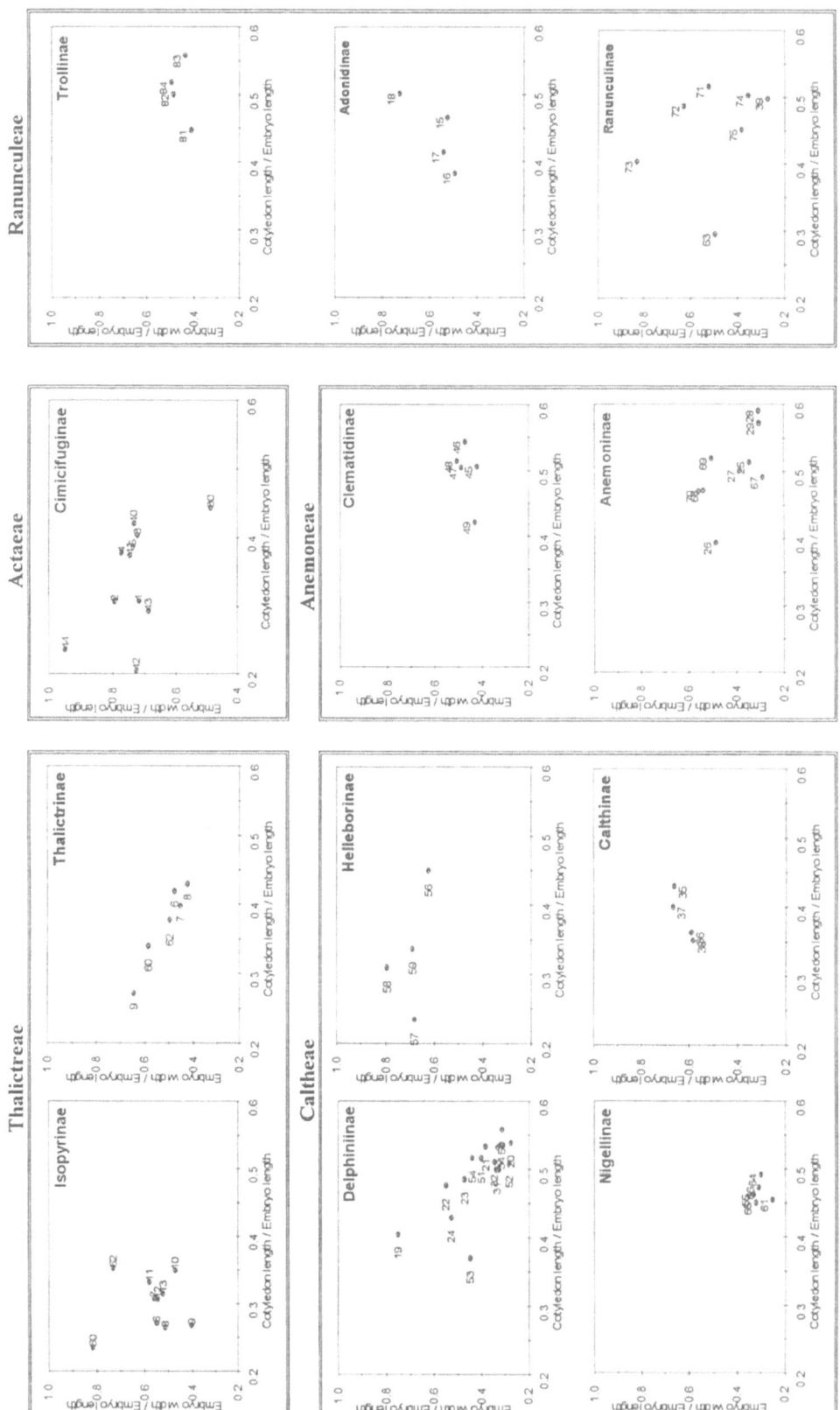

Fig. 3. Correlation between the two differentiation coefficients of the embryos of the *Ranunculaceae*

about one half of the total embryo length, but there are no particular specialisations in the shape of the cotyledons; the differentiation is more advanced, but it is not reasonable to classify it as well-differentiated. Almost the same comments apply to subtribe *Adonidinae* (Fig. 2: 15–18). Here the embryos are bigger and more variable. In all of them the cotyledons are a little leaf-like, and they could be classified as almost well-differentiated. Subtribe *Ranunculinae* in this investigation is represented by *Myosurus*, *Ranunculus* and *Ceratocephalus*. *Myosurus minimus* L. (Fig. 2: 63) is very small and only of weak differentiation, *Ceratocephalus falcatus* Pers. (Fig. 2: 39) is almost well-differentiated, and in *Ranunculus* (Fig. 2: 71–75) the variation in size of the investigated species is great. For most of them, however, the ratio cotyledon length to total embryo length is close to one half, so that despite the variation the differentiation coefficients are alike.

The last tribe *Anemoneae* has two subtribes, *Clematidinae* and *Anemoninae*. The embryos of *Clematis* (Fig. 2: 45–49) are very similar to embryos from most species of *Pulsatilla* (*Anemoninae*) (Fig. 2: 67–70) with a ratio cotyledon length to the embryo length of about one half (Table 1 or Fig. 3). This result also is found in some species of *Anemone*, but variation in *Anemone* (Fig. 2: 25–29) is substantial. The embryos of species with spring seed shedding are small and weak in differentiation; other species (Fig. 2: 28, 29) have the largest differentiation coefficients in the entire family (Fig. 3). This genus probably is among the most variable in the family with respect to embryo structures. Their well-differentiated embryos clearly have leaf-like cotyledons.

Discussion

As regards the structure or differentiation of the embryos, it is obvious that in tribes *Thalictreae* and *Actaeae* the embryos are uniform. They are weakly differentiated, the cotyledons make up about one third of the total embryo length, and the suspensors are of the broad type.

The embryos in tribe *Caltheae* are very variable, and the tribe is not a homogeneous one. Subtribes *Delphiniinae* and *Nigellinae* have in common that the embryos are well-differentiated, and the cotyledons make up about one half of the whole embryo, while subtribes *Helleborinae* and *Calthinae* have embryos of only weak differentiation, and their cotyledons measure about one third of the whole embryo length as in the *Thalictreae* and the *Actaeae*.

Tribes *Ranunculeae* shows great variation. *Myosurus* and some of the species of *Ranunculus* (incl. *Ficaria*) are similar to each other, whereas *Ceratocephalus* is quite similar to the most well-differentiated species of *Ranunculus*.

The situation is quite different in tribe *Anemoneae*. The embryological data give no reason for its division in subtribes, and the embryo structures are very uniform.

The conclusion of this investigation of the embryological data is that the more primitive tribes of the *Ranunculaceae* are quite homogeneous in their embryo structure or differentiation, and their cotyledons measure about one third of the total length of the embryos. In the more advanced tribes the diffferentiation coefficients are about one half, and the embryos are well-differentiated with more or less leaf-like cotyledons. Only in tribe *Caltheae* no clear relationship is found.

The author likes to express her great appreciations to Prof. UWE JENSEN, Bayreuth, for his valuable comments during the work, and to the technical staff at the Botanical Laboratory in Copenhagen.

References

DAVIS, G. L., 1966: Embryology of angiosperms. – New York, London, Sydney: Wiley.

EARLE, T. T.: 1938: Embryology of certain *Ranales*. – Bot. Gaz. **100**: 257–275.

FEDER, N., O'BRIEN, T., 1968: Plant microtechnique: Some principles and new methods. – Amer. J. Bot. **55**: 123–142.

FISCHER, D. B., 1968: Protein staining of ribboned epon sections for light microscopy. – Histochemie **16**: 92–96.

JENSEN, U., 1968a: Serologische Beiträge zur Systematik der *Ranunculaceae*. Teil I. – Bot. Jahrb. **88**: 204–268.

– 1968b: Serologische Beiträge zur Systematik der *Ranunculaceae*. Teil II. – Bot. Jahrb. **88**: 269–310.

JOHANSEN, D. A., 1950: Plant embryology. – Waltham, Mass: Published by the Chronica Botanica Company.

LY THY BA, 1974: Embryogénie comparée et phylogénie des Helléborées. – Rev. Gen. Bot. **81**: 151–191.

– 1980: Embryogénie comparée et phylogénie des Ranales et des Hélobiales II. – Les Ranales herbacées. – Rev. Gen. Bot. **88**: 105–197.

– 1981: Embryogénie comparée et phylogénie des Ranales et des Hélobiales IV. Les conclusions. – Rev. Gén. Bot. **88**: 347–373.

STERCKX, R., 1900: Recherches anatomiques sur l'embryon et les plantules dans la famille des Renonculacées. – Arch. Inst. Bot. Univ. Liège. **2**: 3–94.

TAMURA, M., MIZUMOTO, Y., 1972: Stages of embryo development in ripe seeds or achenes of the *Ranunculaceae*. – J. Japan. Bot.**47**: 225–237.

Address of the author: KIRSTEN ENGELL, Botanical Institute, Botanical Laboratory, University of Copenhagen, Gothersgade 140, DK – 1123 Copenhagen, Denmark.

Accepted December 20, 1994

Pl. Syst. Evol. [Suppl.] 9: 217–227 (1995)

Serological legumin data and the phylogeny of the *Ranunculaceae*

UWE JENSEN

Received December 1, 1994

Key words: *Ranunculaceae*. – Legumin, serology, phylogeny.

Abstract: Legumin, the main seed storage protein in *Ranunculaceae*, was isolated from species of 20 genera (19 belong to *Ranunculaceae*, one to *Papaveraceae*) and used to elicit monospecific antisera. The precipitation bands and the presence of spurs were observed in Ouchterlony two-dimensional double diffusion tests. These data were used to calculate similarity indices as well as neighbor joining and UPGMA phylogenetic trees. Seven groups of closely related genera could be detected, i.e., *Anemone/Hepatica/Clematis*; *Ranunculus/ Ficaria*; *Adonis/Trollius/Trautvetteria*; *Cimicifuga/Actaea/Caltha*; *Aconitum/ Delphinium/ Consolida*; *Aquilegia/Thalictrum*; *Xanthorhiza/Coptis*. They support the classifications based on other molecular data, and correlate with both morphological and chemical data. This strengthens the status as monophyletic groups.

Immunological methods have been used in plant systematics a long time before their molecular background could be interpreted. RELANDER (1911) recommended the precipitin reaction generated by mixtures of an antiserum and antigenic material from different taxa for the identification of seeds, and GOHLKE (1913) used the "Sero-Diagnostik" for relationship studies in higher plants. Such experiments were expanded especially between 1922 and 1929 by the Königsberg group, which developed the "Königsberger sero-diagnostischen Stammbaum" (MEZ & ZIEGENSPECK 1926). In these early experiments neither the structure of antigenic proteins was known, nor the molecular background of the specificity of antibodies produced by an immune system.

During the past 30 years significant amounts of valuable serological data have been included in the interpretation of phylogenetic relationships of plants, e.g., of *Magnoliidae* (FAIRBROTHERS & PETERSEN 1983, JENSEN & GREVEN 1984), *Fabaceae* (CRISTOFOLINI & FEOLI CHIAPELLA 1977, CRISTOFOLINI 1987), or *Poaceae* (ESEN & HILU 1989). All these researchers used complex seed antigenic materials, and immunodiffusion techniques were preferred.

Today the interpretation of the immunoprecipitation bands in gels including the conditions for the forming of bands and spurs are very well understood (EISEN 1974, WEIR 1986); without additional sophisticated experiments, however, no information is obtained about the size and specific structure of the epitopes involved. Nevertheless, those portions of the protein molecules which form epitopes have proved to be valuable parts of the protein chain for use in systematic interpretation. Future knowledge of the gene sequences responsible for the protein epitopes will provide additional information.

In view of this situation the epitopic molecular structures are indirectly compared, e.g., by registering precipitation bands in immuno-electrophoresis or immunodiffusion experiments. Based upon the precipitation patterns, a statement of serological similarity or identity can clearly be reported. For their registration, qualitative methods are used, because serological characters are qualitative and not quantitative entities. Therefore, quantitative methods, e.g., turbidity measurements, are rarely used in taxonomic research.

For the *Ranunculaceae*, HAMMOND (1952, 1955) was the first to use serological methods employing protein mixtures as antigens. He applied a turbidimetric method to measure the amount of precipitate. Also JENSEN (1968) used the complex antigenic materials of the seeds in his comparative work of the *Ranunculaceae*, using immunodiffusion, preabsorption and turbidimetric methods.

In much of the research in the past, the antigenic material of each taxon contained many proteins ("antigenic systems"). Thus, many protein characters were compared automatically and simultaneously in one experiment. Such a procedure was in agreement with the concept of the numerical taxonomic methodology. The question of exact comparability of the antigenic systems, however, was not answered. Our recent investigation of taxa of the *Ranunculaceae* (JENSEN & HÖHN, unpubl.) reveals that obvious qualitative and quantitative differences can be detected already among seed globulins.

Presently the analysis and comparison of molecular characters demonstrate that one protein or gene can be used to produce a phylogenetic tree of considerable significance. JENSEN & PENNER (1980) demonstrated this for serological characters of legumin and vicilin proteins. Therefore, this investigation has been restricted to one antigenic protein, legumin, which is the main seed storage protein in most flowering plants, including the *Ranunculaceae* (JENSEN & HÖHN, unpubl.).

Material and methods

Twenty species were selected for the experiments including *Papaver somniferum* L. as an outgroup. The genera *Eranthis* SALISB. and *Isopyrum* L. had to be excluded because they do not produce detectable amounts of legumin (JENSEN & HÖHN, unpubl.). Only one species per genus was used because earlier experiments have shown that congeneric taxa are serologically identical. The species investigated here are presented in Table 1.

Legumin isolation. The seeds were ground to powder and extracted with 10 mM tris/80 mM glycine buffer, 2% NaCl, pH 8.2. The soluble proteins containing the legumins were extracted from the supernatant after centrifugation (no legumin-like non-soluble proteins have been detected in *Ranunculaceae*). The lipids were removed with 1,1,2-trichlortrifluoroethane. The proteins were separated by gel filtration. The legumins were eluted in the first protein peak, with molecular weights in the range of 300 to 400 kD. In case of incomplete separation from the vicilin-like proteins, an additional separation procedure was applied: The proteins were precipitated with 80% ammonium sulfate, centrifuged, and redissolved in 10 mM tris/80 mM glycine buffer, 0.05 M NaCl, pH 8.2. After the ammonium sulfate was

Table 1. Sources of seed material for the experiments (classification according to
JENSEN & al. 1995, this volume). Some of the seeds were collected in the wild by U.
JENSEN (*Je.*) or K. MCQUEENEY (*McQµ*)

Ranunculaceae
Coptideae LANGLET
    Xanthorhiza simplicissima MARSH., leg. Je: USA; Blue Mts 89
    Coptis trifolia (L.) SALISB., leg. McQu: USA; Marinette/WI 94
Isopyreae SCHRÖDINGER
    Aquilegia vulgaris L., Hort. Bot. BT.
    Thalictrum minus L., Hort. Bot. BT.
Cimicifugeae TORREY & A. GRAY
    Actaea rubra (AIT.) WILLD., Fa. Bornträger
    Cimicifuga racemosa (L.), NUTT., Fa. Bornträger
Caltheae REICHB.
    Caltha palustris L., leg. Je: Bohemia 90
Delphinieae WARM.
    Aconitum napellus L., Fa. Bornträger
    Delphinium halteratum SIBTH. & SM., Hort. Bot. Cologne (ex Sardinia)
    Consolida regalis S. F. GRAY, Hort. Bot. BT.
    Nigella damascena L., Fa. Bornträger
Helleboreae DC
    Helleborus niger L., Fa. Bornträger
Adonideae KUNTH
    Trollius chinensis BUNGE, Fa. Bornträger
    Adonis autumnalis L., Fa. Bornträger
Ranunculeae DC.
    Trautvetteria carolinensis (WALT.) VAIL., leg. Je: USA/Mt St. Helens 93
    Ranunculus auricomus agg., leg. Je: D/Bayern/Wüstenstein 93
    Ficaria verna HUDS. subsp. *verna*, leg. Je: DK/Jylland 93
Anemoneae
    Clematis vitalba L., Fa. Bornträger
    Anemone narcissiflora L., leg. Je: Schweiz 90, and Fa. Jelitto
    Hepatica nobilis SCHREB., leg. Je: DK/Sjelland 91
Papaveraceae
    Papaver atlanticum (BALL) COSS., Fa. Jelitto

---

removed by using a Sephadex G 25 column, the protein mixture was separated by Sephadex
A50 ion exchange chromatography using an NaCl gradient between 0.05 and 1 M. The
electrophoretically detected legumin fractions were stored at $-20\,°C$ until used for the
immunization procedure.

    Immunization. Rabbits from commercial sources were injected subcutaneously or intra-
muscularly to elicit legumin specific monospecific antibodies. The injection procedure
included the initial application of 1 ml (ca. 1–2 mg protein each) legumin solution plus
incomplete Freund's adjuvant, and a cycle of one injection weekly over four weeks beginning
one month after the initial injection. Two to five such injection series were performed. One
week after the last injection of each cycle ca. 30 ml blood were taken. The serum was stored at
$-20\,°C$ until used in the experiment, without isolation of the IgG.

Preparation of the reacting antigen materials. The antigenic materials were obtained by extracting seed materials with 10 mM tris/80 mM glycine buffer, 2% NaCl, pH 8.2 (15 ml buffer/10 g seed material). The solution was delipified and frozen in aliquots until used in the experiments.

Immunological techniques. For the experiments antigen-antibody-precipitations were performed in agar gels (1% in 0.4 M borate buffer, 1% NaCl, pH 8.0). A two-dimensional double diffusion arrangement (Ouchterlony-technique) was used.

The Ouchterlony test was used because this technique has proven to be adequate for our experiments. Only one precipitation band appeared, and the spurs could be easily identified. They indicate whether a protein has or has not one determinant or determinant group more in common with the reference protein than the adjacent protein. A quantitative method like the ELISA test would measure different absorption or radioactivity values in a more refined graduation; the differences found using such a technique, however, do not unequivocally represent character differences.

The assays filled with antisera and antigens were stored in humid boxes at room temperature. The precipitates were recorded after 2–3 days, and the gels were dried and stained with Coomassie Brilliant Blue. Fusion of bands between two adjacent precipitations indicated serological identity, the formation of spurs of double spurs indicated serological differences (Fig. 1).

Serological similarity calculations. For each of the 20 antisera a data matrix was prepared. As an example the matrix for the *Adonis autumnalis* antiserum is shown in Fig. 2a. For this antiserum, each test antigen was placed side by side with another test antigen in the gel. It was recorded whether the two serological reactions were identical or a spur or double spur was formed. The data included in Fig. 2a are used to produce the scheme of Fig. 2b. This figure represents an exact and parsimonious presentation of the results, e.g., *Adonis* spurs all antigens; *Helleborus* spurs *Aquilegia, Thalictrum, Nigella, Anemone, Hepatica,* and *Papaver,* but is spurred by *Adonis, Trollius, Trautvetteria, Xanthorhiza, Cimicifuga, Actaea, Caltha, Ranunculus, Ficaria, Aconitum, Consolida;* e.g., *Aquilegia/Thalictrum* and *Anemone/Hepatica* are identical in their reactions with the *Adonis* antiserum. By such a parsimonious calcula-

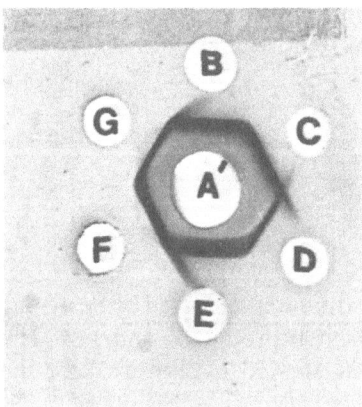

Fig. 1. Two-dimensional double diffusion (Ouchterlony technique) on agar plates; one example of the experiments performed, showing the precipitation line and spurs between an *Adonis autumnalis* antiserum (*A'*, central well) and six antigens (outer wells). The six antigens are *B Nigella hispanica, C Garidella nigellastrum, D Hepatica nobilis, E Hepatica nobilis,* another origin, *F Aquilegia vulgaris, G Thalictrum aquilegifolium.* Identical reactions between *B* and *C,* and between *F* and *G; G* spurs *B,* and *F* spurs *E;* double spurs between *C* and *D*

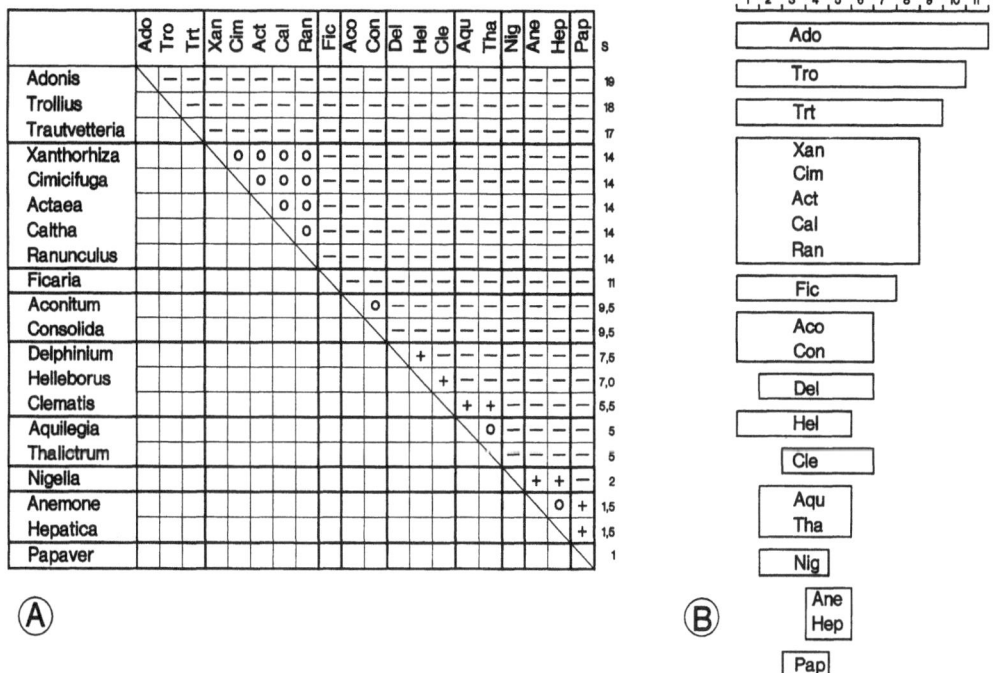

Fig. 2. Serological reaction data for an *Adonis autumnalis* antiserum with nineteen *Ranunculaceae* genera plus one *Papaveraceae* legumin antigens. *A* Matrix of the serological data, and S-, IS-, and ID values. – the left antigen spurs the upper antigen, + double spur, o identical reaction. For plant species used see Table 1. For the calculation of *S, IS* and *ID* see text. *B* Distribution of determinant groups (11 groups)

tion, a minimum number of determinant groups can be estimated (eleven in the example cited above).

The immunological similarity and immunological distance were calculated according to the formula of JENSEN & GREVEN (1984), see also FISCHER & JENSEN (1992) and LESTER & al. (1993):

Immunological similarity IS (%) = $\dfrac{S_{het}}{S_{ref}} \times 100$; immunological distance ID (%) = 100-IS

$S = \sum (no._{spurs}) + 1/2 \sum (no._{identities} + no._{double\ spurs}$
het = heterologous taxon, ref = reference taxon

Example (see Fig. 2a):
  $S_{ref} = S_{Adonis} = 19$ (19 spurs + 0 identities + 0 double spurs)
  $S_{het1} = S_{Trollius} = 18 + 0 + 0 = 18$
  $S_{het_2} = S_{Anemone} = 0 + 1/2 \times 1 + 1/2 \times 2 = 1.5$

The ID values for all 20 experiments were used to establish a taxon/taxon matrix. For the two data pairs (e.g., *Adonis* in the *Thalictrum* experiment and *Thalictrum* in the *Adonis* experiment) the mean was calculated (MAXSON & MAXSON 1990). The resulting ID values are shown in Table 2. The taxa in this table are ordered in that way that the lowest distance values are arranged along the diagonal (see HEYWOOD 1971); values of least distant taxa are framed.

All data of similarity revealed in a serological experiment are referred to each reference system/antibody system. Because in a serological experiment information about high similarities is a more important information than low similarities, the data at the top of Fig. 2a and b are more valuable than those at the bottom. The latter only indicate that small and

Table 2. Legumin distance matrix; for calculation see text. Low distance values are indicated by frames

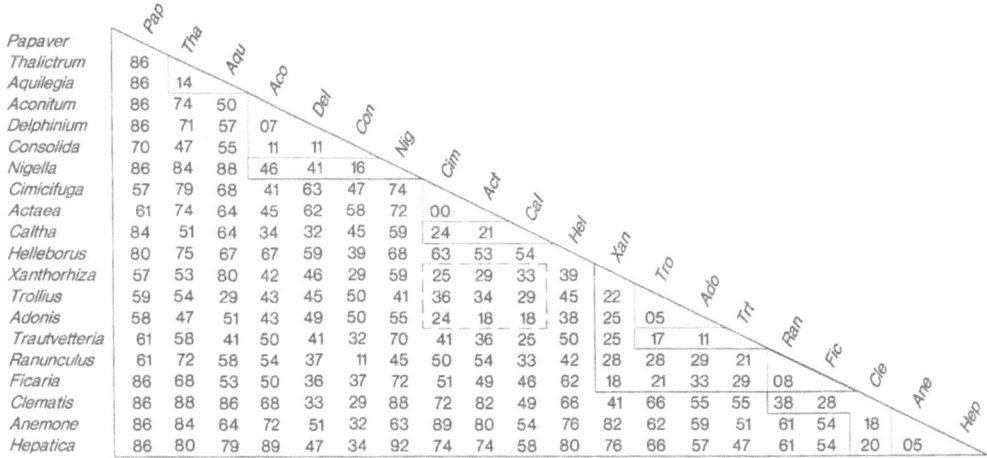

	Pap	Tha	Aqu	Aco	Del	Con	Nig	Cim	Act	Cal	Hel	Xan	Tro	Ado	Trt	Ran	Fic	Cle	Ane	Hep
Papaver																				
Thalictrum	86																			
Aquilegia	86	14																		
Aconitum	86	74	50																	
Delphinium	86	71	57	07																
Consolida	70	47	55	11	11															
Nigella	86	84	88	46	41	16														
Cimicifuga	57	79	68	41	63	47	74													
Actaea	61	74	64	45	62	58	72	00												
Caltha	84	51	64	34	32	45	59	24	21											
Helleborus	80	75	67	67	59	39	68	63	53	54										
Xanthorhiza	57	53	80	42	46	29	59	25	29	33	39									
Trollius	59	54	29	43	45	50	41	36	34	29	45	22								
Adonis	58	47	51	43	49	50	55	24	18	18	38	25	05							
Trautvetteria	61	58	41	50	41	32	70	41	36	25	50	25	17	11						
Ranunculus	61	72	58	54	37	11	45	50	54	33	42	28	28	29	21					
Ficaria	86	68	53	50	36	37	72	51	49	46	62	18	21	33	29	08				
Clematis	86	88	86	68	33	29	88	72	82	49	66	41	66	55	55	38	28			
Anemone	86	84	64	72	51	32	63	89	80	54	76	82	62	59	51	61	54	18		
Hepatica	86	80	79	89	47	34	92	74	74	58	80	76	66	57	47	61	54	20	05	

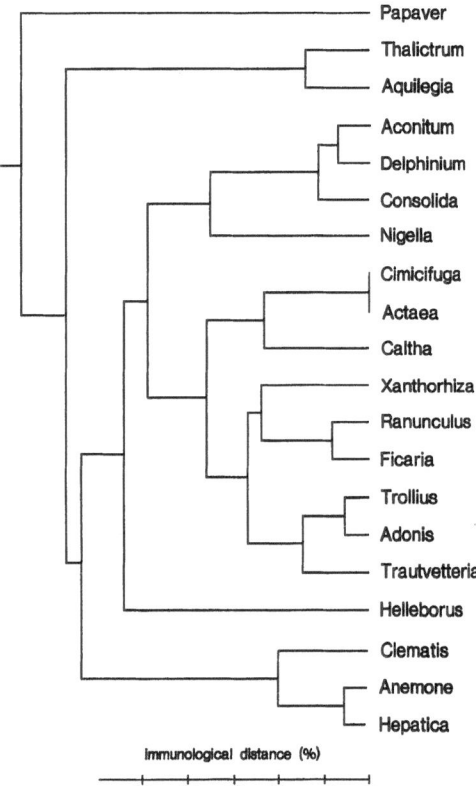

immunological distance (%)

50   40   30   20   10   0

Fig. 3. UPGMA tree, generated by using the PHYLIP program version 3.5c (FELSENSTEIN 1993)

Fig. 4. Neighbor joining tree, generated by using the PHYLIP program version 3.5c. The negative branch length of *Consolida* has been reduced to 0

ubiquitous legumin determinants are shared with the reference legumin. Therefore the data matrix of Table 2 can be used for calculating the UPGMA and neighbor joining trees (Figs. 3 and 4). In both procedures the calculation process starts with the serologically most evident statements, i.e., pairs of highest similarity.

Results and discussion

The UPGMA and neighbor joining algorithms reveal trees of similar topology. The branch lengths of the neighbor joining tree additionally provide information about the phylogenetic distance of taxa. For example, *Papaver*, the *Aquilegia-Thalictrum* group, *Helleborus*, the *Clematis-Anemone-Hepatica* group, and *Nigella* are placed on extended branches.

Seven monophyletic groups of genera can be distinguished. They are supported by other molecular data (JENSEN & al. 1995) obtained from *atp*B/*rbc*L/18SrDNA (HOOT 1995, this volume) and *adh* 1 sequences (KOSUGE & al. 1995, this volume) and from a restriction site variation analysis of cpDNA (JOHANSSON 1995, this volume). These groups are: (1) *Anemone-Hepatica* (-*Clematis*), (2) *Ranunculus-Ficaria*, (3) *Trollius-Adonis* (-*Trautvetteria*), (4) *Cimicifuga-Actaea*, (5) *Aconitum-Delphinium-Consolida* (-*Nigella*), (6) *Aquilegia-Thalictrum*, (7) *Xanthorhiza* (-*Coptis*).

Group 1: *Clematis, Anemone, Hepatica*. The legumin data indicate *Clematis, Anemone* and *Hepatica* to be a monophyletic group. This result, which already had been observed in the serological investigation of JENSEN (1968), is in agreement with other molecular data (JENSEN & al. 1995), and correlates with the distribution of diagnostic morphological and chemical characters (achenes, R-type chromosomes, ranunculin, etc.). In most classifications (e.g., TAMURA 1993, JENSEN & al. 1995, this volume) these genera are included in the tribe *Anemoneae*.

Group 2: *Ranunculus, Ficaria*. These genera are also perceived to form a mono-phyletic group, which is confirmed by former serological data of JENSEN (1968, which included the genus *Myosurus*), all other molecular data available (JENSEN & al. 1995, this volume), and karyological, anatomical, morphological and chemical characters.

Group 3: *Trollius, Adonis, (Trautvetteria)*. The high serological similarity between *Adonis* and *Trollius* has already been reported by JENSEN (1968) and is confirmed by other molecular (JENSEN & al. 1995, this volume), karyological (BHANDARI 1966, BHANDARI & KAPIL 1964, JENSEN & al. 1995, this volume) and chemical characters (JENSEN 1995, this volume). This relationship, however, is not evident from classical systematic data. Probably taxonomists have concentrated for too long on the fruit type as a major character within the *Ranunculaceae*: The old concept of PRANTL (1891) with a separation of follicle-bearing *Helleboroideae* and achene-bearing *Anemonoideae* has dominated the classification of the family, and it is still the concept of the recent classification of TAMURA (1993). When evaluating the number of carpels per flower, *Trollius* is the only polycarpous helleboroid genus which is similar to, e.g., the anemonoid genus *Adonis*. Present knowledge indicates that *Trollius* might represent the link between oligomeric follicle-bearing genera and polymeric achene-bearing genera. Therefore, the designation of subtribe *Trolliinae* close to the *Adonidinae* as proposed by JENSEN (1968), or a tribe *Adonideae* including both *Adonis* and *Trollius* (JENSEN & al. 1995, this volume) best reflects the phylogenetic relationships of these genera.

Trautvetteria serologically clusters with the *Adonis-Trollius* group, while other molecular and non-molecular data indicate a relationship with the *Ranunculus* group (JENSEN & al. 1995, this volume).

In most classifications, groups 1, 2 and 3 form a cluster and are considered to represent a monophyletic lineage which evolved–in parallel to *Thalictrum*- polycarpous gynoecia and achene fruits. Our serological and other molecular data (JENSEN & al. 1995, this volume) do not support this monophyly unequivocally. At least the *Trollius-Adonis* group is indicated (Fig. 4) to be separated from the two other groups by serological data.

Group 4: *Cimicifuga, Actaea*. Among all genera included in this investigation, *Cimicifuga* and *Actaea* always revealed serological identity. This observation had been also reported by JENSEN (1968) from serological data obtained from the main seed globulins. Between other closely related genera such as *Anemone* and *Hepatica*, *Ranunculus* and *Ficaria*, or *Delphinium* and *Consolida*, always serological differences could be detected. Therefore, it is important to consider whether the difference in a few conspicuous morphological characters (follicle versus "follicle-berry" sensu TAKHTAJAN 1959) justifies the separation of both taxa which, e.g., PRANTL (1891) included in the genus *Actaea*.

Within the *Ranunculaceae* with R-type chromosomes, the *Cimicifuga* group has relative few advanced morphological characters and seems to represent a primitive taxon: petal nectaries are absent or non-specialized in some *Cimicifuga* species (HIEPKO 1965); also lacking are zygomorphic flowers, a clear separation between sepals and petals, polycarpous gynoecia, or annual species.

It was not possible to include *Eranthis* in this investigation because no legumin is detectable (JENSEN & HÖHN, unpubl.). Using an *Eranthis* vicilin antiserum, *Actaea* and *Cimicifuga* have been shown to be the most closely related genera. The same relationship has been reported using other molecular data (JENSEN & al. 1995, this volume).

Group 5: *Aconitum, Delphinium, Consolida.* In this group, vicilins and not legumins are the main seed globulins (JENSEN & HÖHN, unpubl.). Especially for *Consolida*, the precipitations were difficult to interprete because of the low legumin production. This is why *Consolida* has only been used as reacting antigen, and not as antiserum (reference system). Therefore serological data for *Consolida* are limited. Close serological relationships between all three genera, however, have been reported (see also JENSEN 1968). All other published molecular (JENSEN & al. 1995, this volume) and non-molecular data (JENSEN 1968, ZIMMERMANN 1965, HEGNAUER 1973, TAMURA 1993) also support this group.

Nigella is also connected with this group of zygomorphic *Ranunculaceae* (Fig. 4), although, with low similarity values (Table 2). Other criteria support such a connection (ZIMMERMANN 1965). JENSEN & al. (1995, this volume) include *Nigella* in tribe *Delphinieae* on the basis of various molecular data.

As regards the position of *Helleborus*, which sometimes is associated with the *Aconitum* group (KORDJUM 1959), our legumin data give no clear evidence.

Group 6: *Aquilegia, Thalictrum.* LANGLET (1932) was the first to demonstrate the close alliance between the achene-bearing genus *Thalictrum* and the follicle-bearing genus *Aquilegia* and other taxa characterized by seven (or 14) small chromosomes ("T-type") versus a connection with other genera with achenes but "R-type" chromosomes. This has been proved by our present data (as well as by JENSEN 1968) and all other molecular (JENSEN & al. 1995, this volume) and chemical data (HEGNAUER 1973, 1990; JENSEN 1995, this volume). Therefore, tribe *Isopyreae* sensu JENSEN & al. (1995, this volume) including *Aquilegia* and *Thalictrum* should represent a monophyletic group.

Group 7: *Xanthorhiza, (Coptis).* Although a *Coptis* antisystem has not been used, the exceedingly high heterologous reactivity of *Coptis* with *Xanthorhiza* antibodies in the *Xanthorhiza* reference system indicates close serological relationships between these two genera. This corresponds to reported systematic data including the recent molecular evidence (JENSEN & al. 1995, this volume).

The woody, monotypic genus *Xanthorhiza* is often perceived to represent a relatively primitive type within the *Ranunculaceae*: It has many small and inconspicuous flowers in an racemose inflorescence, no specialized nectaries (HIEPKO 1965), accumulates benzylisoquinoline alkaloids, etc. The molecular data (JENSEN & al. 1995, this volume) suggest that the *Coptidoideae* are a sister group to the other *Ranunculaceae*.

This raises the question whether a common origin can be postulated for the x = 7 and the x = 9 T-type taxa (JENSEN 1968); in other words, are *Isopyreae* and *Coptideae*, both possessing small T-chromosomes, sister groups, or do they have a different origin in the *Ranunculaceae*? Although the serological investigation using seed protein mixtures (JENSEN 1968) supports a monophyletic origin for the *Isopyreae* plus *Coptideae*, the present legumin data do not augment such a statement. For the time being this problem must remain unanswered.

Significance of serological characters. Serological characters represent those parts of the protein chain which elicid an antigenic response. This response is expressed by the specific antibodies which constitute the used monospecific antisera. The observed precipitation bands, their identity or formation of spurs reflect the similarities or dissimilarities of the molecular structures of the epitopes involved. Accordingly, serologically revealed data comparing sequence sections of proteins (i.e. epitopes) can be used to elucidate relationships. It is not essential that such a serological investigation is based on one specific protein (i.e. legumin) resp. its monospecific antiserum because the results based on protein mixtures (obtained e.g., by JENSEN 1968) are similar to those obtained here. The use of one specific protein, however, makes the procedure comparable with other molecular data and leads to comparable results (see JENSEN & al. 1995, this volume).

ELKE HÖHN has conducted the serological experiments. NIKOLAUS SCHÜLER has provided the UPGMA and neighbor joining trees plus advice for PHYLIP and PAUP tree constructions. KATRIN MCQUEENEY collected the *Coptis trifolia* seed material. I thank them all for their valuable support.

References

BHANDARI, N. N., 1966: Studies in the family *Ranunculaceae* IX. Embryology of *Adonis*. – Phytomorphology **16**: 578–587.

– KAPIL, R. N., 1964: Studies in the family *Ranunculaceae* VII. Two types of embryo sacs in *Trollius* L. – Beitr. Biol. Pfl. **40**: 113–120.

CRISTOFOLINI, G., 1987: Serological relationships among *Sophoreae*, *Thermopsideae* and *Genisteae* (*Fabaceae*). – Bot. J. Linn. Soc. **94**: 421–432.

– FEOLI CHIAPELLA, L., 1977: Serological systematics of the tribe *Genisteae* (*Fabaceae*). – Taxon **26**: 43–56.

EISEN, H. N., 1974: Immunology. – Hagerstown: Harper & Row.

ESEN, A., HILU, K. W., 1989: Immunological affinities among subfamilies of the *Poaceae*. – Amer. J. Bot. **76**: 196–203.

FAIRBROTHERS, D. E., PETERSEN, F. P., 1983: Serological investigation of the *Annoniflorae* (*Magnoliiflorae*, *Magnoliidae*). – In JENSEN, U., FAIRBROTHERS, D. E., (Eds): Proteins and nucleic acids in plant systematics, pp. 301–310. – Berlin, Heidelberg, New York, Tokyo: Springer.

FELSENSTEIN, J., 1993: PHYLIP (Phylogeny Inference Package), version 3.5c. – Computer program distributed by the author. – Washington: Department of Genetics, University of Washington, Seattle, USA.

FISCHER, H., JENSEN, U., 1992: Utilization of proteins to estimate relationships in plants: serology. A discussion based on the *Asteraceae-Cichorioideae*. – Belg. J. Bot. **125**: 243 255.

GOHLKE, K., 1913: Die Brauchbarkeit der Serum-Diagnostik für den Nachweis zweifelhafter Verwandtschaftsverhältnisse im Pflanzenreiche. – Stuttgart, Berlin: Fr. Grub.

HAMMOND, H. D., 1952: Serology applied to systematic studies in the *Ranunculaceae* JUSS. – Dissertation, University of Pennsylvania.

– 1955: Systematic serological studies in *Ranunculaceae*. – Serol. Mus. Bull. **14**: 1–3.

HEGNAUER, R., 1973: Chemotaxonomie der Pflanzen, **6**: *Dicotyledoneae: Rafflesiaceae-Zygophyllaceae*. – Basel, Stuttgart: Birkhäuser.

– 1990: Chemotaxonomie der Pflanzen, **9**: *Dicotyledoneae:Magnoliaceae-Zygophyllaceae*. – Basel, Boston, Berlin: Birkhäuser.

HEYWOOD, V. H., 1971: Taxonomie der Pflanzen. – Stuttgart: G. Fischer.

HIEPKO, P., 1965: Vergleichend-morphologische und entwicklungsgeschichtliche Untersuchungen über das Perianth bei den *Polycarpicae*. – Bot. Jhrb. **84**: 359–426.

HOOT, S. B., 1995: Phylogeny of the *Ranunculaceae* based on *atp*B, *rbc*L and 18S nuclear ribosomal DNA sequence data. – Pl. Syst. Evol. [Suppl.] **9**: 241–251.

JENSEN, U., 1968: Serologische Beiträge zur Systematik der *Ranunculaceae*. – Bot. Jahrb. **88**: 204–268.

– 1995: Secondary compounds of the *Ranunculiflorae*. – Pl. Syst. Evol. [Suppl.] **9**: 85–97.

– PENNER, R., 1980: Investigation of serological determinants from single storage plant proteins. – Biochem. Syst. Ecol. **8**: 161–170.

– GREVEN, B., 1984: Serological aspects and phylogenetic relationships of the *Magnoliidae*. – Taxon **33**: 563–577.

– HOOT, S. B., JOHANSSON, J. T., KOSUGE, K., 1995: Systematics and phylogeny of the *Ranunculaceae* – a revised family concept on the basis of molecular data. – Pl. Syst. Evol. [Suppl.] **9**: 273–280.

JOHANSSON, J. T., 1995: A revised chloroplast DNA phylogeny of the *Ranunculaceae*. – Pl. Syst. Evol. [Suppl.] **9**: 253–261.

KORDJUM, E. L., 1959: Comparative embryological investigation of the family *Ranunculaceae* DC. – Ukr. Bot. Ž. **16**: 32–43.

KOSUGE, K., SAWADA, K., DENDA, T., ADACHI, J., WATANABE, K., 1995: Phylogenetic relationships of some genera in the *Ranunculaceae* based on alcohol dehydrogenase genes. – Pl. Syst. Evol. [Suppl.] **9**: 263–271.

LANGLET, O., 1932: Über Chromosomenverhältnisse und Systematik der *Ranunculaceae*. – Svensk. Bot. Tidskr. **26**: 381–400.

LESTER, R. N., ROBERTS, P. A., LESTER, C., 1983: Analysis of immunotaxonomic data obtained from spur identification and absorption techniques. – In JENSEN, U., FAIRBROTHERS, D. E., (Eds): Proteins and nucleic acids in plant systematics. – Berlin, Heidelberg, New York, Tokyo: Springer.

MAXSON, L. R., MAXSON, R. D., 1990: Immunological techniques. – In HILLIS, D. M., MORITZ, C., (Eds): Molecular systematics. – Sunderland, Mass.: Sinauer.

MEZ, C., ZIEGENSPECK, H., 1926: Der Königsberger serologische Stammbaum. – Bot. Arch. **13**: 483–485.

PRANTL, K., 1891: *Ranunculaceae*. – In ENGLER, A., PRANTL, K., (Eds): Die natürlichen Pflanzenfamilien 3, pp. 43–66. – Leipzig: Engelmann.

RELANDER, L. K., 1911: Studien über die Verwendbarkeit der Präzipitinreaktion in der Samenprüfung. – Abh. Agrikulturwiss. Ges. Finland, 1, SANA, Helsinki.

TAKHTAJAN, A., 1959: Die Evolution der Angiospermen. – Jena: G. Fischer.

TAMURA, M., 1993: *Ranunculaceae*. – In KUBITZKI, K., ROHWER, J. G., BITTRICH, V., (Eds): The families and genera of vascular plants, **II**. – Berlin, Heidelberg, New York, London, Paris, Tokyo, Hongkong, Barcelona, Budapest: Springer.

WEIR, D. M., 1986: Handbook of experimental immunology. – Oxford: Blackwell.

ZIMMERMANN, W., 1965: Familie *Ranunculaceae*. – In RECHINGER, K.-H., DAMBOLDT, J., (Eds): HEGI, Illustrierte Flora von Mitteleuropa, III/3, 2nd edn. – München: Hanser.

Address of the author: Dr UWE JENSEN, Lehrstuhl Pflanzensystematik, Universität Bayreuth, Universitätsstrasse 30, D-95440 Bayreuth, Federal Republic of Germany.

Accepted January 19, 1995

Pl. Syst. Evol. [Suppl.] 9: 229–240 (1995)

Fatty acid patterns of *Ranunculaceae* seed oils: phylogenetic relationships

KURT AITZETMÜLLER

Received September 27, 1994

Key words: *Ranunculaceae.* – Seed fatty acids, seed oils, capillary gas chromatography, fatty acid fingerprints, phylogeny.

Abstract: The current interest in renewable resources has led to research regarding the occurrence and distribution of unusual and technically interesting fatty acids. Their sporadic occurrence in seed fats is genetically determined and they are highly significant indicators of phylogenetic relationships. Capillary gas chromatographic „fingerprints" of unusual fatty acids can be used in chemotaxonomic studies. In the family *Ranunculaceae* a large number of unusual fatty acids occurs and the seed fatty acid patterns observed are highly correlated to plant genus. Differences in fatty acid patterns illustrate a separate position of *A. rivularis* within the genus *Anemone* and show similarities between this species on the one hand and *Caltha, Cimicifuga* and *Actaea* on the other hand.

Renewable resources are of great interest in long term oilseed research. In this context, research is going on regarding the occurrence and distribution of technically interesting fatty acids in seed oils. Phylogenetic and taxonomical relations are also important in this context, and we are looking for the distribution of the genetic potential in the plant kingdom to produce certain technically interesting fatty acids. With further advances in gene transfer, it may be possible in the future to transplant genes for the production of such fatty acids into agronomically more developed species such as rapeseed. Transgenic rapeseed has been produced, for example, by transplanting genes from *Umbellularia californica* NUTT. into rapeseed (DAVIES & FLIDER 1994).

Many technically interesting fatty acids have been found in the plant kingdom (SMITH 1970, BADAMI & PATIL 1981). Fatty acids with unusual structures often accumulate in seed oils. They are usually not present or present in much lower concentrations only, in the lipids of green leaves (COLOMBO & al. 1991, AITZETMÜLLER 1993).

Seed fatty acids of *Ranunculaceae* JUSS. had been investigated by many authors [see literature quoted in HEGNAUER (1990) and AITZETMÜLLER & TSEVEGSÜREN (1994 a, b)]. In the family *Ranunculaceae*, there is a situation that differs from that in most other plant families. For one, the observed fatty acid patterns seem to be correlated

Table 1. Numbers used for diagnostically important fatty acids in the gas chromatograms shown (this paper and AITZETMÜLLER 1993)

No.	Fatty acid	No.	Fatty acid
1	16:1 Δ 5 *trans*	12	20:2 Δ 11c, 14c
2	16:1 Δ 5 *cis*	13	20:3 Δ 5c, 11c, 14c
3	16:2 Δ 7c, 10c	14	20:3 Δ 11c, 14c, 17c
4	18:1 Δ 5 *trans*	15	20:4 Δ 5c, 11c, 14c, 17c
5	18:1 Δ 5 *cis*	16	18:2 Δ 5, 6 (allene)
6	18:2 Δ 5t, 9c	17	18:2 Δ 5c, 9c
7	18:3 Δ 5t, 9c, 12c	18	18:3 Δ 5c, 9c, 12c
8	18:3 Δ 6c, 9c, 12c	19	18:4 Δ 6c, 9c, 12c, 15c
9	18:4 Δ 5c, 9c, 12c, 15c	20	20:1 Δ 5 *cis*
10	20:1 Δ 11 *cis*	21	22:1 Δ 13 *cis*
11	20:2 Δ 5c, 11c	22	22:2 Δ 13c, 16c
		23	24:1 Δ 15 *cis*

to plant genus in a rather strict way. Usually, all the species of one genus that were investigated showed exactly the same pattern of unusual fatty acids. Only the levels present (in % of total fatty acids) varied slightly (AITZETMÜLLER & TSEVEGSÜREN 1994b). This has been illustrated in terms of capillary gas chromatographic „fatty acid fingerprints" (AITZETMÜLLER 1993).

Secondly, the differences in seed fatty acid patterns between genera are very large. In terms of seed fatty acid patterns, the genus-to-genus differences in the *Ranunculaceae* are larger than most family-to-family differences elsewhere in the plant kingdom. The combination of very little species-to-species variation, combined with very large genus-to-genus differences, is rather unique. Thirdly, the occurrence of individual unusual fatty acids seems related to genus or tribe. This was already noted by HEGNAUER (1990), who mentioned that seed oils in the *Ranunculaceae* were particularly characteristic for genus and tribe. A better knowledge of the seed fatty acid patterns could yield taxonomically useful results (HEGNAUER 1990).

Material and methods

Seed material, oil extraction, saponification and methyl ester preparation were as previously described (TSEVEGÜREN & AITZETMÜLLER 1993, AITZETMÜLLER & TSEVEGSÜREN 1994b). Samples originally received from the Botanical Institute, Ulan Bator, were labelled (in Figs. 1–4 and Tables 2 and 3) as received (e.g., *Batrachium, Atragene*). It is recognized that most authors now list *Batrachium* (DC.) S. F. GRAY as a section of *Ranunculus* L. and *Atragene* L. as a section of *Clematis* L. (TAMURA 1993).

The gas chromatograms were obtained as described (AITZETMÜLLER & al. 1993) using a Hewlett-Packard 5890 gas chromatograph equipped with a 50 m Silar 5 CP column (0.25 mm i.d.) and an FID detector. The fatty acid methyl ester peaks were identified via their relative retention times and ECL-values and by comparison with known samples and standards, incl. co-chromatography of mixtures on two columns of different selectivity

(AITZETMÜLLER & al. 1993). In the GLC fingerprints shown, the large fatty acid peaks are cut off at the top.

In the fingerprints shown in Figs. 1–3 in this paper, only those fatty acid peaks are marked or labelled, that are considered to be chemotaxonomically significant. In every seed oil, „normal" fatty acids, such as palmitic, stearic, oleic, linoleic and α-linolenic, are found. These are not labelled at all in the figures shown here, because they lack chemotaxonomic significance (AITZETMÜLLER 1993). The numbers given to the unusual fatty acids are illustrated in Table 1.

Results

Figure 1 shows examples for typical seed oil fatty acid fingerprints for several genera of the family *Ranunculaceae*. In the 5 tested species of *Aconitum*, there were no really unusual fatty acids present, so that *Aconitum* L. has a „normal" fatty acid composition. However, three of the five investigated *Aconitum* species contained elevated levels of 14:0 (TSEVEGSÜREN & AITZETMÜLLER, unpubl.).

In the fatty acid pattern of *Delphinium* L. species, we have the occurrence of 20:1 Δ11 *cis* (fatty acid no. 10 is the most significant peak). This fatty acid as such is not a highly unusual fatty acid and it can occur in several other genera and plant species. In the *Ranunculaceae*, it is typical for *Delphinium* and *Consolida* (DC.) S. F. GRAY, but it occurs in a few other genera, too. Whereas *Delphinium* contains about 10 to 15% of it, in *Consolida* this can go up to between 25 and 40%. In *Delphinium* (and *Consolida*) peak 10 (cut off at the top) is more than twenty times the size of peak 12.

Adonis L. again has a normal fatty acid pattern, and the same applies to *Trollius* L., *Pulsatilla* MILL. and several species of *Anemone* L.

In Fig. 1 the fatty acid pattern of *Helleborus* L. is shown. Typically, in *Helleborus*, the fatty acid 20:1 Δ11 *cis* (no. 10) is accompanied by 20:2 Δ11 *cis*, 14 *cis* (no. 12) and a smaller amount of 20:3 Δ11c, 14c, 17c (no. 14).

Figure 1 also shows the typical fatty acid pattern of *Ranunculus*. In all *Ranunculus* species investigated, there is an unusual 16:2 fatty acid (no. 3). The double bond positions of this fatty acid have not yet been finally determined. However, this fatty acid was present in notable amounts – between 2 and 10 percent of total fatty acids – in all species of *Ranunculus* investigated and it is usually accompanied by elevated levels of 16:1 Δ7 *cis*. Recently we have also found both these fatty acids in *Batrachium* – a finding that supports the inclusion of *Batrachium* in *Ranunculus*. This is a good example for the fact that an unusual fatty acid can still be highly significant chemotaxonomically, even though it occurs consistently only in small percentages in a seed oil. Except for one older report on *R. falcatus* L. it was found in every *Ranunculus* species investigated. It was not found in any other *Ranunculaceae* genus or species with the exception of trace amounts in one species each of *Trollius* and *Halerpestes* E. Greene. *R. falcatus* should be re-investigated using modern high-resolution capillary GLC techniques.

Figure 2 shows the fatty acid pattern of *Caltha palustris* L. One can immediately recognize that in this seed oil a great variety of unusual fatty acids is present (SMITH & al. 1968). This complicated „fingerprint" contains peaks for the fatty acids no. 2, 5, 9, 10, 11, 12, 13, 14, 15, 17 and 20.

Figure 2 also shows the typical fatty acid pattern of *Nigella* L. In this genus, the occurrence of 20:2 Δ11 *cis*, 14 *cis* (no. 12) should be noted which is rather isolated and

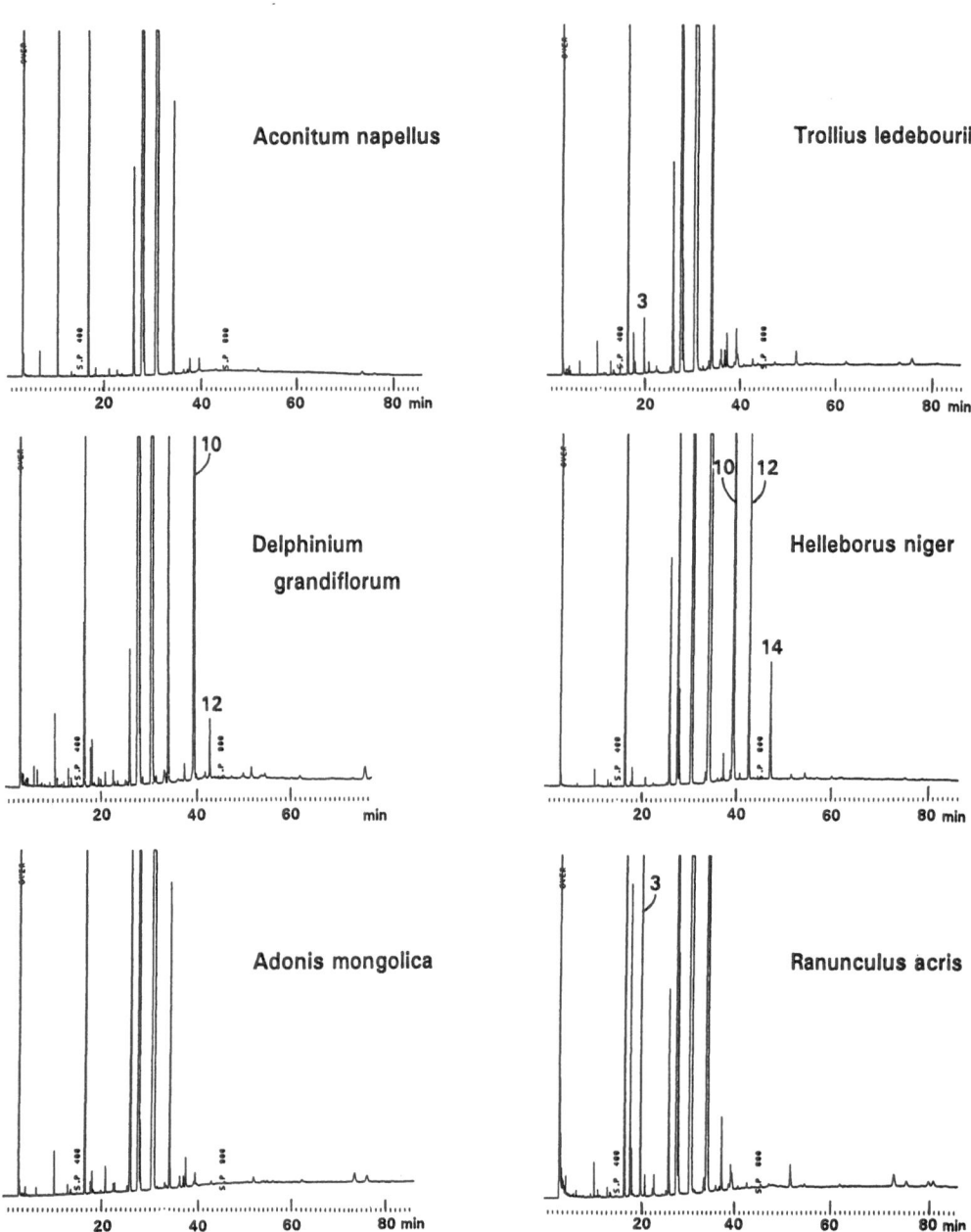

Fig. 1. Seed oil fatty acid fingerprints typical for *Aconitum*, *Delphinium*, *Adonis*, *Trollius*, *Helleborus* and *Ranunculus*. Shown are fingerprints of *Aconitum napellus* L., *Delphinium grandiflorum* L., *Adonis mongolica* SIMONOVICH, *Trollius ledebourii* REICHB., *Helleborus niger* L. and *Ranunculus acris* L.

is accompanied by very little 20:1. (Here, peak 12, cut off at the top, is 10 times the size of peak 10–and „12 ≫ 10“ is typical for *Nigella*.) The fatty acid no. 12, however, also occurs in several other genera where it is usually accompanied by 20:1 Δ11 *cis* and 20:3 Δ5c, 11c, 14c.

Figure 2 illustrates the fatty acid pattern of *Cimicifuga* WERNISCH. Again, in all *Cimicifuga* species there is a highly complex fatty acid pattern, not unlike that of

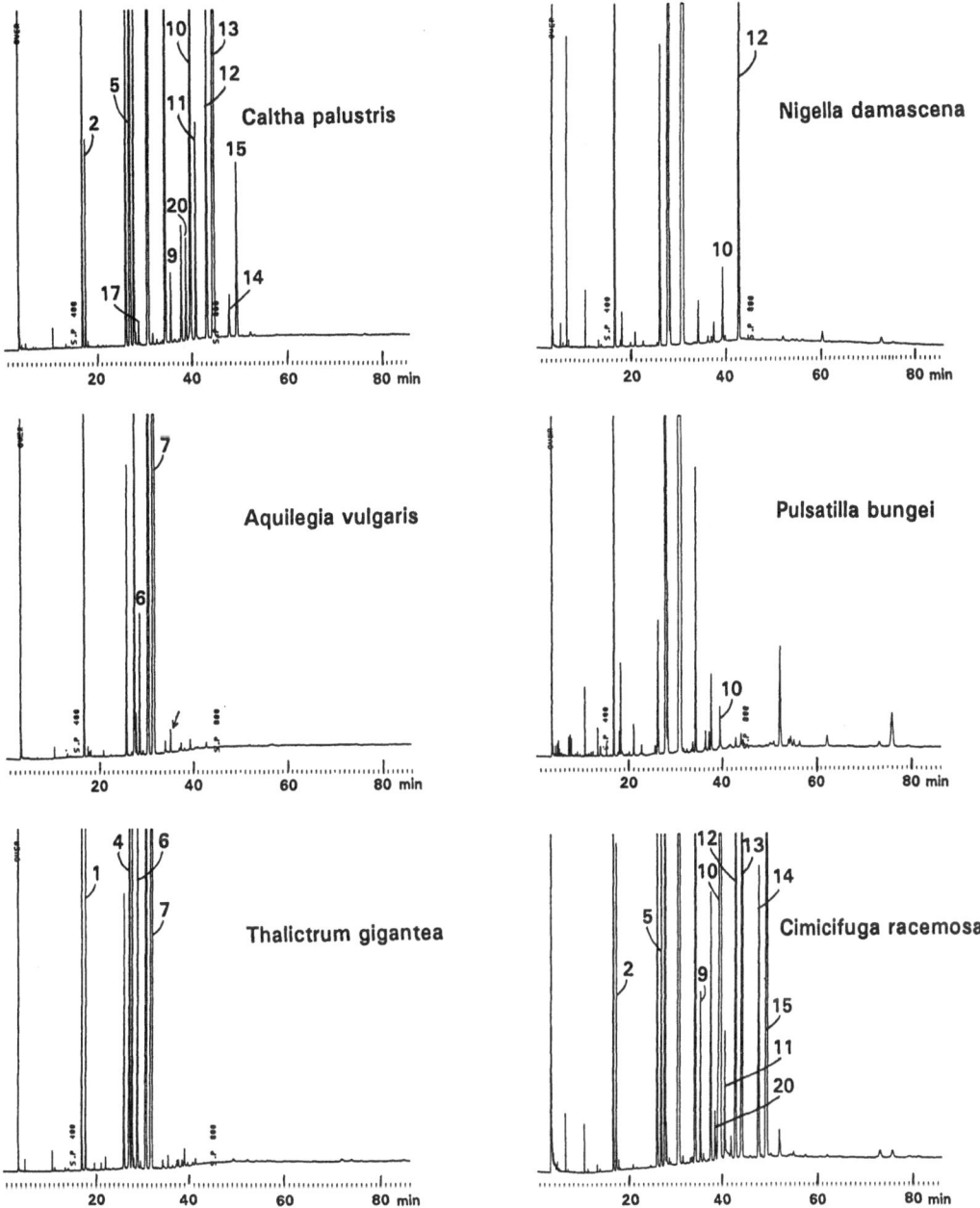

Fig. 2. Seed oil fatty acid fingerprints typical for *Aquilegia, Thalictrum, Nigella, Pulsatilla, Cimicifuga* and *Caltha*. Shown are fingerprints of *Caltha palustris* L., *Aquilegia vulgaris* L., *Thalictrum aquilegifolium* L. cv. 'Gigantea', *Nigella damascena* L., *Pulsatilla bungei* C. A. MEYER and *Cimicifuga racemosa* (L.) NUTT. var. *cordifolia* (PURSH) A. GRAY

Caltha (AITZETMÜLLER, unpubl.) All three *Cimicifuga* spp. investigated contain fatty acids no. 2, 5, 9, 10, 11, 12, 13, 14, 15 and 20.

Figure 2 also shows the fatty acid fingerprints of *Pulsatilla, Aquilegia* L. and *Thalictrum* L. The latter two contain the highly unusual 18:3 *trans* fatty acid, columbinic acid (18:3Δ5 *trans*, 9 *cis*, 12 *cis*, peak no. 7).

Figure 3 shows the fatty acid pattern of several species of *Anemone*. Obviously, a typical *Anemone* contains mostly normal fatty acids. However, sometimes there is

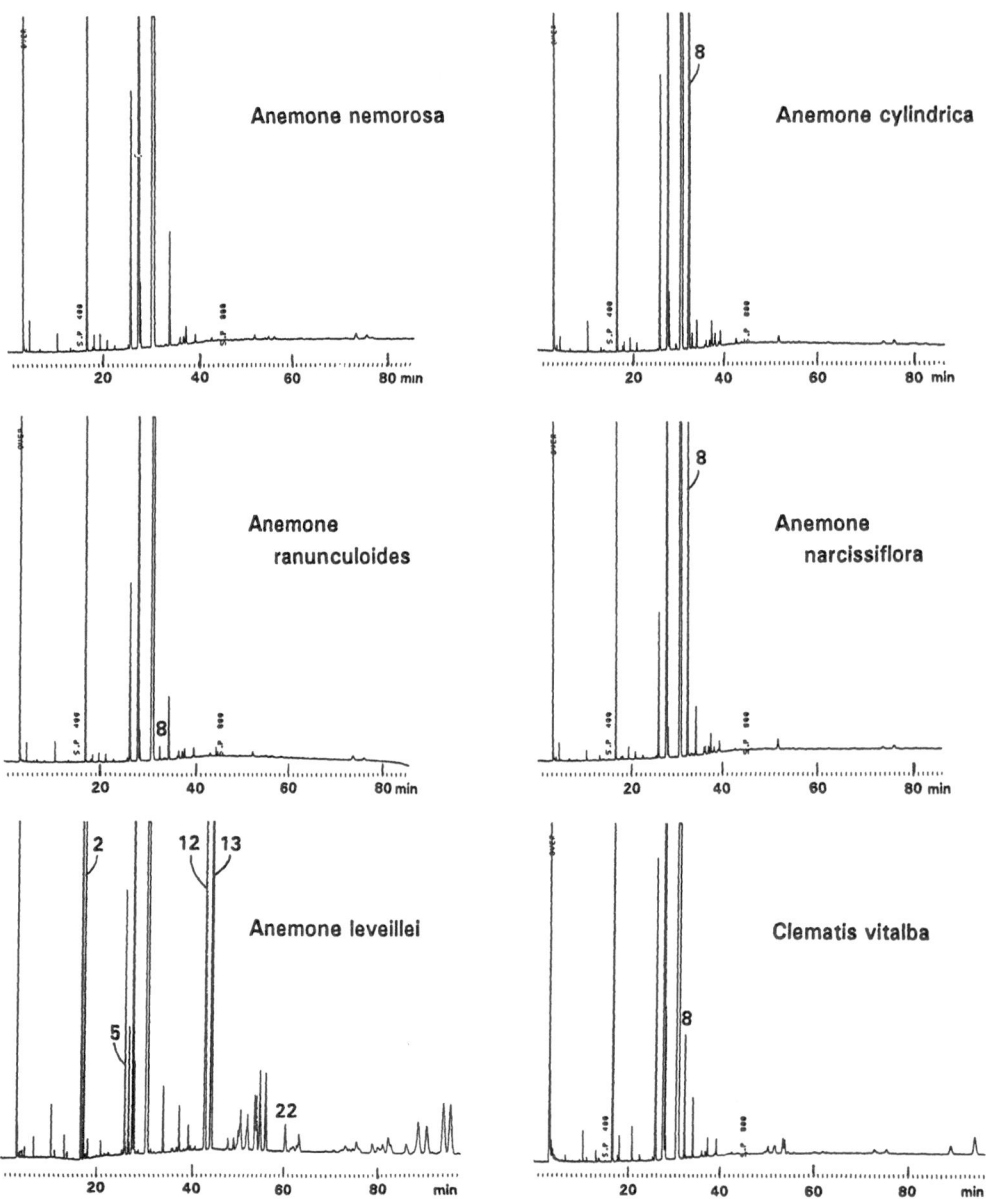

Fig. 3. Seed oil fatty acid fingerprints typical for *Anemone* and *Clematis*, and deviating fingerprint for *Anemone leveillei*. Shown are fingerprints of *A. nemorosa* L., *A. ranunculoides* L., *A. leveillei*, *A. cylindrica* GRAY, *A. narcissiflora* L. and *C. vitalba* L.

a peak (labelled no. 8) which is γ-linolenic acid (TSEVEGSÜREN & AITZETMÜLLER, 1993). γ-Linolenic acid is a fatty acid that is of great interest for the pharmaceutical industry (GUNSTONE 1992, HORROBIN 1992), and so we have investigated several more species of *Anemone*. All the *Anemone* species investigated exhibited essentially the same fatty acid fingerprints: normal fatty acids only, or combined with the occurrence of γ-linolenic acid. The latter fatty acid (no. 8) was also found in species of *Clematis* and *Atragene*. One should note the great similarity of all these fatty acid patterns which are typical for the genera *Anemone*, *Clematis* and *Atragene*. Recent investigations of various *Anemone* species, however, showed an anomaly. One species was found which differs greatly from all the other *Anemone* species, and this was *Anemone rivularis* BUCH.-HAM. ex DC. (TSEVEGSÜREN & AITZETMÜLLER 1993). This was rather surprising and so we looked at other representatives of the same section. *A. rivularis* belongs to sect. *Rivularidium* JANCZEWSKI. We managed to get seeds of another species of sect. *Rivularidium*, namely from *A. leveillei* E. ULBRICH. The analysis of the seed oil fatty acids of *A. leveillei* gave the same pattern as *A. rivularis* (AITZETMÜLLER, unpubl.).

Figure 3 also shows the fatty acid fingerprint for *Anemone leveillei*. One should note that this is a totally different fatty acid pattern. Peak 8 (γ-linolenic acid) does not occur. On the other hand, the fingerprint shows the occurrence of peaks 2, 5, 12 and 13. In the *Ranunculaceae* this was the first case that two species belonging to a certain genus should deviate so strongly from all the other representatives of that genus. The fatty acids number 2, 5, 12 and 13 which are found in *A. rivularis* and *A. leveillei* were also found in *Caltha palustris* and in various *Cimicifuga* species.

Discussion

The seed oil of *Aquilegia* has been already investigated in the late 50's and early 60's. It contains columbinic acid as the major fatty acid (KAUFMANN & MANKEL 1964). This is 18:3Δ5 *trans*, 9*cis*, 12*cis*; i.e. a highly unusual 18:3-fatty acid with the first double bond in Δ-5 position and in *trans* configuration. Now, the interesting thing is that all *Aquilegia* species investigated so far contain this fatty acid. They also all contain a lesser amount of 18:2Δ5 *trans*, 9*cis* (TAKAGI & al. 1983, AITZETMÜLLER & TSEVEGSÜREN 1994 b). The occurrence of these two fatty acids is characteristic for the genus *Aquilegia*, and has been documented at the same time only for one other genus, namely *Thalictrum*. So far these two fatty acids have not been found in any other *Ranunculaceae* seed oils outside *Aquilegia* and *Thalictrum* and, as most recent research (AITZETMÜLLER, unpubl.) has shown, in *Isopyrum* L. They do not occur in other plant seed oils. So in our labeling system, all *Aquilegia* (and *Isopyrum*) species have „unusual fatty acid pattern 6, 7" (AITZETMÜLLER 1993, AITZETMÜLLER & TSEVEGSÜREN 1994 b).

Thalictrum contains the same unusual fatty acids as *Aquilegia* – numbers 6 and 7 – but in addition to these also the fatty acids number 1 and 4. Again all *Thalictrum* species investigated so far contain these four unusual fatty acids, although some of them only in low amounts. Again, in the whole family *Ranunculaceae* the two fatty acids number 1 and 4 occur only in the genus *Thalictrum*, which therefore has the „unusual fatty acid pattern 1, 4, 6, 7". In this way, several of the fatty acids listed in Table 1 can be identified as being typical for certain genera – and usually they are present in all species belonging to that particular genus. Table 2 indicates their presence in the various genera and their postulated chemotaxonomic significance.

In most *Anemone* spp. as well as in *Clematis* and *Atragene* only the usual fatty acids, with or without peak 8 are found (AITZETMÜLLER & TSEVEGSÜREN 1994a). In conclusion, it seems that *Anemone* sect. *Rivularidium* deviate strongly, in their fatty acid pattern 2, 5, 12, 13 from all the other *Anemone* species investigated so far which all have fatty acid pattern 0 or 8. In terms of their fatty acid patterns, *A. rivularis* and *A. leveillei* show similarity to *Caltha* and *Cimicifuga*. This fact, the enzymes that may be necessary to produce these different fatty acid patterns, and some possible evolu-

Table 2. Chemotaxonomically significant fatty acids in the seed lipids of *Ranunculaceae*

Fatty acid	Typical for
16:1 Δ5 *trans*	*Thalictrum*
16:1 Δ5 *cis*	*Caltha* I, *Caltha* II, *Cimicifuga*, *Anemone* III, *Actaea*
16:2 (Δ7c, 10c?)	**Ranunculus, Batrachium**, (*Trollius*), (*Halerpestes*)
18:1 Δ5 *trans*	**Thalictrum**, *Caltha* II
18:1 Δ5 *cis*	**Caltha** I, **Caltha** II, *Cimicifuga*, *Anemone* III, *Actaea*
18:2 Δ5t, 9c	**Thalictrum**, *Isopyrum*, *Aquilegia*
18:2 Δ5c, 9c	(*Caltha* I)
18:3 Δ5t, 9c, 12c	**Aquilegia, Thalictrum, Isopyrum**
18:3 Δ6c, 9c, 12c	**Anemone** I, *Clematis*, *Atragene*
18:4 Δ5t, 9c, 12c, 15c	(*Isopyrum*), (*Aquilegia*), (*Thalictrum*)
18:4 Δ5c, 9c, 12c, 15c	*Cimicifuga*, *Actaea*, (*Caltha* I)
20:1 Δ5 *cis*	(*Caltha* I), (*Caltha* II), (*Actaea*)
20:1 Δ5 *trans*	(*Thalictrum*)
20:1 Δ11 *cis*	**Delphinium, Consolida, Caltha, Actaea, Cimicifuga**, *Helleborus*, *Nigella*
20:2 Δ5c, 11c	*Caltha* I, *Actaea*
20:2 Δ11c, 14c	**Anemone** III, **Nigella**, *Actaea*, *Caltha* I, *Cimicifuga*, *Helleborus*
20:3 Δ5c, 11c, 14c	**Caltha** I, **Anemone** III, *Cimicifuga*, *Actaea*
20:3 Δ11c, 14c, 17c	*Helleborus*, *Cimicifuga*, *Actaea*, (*Caltha* I)
20:4 Δ5c, 11c, 14c, 17c	**Actaea, Cimicifuga**, *Caltha*
22:2 Δ13c, 16c	(*Caltha* I), (*Nigella*)

Bold face: exceeds 5% of total fatty acids in at least one species.
In brackets: below 1%, but considered a taxonomically highly significant fatty acid

tionary or taxonomic aspects of this observation, have already been discussed (AITZETMÜLLER & TSEVEGSÜREN, 1994b). Based on the distribution and occurrence of these unusual fatty acids, and the patterns that are present in the various genera, Table 3 can be constructed:

Table 3 is an updated version of a similar table published by AITZETMÜLLER & TSEVEGSÜREN (1994b) and it includes the latest results. Table 3 shows these patterns in a symbolic way. The numbers in Table 3 mean the occurrence of the fatty acid of this number in larger than trace amounts; however, a „weighted limit" was applied to construct Table 3. For the highly significant fatty acids a limit of 0.3% was applied, whereas for the less significant fatty acids (no. 10 and 12) a limit of 1% was used. The species of *Anemone* sect. *Rivularidium* are labelled „*Anemone* III". Generally, all representatives of the genera mentioned here contained, in their seed oils, the fatty acids with the numbers shown here in the Table. (The single exception was *R. falcatus*.)

What follows from all this? A chemotaxonomic classification of *Ranunculaceae* was published by JENSEN and illustrates the closer or less close relation of various

Table 3. Unusual fatty acid patterns in various genera of *Ranunculaceae*. Numbers as in Table 1

Genus	Species*	1	2	3	4	5	6	7	8	9	10	11	12	13	14	15
*Pulsatilla***	3	—	—	—	—	—	—	—	—	—	—	—	—	—	—	—
Anemone II[1]	2	—	—	—	—	—	—	—	—	—	—	—	—	—	—	—
Anemone I[2]	4	—	—	—	—	—	—	—	8	—	—	—	—	—	—	—
Clematis	2	—	—	—	—	—	—	—	8	—	—	—	—	—	—	—
Atragene	1	—	—	—	—	—	—	—	8	—	—	—	—	—	—	—
Aquilegia	4	—	—	—	—	—	6	7	—	—	—	—	—	—	—	—
*Isopyrum****	1	—	—	—	—	—	6	7	—	—	—	—	—	—	—	—
Thalictrum	13	1	—	—	4	—	6	7	—	—	—	—	—	—	—	—
Adonis	2	—	—	—	—	—	—	—	—	—	—	—	—	—	—	—
Halerpestes	1	—	—	—	—	—	—	—	—	—	—	—	—	—	—	—
Trollius	2	—	—	—	—	—	—	—	—	—	—	—	—	—	—	—
Batrachium	1	—	—	3	—	—	—	—	—	—	—	—	—	—	—	—
Ranunculus	5	—	—	3	—	—	—	—	—	—	—	—	—	—	—	—
Caltha II[3]	1	—	2	—	4	5	—	—	—	—	—	—	12	13	—	—
Anemone III[4]	2	—	2	—	—	5	—	—	—	—	—	—	12	13	—	—
Caltha I[5]	1	—	2	—	—	5	—	—	—	9	10	11	12	13	14	15
Actaea	1	—	2	—	—	5	—	—	—	9	10	11	12	13	14	15
Cimicifuga	3	—	2	—	—	5	—	—	—	9	10	—	12	13	14	15
Helleborus	1	—	—	—	—	—	—	—	—	—	10	—	12	—	14	—
Nigella	2	—	—	—	—	—	—	—	—	—	—	—	12	—	—	—
Consolida	1	—	—	—	—	—	—	—	—	—	10	—	—	—	—	—
Delphinium	6	—	—	—	—	—	—	—	—	—	10	—	—	—	—	—
Aconitum[6]	5	—	—	—	—	—	—	—	—	—	—	—	—	—	—	—

* Number of species for which data are available

** One *Pulsatilla* species (*P. turczaninovii* AICHELE & SCHWEGLER) contained one as yet unidentified unusual fatty acid – different from all the ones mentioned here; some spp. contained elevated levels of 20:0

*** *I. thalictroides* L. also contained 0.4% of 18:4 Δ 5t, 9c, 12c, 15c

[1] *A. nemorosa, A. ranunculoides*
[2] *A. narcissiflora, A. altaica* FISCHER, *A. cylindrica, A. crinita* JUZ.
[3] *C. natans* PALL.
[4] *A. rivularis, A. leveillei.*
[5] *C. palustris.*
[6] three (of five) *Aconitum* species contained elevated levels of 14:0

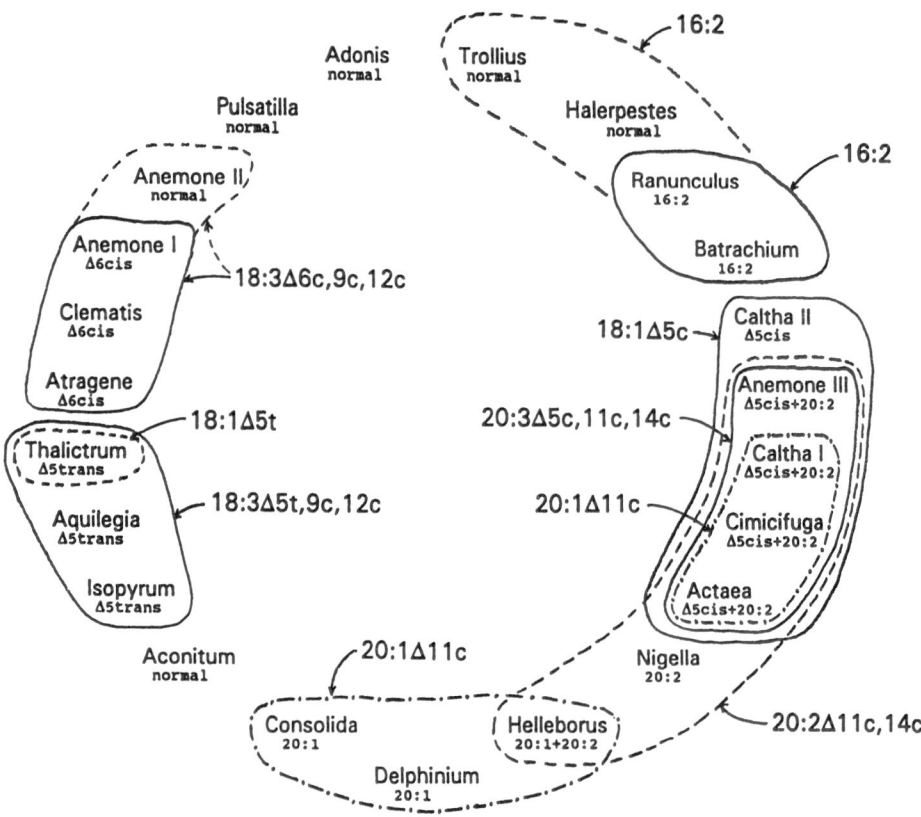

Fig. 4. Chemotaxonomical relation of genera in the family *Ranunculaceae*, based solely on information obtained from seed oil fatty acid fingerprints

genera in the *Ranunculaceae* (JENSEN 1968, FROHNE & JENSEN 1979). From our fatty acid patterns we can construct the following purely chemical relationship:

Aquilegia and *Thalictrum* (and *Isopyrum*) are closely related because both contain the fatty acids 6 and 7. Fatty acids with the *trans* double bond in position Δ5 occur only in the lower left hand part of the scheme. On the other hand, γ-linolenic acid (Δ6 *cis*) occurs only in the upper left part of the scheme. C_{20} unsaturated fatty acids occur in the lower right hand areas, and also in *Anemone* III. The Δ5-monoenoic fatty acids, which are in the *trans* form in *Thalictrum* do occur in *cis* form in the C_{20} region, and we see the occurrence of 20:3 Δ5 *cis*, 11 *cis*, 14 *cis* in *Caltha* and *Cimicifuga*, as well as in *Anemone* III, and also in *Actaea* L. (AITZETMÜLLER, unpubl.). Other C_{20} fatty acids, 20:1 Δ11 *cis* and 20:2 Δ11 *cis*, 14 *cis* also occur in *Delphinium*, *Consolida*, *Helleborus* and *Nigella*. These areas overlap only partially with the areas for 20:3 Δ5c, 11c, 14c.

In conclusion, from the point of view of a fatty acid chemist, and the observed seed oil fatty acid patterns, there must be a close relationship between *Aquilegia* and *Thalictrum*, as postulated by JENSEN (1968) and TAKHTAJAN (1987). Main differences are in the position of *Caltha*, here closer to *Cimicifuga*, and in the position of *Anemone* III. Also the present findings on seed fatty acids fit better to the systems of JENSEN and TAKHTAJAN than to TAMURA's system (TAMURA, 1993). This is not surprising if one considers that JENSEN, in his serological work (1968), produced antibodies reacting with the proteins present in the seeds. These proteins may be enzymes necessary for the production of these fatty acids and triglycerides, and for their catabolism during seed germination. To produce the seed oils with the more complicated fatty acid patterns, such as *Caltha*, *Cimicifuga* and *Anemone* III, a whole range of enzymes is required (elongases, desaturases, acyltransferases etc.) – and some aspects regarding the required desaturases have been discussed (AITZETMÜLLER & TSEVEGSÜREN 1994 b). Other different enzymes may be required for the catabolism of these unusual fatty acids during seed germination. JENSEN (1968), in his serological work, may have measured the enzymes. The present paper measures the products of these enzymes – the fatty acids.

To shed more light on the phylogeny of various genera in the *Ranunculaceae* and to further discuss the importance of unusual seed fatty acids for chemotaxonomical considerations, however, many more species and genera from this plant family should be investigated.

The author is indebted to Dr N. TSEVEGSÜREN (Chemistry Institute, Mongolia) and to the Botanical Institute, Ulan Bator for seed samples and to Prof. Dr U. JENSEN, University of Bayreuth, for helpful discussions. The author wishes to thank Ms G. WERNER for her skilful assistance with the GLC work, and Mr KLAUS R. JELITTO, Schwarmstedt, for gifts of seeds.

References

AITZETMÜLLER, K., 1993: Capillary GLC fatty acid fingerprints of seed lipids – A tool in plant chemotaxonomy? – J. High Resol. Chromatogr. **16**: 488–490.
– TSEVEGSÜREN, N., 1994 a: Occurrence of γ-linolenic acid in *Ranunculaceae* seed oils. – J. Pl. Physiol. **143**: 578–580.
– – 1994 b: Seed fatty acids, „front-end"-desaturases and chemotaxonomy – a case study in the *Ranunculaceae*. – J. Pl. Physiol. **143**: 538–543.
– WERNER, G., TSEVEGSÜREN, N., 1993: Screening of seed lipids for γ-linolenic acid: capillary gas-liquid chromatographic separation of 18:3 fatty acids with Δ 5 and Δ 6 double bonds. – Phytochem. Anal. **4**: 249–255.
BADAMI, R. C., PATIL, K. B., 1981: Structure and occurrence of unusual fatty acids in minor seed oils. – Progr. Lipid Res. **19**: 119–153.
COLOMBO, M. L., TOME, F., BUGATTI, C., 1991: Lipid content and fatty acid composition in hypogeous organs of *Helleborus* species (*Ranunculaceae*). – Pl. Syst. Evol. **178**: 55–63.
DAVIES, H. M., FLIDER, F. J., 1994: Designer oils. – Chemtech (Washington) **24**: 33–37.
FROHNE, D., JENSEN, U., 1979: Systematik des Pflanzenreichs, 3rd edn. – Stuttgart: Fischer.
GUNSTONE, F. D., 1992: γ-Linolenic acid – occurrence and physical and chemical properties. – Progr. Lipid Res. **31**: 145–161.
HEGNAUER, R., 1990: Chemotaxonomie der Pflanzen, **IX** – Basel: Birkhäuser.
HORROBIN, D. F., 1992: Nutritional and medical importance of γ-linolenic acid. – Progr. Lipid Res. **31**: 163–194.
JENSEN, U., 1968: Serologische Beiträge zur Systematik der *Ranunculaceae*. – Bot. Jahrb. **88**: 204–310.
KAUFMANN, H. P., MANKEL, G., 1964: Über das Vorkommen von *trans*-Fettsäuren. – Fette, Seifen, Anstrichm. **66**: 6–13.
SMITH, C. R. Jr., KLEIMAN, R., WOLFF, I. A., 1968: *Caltha palustris* L. seed oil. A source of four fatty acids with *cis*-5-unsaturation. – Lipids **3**: 37–42.

SMITH, C. R., 1970: Occurrence of unusual fatty acids in plants. – In HOLMAN, R. T., (Ed.): Progr. chem. fats other lipids, **XI**, pp. 137–177. Oxford, UK: Pergamon.

TAKAGI, T., ITABASHI, Y., KANENIWA, M., MIZUKAMI, M., 1983: *Trans*-5-olefinic unusual fatty acids in seed lipids of *Aquilegia*. – Yukagaku **32**: 367–374.

TAKHTAJAN, A., 1987: Systema Magnoliophytorum. – Leningrad: Nauka.

TAMURA, M., 1993: *Ranunculaceae*. – In KUBITZKI, K., (Ed.): The families and genera of vascular plants **2**. – Springer: Berlin.

TSEVEGSÜREN, N., AITZETMÜLLER, K., 1993: γ-Linolenic acid in *Anemone* spp. seed lipids. – Lipids **2**: 841–846.

Address of the author: KURT AITZETMÜLLER, Institut für Chemie und Physik der Fette, Bundesanstalt für Getreide-, Kartoffel- und Fettforschung, Piusallee 76, D-48147 Münster, Federal Republic of Germany.

Accepted January 2, 1995

Pl. Syst. Evol. [Suppl.] 9: 241–251 (1995)

Phylogeny of the *Ranunculaceae* based on preliminary *atp*B, *rbc*L and 18S nuclear ribosomal DNA sequence data

SARA B. HOOT

Received December 19, 1994

Key words: *Ranunculaceae,–atp*B, *rbc*L, nuclear ribosomal 18S DNA, phylogeny.

Abstract: The chloroplast genes *atp*B and *rbc*L and nuclear ribosomal 18S DNA were sequenced for 23 genera of the *Ranunculaceae* and two outgroup taxa (*Hydrastis* and *Glaucidium*). The three sequence data sets were combined and the resulting preliminary phylogenetic tree used to assess relationships within the *Ranunculaceae*. The phylogeny strongly supports the monophyly of the family, with 26 substitutions, a bootstrap value of 98% and a decay index of > 7. Within the family, the T-type chromosome group is basal and paraphyletic with respect to the larger R-type chromosome group. Within the T-type chromosome group, *Coptis* and *Xanthorhiza* from a monophyletic group and are basal to all other *Ranunculaceae*. Other alliances previously proposed by taxonomists are confirmed: *Anemonella/Thalictrum/ lsopyrum/Aquilegia/Semiaquilegia*; *Anemone/Clematis, Trautvetteria/ Myosurus/Ranunculus*; *Aconitum/Delphinium*; and *Anemonopsis/Cimicifugal/Actaea*. Other groupings that could not have been predicted on the basis of traditional data include a clade consisting of *Adonis* and *Trollius* and the inclusion of *Eranthis* in a clade with *Anemonopsis, Cimicifuga*, and *Actaea*. *Nigella* is weakly allied with a clade consisting of *Aconitum* and *Delphinium*. The molecular sequence data are largely congruent with results based on cytology, phytochemistry, and micromorphology. Flower and fruit characters are homoplastic in relation to the cladogram based on sequence data. The pattern of relationships based on sequence data supports the view that staminodia/petals and achenes have evolved independently several times within the *Ranunculaceae*.

Homoplasy in the *Ranunculaceae* has always been a limiting factor in the formation of a reliable evolutionary hypothesis and taxonomic treatment of the family. Not only are there the problems intrinsic to such a large and varied family, but also the probable occurrence of widespread parallel and convergent evolution. This is clearly exemplified in the present symposium volume, where widely differing phylogenetic hypotheses can be proposed according to individual data sets: e.g., embryology (ENGEL 1995), morphology (LOCONTE & al. 1995, this volume), and molecular data (JOHANSSON 1995, KOSUGE & al. 1995, both in this volume).

The karyological research of LANGLET (1932) and GREGORY (1941) was instrumental to a major reassessment of evolutionary relationships in the *Ranunculaceae*.

This work resulted in the recognition of two major groups according to chromosome type: the large *Ranunculus*-type (R-type) and the small *Thalictrum*-type (T-type) groups. Subsequent work based on serology (JENSEN 1968, 1971), phytochemistry (HEGNAUER 1966, 1973, 1990; RUIJGROK 1966), amino acid sequences (GRUND & al. 1981), micromorphology (HOOT 1991), and chloroplast DNA restriction site variation (JOHANSSON & JANSEN 1993) have produced varying results and failed to resolve many of the basal branching patterns within the family.

While most systematists since the time of LANGLET (1932) and GREGORY (1941) have treated the family as monophyletic, I know of no morphological, anatomical, or chemical character that is unique to the *Ranunculaceae*. Further debates center on the familial inclusion or exclusion of such genera as *Paeonia*, *Glaucidium*, *Hydrastis*, and *Kingdonia* (sometimes allied with the genus *Anemone*). In addition, the relative placement of several genera (e.g., *Eranthis*, *Adonis*, and *Nigella*) within the family has been especially problematic.

There have also been differing views concerning the position of the *Ranunculaceae* within the *Ranunculidae*. Many past workers considered the family most closely allied to the *Berberidaceae*, *Hydrastidaceae*, and less frequently, *Glaucidiaceae* (TAKHTAJAN 1980, CRONQUIST 1981, HOOT 1991). A recent phylogenetic analysis based on traditional data places the family as most closely allied to the *Papaveraceae* (LOCONTE & STEVENSON 1991; LOCONTE & al. 1995, this volume).

In an attempt to overcome the problems posed by homoplasy, resolve basal branching patterns within the family, and clarify the family's placement within the *Ranunculidae*, three conservative genes were sequenced. This work is part of an intensive systematic study of phylogenetic relationships among basal eudicots (*Ranunculidae* and 'lower' *Hamamelididae*) based on *atp*B and *rbc*L sequences from the chloroplast genome and 18S nuclear ribosomal DNA (nrDNA) sequences. The resulting gene sequences were analyzed separately and as a combined data set. Because of the preliminary nature of the data and the analyses, this paper discusses the evolutionary relationships resulting from the combined analysis only. The resulting cladogram is compared to some important morphological, cytological, and phytochemical characters to assess homoplasy and homology.

Material and methods

Sequence data. The species studied, source of sequences, and voucher information are given in Table 1. The same genera were sequenced for all three genes with one exception: *Aquilegia* has not yet been sequenced for 18S. The taxa included represent all the major subfamilies and tribes found in most classification schemes of the family.

Total cellular DNA was isolated from fresh, silica-dried, or herbarium material according to the miniprep method of DOYLE & DOYLE (1987). In some cases, DNA was further purified after extraction by cesium chloride/ethidium bromide gradient centrifugation (*Actaea*, *Adonis*, *Anemone*, *Caltha*, *Clematis*, *Myosurus*, *Ranunculus*, and *Xanthorhiza*).

The amplification primers and PCR protocol for the chloroplast genes, *atp*B and *rbc*L, are described in HOOT & al. (1995). In most cases, the amplification primers for the chloroplast gene *rbc*L were those described in OLMSTEAD & al. (1992). 18S nrDNA was amplified by using either the primers and protocols of NICKRENT (1993), NICKRENT & STARR (1993), or those of HAMBY & al. (1988).

Table 1. Taxa sampled, sources of *rbc*L sequences and/or leaf material, and voucher information (the source of all *atp*B and 18S nrDNA sequences is S. Hoot)

	*rbc*L	Source/voucher
Aconitum napellus L.	Hoot	S. Hoot 926, UWM
Actaea pachypoda Ell.	,,	M. W. Chase 550, K
Adonis amurensis Regal & Radde	,,	S. Hoot 933, UWM
Anemone canadensis L.	,,	S. Hoot 867, MICH
Anemonella thalictroides Spach.	,,	A. Reznicek, Hoot 9223, UWM
Anemonopsis macrophylla Siebold & Zucc.	,,	A. Reznicek 9977, MICH
Aquilegia vulgaris L.	,,	A. Reznicek 1108, MICH
A. ecalcarata Maxim.	Hoot	A. Reznicek 9996, MICH
Caltha palustris L.	Olmstead	Rolmstead, none; S. Hoot 299, MICH
Caulophyllum thalictroides (L.) Michx.	Les	Les. s.n., CONN
Cimicifuga simplex Wormsk.	Hoot	A. Reznicek 9238, MICH
Clematis hexapetala Pall.	,,	S. Hoot 9150, MICH
Coptis trifolia (L.) Salisb.	,,	E. Voss, MICH
Delphinium tricorne Michaux	,,	A. Reznicek 9997, MICH
Eranthis hyemalis (L.) Salisb.	,,	S. Hoot 920, UWM
Glaucidium palmatum Siebold & Zucc.	,,	S. Hoot 924, UWM
Helleborus orientalis Lam.	,,	A. Reznicek, Hoot 9224, UWM
Hydrastis canadensis L.	,,	R. Naczi 2883, MICH
Isopyrum biternatum (Raf.) T. & G.	,,	S. Hoot 9214, UWM
Myosurus minimus L.	,,	M. W. Chase 532, K
Nigella damascena L.	,,	S. Hoot 9211, UWM
Nandina domestica Thunb.	,,	S. Hoot 922, UWM
Ranunculus trichophyllus Chaix	Les	D. Les s.n., CONN
R. hispidus Michx.		S. Hoot 901, MICH
Thalictrum adiantifolium Bess.	,,	A. Reznicek 10107, MICH
Trautvetteria carolinensis (Walt.) Vail	,,	S. Hoot 9218, UWM
Trollius ledebourii Reichb.	,,	A. Reznicek 100035, MICH
Xanthorhiza simplicissima Marshall	Qui	Qui 91030, NCU

Particular attention was given to purification of PCR products to avoid superimposed sequences that can result from priming by the amplification primers as well as the internal sequencing primer during double-stranded sequencing. Samples were run on a 2% low melt agarose gel (NuSieve GTG) with 1% TAE buffer and Et Br, bands were visualized by means of UV illumination, then removed as gel plugs. To remove agarose gel and concentrate the PCR product, gel plugs were melted at 65–75 °C for approximately 10 min, then further purified and concentrated either with glass milk (GeneClean, Bio 101, Inc.) or Wizard Columns (Promega).

Double stranded sequencing. Most of the purified dsPCR products (Table 1) were sequenced directly with the dideoxy-termination method and sequenase T7 DNA polymerase (US Biochemical) using the method described in Hoot & al. (1995) or a method allowing for simultaneous reactions of 10 samples as follows: Annealing reaction mix and labelling cocktail were premixed – annealing reaction mix for each $10\,\mu l$ sample contained $2.0\,\mu l$ reaction buffer (US Biochemical), $1.0\,\mu l$ 50% acetamide, $1.0\,\mu l$ 5uM primer, $6.0\,\mu l$ ds DNA; – labelling cocktail for 10 samples ($90\,\mu l$) contained $35.0\,\mu l\,T_{10}E_{0.1}$, $10.0\,\mu l\,0.1$ M DTT, $20.0\,\mu l$ dilute labelling mix (0.4%), $5.0\,\mu l$ ^{35}S-dATP (12.5 mCi/ml), and $20.0\,\mu l$ diluted sequenase (0.125%, US Biochemical). The annealing reactions were heated at 98–99 °C for six min., cooled and centrifuged at room temperature for 1 min, then placed on ice. The annealing reactions were then divided evenly between the termination tubes placed in four centrifuge racks containing 10 tubes each (Brinkman, racks clearly labelled for each termination mix). The racks were spun down and placed on ice. The labelling cocktail was spot pipetted into each termination tube using a $2.0\,\mu l$ repeat pipettor (Hamilton), spun down, and returned to ice. $2.0\,\mu l$ of each ddNTP were spot pipetted into the appropriate termination tubes using the repeat pipettor, spun down, and incubated at 46–47 °C for three min., then returned to ice. $4.0\,\mu l$ of stop mix were spot pipetted into all tubes, spun, and frozen.

Internal primers for sequencing of *atp*B are listed in Hoot & al. (1995). Sequencing of *rbc*L used a combination of internal primers, kindly provided by G. Zurawski (DNAX Research Institute, Palo Alto, California) and a few primers designed specifically for this study (positions and sequences available from S. Hoot). The 18S internal primers used to date were generously furnished by D. Nickrent (Nickrent 1993) or E. A. Zimmer (Hamby & al. 1988). Electrophoresis using polyacrylamide gels (6%), subsequent gel drying, and exposure to x-ray film are described in Hoot & al. (1995).

Data collection and analysis. Sequence comparisons for the genes *atp*B, *rbc*L, and nr18S included 1468, 1397, and 1671 bp, respectively. Both strands of DNA were sequenced for both *atp*B and *rbc*L with approximately 80% overlap. Both strands were also sequenced for 18S, but with less overlap between the two directions (40–50%). The sequences were read from the autoradiographs, recorded on a data sheet, then entered into MacClade (Maddison & Maddison 1992). All phylogenies are preliminary, pending a recheck of the data sets with the original autoradiographs.

Alignment problems (caused by compressions) for *atp*B and *rbc*L often could be rectified by reading the opposite strand. There were several regions in the 18S nrDNA sequences where alignment was impossible because of compressions or base insertion/deletion events. These regions were deleted from the data matrix and are located at the following positions in relation to the soybean 18S sequence (Eckenrode & al. 1985): 224–231, 236, 270–276, 666–670, 1180–1181, and the very end of the amplified region, 1711–1761. The possibility of PCR-generated anomalous sequences was checked by comparison of sequences from closely related taxa.

Phylogenetic analyses were performed using PAUP 3.0 (Swofford 1993) using the heuristic search option with 100 random additions, and Mulpars in effect (with collapse of zero-length branches). PAUP was also used to perform bootstrap analysis with 1000 replications using the heuristic search option on the combined data set (Felsenstein 1985). The decay indices (the number of steps that must be added to the minimal-length tree before a clade collapses) were computed using the heuristic search option (Donoghue & al. 1992). Alternative tree topologies and resultant changes in tree length were explored using MacClade 3.0 (Maddison & Maddison 1992).

Results

Repeated cladistic analyses of the *Ranunculidae* for each gene separately and combined, with considerable variation in the genera and families included, consis-

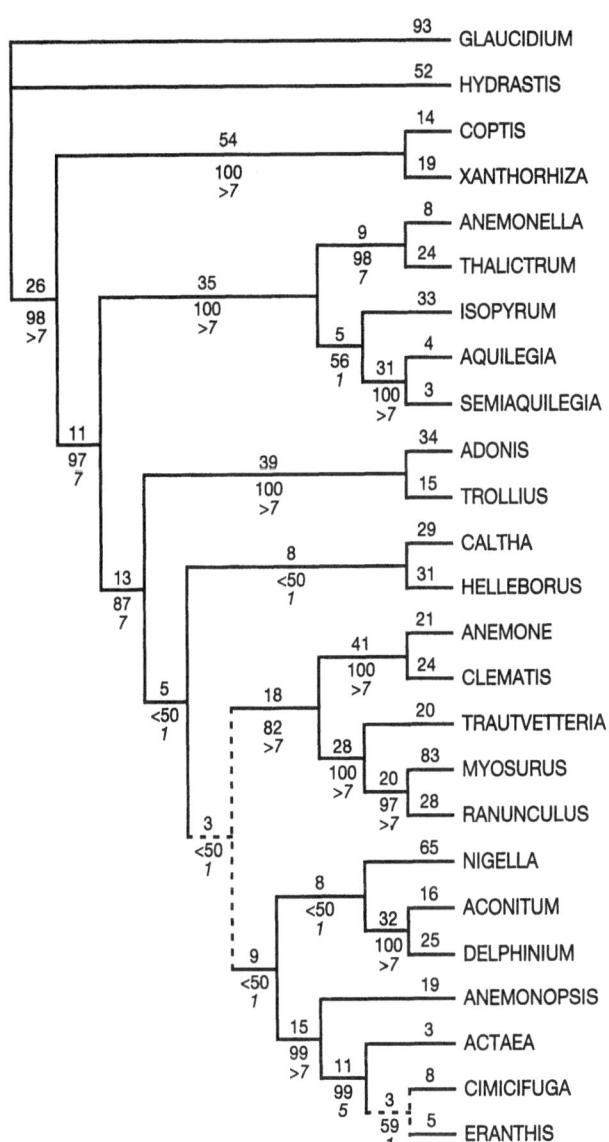

Fig. 1. One of the four most parsimonious trees resulting from a preliminary cladistic analysis based on a combination of *atp*B, *rbc*L, and 18S nrDNA sequence data for the genera indicated in Table 1. Dotted lines indicate branches which collapsed in the strict consensus tree derived from the four shortest trees. Tree length = 824 steps, CI excluding autapomorphies = 0.55, RI = 0.62. Numerals above branches indicate the number of nucleotide changes supporting each branch. Numerals immediately below indicate the percentage of times that the branch was recovered in 1000 bootstrap replications. Numerals in italics are decay indices, computed for trees up to seven steps longer than the shortest tree

tently result in a monophyletic *Ranunculaceae*, with *Hydrastis* and *Glaucidium* basal to the *Ranunculaceae* s. str. The *Berberidaceae* (*Nandina* and *Caulophylum*) are a sister group to a clade consisting of *Hydrastis*, *Glaucidium*, and the *Ranunculaceae* (HOOT & CRANE 1995, this volume; and unpubl.). Therefore, *Hydrastis* and *Glaucidium* were chosen as the outgroups for all the following analyses of the *Ranunculaceae*.

Each of the *Ranunculaceae* sequence data sets (*atp*B, *rbc*L, and 18S nrDNA) when analysed separately yielded largely similar phylogenetic estimates, indicating a common evolutionary history. Therefore, the individual data sets were combined and a further analysis was performed. This resulted in four shortest trees derived from 705 variable sites (357 informative sites) with a tree length = 824 (calculated for informative characters only), a consistency index excluding autapomorphies (CI) = 0.55 (KLUGE & FARRIS 1969), and a retention index (RI) = 0.62 (FARRIS 1989).

One of the four shortest trees is illustrated in Fig. 1; the dotted lines indicate the branches which collapsed in a strict consensus tree of the four shortest trees. The consensus tree is highly resolved and branch support (as indicated by number of substitutions, bootstrap values, and decay indices) are considerably higher than was found in the trees resulting from analysis of each gene separately (trees not presented here). Using 75% as a conservative cutoff point for the accuracy of bootstrap values (HILLIS & BULL 1993), many branching patterns on the total evidence tree are well supported (as indicated with bold lines in Fig. 2). As in the analyses of the *Ranunculidae*, the monophyly of the *Ranunculaceae* is confirmed, with 26 base substitutions, a bootstrap value of 98%, and a decay indice of > 7. The T-Type taxa consisting of *Coptis* and *Xanthorhiza* and a clade composed of ((*Anemonella*, *Thalictrum*) (*Isopyrum* (*Aquilegia*, *Semiaquilegia*))) are basal to the R-type members of the *Ranunculaceae*. The R-type chromosome group is a monophyletic clade, supported by 13 substitutions, a bootstrap value of 87%, and a decay index of 7. Within the R-type chromosome group, several clades are extremely well supported: (*Adonis*, *Trollius*), (*Anemone*, *Clematis*), (*Trautvetteria*, *Myosurus*, *Ranunculus*), (*Aconitum*, *Delphinium*), and (*Anemonopsis*, *Actaea*, *Cimicifuga*, *Eranthis*). *Anemone* and *Clematis* are found in a moderately well supported clade with *Trautvetteria*, *Myosurus*, and *Ranunculus* (bootstrap value = 82%). *Nigella* is weakly allied with *Delphinium* and *Aconitum* (bootstrap value < 50%).

Discussion

Phylogeny of the *Ranunculidae*. Numerous cladistic analysis of the *Ranunculidae* using three gene sequences (*atp*B, *rbc*L, and nr18S), analysed separately and as a combined data set, established that *Hydrastis* and *Glaucidium* are most closely related to the *Ranunculaceae*. Because of their consistent placement basal to the *Ranunculaceae*, *Hydrastis* and *Glaucidium* make appropriate outgroups and placeholders for the family in broader systematic studies (HOOT & CRANE 1995, this volume; and unpubl.). These results are consistent with the presence of T-type chromosomes in *Hydrastis*, *Glaucidium*, and the basal members of the *Ranunculaceae* (GREGORY 1941). In addition, *Hydrastis* shares the presence of berberin and the correlated character, yellow rhizomes, with the two most basal members of the *Ranunculaceae*, *Coptis* and *Xanthorhiza*. None of these analyses support the sister group relationship between the *Ranunculaceae* and the *Papaverales* as previously suggested (LOCONTE & STEVENSON 1991; LOCONTE & al. 1995, this volume).

Some taxonomists have speculated about the inclusion of *Glaucidium* and/or *Hydrastis* within the *Ranunculaceae* (CRONQUIST 1981, TAMURA 1962, HOOT 1991). Preliminary analyses of sequence data consistently place *Glaucidium* and *Hydrastis* basal to the *Ranunculaceae*, but never nested within the family (trees not shown). Furthermore, the large number of autapomorphies characteristic of both genera indicate that in addition to a history of genetic evolution in common with the *Ranunculaceae*, there has been an extended period of independent evolution for both *Glaucidium* and *Hydrastis*. There is some evidence (*atp*B sequence data analyzed alone and in combination with *rbc*L data) supporting the inclusion of *Glaucidium* and *Hydrastis* in the same family. However, pending further investigation, I recommend that *Glaucidiaceae* and *Hydrastidaceae* be kept as separate monotypic families, neither included within the *Ranunculaceae*.

Although there are no traditional characters that are unique to the *Ranunculaceae*, it is clear from previous analyses of both the *Ranunculidae* and the *Ranunculaceae* that the *Ranunculaceae* s. str. are a monophyletic family (DRINNAN & al. 1994; HOOT & CRANE 1995, this volume; unpubl.). The monophyly of the family is well-supported by the tree based on a combination of all the data (Fig. 1).

Phylogeny within the *Ranunculaceae*. The tree based on a combination of the three sequence data sets provides additional well-supported hypotheses about phylogenetic relationships within the *Ranunculaceae*, including some of the more elusive basal branching patterns.

The results are congruent with previous divisions of the family based on chromosome size, recognizing a separate R-type chromosome group (LANGLET 1932, GREGORY 1941). The T-type chromosome group is paraphyletic and basal to a monophyletic group consisting of the R-type taxa (Fig. 1), consistent with previous phylogenetic analyses based on micromorphology (HOOT 1991).

In addition, some of the terminal clades found in the molecular tree (Fig. 1) have long been recognized by workers using nonmolecular data (LANGLET 1932; GREGORY 1941; TAMURA 1966, 1967, 1968, 1993): (*Coptis Xanthorhiza*), ((*Anemonella Thalictrum*) (*Isopyrum* (*Aquilegia Semiaquilegia*))), ((*Anemone Clematis*) (*Trautvetteria* (*Myosurus Ranunculus*))), (*Aconitum Delphinium*), and (*Anemonopsis* (*Cimicifuga Actaea*)).

However, other well-supported alliances were not predicted based on traditional data. *Adonis* and *Trollius* form a well supported clade basal to all others in the R-type chromosome group. Although the close relationship of *Adonis* and *Trollius* is surprising from a morphological point of view (e.g., *Adonis* has achenes, *Trollius* follicles), similar results were found based on phytochemistry (JENSEN 1995, this volume), serology (JENSEN 1968), and restriction site variation (JOHANSSON & JANSEN 1993; JOHANSSON 1995, this volume). *Trollius* has been considered most closely related to *Caltha* by many workers (e.g., LANGLET 1932; TAMURA 1966, 1993). Moving *Trollius* together with *Caltha* (using McClade) costs an additional 37 steps in tree length.

Another surprising alliance is the inclusion of *Eranthis* in a clade with *Anemonopsis*, *Actaea*, and *Cimicifuga* (Fig. 1). This clade is extremely well-supported, with a bootstrap value of 99% and decay index of > 7. There are few if any synapomorphies among traditional data to support this relationship, although JENSEN (1971, 1995, this volume) reported that serologically *Eranthis* is more similar to *Actaea* and *Cimicifuga* than to *Helleborus*. *Eranthis* is very different from *Actaea* and *Cimicifuga* morphologically. It is a low-growing spring ephemeral; the flowers have peltate staminodia. In contrast, *Actaea* and *Cimicifuga* are robust herbaceous or woody plants and the flowers have flat staminodia. Mainly on the basis of floral morphology, *Eranthis* had previously been placed with *Helleborus* (e.g., LANGLET 1932; TAMURA 1966, 1993). Placing *Eranthis* and *Helleborus* together as sister taxa costs a minimum of 29 additional steps in tree length.

There is weak evidence for the inclusion of *Nigella* in the clade consisting of *Aconitum* and *Delphinium* (8 substitutions, bootstrap value < 50%, decay index = 1; Fig. 1). Similar results were obtained in earlier cladistic analyses based on morphol-

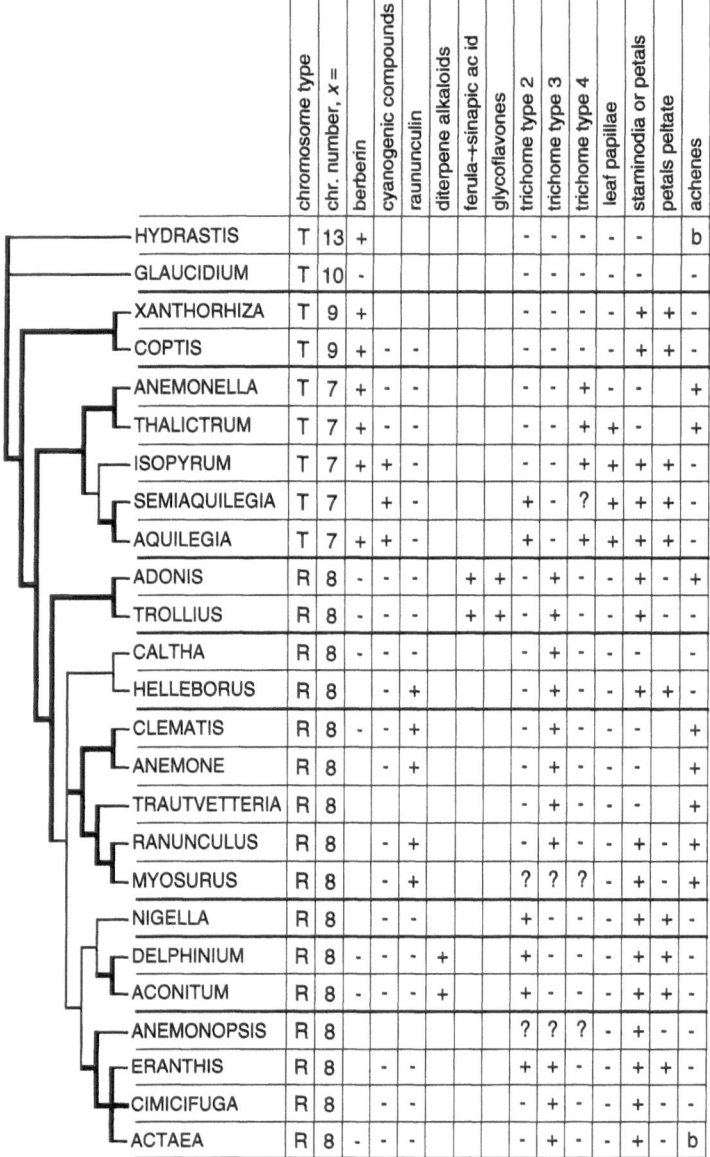

Fig. 2. Consensus tree of the four shortest trees derived from a preliminary cladistic analysis based on a combination of *atp*B, *rbc*L and 18S data (as in Fig. 1) showing the distribution of selected character states. Branches in bold had bootstrap values higher than 75%. Key to chemical data (columns 3–8): "+" = present, "–" = missing, blank indicates not tested, information not available, or present in trace amounts only; "?" indicates missing data, "b" = berries. Characters 1–2 from Langlet (1932) and Gregory (1941), characters 3–8 from Jensen (1995), characters 9–12 from Hoot (1991)

ogy and other traditional characters (Hoot 1991). Morphological characters which support this clade are presence of trichomes with swollen bases and reflexed tips, staminodia with nectaries and spurs (spurs weakly developed in *Nigella*), highly dissected leaves, tendency toward the annual habit, and three-angled seeds, frequently with rough surfaces.

Character correlation. The strict consensus tree based on a combination of all sequence data correlated with some key traditional characters is presented in Fig. 2. It is evident from this figure that certain kinds of characters are more congruent with the molecular phylogeny than others. For example, type and base number of chromosomes is completely congruent with the molecular phylogeny.

Some chemical characters, such as presence of berberin, cyanogenic compounds, ranunculin, diterpene alkaloids, ferulic-sinapic acid, and glycoflavones are seemingly largely congruent with the molecular and cytological data (Fig. 2). However, phytochemical data are difficult to evaluate because of several problems inherent in the data collection procedures and reporting. In many cases, it is difficult to determine from the literature which taxa were investigated and the compound of interest not found; frequently only positive results are reported. Secondly, the compounds can be present in varying amounts, complicating the scoring of presence or absence. And lastly, related compounds may be grouped into one class or recorded each as a separate compound. Until further work on phytochemistry in the family, it is premature to assess phytochemical characters in general as highly congruent with molecular characters.

Some micromorphological characters are also largely congruent with the molecular tree, for example, the presence of certain trichome types on leaf and stem surfaces and papillae on the periclinal cell walls of leaves (Fig. 2). Others, such as presence of striations on the periclinal cell walls or stomata on the adaxial leaf surface (characters not presented in Fig. 2), are homoplastic (Hoot 1991).

Other morphological characters are not congruent with molecular, phytochemical, and cytological data. Some of these homoplastic characters are those most frequently used in past taxonomic treatments of the family (i.e., Janchen 1949, Tamura 1993). They include floral characters, such as presence and type of staminodia/petals and fruit type, such as follicles, achenes, or berries (Fig. 2). Consequently, the molecular phylogeny supports the view that staminodia/petals have evolved independently many times (Janchen 1949, Leppik 1964, Tamura 1965, Cronquist 1981, Hoot 1991). Similarly, assuming that follicles are primitive (the outgroup condition), achenes have evolved independently at least three times. These molecular results, in conjunction with cytology, phytochemistry, micromorphology, and other sources of molecular data (e.g., restriction site variation; Johansson 1995, this volume) place into question the traditional emphasis given to floral and fruit morphology in the systematics of the family.

I am indebted to Dr Uwe Jensen and Dr Joachim Kadereit for the opportunity to participate in the international conference entitled "The Systematics and Evolution of the *Ranunculiflorae*". I thank Dr Peter Crane for his invaluable advice and both Dr Crane and the Field Museum, Chicago, for the opportunity and facilities to carry out most of this work. I also thank Dr Daniel Nickrent and Dr Elizabeth Zimmer for ribosomal DNA amplification and sequencing primers, and Dr Gerald Zurawski for *rbc*L sequencing primers. I am grateful to the following for providing leaf material or DNA: Dr Anton Reznicek and Dr Edward Voss, University of Michigan; Dr Mark Chase, Royal Botanic Garden, Kew; and Dr Rob Naczi, North Kentucky University. I also thank Dr John Hall for technical advice and Ernest Mui for laboratory assistance. This work was supported in part by NSF grants DEB-9020237 to Peter R. Crane and DEB-9306533 to Sara B. Hoot.

References

CRONQUIST, A., 1981: An integrated system of classification of flowering plants. – New York: Columbia University Press.

DONOGHUE, M. J., OLMSTEAD, R. G., SMITH, J. F., PALMER, J. D., 1992: Phylogenetic relationships of *Dipsacales* based on *rbc*L sequences. – Ann. Missouri Bot. Gard. **79**: 333–345.

DOYLE, J. J., DOYLE, J. L., 1987: A rapid DNA isolation procedure for small quantities of fresh tissue. – Phytochem. Bull. **19**: 11–15.

DRINNAN, A. N., CRANE, P. R., HOOT, S. B., 1994: Patterns of floral evolution in the early diversification of non-magnoliid dicotyledons (eudicots). – Pl. Syst. Evol. [Suppl]. **8**: 93–122.

ECKENRODE, V. K., ARNOLD, J., MEAGHER, R. B., 1985: Comparison of the nucleotide sequence of soybean 18S rRNA with the sequences of other small-subunit rRNA's. – J. Mol. Evol. **21**: 259–269.

ENGELL, K., 1995: Embryo morphology of the *Ranunculaceae*. – Pl. Syst. Evol. [Suppl]. **9**: 207–216.

FARRIS, J. S., 1989: The retention index and rescaled consistency index. – Cladistics **5**: 417–419.

FELSENSTEIN, J., 1985: Confidence limits on phylogenies: an approach using the bootstrap. – Evolution **39**: 783–791.

GREGORY, W. C., 1941: Phylogenetic and cytological studies in the *Ranunculaceae*. JUSS. – Trans. Amer. Phil. Soc., n.s., **31**: 441–520.

GRUND, C., GILROY, J., GLEAVES, T., JENSEN, U., BOULTER, D., 1981: Systematic relationships of the *Ranunculaceae* based on amino acid sequence data. – Phytochemistry **20**: 1559–1565.

HAMBY, R. K., SIMS, L. E., ISSEL, L. E., ZIMMER, E. A., 1988: Direct ribosomal RNA sequencing: optimization of extraction and sequencing methods for work with higher plants. – Pl. Mol. Biol. Rep. **6**: 175–192.

HEGNAUER, R., 1966: Comparative phytochemistry of alkaloids. – In SWAIN, T., (Ed.): Comparative phytochemistry, pp. 175–186. – New York: Academic Press.

– 1973: Chemotaxonomie der Pflanzen **6**: *Dicotyledoneae: Rafflesiaceae-Zygophyllaceae*. – Stuttgart: Birkhäuser.

– 1990: Chemotaxonomie der Pflanzen **9**: *Dicotyledoneae: Magnoliaceae-Zygophyllaceae*. – Stuttgart: Birkhäuser.

HILLIS, D. M., BULL, J. J., 1993: An empirical test of bootstrapping as a method for assessing confidence in phylogenetic analysis. – Syst. Biol. **42**: 182–192.

HOOT, S. B., 1991: Phylogeny of the *Ranunculaceae* based on epidermal microcharacters and macromorphology. – Syst. Bot. **16**: 741–755.

– CRANE, P. R., 1995: Inter-familial relationships in the *Ranunculidae* based on molecular systematics. – Pl. Syst. Evol. [Suppl]. **9**: 119–131.

– CULHAM, A., CRANE, P. R., 1995: The utility of *atp*B gene sequences in resolving relationships in the *Lardizabalaceae*, including comparisons with *rbc*L and 18S ribosomal DNA sequences. – Ann. Missouri Bot. Gard. **82**: 194–207.

JANCHEN, E., 1949: Die systematische Gliederung der Ranunculaceen and Berberidaceen. – Denkschr Österr. Akad. Wiss., Math. – Naturwiss. Kl. **108**: 1–82.

JENSEN, U., 1968: Serologische Beiträge zur Systematik der *Ranunculaceae*. – Bot. Jahrb. Syst. **88**: 204–268.

– 1971: Zur systematischen Stellung der *Helleborinae* (*Ranunculaceae*). – Taxon **20**: 747–758.

– 1995: Secondary compounds of the *Ranunculiflorae*. – Pl. Syst. Evol. [Suppl]. **9**: 85–97.

JOHANSSON, J. T., 1995: A revised chloroplast DNA phylogeny of the *Ranunculaceae*. – Pl. Syst. Evol. [Suppl]. **9**: 253–261.

– JANSEN, R. K., 1993: Chloroplast DNA variation and phylogeny of the *Ranunculaceae*. – Pl. Syst. Evol. **187**: 29–49.

KLUGE, A. G., FARRIS, J. S., 1969: Quantitative phyletics and the evolution of the anurans. – Syst. Zool. **18**: 1–32.

KOSUGE, K., SAWADA, K., DENDA, T., ADACHI, J., WATANABE, K., 1995: Phylogenetic relationships of some genera in the *Ranunculaceae* based on the alcohol dehydrogenase gene. – Pl. Syst. Evol. [Suppl]. **9**: 263–271.

LANGLET, O., 1932: Über Chromosomenverhältnisse und Systematik der *Ranunculaceae*. – Svensk Bot. Tidskr. **26**: 381–401.

LEPPIK, E. E., 1964: Floral evolution in *Ranunculaceae*. – Iowa State Coll. Sci. **39**: 1–101.

LOCONTE, H., STEVENSON, D. W., 1991: Cladistics of the *Magnoliidae*. – Cladistics **7**: 267–296.

– CAMPBELL, L. M., STEVENSON, D. W., 1995: Ordinal and familial relationships of Ranunculidy genera. Pl. Syst. Evol. [Suppl]. **9**: 99–118.

MADDISON, W. P., MADDISON, D. R., 1992: MacClade: interactive analysis of phylogeny and character evolution, version 3.0. – Sunderland, Mass.: Sinauer.

NICKRENT, D. L., 1993: From field to film: rapid sequencing methods for field-collected plant species. – Biotechniques **16**: 470–475.

– STARR, E. M., 1993: High rates of nucleotide substitution in nuclear small-subunit (18S) rDNA from holoparasitic flowering plants. – J. Mol. Evol. **39**: 62–70.

OLMSTEAD, R. G., MICHAELS, H. J., SCOTT, K. M., PALMER, J. D., 1992: Monophyly of the *Asteridae* and identification of their major lineages inferred from DNA sequences of *rbc*L. – Ann. Missouri Bot. Gard. **79**: 249–265.

RUIJGROK, H. W. L., 1966: The distribution of ranunculin and cyanogenetic compounds in the *Ranunculaceae*. – In SWAIN, T., (Ed.): Comparative phytochemistry, pp. 175–186. – New York: Academic Press.

SWOFFORD, D. L., 1993: PAUP: Phylogenetic analysis using parsimony, version 3.1. – Computer program distributed by Illinois Natural History Survey., Champaign, Illinois.

TAKHTAJAN, A., 1980: Outline of the classification of flowering plants (*Magnoliophyta*). – Bot. Rev. **46**: 226–359.

TAMURA, M., 1962: Morphology, ecology and phylogeny of the *Ranunculaceae* I. – Sci. Rep. Osaka Univ. **11**: 115–126.

– 1965: Morphology, ecology and phylogeny of the *Ranunculaceae* IV. – Sci. Rep. Osaka Univ. **14**: 53–71.

– 1966: Morphology, ecology and phylogeny of the *Ranunculaceae* VI. – Sci. Rep. Osaka Univ. **15**: 13–35.

– 1967: Morphology, ecology and phylogeny of the *Ranunculaceae* VII. – Sci. Rep. Osaka Univ. **16**: 21–43.

– 1968: Morphology, ecology and phylogeny of the *Ranunculaceae* VIII. – Sci. Rep. Osaka Univ. **17**: 41–56.

– 1993: *Ranunculaceae*. – In KUBITZKI, K., ROHWER, J. G., BITTRICH, V., (Eds): The families and genera of vascular plants, **VII**, pp. 563–583. – Berlin, Heidelberg, New York: Springer.

Address of the author: Dr SARA B. HOOT, Department of Biological Sciences, Lapham Hall, PO Box 413, University of Wisconsin, Milwaukee, WI 53201, U.S.A., email: hoot@csd.uwm.edu.

Accepted February 6, 1995

Pl. Syst. Evol. [Suppl.] 9: 253–261 (1995)

A revised chloroplast DNA phylogeny of the *Ranunculaceae*

JAN T. JOHANSSON

Received September 11, 1994

Key words: Angiosperms, *Ranunculaceae*. – Chloroplast DNA, phylogeny.

Abstract: Phylogenetic analyses of chloroplast DNA restriction site variation among thirty-one genera of the *Ranunculaceae* were performed using three species of *Berberidaceae* and *Lardizabalaceae* as outgroups. 598 phylogenetically informative restriction sites were analysed cladistically. Two most-parsimonious trees were obtained using Wagner parsimony and a single most-parsimonious tree using Dollo parsimony. Many strongly supported monophyletic groups were found, although most of the basal clades varied depending on the parsimony method used. *Hydrastis* is positioned outside of the *Ranunculaceae*, whereas *Coptis* and *Xanthorhiza* form a clade that is sister group to the remaining genera. Several basal clades are weakly supported, and the analyses do not suggest an unambiguous subdivision of the family. A comparison between chloroplast DNA phylogeny and the distribution of morphological, cytological, and phytochemical characters is made.

This investigation of the chloroplast DNA (cpDNA) phylogeny of the *Ranunculaceae* is a continuation of a project which was presented by JOHANSSON & JANSEN (1991, 1993). Intergeneric studies based on restriction site mapping of cpDNA were briefly reviewed in JOHANSSON & JANSEN (1993). Molecular phylogenetic investigations of genera within the *Ranunculaceae* are still very few. An analysis of *Anemone* s.l. using cpDNA restriction site mapping was recently performed (HOOT & al. 1994, HOOT & PALMER 1994).

A brief survey of previous systematic treatments of the *Ranunculaceae* was given by JOHANSSON & JANSEN (1993). The most recent concise taxonomic outline of this family was published by TAMURA (1993). He subdivided the *Ranunculaceae* into five subfamilies: *Helleboroideae* (including the tribes *Helleboreae*, *Cimicifugeae*, *Nigelleae*, and *Delphinieae*), *Ranunculoideae* (comprising *Adonideae*, *Anemoneae*, and *Ranunculeae*), *Isopyroideae* (including *Isopyreae*, *Dichocarpeae* and *Coptideae*), *Thalictroideae* and *Hydrastidoideae*. His classification is based on morphology and cytology.

The aims of this study are to (1) evaluate the robustness of the previous cladistic analyses of the *Ranunculaceae* cpDNA (JOHANSSON & JANSEN 1993) by including additional taxa and (2) find molecular evidence for classification of the family. *Glaucidium* and *Paeonia* were not included in the present study, since their systematic relationships are outside of the *Ranunculales* (TAMURA 1972, 1993; TOBE 1981; THOME 1992; CHASE & al. 1993).

Material and methods

Sources of leaf material for 34 of the species were given by JOHANSSON & JANSEN (1993). Seven additional species were analysed in the present study: *Adonis brevistyla* FRANCH. (cult. Gothenburg Bot. Garden), *A. cyllenea* BOISS., HELDR. & ORPH. (cult. Gothenburg Bot. Garden), *Anemonopsis macrophylla* SIEB. & ZUCC. (cult. Lund Bot. Garden), *Callianthemum anemonoides* (J. ZAHLBR.) ENDL. ex HEYNH. (cult. Linz Bot. Garden), *Eranthis hyemalis* (L.) SALISB. (cult. Lund Bot. Garden), *Paraquilegia grandiflora* (FISCH. ex DC.) DRUMM. & HUTCH. (cult. Gothenburg Bot. Garden) and *Paropyrum anemonoides* (KAR. & KIR.) ULBR. (cult. Gothenburg Bot. Garden).

Isolation of total DNAs, restriction endonuclease digestions, agarose gel electrophoresis, bidirectional transfer of DNA to nylon filters, labelling of tobacco cpDNA probes by nick-translation, filter hybridizations and autoradiography were performed according to methods used by JOHANSSON & JANSEN (1993) and described by, e.g., PALMER (1986). Restriction site maps were constructed for 38 species representing 31 genera of *Ranunculaceae*, and three species of the two closely allied families *Berberidaceae* and *Lardizabalaceae*.

The choice of outgroups (WATROUS & WHEELER 1981) – *Caulophyllum thalictroides* (*Berberidaceae*), *Akebia quinata* (*Lardizabalaceae*) and *Decaisnea fargesii* (*Lardizabalaceae*) – follows the same principles as in JOHANSSON & JANSEN (1993). Restriction sites that vary among the outgroups were excluded from the analyses. Sites occurring in cpDNA regions adjacent to inversion end points and in regions containing extensive length variation were excluded, since such mutations make the alignment of sites too unreliable. The Wagner and Dollo parsimony algorithms in PAUP version 3.0r (SWOFFORD 1990) were applied using the heuristics search option (SWOFFORD & OLSEN 1990). The ACCTRAN option was in effect for character-state optimization. Decay analysis (DONOGHUE & al. 1992) was carried out, in order to evaluate the strength of the monophyletic groups. Only groups present in clado-grams one to four steps longer than the shortest trees (up to 1,143 and 1,353 steps, respectively) were analysed due to the large data set. One-hundred bootstrap replications were run using the Wagner and Dollo parsimonies in PAUP, and bootstrap values were placed on the monophyletic groups (FELSENSTEIN 1985). The cpDNA inversions in *Adonis, Anemone, Clematis, Clematopsis, Hepatica, Knowltonia* and *Pulsatilla* were inverted to be co-linear with the tobacco cpDNA before the data matrix was constructed (JOHANSSON & JANSEN 1993).

Chloroplast DNA restriction site maps and data matrix are available from the author upon request.

Results

598 phylogenetically informative restriction sites were used as characters in the phylogenetic analyses. The Wagner parsimony analysis resulted in two most-parsimonious trees with a length of 1,139 steps, including 465 gains and 133 losses, a consistency index of 0.52 and a retention index (FARRIS 1989) of 0.77. Forty-seven percent of the characters were homoplastic, including 157 parallel site gains, 267 parallel site losses, 37 loss/gains, and 80 gain/losses. A strict consensus tree is shown in Fig. 1.

The cladogram is resolved except for one trichotomy comprising *Eranthis, Anemonopsis,* and *Actaea-Cimicifuga*. Strongly supported major clades which have a bootstrap value (BV) of 100% and a decay index (DI) of more than 3 include the monophyletic groups: *Trollius-Adonis*; *Aconitum-Delphinium-Consolida*; *Anemone-Pulsatilla-Knowltonia-Heptatica-Clematis-Clematopsis*; *Anemonella-Thalictrum-Paropyrum-Paraquilegia-Isopyrum-Aquilegia*, and *Coptis-Xanthorhiza*. A mono-phyletic lineage which includes *Anemonella, Thalictrum, Paropyrum, Paraquilegia* and *Isopyrum* and has *Aquilegia* as sister group is also strongly supported (BV = 100%). On the other hand, the *Caltha-Callianthemum* clade (BV = 82%; DI > 3) and the *Eranthis-Anemonopsis-Actaea-Cimicifuga* clade (BV = 80%; DI = 3) are less strongly supported. The monophyletic group which comprises all the *Ranunculaceae* with the exception of *Hydrastis* has a bootstrap value of only 85% (DI > 3). The remaining major clades are weakly supported, e.g., the following monophyletic lineages: *Trollius-Adonis-Helleborus* (BV = 24%; DI = 1); *Ranun-culus-Myosurus-Trautvetteria* (BV = 56%; DI = 2); and the clade comprising the *Anemone-Clematis* and the *Thalictrum-Isopyrum-Aquilegia* groups (BV = 37%; DI = 1). The sister group relationship *Myosurus-Ranunculus* is likewise weakly supported (BV = 60%; DV = 1). The *Coptis-Xanthorhiza* clade is positioned outside of the remaining *Ranunculaceae* which form a monophyletic group (BV = 39%; DI = 2).

A single most-parsimonious tree was detected by the Dollo parsimony and this had a length of 1,349 steps, including 446 gains and 151 losses, a consistency index of 0.44 and a retention index of 0.95. Fifty-six percent of the characters were homoplas-tic, including 640 parallel losses, and 112 gain/losses. The Dollo tree is presented in Fig. 2. The phylogenetic tree is completely resolved. *Hydrastis* is positioned outside of the *Ranunculaceae*, which form a strongly supported monophyletic group (BV = 92%; DI > 3). Other strongly supported clades include: *Trollius-Adonis* (BV = 100%; DI > 3); *Thalictrum-Isopyrum-Aquilegia* (BV = 99%; DI > 3); *Acon-itum-Delphinium-Consolida* (BV = 100%; DI > 3); *Anemone-Clematis* (BV = 100%; DI > 3); and *Coptis-Xanthorhiza* (BV = 100%; DI > 3). The *Eranthis-Anemonopsis-Actaea-Cimicifuga* clade (BV = 85%; DI > 3) is less strongly supported. However, most of the monophyletic lineages are weakly supported, e.g., the following clades: *Caltha-Callianthemum* (BV = 40%; DI = 1); *Trollius-Adonis-Helleborus* (BV = 66%; DI = 3); *Aconitum-Delphinium-Consolida-Nigella* (BV = 40%; DI = 3); *Anemonopsis-Actaea-Cimicifuga* (BV = 56%; DI = 1); and *Ranunculus-Myosurus-Trautvetteria-Anemone-Clematis* (BV = 50%; DI = 3). The sister group relationship *Myosurus-Ranunculus* is supported in 70% of the bootstrap replicates (DI > 3), and the monophyletic group consisting of *Ranunculus, Myosurus*, and *Trautvetteria* has BV = 67% (DI > 3).

Discussion

There are numerous topological similarities between the Wagner and Dollo trees, although many differences can be found. The *Caltha-Callianthemum* and *Trollius-Adonis-Helleborus* clades appear as sister groups in both of the analyses. The monophyletic group *Aconitum-Delphinium-Consolida* is stable and strongly sup-ported. The *Anemone-Pulsatilla-Knowltonia-Heptatica-Clematis-Clematopsis* and *Anemonella-Thalictrum-Paropyrum-Paraquilegia-Isopyrum-Aquilegia* lineages are likewise monophyletic in both analyses. The clade *Coptis-Xanthorhiza* is sister

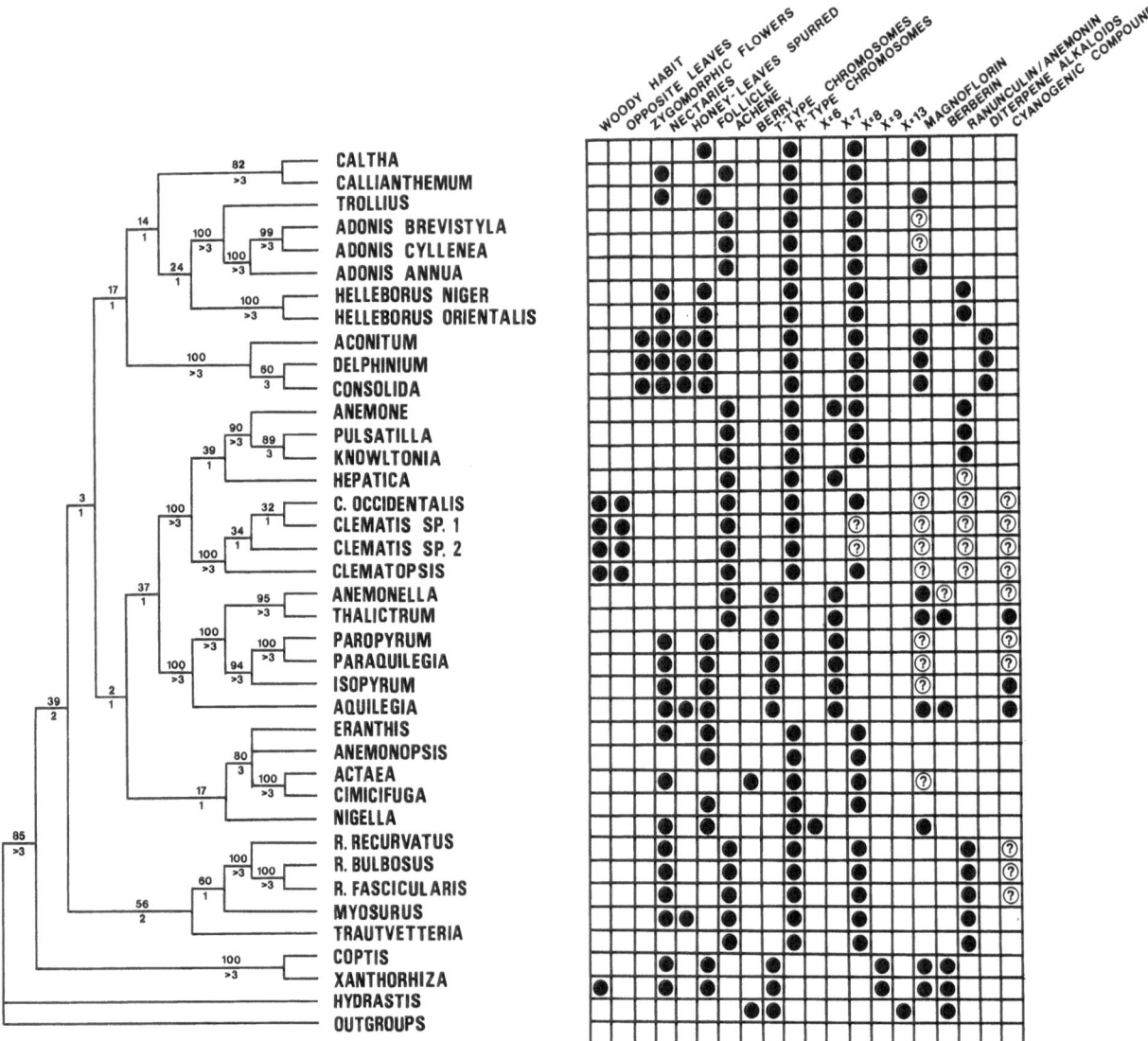

Fig. 1. Strict consensus Wagner tree for the two most-parsimonious cladograms of the *Ranunculaceae* chloroplast DNA. Numbers above the lines indicate bootstrap values given as the number of times that a monophyletic group occurred in 100 bootstrap replicates. Numbers below the lines indicate decay analysis values. The table at the right shows the distribution of 20 features on the Wagner tree. Dot denotes presence of the character; question-mark within circle denotes character state unknown. Characters 9–15 are from Langlet (1927, 1928), Gregory (1941) and Tamura (1993), and characters 16–20 are from Hegnauer (1964, 1973)

group to the remaining *Ranunculaceae* in the Wagner as well as the Dollo analysis. *Eranthis, Anemonopsis, Actaea,* and *Cimicifuga* form a relatively strongly supported clade, whereas the monophyletic group *Ranunculus-Myosurus-Trautvetteria* is more weakly supported by the Wagner than the Dollo parsimony. The strongest groups largely correspond to those found by JOHANSSON & JANSEN (1993). One exception is *Coptis* and *Xanthorhiza* which appear as sister group to the other *Ranunculaceae* in the present analyses, whereas they are allied to, e.g., *Trautvetteria, Anemone-Clematis,* and *Ranunculus* in the previous study.

The dissimilarities between the Wagner and Dollo parsimony analyses are found mainly among the basal clades. The *Ranunculus-Myosurus-Trautvetteria* lineage is placed in the Wagner analysis as sister group to the remaining *Ranunculaceae* with the exception of *Coptis-Xanthorhiza.* On the other hand, in the Dollo analysis the *Ranunculus* clade is closely allied to the *Anemone-Clematis* group. The *Aquilegia-Thalictrum* clade is nested together with the *Anemone-Clematis* lineage in the Wagner analysis, whereas in the Dollo analysis the *Aquilegia-Thalictrum* group is positioned close to the *Aconitum-Delphinium-Nigella* clade. On the other hand, *Nigella* is allied to *Eranthis, Anemonopsis* and *Actaea-Cimicifuga* in the Wagner analysis, and the *Aconitum-Delphinium* clade to the *Caltha- Callianthemum-Trollius-Adonis-Helleborus* lineage. However, the differences in results from the two parsimony methods are found only in weakly supported clades.

Most of the basal clades in both analyses are too weakly supported to allow a thorough classification of the *Ranunculaceae*. However, many groups are strongly supported by both Wagner and Dollo parsimony. *Hydrastis* is placed outside the *Ranunculaceae* s. str. in the two analyses and might well constitute a family of its own, *Hydrastidaceae* (THORNE 1992). *Coptis* and *Xanthorhiza,* which are included in the tribe *Coptideae* (TAMURA 1993), are separated from the remaining family, and this argues in favour of keeping them in a subfamily of their own, *Coptidoideae* (TAMURA 1968). A strongly supported monophyletic group, which may deserve tribal status, comprises *Aquilegia, Isopyrum, Paraquilegia, Paropyrum, Thalictrum* and *Anemonella*. This group corresponds to TAMURA's subfamilies *Isopyroideae* and *Thalictroideae* (TAMURA 1993), although several of the genera that were included by him were not investigated by me. Three genera of the *Ranunculeae* (TAMURA 1993) were included in the present analyses: *Ranunculus, Myosurus* and *Trautvetteria*. This tribe is now being subjected to a thorough molecular systematic study (JOHANSSON, unpubl.). The genera *Anemone, Pulsatilla, Knowltonia, Hepatica, Clematis* and *Clematopsis* form a well supported clade. They belong in the tribe *Anemoneae* (TAMURA 1993) together with several other genera which are not yet investigated molecularly. *Aconitum, Delphinium* and *Consolida* are closely allied, and TAMURA (1993) included them in the tribe *Delphinieae,* which is well supported by cpDNA data. TAMURA's subfamily *Helleboroideae* is not supported by the results from the present analyses. *Trollius,* which was placed in *Helleboreae* by TAMURA, is allied to *Adonis,* according to cpDNA information. *Adonis* is placed together with *Callianthemum* in the tribe *Adonideae* (*Ranunculoideae*) by TAMURA, whereas cpDNA data suggest that *Callianthemum* is close to *Caltha* (although this is weakly supported by the Dollo analysis). The tribe *Helleboreae* includes, e.g., *Caltha, Trollius, Helleborus* and *Eranthis,* according to TAMURA (1993). However, *Eranthis* is here positioned together with *Anemonopsis, Actaea* and *Cimicifuga,* the last three genera being

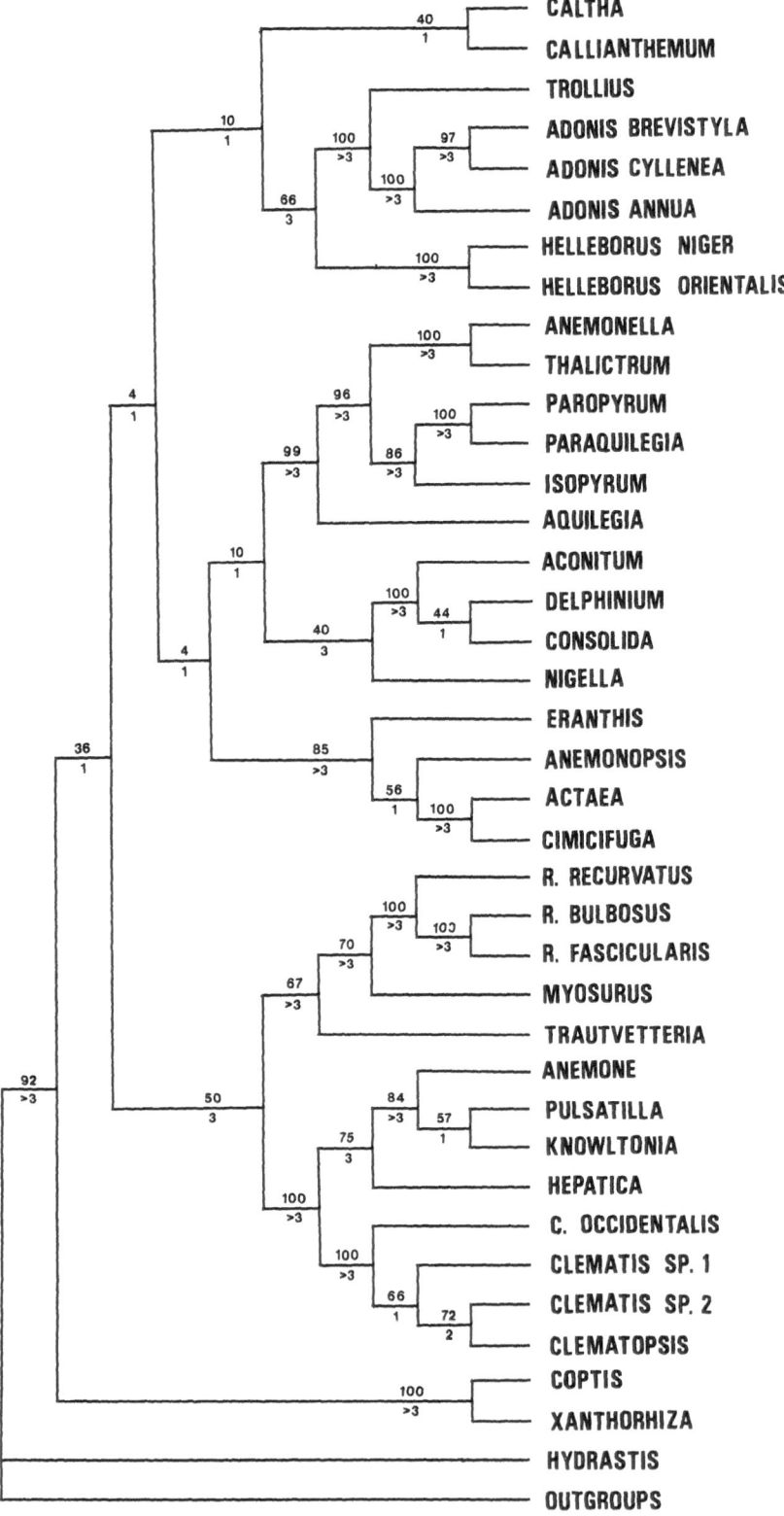

Fig. 2. Single most-parsimonious cladogram of the *Ranunculaceae* chloroplast DNA obtained using Dollo parsimony analysis. Numbers above the lines indicate bootstrap values. Numbers below the lines indicate decay analysis values

placed in *Cimicifugeae* by TAMURA. *Nigella*, finally, is included by TAMURA in a separate tribe, *Nigelleae* (*Helleboroideae*), in addition to *Komaroffia* and *Garidella*. Chloroplast DNA data do not unambiguously suggest any phylogenetic relationship of *Nigella*.

The occurrence of morphological, cytological, and phytochemical features in the *Ranunculaceae* is shown in Fig. 1. Twenty such traits are plotted here against the cpDNA Wagner tree. Information on morphological features were taken from various literature on the *Ranunculaceae*, cytological characters from LANGLET (1927, 1928), GREGORY (1941), and TAMURA (1993), and phytochemical traits from HEGNAUER (1964, 1973). The distribution of these characters corresponds fairly well with groups resulting from analyses of molecular data. Woody stems are restricted to *Clematis* (incl. *Clematopsis*) and *Xanthorhiza*, and opposite leaves are an autapomorphy of *Clematis* (incl. *Clematopsis*). Zygomorphic flowers and diterpene alkaloids are synapomorphies of the tribe *Delphinieae*, whereas spurred honey-leaves also occur in *Aquilegia* and *Myosurus*. Nectaries are likewise present in *Callianthemum, Trollius, Helleborus, Paropyrum, Paraquilegia, Isopyrum, Eranthis, Actaea, Nigella, Ranunculus, Coptis* and *Xanthorhiza*, and are thus not a useful feature at the tribal level. Berries occur only in *Actaea* and *Hydrastis*, whereas achenes are widely distributed: *Callianthemum, Adonis*, the *Anemone-Clematis* group, *Thalictrum* (incl. *Anemonella*), *Ranunculus, Myosurus* and *Trautvetteria*. T-type chromosomes are present in the *Aquilegia-Isopyrum-Thalictrum* and the *Coptis-Xanthorhiza* lineages and also in *Hydrastis*. *Nigella* has the basic chromosome number x = 6, *Coptis* and *Xanthorhiza* have x = 9, and *Hydrastis* x = 13. *Hepatica* and the *Aquilegia-Isopyrum-Thalictrum* group have x = 7, whereas x = 8 is predominant in the *Ranunculaceae*. In *Anemone* both x = 7 and x = 8 occur. The distribution of secondary metabolites in the *Ranunculaceae* is very incompletely known. Cyanogenic compounds have been found in *Clematis, Thalictrum, Isopyrum, Aquilegia*, and *Ranunculus*. Ranunculin and anemonin seem to be restricted to *Helleborus*, the *Anemone-Clematis* group, *Ranunculus, Myosurus* and *Trautvetteria*. Berberin has been found in *Thalictrum, Aquilegia, Coptis, Xanthorhiza* and *Hydrastis*, whereas magnoflorin occurs in a large number of genera. Thus, many of these features support the monophyletic groups that were suggested by the results from the cpDNA analyses.

Restriction site mapping of the chloroplast genomes of further taxa in the *Ranunculaceae* is being performed. Many problems remain to be solved, especially concerning the main branches of the phylogenetic tree. Sequencing of one or more chloroplast and nuclear genes might give a clue to answering at least some of these questions.

I am indebted to the directors and curators of botanical gardens for providing seeds and leaf material. The investigation was supported by grants from the Swedish Natural Science Research Council (B-BU 09277–313, B-BU 09277–314).

References

CHASE, M. W., SOLTIS, D. E., OLMSTEAD, R. G., MORGAN, D., LES, D. H., MISHLER, B. D., DUVALL, M. R., PRICE, R. A., HILLS, H. G., QIU, Y. -L., KRON, K. A., RETTIG, J. H., CONTI,

E., PALMER, J. D., MANHART, J. R., SYTSMA, K. J., MICHAELS, H. J., KRESS, W. J., KAROL, K.
   G., CLARK, W. D., HEDRÉN, M., GAUT, B. S., JANSEN, R. K., KIM, K. -J., WIMPEE, C. F.,
   SMITH, J. F., FURNIER, G. R., STRAUSS, S. H., XIANG, Q. -Y., PLUNKETT, G. M., SOLTIS, P. S.,
   SWENSEN, S. M., WILLIAMS, S. E., GADEK, P. A., QUINN, C. J., EQUIARTE, L. E., GOLENBERG,
   E., LEARN, G. H. Jr., GRAHAM, S. W., BARRETT, S. C. H., DAYANANDAN, S., ALBERT, V. A.,
   1993: Phylogenetics of seed plants: an analysis of nucleotide sequences from the plastid
   gene rbcL. – Ann. Missouri Bot. Gard. **80**: 528–580.
DONOGHUE, M. J., OLMSTEAD, R. G., SMITH, J. F., PALMER, J. D., 1992: Phylogenetic
   relationships of Dipsacales based on rbcL sequences. – Ann. Missouri Bot. Gard. **79**:
   333–345.
FARRIS, J. S., 1989: The retention index and the rescaled consistency index. – Cladistics **5**:
   417–419.
FELSENSTEIN, J., 1985: Confidence limits on phylogenies. An approach using the bootstrap. –
   Evolution **39**: 783–791.
GREGORY, W. C., 1941: Phylogenetic and cytological studies in the Ranunculaceae JUSS. –
   Trans. Amer. Phil. Soc., n.s., **31**: 443–521.
HEGNAUER, R., 1964: Chemotaxonomie der Pflanzen 3. Dicotyledoneae: Acanthaceae- Cyril-
   laceae. – Stuttgart, Basel: Birkhäuser.
– 1973: Chemotaxonomie der Pflanzen 6. Dicotyledoneae: Rafflesiaceae-Zygophyllaceae. –
   Stuttgart, Basel: Birkhäuser.
HOOT, S. B., PALMER, J. D., 1994: Structural rearrangements, including parallel inversions,
   within the chloroplast genome of Anemone and related genera. – J. Mol. Evol. **38**: 274–281.
– REZNICEK, A. A., PALMER, J. D., 1994: Phylogenetic relationships in Anemone (Ranunculaceae) based on morphology and chloroplast DNA. – Syst. Bot. **19**: 169–200.
JOHANSSON, J. T., JANSEN, R. K., 1991: Chloroplast DNA variation among five species of
   Ranunculaceae: structure, sequence divergence, and phylogenetic relationships. – Pl.
   Syst. Evol. **178**: 9–25.
– – 1993: Chloroplast DNA variation and phylogeny of the Ranunculaceae. – Pl. Syst. Evol.
   187: 29–49.
LANGLET, O. F. I., 1927: Beiträge zur Zytologie der Ranunculazeen. – Svensk Bot. Tidskr. **21**:
   1–17.
– 1928: Einige Beobachtungen über die Zytologie der Berberidazeen. – Svensk Bot. Tidskr.
   22: 169–184.
PALMER, J. D., 1986: Isolation and structural analysis of chloroplast DNA. – Meth. Enzymol.
   118: 167–186.
SWOFFORD, D. L., 1990: PAUP. Phylogenetic analysis using parsimony, ver. 3.0r for
   Macintosh. Computer package. – Champaign, Ill.: Illinois Natural History Survey.
– OLSEN, G. J., 1990: Phylogeny reconstruction. – In HILLIS, K. M., MORITZ, C., (Eds):
   Molecular systematics, pp. 411–501. – Sunderland, Mass.: Sinauer.
TAMURA, M., 1968: Morphology, ecology, and phylogeny of the Ranunculaceae VIII. – Sci.
   Rep. **17**: 41–56.
– 1972: Morphology and phyletic relationship of the Glaucidiaceae. – Bot. Mag. (Tokyo) **85**:
   29–41.
– 1993: Ranunculaceae. – In KUBITZKI, K., (Ed.): The families and genera of vascular plants.
   II. Flowering plants. Dicotyledons. Magnoliid, hamamelid and caryophyllid families, pp.
   563–583. – Berlin: Springer.
THORNE, R. F., 1992: An updated phylogenetic classification of the flowering plants. – Aliso
   13: 365–389.
TOBE, H., 1981: Embryological studies in Glaucidium palmatum SIEB. et ZUCC. with
   a discussion on the taxonomy of the genus. – Bot. Mag. (Tokyo) **94**: 207–224.

WATROUS, K. E., WHEELER, Q. D., 1981: The out-group comparison method of character analysis. – Syst. Zool. **30**: 1–11.

Address of the author: JAN T. JOHANSSON, Department of Systematic Botany, University of Lund, Ö. Vallgatan 18–20, S–223 61 Lund, Sweden.

Accepted November 23, 1994

Pl. Syst. Evol. [Suppl.] 9: 263–271 (1995)

Phylogenetic relationships of some genera in the *Ranunculaceae* based on alcohol dehydrogenase genes

K. Kosuge, K. Sawada, T. Denda, J. Adachi, and K. Watanabe

Received October 27, 1994

Key words: Angiosperms, *Ranunculaceae*. – Alcohol dehydrogenase gene, PCR, phylogeny.

Abstract: Fragments of the nuclear-encoded gene for alcohol dehydrogenase (*adh*) were amplified from 23 species representing 17 genera in the *Ranunculaceae* and one in the *Berberidaceae* using polymerase chain reaction method. Homology of sequences among the 26 amplified fragments suggests that the *Ranunculaceae* have two *adh* genes, *adh*1 and *adh*2, which are obviously separated into two clusters in both the neighbor joining and maximum parsimony trees, respectively. The cluster of *adh*1 genes is divided into four clades which support the classification based on cytological and morphological data. The phylogenetic analysis of T-type chromosome group clearly suggests that *Coptis* and *Xanthoriza* are a monophyletic group, and that *Thalictrum* with achenes were derived from *Enemion* and its allied genera with follicles.

In the past decade, the molecular systematics on angiosperms has made great strides with attention focusing on chloroplast gene data, especially from *rbc*L gene (Chase & al. 1993). These sequences, however, are conservative and the usages are restricted to the resolving of phylogenetic relationships in the higher taxonomic rank (Clegg 1993). Therefore, sequences with faster evolutionary rates have been explored to estimate the relationships of closely related taxa. The evolutionary rates of nuclear genes have been shown to be twice or more faster than those of the chloroplast (Wolfe & al. 1987, 1989).

In the present study, we investigated the potential utility of the nuclear-encoded gene for alcohol dehydrogenase (*adh*) to estimate the intergeneric relationships within the *Ranunculaceae*.

Amplification and sequencing of the *adh* fragments

Alcohol dehydrogenase (ADH; EC 1.1.1.1) catalyze the NAD^+-depending oxidation of alcohols. Sequences of *adh* gene have been analyzed in many higher plants, including seven dicot species (Chang & Meyerowitz 1986, Llewellyn & al. 1987, Matton & Brisson 1989, Ellison & al. 1990, Wolyn & Jelenkovic 1990, Gregerson & al. 1991) and five monocot species (Dennis & al. 1984, 1985; Trick & al. 1988; Xie & Wu 1989; Gaut & Clegg 1991, 1993). The protein coding sequences are generally interrupted by nine introns. Primers were designed specific to sequences in the second and fourth exons. We amplified 26 *adh* fragments from 23 species representing 17 genera in the *Ranunculaceae* and one in the *Berberidaceae* (as an outgroup) using the polymerase chain reaction (PCR) method. Voucher specimens are deposited at the Kobe University. The sequenced fragments contain exon 3, intron 3 and ca. three quarters of exon 4. The exon sequences are conserved and identical in length, while the introns differ in sequences and length, ranging from 81 bp to 167 bp.

Analysis of exon sequences

Based on the 273 bp of exon sequences, the Kimura two-parameter distances were calculated (Table 1) and the neighbor joining (NJ) tree (Fig. 1) was constructed using the MEGA program package (Kumar & al. 1993). The NJ tree shows that the *adh* fragments in the *Ranunculaceae* separate into two clusters at the base. In *Anemone*, *Caltha* and *Cimicifuga*, two fragments which differ in sequences and intron length were obtained from the same individuals and were placed in the different clusters of the phylogenetic tree, respectively. Most higher plants have two *adh* genes (Gottlieb 1982) and in maize both genes are considered to be derived from the duplication of a single ancestral gene (Dennis & al. 1985). Although they have not been sequenced yet, two fragments differing in length have been also obtained from each individual of all other species examined through the improvement of PCR primers and methods. Considering these facts, the *Ranunculaceae* have at least two *adh* loci, i.e., *adh*1 and *adh*2, which are separated into two distinct clusters in the phylogenetic tree.

The *adh*1 cluster is divided into four clades and members in each clade roughly correspond to tribes based on cytological characters and fruit morphology. The first clade consists of *Anemone* and *Ranunculus*, which possess R (anunclus)-type chromosomes and achenes. They are classified in the tribe *Anemoneae* by Gregory (1941) and Tamura (1967). The second clade includes *Actaea*, *Anemonopsis*, *Cimicifuga* and *Eranthis*, and *Caltha* is sister to them with relatively low bootstrap support. These genera possess R-type chromosomes and follicles (or many-seeded berries), and are classified into the tribe *Helleboreae* (Gregory 1941). *Aquilegia*, *Semiaquilegia*, *Enemion*, *Isopyrum*, *Dichocarpum* and *Thalictrum*, having T (halictrum)-type chromosomes with the basic chromosome number x = 7 or 6, form the third clade and are included in the tribe *Thalictreae* (Langlet 1932, Gregory 1941). The fourth clade consists of *Coptis* and *Xanthorhiza*, which possess T-type chromosomes with x = 9 and are classified into the tribe *Coptideae* (Langlet 1932, Gregory 1941, Tamura 1968, Tamura & Kosuge 1989, Fu 1990).

Although the sequences of *adh*2 have been analyzed on only a small number of genera, *Anemone* and *Clematis*, which are included in tribe *Anemoneae* (Langlet 1932, Gregory 1941, Tamura 1967), are clustered with a high bootstrap value. *Hydrastis* is weakly clustered with the genera of *Caltha* and *Cimicifuga*.

The maximum parsimony (MP) tree was also constructed using the same data set of the NJ tree by the branch-and-bound procedure of the MEGA program (Kumar & al. 1993). Except for *Caltha*, the members of each clade in both trees are consistent (Fig. 1). In the NJ tree, *Caltha* is clustered with other genera of the *Helleboreae*, but

Table 1. Pairwise divergence (x 100) for exon and intron sequences (273 bp) appear in the upper-right half of the matrix and those of intron 3 in the lower-left half of the matrix. 1 *Anemone flaccida* Fr. Schm. 1, 2 *Ranunculus silerifolius* Lév. 3 *Ranunculus sceleratus* L. 4 *Caltha palustris* L. var. *nipponica* Hara, 5 *Eranthis pinnatifida* Maxim, 6 *Cimicifuga simplex* Wormsk. 1, 7 *Actaea asiatica* Hara, 8 *Anemonopsis macrophylla* Sieb. & Zucc., 9 *Aquilegia buergeriana* Sieb. & Zucc., 10 *Semiaquilegia adoxoides* (DC.) Makino, 11 *Enemion raddeanum* Regel, 12 *Isopyrum thalictroides* L., 13 *Dichocarpum trachyspermum* (Maxim.) W. T. Wang & Hsiao, 14 *Thalictrum filamentosum* Maxim., 15 *Thalictrum minus* L. var. *hypoleucum* (Sieb. & Zucc.) Miq., 16 *Thalictrum actaefolium* Sieb. & Zucc., 17 *Coptis trifolia* (L.) Salisb., 18 *Xanthorhiza simplicissima* Marshall, 19 *Anemone flaccida*, 2, 20 *Anemone keiskeana* Maxim., 21 *Clematis terniflora* DC., 22 *Clematis japonica* Thunb., 23 *Clematis stans* Sieb. & Zucc., 24 *Cimicifuga simplex*, 25 *Caltha palustris* 2, 26 *Hydrastis canadensis* L., 27 *Mahonia japonica* (Thunb.) DC.

	1	2	3	4	5	6	7	8	9	10	11	12	13	14	15	16	17	18	19	20	21	22	23	24	25	26	27
1		8.2	6.7	13.4	12.3	11.6	12.7	11.9	12.7	13.1	13.8	12.7	12.3	13.1	13.4	13.1	14.6	13.8	20.5	20.2	19.0	18.7	18.7	20.9	23.1	21.3	18.7
2	36.0		7.1	13.8	13.8	14.6	14.2	14.2	15.7	15.7	14.9	14.2	13.1	13.1	13.4	13.1	16.4	17.2	21.6	20.2	21.3	21.6	21.6	21.0	23.9	21.3	16.8
3	37.3	28.0		13.4	12.7	11.9	12.3	12.3	12.7	13.1	13.8	13.1	12.3	12.3	12.7	12.3	14.2	13.8	20.9	20.5	20.5	20.9	20.9	21.3	22.4	20.5	19.0
4					9.3	9.3	9.7	10.5	11.9	12.7	11.2	10.1	11.6	11.2	13.1	12.7	12.3	11.9	22.4	22.4	21.6	22.8	22.8	20.9	23.9	23.5	19.4
5						3.7	5.6	5.2	11.6	12.7	10.1	8.6	9.3	10.5	12.3	11.9	11.2	10.8	19.8	20.2	20.5	21.6	21.6	20.5	23.9	20.5	18.3
6					24.0		3.4	3.0	10.5	11.2	9.3	7.8	9.3	9.7	11.6	11.2	9.0	9.3	13.3	19.4	19.0	20.2	20.2	20.5	22.4	20.2	17.9
7					18.7	14.7		1.1	13.1	13.8	11.2	9.7	11.2	11.6	13.8	13.4	10.5	10.8	23.9	21.3	21.6	21.6	21.6	21.3	22.8	20.9	18.3
8					14.7	22.7	13.3		13.1	13.4	10.5	9.0	10.5	10.8	13.1	12.7	9.7	10.1	23.9	21.3	21.3	21.6	21.6	21.3	22.8	20.5	19.0
9										1.5	6.7	6.0	6.7	6.7	8.6	6.3	10.8	10.1	22.0	22.8	20.5	22.4	22.4	19.4	20.9	20.2	19.4
10									1.3		6.7	6.7	7.8	7.8	7.5	7.5	10.8	11.2	22.8	23.5	20.9	21.6	22.4	20.2	21.6	20.2	19.4
11									21.5	21.3		2.2	4.5	4.1	6.3	6.0	9.3	10.1	20.9	22.0	20.2	21.6	21.6	19.8	21.6	18.7	18.7
12									23.9	22.4	5.3		3.0	2.6	5.2	4.9	7.8	8.6	20.5	21.6	20.2	20.9	20.9	19.0	20.5	17.9	17.9
13									20.5	20.5	5.5	3.0		4.5	5.2	5.6	9.7	9.7	21.6	22.0	22.4	23.1	23.1	20.2	22.0	18.3	18.3
14									21.0	21.0	6.9	11.6	7.7		4.1	3.7	8.6	9.0	20.2	21.3	19.8	20.5	20.5	18.7	20.5	17.5	16.8
15									22.9	22.9	10.9	15.7	11.6	6.0		1.9	10.8	11.2	20.2	21.3	19.8	20.5	20.5	18.3	20.2	17.5	18.3
16									27.0	27.0	13.5	19.2	15.6	7.3	5.5		10.8	11.2	20.5	21.6	20.5	21.3	21.3	19.4	21.3	18.3	18.7
17																		3.7	18.7	20.5	18.3	17.5	17.5	19.8	22.0	19.8	17.2
18																	9.4		17.5	19.4	18.3	17.5	17.5	19.8	19.4	19.4	16.4
19																				3.4	11.2	11.2	11.2	14.9	17.2	17.9	21.3
20																			3.4		13.4	13.1	13.1	15.7	17.9	19.0	20.9
21																			52.6	31.6		5.6	5.6	17.2	15.7	16.4	23.1
22																			56.1	28.1	7.0		0.8	16.4	13.8	16.4	22.8
23																			54.4	28.1	5.3	3.5		16.4	13.1	16.4	22.4
24																			50.9	28.1					10.8	13.1	20.9
25																								40.9		11.6	22.0
26																											24.3

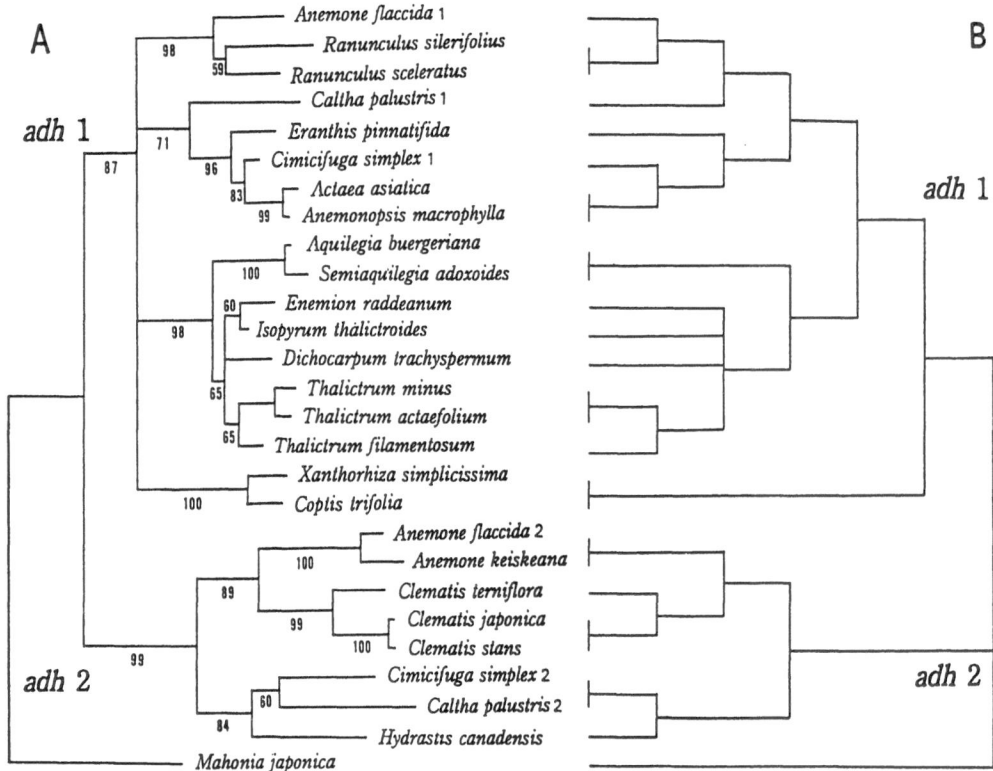

Fig. 1. Phylogenetic trees of *adh* fragments in the *Ranunculaceae*. *A* Neighbor joining (NJ) tree of the 500 bootstrap replicates. *B* 50% majority consensus maximum parsimony (MP) tree

with relatively low bootstrap support. In the MP tree, however, the *adh*1 from *Caltha* is clustered with the *Anemone–Ranunculus* clade and the *Caltha adh*2 with *Hydrastis*. *Caltha* has been considered to be the most primitive genus in subfam. *Helleboroideae*, because of the absence of petals and their distribution (TAMURA 1966) and is similar to some species of *Ranunculus* with achenes in floral and leaf morphologies.

The topologies of both trees are different at the base. In the MP tree of *adh*1, the *Coptis* group is placed basal to all other taxa. In the NJ tree, the relationships among the four major clades are unresolved, for the bootstrap values are low. This is caused mainly by the fact that some sequences in the different clades are relatively more closely related to each other in corrected distance values (Table 1). As both genes seem to share a common ancestor, the NJ trees of *adh*1 were reconstructed using *adh*2 sequences as an outgroup. When we select the *adh*2 from *Anemone* or *Clematis* as the outgroup, *Anemone* groups basal to all other taxa with about 40% bootstrap support. On the contrary, when the *adh*2 sequences from *Hydrastis*, *Caltha* or *Cimicifuga* are used as an outgroup, *Coptis* groups basal to all other taxa with 60% bootstrap support. As the basal topologies change flexibly depending on the choice of an outgroup, we need the complete data sets for both *adh*1 and *adh*2 to clarify the intersubfamiliar relationships in the *Ranunculaceae*.

Analysis of intron sequences

Since sequences of introns evolve more rapidly than those of exons, homology of intron sequences can be detected usually only between pairs with the high exon similarity. The intron 3 sequences were aligned with the CLUSTAL V program (HIGGINS 1991) and were adjusted manually to increase overall similarities. A distance for each pair of sequences was calculated ignoring those gaps that were directly involved in the comparison (Table 1). The intron homologies also corroborate the exon clades and thus these sequences may be useful in analyzing closely related taxa.

Phylogeny of T-type chromosome group

Cytological characters have been most important in the phylogeny of the *Ranunculaceae*. LANGLET (1932) recognized two types of chromosomes in this family and proposed a subdivision into two subfamilies, *Ranunculoideae* and *Thalictroideae*. *Thalictroideae* with T-type chromosomes was divided into two tribes based on chromosome number: tribe *Thalictreae* with x = 7 and tribe *Coptideae* with x = 9. TAMURA (1966) distinguished *Isopyrum* and its allied genera with follicles from *Thalictrum* with achenes as subfamily *Isopyroideae*. After that, new basic chromosome numbers were reported in follicle-bearing groups, *Isopyroideae* and *Coptidoideae*: x = 6 in *Dichocarpum* (LI & HSU 1986, FU 1988, KOSUGE & OKADA 1989), and x = 8 in *Asteropyrum* (CHANG 1982), respectively. Thus, TAMURA and KOSUGE (1989) revised the classification of the follicle-bearing group. Four different systematic treatments of this group were outlined in Table 2.

The *adh*1 genes from the T-type chromosome group are divided into two clades (Fig. 1). *Coptis* and *Xanthorhiza* with x = 9 form a distinct clade from the remaining members, namely *Aquilegia*, *Semiaquilegia*, *Enemion*, *Isopyrum*, *Dichocarpum*, and *Thalictrum*. These two clades are also supported by restriction site mapping of chloroplast DNA data (JOHANSSON & JANSEN 1993, JOHANSSON 1995) and the *atp*B, *rbc*L and 18S nuclear ribosomal DNA sequence data set (HOOT 1995), and correspond to LANGLET's (1932) two tribes *Coptideae* and *Thalictreae*.

Coptis seems to have retained a morphologically primitive conditions, such as unsealed carpels and double bundled petioles (TAMURA 1981). Based on the molecular, cytological and morphological data, *Coptis* and *Xanthorhiza* seem to comprise a relatively primitive monophyletic clade in the T-type chromosome group.

In the clade of *Thalictreae*, the *Aquilegia* and *Semiaquilegia* clade splits from others with high bootstrap support but the relationships among other genera are not resolved as clearly. As the exon sequences are quite similar for these taxa, we compared the sequences of intron 3. Although those introns are generally different in length, their sequences are quite homologous to each other (Fig. 2, Table 1). Except for a short insertion and deletion (10–23, 61–65), the intron sequences of *Aquilegia* and *Semiaquilegia* are nearly the same. The monophyly of both genera with specialized cyclic flowers and spurred petals is well supported by the intron sequences as well as the exon ones. The relationships of the intron sequences among *Dichocarpum*, *Enemion* and *Isopyrum* do not agree with those implied by exon sequences. This may be due to a lack of sufficient variation in the sequenced exon. The intron sequences of three genera are relatively closely related each other (Table

Table 2. Classifications within the T-type chromosome group in the *Ranunculaceae* (excluding *Hydrastis*)

LANGLET (1932)	TAMURA (1966, 1968)	TAMURA & KOSUGE (1989)	FU (1990)
Subfam. *Thalictroideae*	Subfam. *Thalictroideae*	Subfam. *Thalictroideae*	Subfam. *Thalictroideae*
Tribus *Thalictreae*	Tribus *Thalictreae*	Tribus *Thalictreae*	Tribus *Thalictreae*
Thalictrum	*Thalictrum*	*Thalictrum*	*Thalictrum*
Anemonella			Tribus *Calathodeae*
Isopyrum			*Calatodes*
Leptopyrum	Subfam. *Isopyroideae*	Subfam. *Isopyroideae*	Tribus *Isopyreae*
Aquilegia	Tribus *Isopyreae*	Tribus *Isopyreae*	*Enemion*
	Enemion	*Enemion*	*Isopyrum*
	Isopyrum	*Isopyrum*	*Leptopyrum*
Tribus *Coptideae*	*Leptopyrum*	*Leptopyrum*	*Paraquilegia*
Coptis	*Paraquilegia*	*Paraquilegia*	*Urophysa*
Xanthorhiza	*Urophysa*	*Urophysa*	*Semiaquilegia*
	Semiaquilegia	*Semiaquilegia*	*Aquilegia*
	Aquilegia	*Aquilegia*	Tribus *Dichocarpeae*
	Dichocarpum	*Dichocarpum*	Tribus *Asteropyreae*
	Subfam. *Coptidoideae*	Tribus *Coptideae*	*Asteropyrum*
	Tribus *Coptideae*	*Asteropyrum*	Tribus *Coptideae*
	Asteropyrum	*Coptis*	*Coptis*
	Coptis	*Xanthorhiza*	*Xanthorhiza*
	Xanthorhiza		

```
          1                                                                    70
Aquilegia        GTATGTACATTT-CAAGTACATTCGTTTTTATTGTCTTTGCTCACAGCTATTTAATATTA---------
Semiaquilegia    GTATGTACAT------------CGTTTTTACTGTCTTTGCTCACAGCTATTTAATATTAGGGGA-----
Enemion          GTATGTACATTTCCATGTACATTCATTTTTACTTTCATTGCCCGCAACTGTTC---------GATGTTT
Isopyrum         GTATGTACATT-CCAAACACATTCATTCTTACTTTCAT-GCCTGCAACTATTC---------GATGTTT
Dichocarpum      GTATGTACATTTCCAACTACATTCATTCTTACTTTCATTGCCTGCAACTATTC---------GGATGTTT
Thalictrum minus GTATGTACATTTCCAAGTACATGCATTTTTACTTTCATTGGGCGCAACTGTT---------GGGATGTTT
T. actaefolium   GTATGTACATTTCCAAGTACATGCATTTTTACTTTTATTACACGCAACTGTT---------GGGATGTGT
T. filamentosum  GTATGTACATTTCCAAGTACATTCATTTTTAC---------CCGCAACTGTT---------GGGATGTTT

          71                                                                   140
Aquilegia        TGTAAC--GACAATAATATAGATTTCA-TTTATCTTTCCTCTGTTCATTTCATTTATCTCTTTTC-TGTT
Semiaquilegia    TGTAAC--GACAATAATATAGATTTCA-TTTATCTTTCCTCTGTTCATTTCATTTATCTCTCTTC-TGTT
Enemion          TGTAAT--GACC-TAAA-TAGATTTCACTT-ATCTCTCTTCTGTTAACT--------------------
Isopyrum         TGTAATCAGACC-TAAAATAGATTTCACTT-ATCTCTCTTCTGTTAACT--------------------
Dichocarpum      -GTAAT--GACC-TAAAAGAGATTTCACTT-ATCT-TCTTCTGTT-----------------------
Thalictrum minus -ATAA---GACC-TAAAATAGATTTCACTTTATCAATCATTTGT-CATT-CATATATT--CGCTTATGTT
T. actaefolium   TATAAC--GA----------------------------------------------------------
T. filamentosum  TGTAAT--GACC-TAAAATAGATTTCACTT-ATCTTTCATTTCTTCATTTCATATATTTCTGCTTATGTT

          141                                        186
Aquilegia        TACTCAA---ACCATTTCAATCTGTTTGATGGTGTTTA-TTTGTAG          (167 bp)
Semiaquilegia    TACTCAA---ACCATTTCAATCTGTTTGATGGTGTTTA-TTTGTAG          (160 bp)
Enemion          ---------AAC-AGTTCAATTTGTTTGATGTTGTTGGGTCTGTAG          (140 bp)
Isopyrum         ---------AAC-AGTTCAATGTGTTTGATGTTGTTGG-TCTGTAG          (140 bp)
Dichocarpum      ----TAACTAA-CAGTTCAATGTGTTTGATGTTGTTGG-TCTGTAG          (142 bp)
Thalictrum minus TACATAACTAACCAGTTCGATTTGTTGGATGTTGTTGG-TCTTTAG          (167 bp)
T. actaefolium   --------TAACCAGTTCGAT----------TTGTTGG-TCTTTAG          ( 96 bp)
T. filamentosum  TACATAACTAACCAGTTCAATTGGTTGGATGTTGTTGG-TCTTTAG          (163 bp)
```

Fig. 2. Aligned intron 3 sequences of the *adh*1 genes in the *Thalictroideae*. Numbers in parentheses are the total length of intron 3 from each taxon and the hyphens are gaps

1, Fig. 2). Comparing distance values alone, *Dichocarpum* is most related to *Isopyrum*. Many species of *Dichocarpum* were first described under *Isopyrum*. WANG & HSIAO (1964) separated them as a new genus and considered this genus to be related to *Coptis*. The sequence data of *adh* genes, however, reveal that *Dichocarpum* (with x = 6) is closer to *Isopyrum* (with x = 7) than *Coptis* (with x = 9).

The similarity of sequences suggests that *Thalictrum* with achenes is monophyletic but is closely related to *Enemion* with follicles. Anatomical studies have revealed that achenes are derived from follicles by the reduction of ovule numbers and fusion of carpel traces (SMITH 1928, CHUTE 1930, EAMES 1931, KUMAZAWA 1938). In *Isopyrum* and *Enemion*, the number of ovules is reduced and *Enemion* is regarded as a primitive genus without petals like *Thalictrum*. The molecular, chromosomal and morphological data all suggest the genus *Thalictrum* shares a common ancestor with *Enemion*.

Our preliminary results indicate that exon sequences of *adh* contain enough information to resolve intergeneric relationships clearly, despite that the length of sequences examined is very short. The intron sequences are able to resolve intrageneric and some intergeneric relationships more clearly than the exon sequences.

Research was supported by grants from Ministry of Education, Science and Culture Japan (02740349). I would like to thank Prof. MICHIO TAMURA for his advice and for donating material of *Hydrastis canadensis*, and Ms Noriko Seike for sequencing *Mahonia japonica adh*.

References

Chang, C., Meyerowitz, E. M., 1986: Molecular cloning and DNA sequence of the *Arabidopsis thaliana* alcohol dehydrogenase gene. – Proc. Natl. Acad. Sci. USA **83**: 1408–1412.

Chang, C. Y., 1982: Chromosome observations of three ranunculaceous genera in relation to their systematic positions. – Acta Phytotax. Sinica **20**: 402–409.

Chase, M. W., Soltis, D. E., Olmstead, R. G., Morgan, D., Les, D. H., Mishler, B. D., Duval, M. R., Price, R. A., Hills, H. G., Qiu, Y. -L., Kron, K. A., Rettig, J. H., Conti, E., Palmer, J. D., Manhart, J. R., Sytsma, K. J., Michels, H. J., Kress, W. J., Karol, K. G., Clark, W. D., Hedrén, M., Gaut, B. S., Jansen, R. K., Kim, K.-J., Wimpee, C. F., Smith, J. F., Furnier, G. R., Strauss, S. H., Xiang, Q. -Y., Plunkett, G. M., Soltis, P. S., Swensen, S. M., Williams, S. E., Gadek, P. A., Quinn, C. J., Eguiarte, L. E., Golenberg, E., Learn, G. H., Graham, S. W., Barrett, S. C. H., Dayanandan, S., Albert, V. A., 1993: Phylogenetic of seed plants: An analysis of nucleotide sequences from the plastid gene *rbc*L. – Ann. Missouri Bot. Gard. **80**: 528–580.

Chute, H. M., 1930: The morphology and anatomy of the achene. – Amer. J. Bot. **17**: 703–723.

Clegg, M. T., 1993: Chlorophyll gene sequences and the study of plant evolution. – Proc. Natl. Acad. Sci. USA **90**: 363–367.

Dennis, E. S., Gerlach, W. L., Pryor, A. J., Bennetzen, J. L., Inglis, A., Llewelyn, D., Sachs, M. M., Ferl, R. J., 1984: Molecular analysis of the alcohol dehydrogenase (*adh*1) gene of maize. – Nucleic Acids Res. **12**: 3983–4000.

– Sachs, M. M., Gerlach, W. L., Finnergan, E. J., Peacock, W. J., 1985: Molecular analysis of the alcohol dehydrogenase 2 (*adh*2) gene of maize. – Nucleic Acids Res. **13**: 727–743.

Eames, A. J., 1931: The vascular anatomy of the flower with refutation of the theory of carpel polymorphism. – Amer. J. Bot. **18**: 147–188.

Ellison, N. W., Yu, P. L., White, D. W. R., 1990: Nucleotide sequence of a white clover alcohol dehydrogenase cDNA. – Nucleic Acids Res. **18**: 4913.

Fu, D.-Z., 1988: A study on *Dichocarpum* (*Ranunculaceae*). – Acta Phytotax Sinica **26**: 249–264.

– 1990: Phylogenetic considerations on the subfamily *Thalictroideae* (*Ranunculaceae*). – Cathaya **2**: 181–190.

Gaut, B. S., Clegg, M. T., 1991: Molecular evolution of alcohol dehydrogenase 1 in members of the grass family. – Proc. Natl. Acad. Sci. USA **88**: 2060–2064.

– 1993: Molecular evolution of the adh 1 locus in the genus *Zea*. Proc. Natl. Acad. Sci. USA **90**: 5095–5099.

Gottlieb, L. P., 1982: Conservation and duplication of isozymes in plants. – Science **216**: 373–380.

Gregerson, R. M., McLean, M., Beld, M., Gerats, A. G., Strommer, J., 1991: Structure, expression, chromosomal location and product of the gene encoding ADH 1 in petunia. – Pl. Mol. Biol. **17**: 37–48.

Gregory, W. C., 1941: Phylogenetic and cytological studies in the *Ranunculaceae* Juss. – Trans. Amer. Phil. Soc., n.s., **31**: 443–521.

Higgins, D. G., 1991: Clustal V documentation. – EMBL.

Hoot, S. B., 1995: Phylogeny of the *Ranunculaceae* based on *atp*B, *rbc*L and 18S nuclear ribosomal DNA sequence data. – Pl. Syst. Evol. [Suppl]. **9**: 241–251.

JOHANSSON, J. T., 1995: A revised chloroplast DNA phylogeny of the *Ranunculaceae*. – Pl. Syst. Evol. [Suppl]. **9**: 253–261.

– JANSEN, R. K., 1993: Chloroplast DNA variation and phylogeny of the *Ranunculaceae*. – Pl. Syst. Evol. **187**: 29–49.

KOSUGE, K., OKADA, H., 1989: Cytotaxonomical studies on *Dichocarpum* (*Ranunculaceae*) in Japan. – J. Japan. Bot. **64**: 1–7.

KUMAZAWA, M., 1938: On the ovular structure in the *Ranunculaceae* and *Berberidaceae*. – J. Japan. Bot. **14**: 10–25.

KUMAR, S., TAMURA, K., NEI, M., 1993: MEGA: molecular evolutionary genetic analysis, version 1.0 – University Park: The Pennsylvania State University.

LANGLET, O., 1932: Über Chromosomenverhältnisse und Systematik der *Ranunculaceae*. – Svensk Bot. Tidskr. **26**: 381–400.

LI, L. C., HSU, P. S., 1986: Chromosome observations of eight species endemic to China. – Acta Phytotax. Sinica **24**: 157–160.

LLEWELLYN, D. J., FINNEGAN, E. J., ELLIS, J. G., DENNIS, E. S., PEACOCK, W. J., 1987: Structure and expression of an alcohol dehydrogenase 1 gene from *Pisum sativum* (cv. "greenfast"). – J. Mol. Biol. **195**: 115–123.

MATTON, D. P., BRISSON, N., 1989: Nucleotide sequence of two potato alcohol dehydrogenase cDNA. – Nucleic Acids Res. **18**: 3070.

SMITH, G. H., 1928: Vascular anatomy of Ranalian flowers II. *Ranunculaceae* (continued), *Menispermaceae, Calycanthaceae, Annonaceae*. – Bot. Gaz. **85**: 152–177.

TAMURA, M., 1966: Morphology, ecology and phylogeny of the *Ranunculaceae* IV. – Sci. Rep. Osaka Univ. **15**: 13–35.

– 1967: Morphology, ecology and phylogeny of the *Ranunculaceae* VII. – Sci. Rep. Osaka Univ. **16**: 21–43.

– 1968: morphology, ecology and phylogeny of the *Ranunculaceae* VIII. – Sci. Rep. Osaka Univ. **17**: 41–56.

– 1981: Morphology of *Coptis japonica* and its meaning in phylogeny. – Bot. Mag. (Tokyo) **94**: 165–176.

– KOSUGE, K., 1989: Classification of the *Isopyroideae* (*Ranunculaceae*). – Acta Phytotax. Geobot. **40**: 31–35.

TRICK, M., DENNIS, E. S., EDWARDS, K. J. R., PEACOCK, W. J., 1988: Molecular analysis of the alcohol dehydrogenase gene family of barley. – Pl. Mol. Biol. **11**: 147–160.

WANG, W. T., HSIAO, P. K., 1964: A new genus of *Ranunculaceae* – *Dichocarpum* W. T. WANG et HSIAO. – Acta Phytotax. Sinica **9**: 315–333, table 32, 33.

WOLFE, K. H., LI, W. -H., SHARP, P. M., 1987: Rates of nucleotide substitution vary greatly among plant mitochondrial, chloroplast, and nuclear DNAs. – Proc. Natl. Acad. Sci. USA **84**: 9054–9058.

– SHARP, P. M., LI, W. -H., 1989: Rates of synonymous substitution in plant nuclear genes. – J. Mol. Evol. **29**: 208–211.

WOLYN, D. J., JELENKOVIC, G., 1990: Nucleotide sequence of an alcohol dehydrogenase gene in octoploid strawberry (*Fragaria* x *ananassa* DUCH.). – Pl. Mol. Biol. **4**: 855–857.

XIE, Y., WU, R., 1989: Rice alcohol dehydrogenase genes: anaerobic induction, organ specific expression and characterization of cDNA clones. – Pl. Mol. Biol. **13**: 53–68.

Address of the authors: KEIKO KOSUGE, TETSUO DENDA, JUNKO ADACHI and KUNIAKI WATANABE, Department of Biology, Faculty of Science, Kobe University, Kobe 657, Japan. – KAZUTOSHI SAWADA, Biological Science Research Laboratory, Central Research Laboratories, Idemitsu Kosan Co., LTD. 1280 Kami-Izumi, Sodegawra, Chiba 299-02, Japan.

Accepted January 25, 1995

Pl. Syst. Evol. [Suppl.] 9: 273–280 (1995)

Systematics and phylogeny of the *Ranunculaceae* – a revised family concept on the basis of molecular data

U. Jensen, S. B. Hoot, J. T. Johansson, and K. Kosuge

Received January 27, 1995

Key words: *Ranunculaceae.* – *atp*B, *rbc*L, 18S nuclear ribosomal DNA, alcohol dehydrogenase gene, chloroplast restriction site variation, legumin, serology.

Abstract: Phylogenetic trees for the *Ranunculaceae* based on four independent molecular data sets are compared and analyzed. This comparison includes chloroplast DNA restriction site variation; *atp*B (chloroplast), *rbc*L, nuclear ribosomal DNA sequences (analysed as a combined data set), nuclear *adh* sequences, and serological detected characters of the major seed protein, legumin. These trees are highly congruent in terminal branching patterns with high support. This congruency suggests a strong correlation between the evolution of the genes and proteins investigated and the taxa involved and further demonstrates the utility of molecular markers in plant phylogeny. The molecular results are often congruent with non-molecular data as well and are used to develop more reliable systematic classification of the *Ranunculaceae*.

Systematics and phylogenetic relationships within the family *Ranunculaceae* have been considerably improved by exploring new and varied non-molecular data sets: cytogenetics (Langlet 1932, Gregory 1941), palynology (Kumazawa 1936, Wodehouse 1936, Vishnu-Mittre & Sharma 1963), and phytochemistry (Hegnauer 1966, 1973, 1990; Ruijgrok 1966). Molecular methods were first applied to the family with the advent of serologically detectable seed protein characters, reflecting the structures of antigenic molecules ("determinants"; Hammond 1952, 1955; Jensen 1968; Jensen & Penner 1980; Jensen & Greven 1984). Serology indicated the general significance of this class of molecular characters in further understanding the evolutionary relationships of the family. Recent work has applied a spectrum of new molecular methods to the family: chloroplast restriction site variation (RSV) (Johansson & Jansen 1993, Johansson 1995, this volume), the chloroplast gene sequences, *atp*B and *rbc*L (Hoot 1995, this volume), nuclear ribosomal DNA sequences (Hoot 1995, this volume), and nuclear *adh* sequences (Kosuge & al. 1995, this volume).

Although the sampling for the above studies was not identical, many of the same genera were included in all four data sets. For many genera, different researchers used different species in their investigations. This is of minor importance with the genes involved in this study, since the molecular similarities between species within a genus appear to be far greater than those found between genera.

Trees derived from molecular data reflect a phylogeny of the genes involved, not necessarily that of the organisms investigated. Especially genes from different genomes (nuclear versus chloroplast) may undergo different evolutionary histories. Hence, sytematists must consider whether the trees resulting from their molecular data sets are a true indication of an organism's evolutionary pattern. The following comparisons of diverse molecular data sets for the *Ranunculaceae* offer a unique opportunity to assess by tree congruence the appropriateness of each of the genes or molecular markers used in these studies. For example, preliminary separate analyses of the *atp*B, *rbc*L and 18S data sets give highly congruent results, suggesting that there is a correlation between the evolution of the genes and the organisms involved (Hoot 1995, this volume).

Comparative discussion of the molecular data

In this contribution, we will compare and analyze the results from the independent molecular data sets presented at the 1994 *Ranunculiflorae* Conference at Bayreuth, Germany (i.e., Hoot 1995, Jensen 1995b, Johansson 1995, and Kosuge & al. 1995; all this volume). This is the first time to our knowledge that as many as six molecular data sets have been considered in a phylogenetic hypothesis of one plant group (the *Ranunculaceae*). The data will be discussed in relation to the already well known "classical" data and a preliminary revised evolutionary classification for the *Ranunculaceae* will be suggested.

A comparison of the trees derived from the independent molecular data sets leads to the following conclusions: (1) the trees are highly congruent; (2) the level of confidence for the data generally increases from the basal to the terminal branches; (3) the end branches with the highest significance (e.g., bootstrap values) are largely the same in all data sets.

All the molecular data sets universally support seven monophyletic groups of genera (Table 1). High bootstrap values (83% or higher) indicate that the groups are stable in the different tree calculations. Only the *Ranunculus* group (*Trautvetteria*, *Ranunculus*, *Ficaria*, *Myosurus*) in the chloroplast RSV tree had lower bootstrap values (56% in the Wagner tree, 67% in the Dollo parsimony tree; Johansson 1995, this volume). This group is also not sustained by the legumin tree, where *Trautvetteria* is joined with *Adonis/Trollius* (Jensen 1995b, this volume).

Many significant non-molecular characters also support the monophyly of these seven groups. Such characters are found in phytochemistry (Hegnauer 1973, 1990; Jensen 1995a, this volume), karyology (Langlet 1932, Gregory 1941), morphology (Davezac 1957; Tamura 1962, 1965, 1966, 1967, 1968; Hoot 1991), embryology (Kordjum 1959; Kurita 1961; Bhandari 1966; Engell 1995, this volume), and anatomy (Tamura 1962, 1993; Carlquist 1995, this volume).

However, there are also noticeable differences between the phylogenetic patterns derived from "classical" and molecular data. A close relationship between *Eranthis* and *Actaea/Cimicifuga/Anemonopsis* is indicated by the molecular data. This has never been postulated on the basis of morphology (leaf shape, nectary type, and growth habit are very different) and it is also not verified by the morphological tree of Loconte & al. (1995, this volume). Most previous taxonomic treatments place *Eranthis* near *Helleborus* L. because of similar floral features, e.g., similar perigon

Table 1. Groups of *Ranunculaceae* genera significantly separated by the six molecular data sets. Numbers in the legumin column refer to the similarity values (JENSEN 1995b, this volume). Numbers in the remaining columns are bootstrap values (for RSV, an average between the Dollo and Wagner values are given). ( ) Genus not included in the entire experiment,/no data available for this group

Group Genus no.	Legumin, JENSEN 1995b	atpB, rbcL, & 18S rDNA, HOOT 1995	adh1, KOSUGE & al. 1995	RSV, JOHANSSON 1995
1 *Xanthorhiza* MARSH. *Coptis* SALISB.	/[1]	100	100	100
2. *Isopyrum* L. *Aquilegia* L. *Anemonella* SPACH. *Thalictrum* L.	86 (*Isopyrum*) (*Anemonella*)	100	98 (*Anemonella*)	100
3 *Actaea* L. *Cimicifuga* L. *Anemonopsis* SIEB. & ZUCC. *Eranthis* SALISB.	100[2] (*Anemonopsis*)	100	96	83
4 *Aconitum* L. *Delphinium* L. *Consolida* (DC.) S. F. GRAY	90	100 (*Consolida*)	/	100
5 *Adonis* L. *Trollius* L.	95	100	/	100
6 *Ranunculus* L. *Ficaria* HUDS. *Myosurus* L. *Trautvetteria* FISCH & MEY.	89 (*Myosurus*)	100 (*Ficaria*)	/	61 (*Ficaria*)
7 *Clematis* L. *Anemone* L. *Hepatica* MILL.	86	100 (*Hepatica*)	89[3] (*Hepatica*)	100

1 *Coptis* has been added in some legumin experiments only; it proved to be serologically the least distant genus to *Xanthorhiza* (ref.: *Xanthorhiza* legumin antiserum)

2 *Eranthis* has been excluded from the legumin test, because this storage protein is not produced. For vicilin, *Actaea* and *Cimicifuga* are the serologically least distant genera to *Eranthis* (ref.: *Eranthis* vicilin antiserum)

3 Referring to *adh2*

leaves and nectaries (ZIMMERMANN 1965; TAMURA 1962, 1993), characters which may be plesiomorphies.

The sister relationship of *Adonis* and *Trollius* has possibly been obscured by the high value placed on fruit types in the previous systematics of the family (*Adonis* has achenes, *Trollius* follicles). However, both *Trollius* and *Adonis* have a typical polycarpous gynoecium, also characteristic of other genera with achenes (e.g., *Anemone*, *Ranunculus*). *Trollius* is also closest to *Adonis* in some chemical characters (JENSEN 1995a). JENSEN (1968) placed the two genera in the same tribe based on serology. In addition, embryology supports a close relationship between *Adonis* and *Trollius*. Most members of the *Ranunculaceae* investigated have the monosporic, 8-nucleate *Polygonum* type of embryo sac. *Adonis* has the bisporic *Allium* type (BHANDARI 1966), while both the *Polygonum* and *Allium* types have been found in *Trollius* (BHANDARI & KAPIL 1964), *Callianthemum* C. A. MEY. (BHANDARI & VIJAYARAGHAVAN 1970) and *Ceratocephalus* MOENCH. (BHANDARI & ASNANI 1966).

Some molecular data support the inclusion of other taxa in the seven basic groups mentioned in Table 1, but with weak branch support. *Nigella* L. is connected with the *Aconitum* group (group 4) in some of the trees derived from legumin (JENSEN 1995b, this volume), the *atp*B/*rbc*L/18S tree (HOOT 1995, this volume), and the Dollo parsimony RSV tree (JOHANSSON 1995, this volume). Various morphological arguments have been discussed in favour of this relationship (ZIMMERMANN 1965, HOOT 1991).

Caltha L., often connected to *Trollius* in previous classifications, and *Helleborus*, often connected to *Eranthis*, show no clear molecular relationships and do not support the mentioned sister pairs. The clade consisting of *Caltha* and *Helleborus* in the *atp*B/*rbc*L/18S tree does not have significant support (B.S < 50%, D.I. = 1; HOOT 1995, this volume). The legumin and *adh* data connect *Caltha* with the *Cimicifuga*/*Actaea*/*Anemonopsis*/*Eranthis* group (JENSEN 1995b, KOSUGE & al. 1995; both this volume). The monophyly of the *Helleborus* and *Trollius*/*Adonis* clade is only weakly supported by the RSV data (JOHANSSON 1995, this volume).

The position of *Callianthemum* is unresolved by the molecular and morphological data. The serological data of JENSEN (1968) supported a close relationship with *Adonis* which is congruent with the morphological evidence (TAMURA 1962; LOCONTE & al. 1995, this volume). In contrast, both RSV trees place *Callianthemum* as a sister genus to *Caltha* (JOHANSSON 1995, this volume).

Many of the relationships between the seven groups remain unresolved or are not congruent between the individual molecular trees. However, in agreement with a previous phylogenetic hypothesis based on karyology (LANGLET 1932, GREGORY 1941) which divided the family into two major groups based on chromosome size, the tree based on a combination of *atp*B, *rbc*L and 18S sequence data (B.S = 87%; HOOT 1995, this volume) and the MP *adh*1tree (KOSUGE & al. 1995, this volume) support the monophyletic status of the large R-type chromosome group. Similar findings were found in an analysis based on micromorphological characters (HOOT 1991), other morphological characters (TAMURA 1962; LOCONTE & al. 1995, this volume), and phytochemistry (accumulation of benzylisoquinoline alkaloids; HEGNAUER 1973, 1990; JENSEN 1995a, this volume). However, in other molecular trees, the monophyly of the R-type chromosome genera is unresolved (*adh*1 NJ tree; KOSUGE & al. 1995, this volume) or not found: In the legumin tree of JENSEN (1995b, this volume), *Xanthorhiza* (T-type chromosome group) is found nested within the R-type chromosome group, although the seed protein mixtures indicate a close relationship (JENSEN 1968). In addition, the monophyly of the R-type chromosome group is not found in either of the RSV trees, but support for the pertinent basal branching patterns in these trees is weak (JOHANSSON 1995, this volume).

The R-type chromosome group with achenes, i.e. *Ranunculus* complex (*Ranunculus, Myosurus, Trautvetteria*), *Anemone* complex (*Anemone, Hepatica, Pulsatilla* MILL., *Knowltonia* SALISB.), *Clematis* complex (*Clematis, Clematopsis* BOJ.), and *Adonis*, had previously been considered to be a monophyletic group by ZIMMERMMANN (1965) and are included in one subfamily (*Ranunculoideae*) by TAMURA (1993). The molecular data are not unequivocal but most data analyses support the monophyly of a clade consisting of the *Anemone, Clematis* and *Ranunculus* complexes (excluding *Adonis*). Support for this clade can be found in the combination of *atp*B, *rbc*L and 18S data (HOOT 1995, this volume), chloroplast restriction site data (Dollo tree only, JOHANSSON 1995, this volume) and *adh* sequences (KOSUGE & al. 1995, this volume) with bootstrap values of 82, 50, and 99%, respectively. Serological data (JENSEN 1968) and phytochemistry (presence of ranunculin; RUIJGROK 1966) also support the monophyly of this clade. The proposed tribes *Ranunculeae* and *Anemoneae* are therefore probably closely related. *Adonis*, which does not produce ranunculin, appears to have evolved achenes independently.

Systematics of the *Ranunculaceae*

In most of the molecular trees presented, the *Xanthorhiza/Coptis* group branches first and is the sister group to the remaining *Ranunculaceae*. Chemical and karyological characters (T-type chromosomes, n = 9) also separate them clearly. Therefore, the recognition of this group as the subfamily *Coptidoideae* is justified. The inclusion of *Thalictrum* (LOCONTE & al. 1995, this volume) is contradicted by numerous characters, both molecular and traditional.

The *Isopyrum* group (Table 1) is clearly separated from other *Ranunculaceae* by additional chemical and karyological characters (small T-type chromosomes, n = 7) and can be recognized as the subfamily *Isopyroideae*. Within the tribe, the two well-supported subtribes, *Isopyrinae* and *Thalictrinae*, are recognized.

The remaining five groups (Table 1) with R-type chromosomes and some genera not included in these groups (*Caltha, Helleborus*) are less reliably divided into tribes and subtribes. Within this subfamily, we propose the separation of tribes and subtribes on the basis of the above mentioned molecular similarities and phytochemical data.

The following preliminary systematic treatment is proposed:

Subfamily *Coptidoideae* TAMURA
  Tribe *Coptideae* LANGLET (*Coptis* SALISB., *Xanthorhiza* MARSH.)

Subfamily *Isopyroideae* TAMURA
  Tribe *Isopyreae* SCHRÖDINGER
    Subtribe *Isopyrinae* SPACH (*Isopyrum* L., *Enemion* RAF.,
      Dichocarpum W. T. WANG & HSIAO, *Paraquilegia*
      J. R. DRUMM. & HUTCH., *Semiaquilegia* MAKINO,
      Aquilegia L.)
    Subtribe *Thalictrinae* LANGLET
      (*Thalictrum* L., *Anemonella* SPACH)

Subfamily *Ranunculoideae* HUTCH.
   Tribe *Cimicifugeae* TORR. & A. GRAY
      Subtribe *Cimicifuginae* BENTH. (*Cimicifuga* WERNISCHEK,
         Actaea L., *Anemonopsis* SIEB. & ZUCC.)
      Subtribe *Eranthinae,* subtribe nov. (*Eranthis* SALISB.)
   Tribe *Delphinieae* WARM.
      Subtribe *Delphiniinae* BENTH. (*Aconitum* L., *Delphinium* L.,
         Consolida (DC.) S. F. GRAY)
      Subtribe *Nigellinae* SCHRÖDINGER (*Nigella* L.)
   Tribe *Caltheae* REICHB. (*Caltha* L.)
   Tribe *Helleboreae* DC. (*Helleborus* L.)
   Tribe *Adonideae* KUNTH
      Subtribe *Trolliinae* SCHRÖDINGER (*Trollius* L.)
      Subtribe *Adonidinae* REICHB.
        (*Adonis* L., possibly *Callianthemum* C. A. MEYER)
   Tribe *Ranunculeae* DC.
      Subtribe *Ranunculinae* SPACH (*Trautvetteria* FISH. & C. A.
        MEYER, *Ranunculus* L., *Ficaria* HUDS., *Myosurus* L.)
   Tribe *Anemoneae* DC.
      Subtribe *Clematidinae* LOTSY
        (*Clematis* L., *Clematopsis* HUTCH.)
      Subtribe *Anemoninae* SPACH (*Anemone* L., *Hepatica* MILL.,
        Pulsatilla MILL., *Knowltonia* SALISB.)

This classification includes the new **subtribus *Eranthinae* JENSEN, HOOT, JOHANSSON & KOSUGE, subtribus nov.:** herbae perennes; rhizomata tuberifera; folia basalia palmatisecta, caulina verticillata; petala tubularia, carpella stipitata. Typus: *Eranthis* SALISB.

Phylogenetic considerations

Molecular characters, i.e. genes and proteins, have been postulated to be less influenced by selection (KIMURA 1983, STEBBINS 1988). The trees derived from the six molecular data sets (including the separate analyses of *atp*B, *rbc*L, and 18S not published in this volume) for the *Ranunculaceae* show much congruence in the terminal branching patterns with high support. Considering the extensive agreement between these trees, we conclude that they represent not only the evolution of the genes but also of the taxa involved (i.e., genera within the *Ranunculaceae*).

    With this in mind, we can make the following proposals: (a) The *Coptideae* may be the basal group within the *Ranunculaceae*; (b) *Adonis* (and *Callianthemum*?) evolved independently of the *Ranunculeae/Anemoneae* stock and (c) in some branches a profound phenotypic difference may mask close genetic relationships (*Eranthis – Cimicifuga/Actaea, Adonis – Trollius,* and *Nigella – Aconitum/Delphinium/Consolida*).

    This congruence between the independent molecular data sets has allowed us to propose a preliminary reclassification of the family. Future work should focus on additional data acquisition so that the more elusive basal branching patterns may be resolved.

References

BHANDARI, N. N., 1966: Studies in the family *Ranunculaceae* IX. Embryology of *Adonis.* – Phytomorphol. **16**: 578–587.

– ASNANI, S., 1966: Studies in the family *Ranunculaceae* XI. Morphology and embryology of *Ceratocephalus falcatus* PERS. – Beitr. Biol. Pfl. **45**: 271–290.

– VIJAYARAGHAVAN, M. R., 1970: Studies in the family *Ranunculaceae* XII. Embryology of *Aquilegia vulgaris.* – Beitr. Biol. Pfl. **46**: 337–354.

CARLQUIST, S., 1995: Wood anatomy of *Ranunculiflorae*: a summary. – Pl. Syst. Evol. [Suppl.] **9**: 11–24.

DAVEZAC, T., 1957: Recherches ontogéniques sur les Renonculacées. – Bull. Soc. Hist. Nat. Toulouse **92**: 68–82.

ENGELL, K., 1995: Embryo morphology of the *Ranunculaceae.* – Pl. Syst. Evol. [Suppl.] **9**: 207–216.

GREGORY, W. C., 1941: Phylogenetic and cytological studies in the *Ranunculaceae* JUSS. – Trans. Amer. Phil. Soc., n.s. **31**: 441–520.

HAMMOND, H. D., 1952: Serology applied to systematic studies in the *Ranunculaceae* JUSS. – Dissertation, University of Pennsylvania.

– 1955: Systematic serological studies in *Ranunculaceae.* – Serol. Mus. Bull. **14**: 1–3.

HEGNAUER, R., 1966: Comparative phytochemistry of alkaloids. – In SWAIN, I., (Ed.): Comparative Phytochemistry, pp. 175–186. – New York: Academic Press.

– 1973: Chemotaxonomie der Pflanzen **6**: *Dicotyledoneae: Rafflesiaceae – Zygophyllaceae.* – Stuttgart: Birkhäuser.

– 1990: Chemotaxonomie der Pflanzen **9**: *Dicotyledoneae: Magnoliaceae – Zygophyllaceae.* – Stuttgart: Birkhäuser.

HOOT, S. B., 1991: Phylogeny of the *Ranunculaceae* based on epidermal microcharacters and macromorphology. – Syst. Bot. **16**: 741–755.

– 1995: Phylogeny of the *Ranunculaceae* based on *atp*B, *rbc*L and 18S nuclear ribosomal DNA sequence data. – Pl. Syst. Evol. [Suppl.] **9**: 241–251.

JENSEN, U., 1968: Serologische Beiträge zur Systematik der *Ranunculaceae.* – Bot. Jahrb. Syst. **88**: 204–268.

– 1995a: Secondary compounds of the *Ranunculiflorae.* – Pl. Syst. Evol. [Suppl.] **9**: 85–97.

– 1995b: Serological legumin data and the phylogeny of the *Ranunculaceae* – Pl. Syst. Evol. [Suppl.] **9**: 217–227.

– PENNER, R., 1980: Investigation of serological determinants from single storage plant proteins. – Biochem. Syst. Ecol. **8**: 161–170.

– KAPIL, R. N., 1964: Studies in the family *Ranunculaceae* VII. Two types of embryo sacs in *Trollius* L. – Beitr. Biol. Pfl. **40**: 113–120.

– GREVEN, B., 1984: Serological aspects and phylogenetic relationships of the *Magnoliidae.* – Taxon **33**: 563–577.

JOHANSSON, J. T., 1995: A revised chloroplast DNA phylogeny of the *Ranunculaceae.* – Pl. Syst. Evol. [Suppl.] **9**: 253–261.

– JANSEN, R. K., 1993: Chloroplast DNA variation and phylogeny of the *Ranunculaceae.* – Pl. Syst. Evol. **187**: 29–49.

KIMURA, M., 1983: The neutral theory of molecular evolution. – Cambridge: University Press.

KORDJUM, E. L., 1959: Comparative embryological investigation of the family *Ranunculaceae* DC. – Ukrain. Bot. Ž. **16**: 32–43.

KOSUGE, K., SAWADA, K., DENDA, T., ADACHI, J., WATANABE, K., 1995: Phylogenetic relationships of some genera in the *Ranunculaceae* based on alcohol dehydrogenase genes. – Pl. Syst. Evol. [Suppl.] **9**: 263–271.

Kumazawa, M., 1936: Pollen grain morphology in *Ranunculaceae*, *Lardizabalaceae*, and *Berberidaceae*. – Japan. J. Bot **8**: 19–47.

Kurita, M., 1961: Chromosome studies in *Ranunculaceae* XVIII. Karyotypes of several species. – Mem. Ehime Univ., Sect. II (Science), Ser. B (Biology), **4**: 251–261.

Langlet, O., 1932: Über Chromosomenverhältnisse und Systematik der *Ranunculaceae*. – Svensk Bot. Tidskr. **26**: 381–401.

Loconte, H., Campbell, L. M., Stevenson, D. W., 1995: Ordinal and familial relationships of the *Ranunculiflorae*. – Pl. Syst. Evol. [Suppl.] **9**: 99–118.

Ruijgrok, H. W. L., 1966: The distribution of ranunculin and cyanogenic compounds in the *Ranunculaceae*. – In Swain, T., (Ed.): Comparative phytochemistry, pp. 175–186. – New York: Academic Press.

Stebbins, G. L., 1988: Essays in comparative evolution. The need for evolutionary comparisons. – In Gottlieb, L. D., Jain, S. K., (Eds): Plant evolutionary biology, pp. 3–20. – London, New York: Chapman & Hall.

Tamura, M., 1962: Morphology, ecology and phylogeny of the *Ranunculaceae* I. – Sci. Rep. Osaka Univ. **11**: 115–126.

– 1965: Morphology, ecology and phylogeny of the *Ranunculaceae* IV. – Sci. Rep. OsakaUniv. **14**: 53–71.

– 1966: Morphology, ecology and phylogeny of the *Ranunculaceae* VI. – Sci. Rep. Osaka Univ. **15**: 13–35.

– 1967: Morphology, ecology and phylogeny of the *Ranunculaceae* VII. – Sci. Rep. Osaka Univ. **16**: 21–43.

– 1968: Morphology, ecology and phylogeny of the *Ranunculaceae* VIII. – Sci. Rep. Osaka Univ. **17**: 41–56.

– 1993: *Ranunculaceae*. – In Kubitzki, K., Rohwer, J. G., Bittrich, V., (Eds): The families and genera of vascular plants, **7**, pp. 563–583. – Berlin: Springer.

Vishnu-Mittre, Sharma, B. D., 1963: Studies of Indian pollen grains. 2. *Ranunculaceae*. – Pollen & Spores **5**: 285–296.

Wodehouse, R. P., 1936: Pollen grains in the identification and classification of plants. VII. The *Ranunculaceae*. – Bull. Torrey Bot. Club **63**: 495–514.

Zimmermann, W., 1965: Familie *Ranunculaceae*. – In Rechinger, K. -H., Damboldt, J. (Eds): Hegi, Illustrierte Flora von Mitteleuropa, **III/3**, pp. 53–341, 2nd edn. – München: Hanser.

Addresses of the authors: Dr Uwe Jensen, Lehrstuhl Pflanzensystematik, Universität Bayreuth, Universitätsstraße 30, D-95440 Bayreuth, Germany. – Dr Sara B. Hoot, Department of Biological Sciences, Lapham Hall, PO Box 413, University of Wisconsin–Milwaukee, Milwaukee, WI 53201, USA. – Dr Jan T. Johansson, Department of Systematic Botany, University of Lund, Ö. Vallgatan 18–20, S-223 61 Lund, Sweden. – Dr Keiko Kosuge, Department of Biology, Faculty of Science, Kobe University, Rokkodai, Nada-ku, Kobe, 657, Japan.

Accepted January 29, 1995

Pl. Syst. Evol. [Suppl.] 9: 281–282 (1995)

Numerical taxonomy of the *Ranunculaceae*

Toni Nikolić

Received September 20, 1994

Key words: *Ranunculaceae*. – Numerical taxonomy, systematics.

With the aim of determining phenetic relationships among genera of the *Ranunculaceae*, a numerical taxonomic analysis was carried out.

Operational taxonomic units (OTUs) involve 25 genera which represent the diversity of the family to a large extent. The following genera have been processed: *Aconitum* L., *Actaea* L., *Adonis* L., *Anemone* L., *Aquilegia* L., *Callianthemum* C. A. Meyer, *Caltha* L., *Ceratocephalus* Moench, *Cimicifuga* L., *Clematis* L., *Consolida* S. F. Gray, *Coptis* Palisb., *Delphinium* L., *Eranthis* Salisb., *Glaucidium* Sieb. & Zucc., *Helleborus* L., *Hepatica* Mill., *Hydrastis* L., *Isopyrum* L., *Myosurus* L., *Nigella* L., *Pulsatilla* Mill., *Ranunculus* L., *Thalictrum* L., and *Trollius* L.

Altogether 135 characters were taken from the literature. Anatomy is represented by 29, morphology by 56, cytology by 9, embryology by 26 and phytochemistry by 15 characters. For micro- and macromorphological characters numerous herbarium specimens and living material was consulted. All variables were coded binary and weighted equally.

The product-moment correlation coefficient was calculated as a measure of similarity. Cluster analysis was carried out using the UPGMA and WPGMA techniques. For the presentation of relationships of the genera a PCO analysis was done. The MST (Minimum Spanning Tree) was projected onto 3D results of the PCO analysis. The groups detected by the ordination method as well as by the cluster method was tested by K-means clustering (Sneath & Sokal 1973).

The results from the different numerical techniques were highly congruent (Fig. 1). Genera are clearly divided into two clusters that correspond to genera with T and R type chromosomes (Langlet 1932, Gregory 1941). Several groups, which had been observed earlier (Janchen 1949, Tamura 1967, Jensen 1968, Ziman 1985) could be confirmed: *Aconitum – Consolida – Delphinium*, *Ceratocephalus – Myosurus – Ranunculus*, *Actaea – Cimicifuga*, *Caltha – Trollius – Eranthis – Nigella – Helleborus*, *Aquilegia – Isopyrum – Thalictrum* and *Anemone – Hepatica – Pulsatilla – Clematis*. *Hydrastis* and *Glaucidium* were clearly separate. *Coptis* shows an unambiguous close relationship to the genera *Isopyrum – Aquilegia* and the somewhat isolated *Thalictrum*. *Nigella* shows affinities to *Eranthis* and *Helleborus*. *Actaea* and *Cimicifuga* are quite isolated and show weak affinities to the *Helleborus* group. *Clematis* is somewhat isolated in the *Anemone* group. *Adonis* and *Callianthemum* have an ambiguous position and are connected alternatively with the *Helleborus* or *Ranunculus* groups.

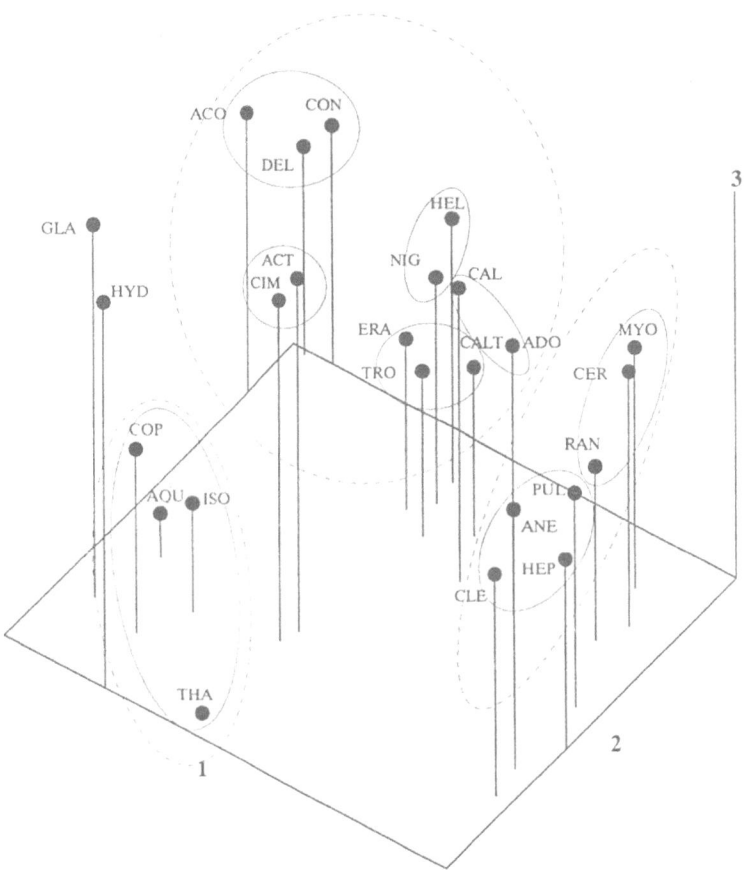

Fig. 1. 3D projection of 25 genera of the *Ranunculaceae* onto the first three principal co-ordinates. K-means clusters superimposed, dividing into three (dotted line) and eight (solid line) initial groups

References

GREGORY, W. C., 1941: Phylogenetic cytological studies in the *Ranunculaceae* JUSS. – Trans. Amer. Phil. Soc. **31**: 443–520.

JANCHEN, E., 1949: Die Systematische Gliederung der Ranunculaceen und Berberidaceen. – Denkschriften Akad. Wiss., math. – naturwiss. Klasse Wien **108**: 1–82.

JENSEN, U., 1968: Serologische Beiträge zur Systematik der Ranunculaceen. – Bot. Jahrb. **88**: 204–310.

LANGLET, O., 1932: Über Chromosomenverhältnisse und Systematik der *Ranunculaceae*. – Svensk Bot. Tidskrift. **26**: 381–400.

SNEATH, P. H. A., SOKAL, R. R., 1973: Numerical taxonomy. – San Francisco: Freeman.

TAMURA, M., 1967: Morphology, ecology and phylogeny of the *Ranunculaceae*. VII. – Sci. Rep. **16**: 21–43.

ZIMAN, S., 1985: Morfologiya i filogeniya semejstva lyutikovyih. – Kiev: Naukova dumka.

Address of the author: Dr TONI NIKOLIĆ, Department of Botany, Faculty of Science, University of Zagreb, Marulićev trg 20/2, HR–41000 Zagreb, Croatia.

Accepted December 22, 1994

Pl. Syst. Evol. [Suppl.] 9: 283–293 (1995)

Evolutionary trends and patterns in the *Anemoninae*

F. Ehrendorfer

Received February 6, 1995

Key words: *Ranunculaceae, Anemoninae, Anemone.* – Diversity, adaptive syndromes, morphological versus molecular data, phylogeny, systematics.

Abstract: On the basis of cpDNA restriction site analyses (Hoot & al. 1994) an improved but still informal hierarchical survey of *Anemoninae* is presented and combined with diversity and distribution data. Several adaptive syndromes are recognized on the basis of interrelated vegetative, reproductive, cytogenetic and group-organizational aspects. These adaptive patterns or types often include superficially similar species groups from different *Anemoninae* clades and apparently are the result of the convergent canalizing effect of similar environments and histories during the evolutionary process of eco-geographical radiation. This and several other reasons are discussed as obstacles in matching morphological and molecular data for taxonomic purposes.

The morphologically and ecologically very diverse and world-wide distributed group of taxa assembled within the *Ranunculaceae* subtribe *Anemoninae* has been the object of phylogenetic speculations for nearly 100 years (Ulbrich 1905/06, Scharfetter 1953, Starodubtsev 1991). Only recently, molecular data obtained by Hoot & al. (1994) and Hoot (1995, this volume) provide a more solid basis for phylogenetic interpretations of *Anemone* L. and its relatives. At the same time considerable discrepancies with former systematic arrangements are revealed. All this and our present multidisciplinary knowledge of the subtribe should give a good opportunity to pursue some more general problems concerning evolutionary trends and patterns in *Anemone* (Ehrendorfer 1988). Thus, I will try to answer the following two questions: (1) What are the main obstacles in coordinating DNA-based genealogies with morphological and other data so far relevant in *Anemoninae* systematics? (2) Which evolutionary processes have shaped today's appearance of this complex plant group of common descent taxonomists call *Anemoninae*?

Phylogeny and systematics

Our present concept of the phylogenetic, i.e. genealogical relationships within *Ranunculaceae – Anemoneae* are shown in a simplified cladogram (Fig. 1). This is primarily based on restriction site analyses of cpDNA, but also considers some other molecular, karyological and morphological markers, as published by Hoot & al.

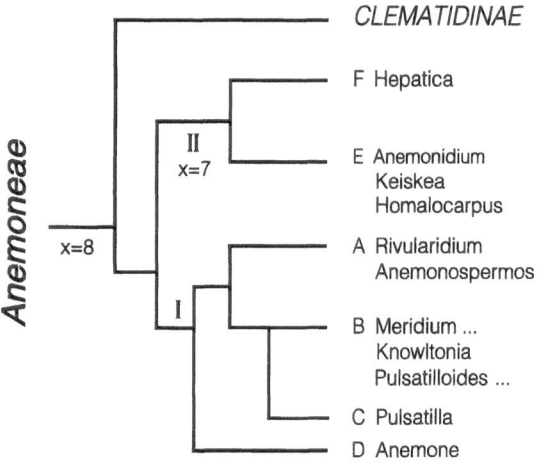

Fig. 1. Simplified tree of major clades in *Ranunculaceae–Anemoneae*, based mainly on restriction site analyses of cp DNA (data from Hoot & al. 1994) and base chromosome numbers. Figures and names correspond to Table 1

(1994) and Hoot (1995, this volume). The comparison of preliminary sequence data from a cpDNA intergene region adjacent to the rbcL gene obtained from ten *Anemone* species (Manen & Ehrendorfer, unpubl.; see also Manen & al. 1994) supports this cladogram. One can therefore assume that the *Anemoninae* consist of several clades (I with A–D and II with E–F) and are a monophyletic sister group to the *Clematidinae* within the *Anemoneae*. It is remarkable that the major phylogenetic branching of *Anemoninae* (I/II), so far overlooked by morphologists, has been foreshadowed by the karyological differentiation into groups with x = 8 and x = 7 chromosomes (Kurita 1958), corresponding to two major karyotypes (Baumberger 1970). All seemingly contradictory chromosome counts in this and other publications are obviously due to erroneous identifications, erroneous counts or both. Furthermore, data by Rothfels & al. (1966) suggest that the nuclear DNA content is higher in clade II compared to clade I.

On the basis of the data provided by Hoot & al. (1994), available literature references, and personal studies, a proposal for an improved and hopefully more "natural" (i.e. phylogenetically orientated) but still very provisional arrangement of the taxa within *Anemoninae* is presented in Table 1.

Discrepancies of this hierarchical arrangement with earlier systems are obvious and have been stressed already by Hoot & al. (1994) and Hoot (1995, this volume), particularly when compared to Ulbrich (1905/06). Whereas most of the present species groups are not in conflict with this classical concept, as many as 15 of the 40 supraspecific taxa recognized by Ulbrich (1905/06) today have to be considered as polyphyletic and only superficially similar (e.g., his sect. *Anemonanthea*, including taxa of the groups I D: m 37–43 and II E: o 46–47, sect. *Pulsatilloides* with taxa from I B: d 12, e 13, g 15–17, and II E: p 48, etc.). Also, Ulbrich has included the usually separated genus *Hepatica* Mill., while maintaining *Pulsatilla* Mill. (I C: k 21–22, l 23–31), *Barneoudia* C. Gay (I B: j 20), *Oreithales* Schldl. (= *Capethia* Britton; I B: i 19) and *Knowltonia* Salisb. (I B: h 18). This leaves *Anemone* L. as an unbalanced and paraphyletic residue.

In his studies on *Ranunculaceae*, Tamura (1967) recognized as additional genera within *Anemoninae Eriocapitella* Nakai (corresponding to our group I A: b 3) and *Miyakea* Miyabe & Tatewaki (group I C: l 27), but in Tamura (1993) these were sunk

Table 1. Provisional survey of *Anemoninae* taxa and their possible phylogenetic hierarchy. Clades I A–D and II E–F correspond to Fig. 1. Species groups are constituted from smallest obvious monophyla; their diversity is indicated by approximate (sub)species numbers

ANEMONINAE

cpDNA affinities	(sub)genera, sections	species groups	(sub)species numbers
I A	a) RIULARDIUM	1) *Rivularis*	(3-6)
		? 2) *Sumatrana*	
	b) ANEMONOSPERMOS	3) *Vitifolia*	(4)
		3a) *Rupicola*	(2)
B	c) MERIDIUM	4) *Mexicana*	(1)
		5) *Sellowii*	(2-3)
		6) *Helleborifolia*	(1-2)
		7) *Rigida*	(2)
		8) *Hepaticifolia*	(1)
		9) *Antucensis*	(1)
		?10) *Tenuicaulis*	(1)
		11) *Crassifolia*	(1)
	d) ANEMOCLEMA	12) *Glaucifolia*	(1)
	? e) BEGONIIFOLIA	13) *Begoniifolia*	(1)
	? f) METANEMONE	14) *Ranunculoides*	(1)
	g) PULSATILLOIDES	15) *Thomsonii*	(1)
		16) *Caffra*	(2)
		17) *Capensis*	(1)
	h) KNOWLTONIA	18) *Vesicatoria*	(8)
	? i) OREITHALES	19) *Integrifolia*	(1)
	? j) BARNEOUDIA	20) *Chilensis*	(3)

C	k) *PREONANTHUS*		
	l) *PULSATILLA*	21) *Alpina*	(2-5)
		22) *Occidentalis*	(1)
		23) *Taraoi*	(1-2)
		24) *Kostyczewii*	(1)
		25) *Chinensis*	(7)
		26) *Albana*	(8-15)
		27) *Integrifolia*	(1)
		28) *Vernalis*	(1)
		29) *Patens*	(1-3)
		30) *Vulgaris*	(5-7)
		31) *Pratensis*	(3-7)
D	m) *ANEMONE*	32) *Parviflora*	(1)
		33) *Baldensis*	(5-8)
		34) *Multifida*	(5-11)
		35) *Tuberosa*	(5-7)
		36) *Coronaria*	(6-15)
		37) *Blanda*	(2-3)
		38) *Stolonifera*	(4)
		39) *Umbrosa*	(4)
		40) *Deltoidea*	(1)
		41) *Ranunculoides*	(6)
		42) *Reflexa*	(1)
		43) *Nemorosa*	(7-11)
II E	n) *ANEMONIDIUM*	44) *Dichotoma*	(2)
		45) *Richardsonii*	(1)
	o) *KEISKEA*	46) *Keiskeana*	(1)
		47) *Flaccida*	(8-13)
	p) *HOMALOCARPUS*	48) *Obtusiloba*	(5-6)
		49) *Atropurpurea*	(1)
		50) *Narcissiflora*	(6-25)
F	q) *HEPATICA*	51) *Transsilvanica*	(4)
		52) *Triloba*	(3-5)

again while accepting as new *Metanemone* W. T. Wang (I B: f 14). Tamura's (1967, 1993) infrageneric classification deviates little from Ulbrich (1905/06), but he correctly recognized the affinities of groups II E: p 48 and 49 within his sect. *Omalocarpus*, all with a monopodial rhizome.

Recently, Starodubtsev (1991) has presented a taxonomic survey of *Anemoninae*, considering fruit anatomy, chromosome numbers and other relevant characters. Following suggestions by Holub (1973) he has split *Anemone* as recognized by Ulbrich (1905/06) and Tamura (1967) in as many as eight genera: *Anemonidium* (Spach) Holub (incl. groups I A: a 1, I B: c 4–11, II E: n 44–45), *Anemone* L. s. str. (I A: b 3–3a, I D: m 32–36), *Anemonastrum* Holub (II E: p 50), *Arsenjevia* Starodub. (II E: o 47), *Tamuria* Starodub. (II E: o 46, II D: m 40), *Hepatica* Mill. (II F: q 51–52), *Anemonoides* Mill. (I D: m 37–39, 41–43), and *Pulsatilloides* (DC.) Starodub. (I B: d 12, e 13, g 15–17, II E: p 48). Inspite of this far-reaching taxonomic splitting, several of these genera still appear to be polyphyletic: in *Anemonidium* conservative groups from different basal branches (I A: a 1, I B: c 4–11, and II E: n 44–45) which share plesiomorphic features are united, whereas those included in *Anemone* s. str. (I A: b 3–3a and I D: m 32–36) share only superficial similarities such as long wooly fruit hairs, etc.

The reverse taxonomic procedure is proposed by Hoot & al. (1994), i.e. lumping all *Anemoninae* into the single genus *Anemone* s.l. This is a taxonomic procedure certainly in line with phylogenetic reasoning and available karyological, molecular and other data. These data clearly do not support the usual partial splitting proposed by Ulbrich (1905/06) and Tamura (1967) and still practiced in current floras. The only other alternative, taxonomically and phylogenetically acceptable scenario one could imagine, would be an even more excessive generic splitting beyond that advocated by Starodubtsev (1991). Nevertheless, for such a taxonomic approach the information available certainly is not yet sufficient.

The preliminary classification of *Anemone* s.l. proposed by Hoot & al. (1994: appendix 2) differs in some respect from the one presented here. Group I (with x = 8) is put before II (with x = 7), because the trend in *Anemoninae* certainly is x = 8 → 7 (Baumberger 1970). The monotypic *Metanemone* described by Wang (1980: Fl. Rep. Pop. Sin. 28: 72, 351) from SW China (Yunnan) is lacking in the survey by Hoot & al. (1994), and may best fit into clade I B. Instead of lumping such extremely divergent groups into sect. *Pulsatilloides* s.l. as I B: d 12, g 15–17, i 19 and j 20 (*Caffra* group) and of c 4–9 and h 18 (*Knowltonia* group) as proposed by Hoot & al. (1994), a more discrete placement at a higher rank is preferred here. Generally, informal "species groups" are much more narrowly circumscribed in Table 1 than in their taxonomic survey.

In respect to chromosome numbers, references for *A. thomsonii* (2n = 16: Hedberg & Hedberg 1977) and *Oreithales integrifolia* (2n = 48: Duncan & Perez 1979) substantiate their placement in clade I B: g 15 and i 19. In contrast, a reliable chromosome count for *A. tenuicaulis* (2n = 28, 4x on x = 7 ?!; Hair 1963), not considered by Hoot & al. (1994), makes the inclusion of this New Zealand endemic (and of the probably related Tasmanian *A. crassifolia*) in subg. *Anemone* sensu Hoot and our clade I B (*Meridium*) doubtful. It is remarkable that from the few chromosome counts available from S Hemisphere *Anemoninae* not only the former two but also those for several species of the African *Knowltonia* clade (2n = 48: Rasmussen 1979) and for the isolated S American *A. helleborifolia* (I B: c 6, 2n = 48: HUYNH 1965) apparently are paleopolyploid and thus suggest an ancient evolutionary status for these taxa (see also LOURTEIG 1951, 1956 a, b).

A remarkable incongruence has been documented by HOOT & al. (1994: fig. 5; tables 1, 3) between their cpDNA and morphology based cladograms. Furthermore, their selection of *Clematis* as an outgroup has far-reaching consequences for the classification of plesio- and apomorphic character states within *Anemoneae*. This procedures results in a phylogenetically incorrect affiliation of *Clematis*, *Pulsatilla* and *Pulsatilloides*, and does not reflect the x = 7/8 divergence within *Anemoninae*. In my opinion this is due to the highly advanced morphological status of *Clematis* which makes it quite unsuitable as an outgroup and starting point for character state evaluations. If the wider and obviously monophyletic assembly of *Ranunculoideae* s. str. (*Ranunculeae* + *Anemoneae*) is screened for a more suitable outgroup candidate, the genus *Trautvetteria* may prove to be a much better candidate than *Clematis*. Thus, it is suggested to accept i. a. sympodial polycorm rootstock, herbaceous (not woody) habit, compound (not simple) basal and involucral leaves, compound (not simple) inflorescences, sepal numbers of 5–8 (not 4), lack of petals, tricolpate pollen grains, and achenes with a hooked and rather short style (not more than 5 mm) and a thick wall (inner layers sclerenchymatous) as plesiomorphic character states for *Anemoninae*. This together with the many other features already used by HOOT & al. (1994) will result in more congruence between morphological and molecular cladograms.

To the other obstacles already referred to in coordinating molecular and morphological data in *Anemoninae*, we will return after a discussion of the adaptive syndromes apparent in *Anemoninae*.

Diversity and adaptive syndromes

The amount of solid data available on the diversity, eco-geographical differentiation and phylogeny of *Anemoninae* offers a basis for inquiries of a more general nature. Can we recognize certain regularities and causal links (i.e. syndromes) with respect to: (A) the diversity of vegetative, (B) reproductive and (C) cytogenetic structures and functions, the integration of this diversity into (D) the hierarchy from populations to groups of different nature and rank, and into (E) different habitats, ecological niches, and distribution areas, and (F) the historical aspects of the evolutionary processes involved?

A few examples from *Anemoninae* will illustrate these points. There are very obvious trends in this subtribe towards anemochory, either by the development of an extensive, often wooly indumentum on the surface of the achenes or the excessive postfloral elongation of hairy styles. This is linked to elongated and stiffly upright pedicels in the fruiting stage. Understandably, this trend has been favoured in taxa growing in open grassland or pioneer habitats. It is often linked to polycorm hemicryptophytes relying more on sexual than on vegetative reproduction. An opposite trend is that towards myrmecochory, linked to the development of an elaiosome from the achene basis and pedicels bending to the ground after flowering. Relevant taxa mostly grow in deciduous forests and have excessive vegetative reproduction.

Centres of taxonomic diversity often can be linked to centres of distribution areas, which sometimes reach considerable, transcontinental dimensions. The

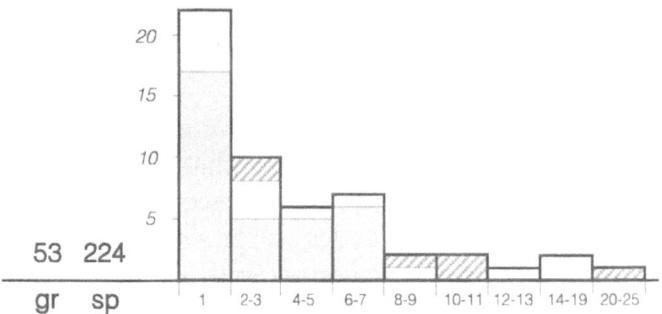

Fig. 2. Diversity spectrum of 53 species groups within *Anemoninae*: Block diagrams of frequencies of gr = groups (on vertical axis) with a maximum of 1, 2–3, 4–5 ··· 20–25 sp = (sub)species (on horizontal axis). Groups with local (regional) distribution shaded, groups with transcontinental distribution hatched. Data from Table 1

Homalocarpus clade (II E: p 48–50) is centred with its most diverse and most plesiomorphic taxa in the Himalayas and SW Chinese mountains and extends with the more apomorphic *Narcissiflora* group to the Iberian Peninsula in the W and to the Rocky Mountains in the E. The related and interconnected *Baldensis* and *Multifida* groups (I D: m 33, m 34), have their centre of diversity in N America and reach with *A. sylvestris* to W Europe and with *A. multifida* to Patagonia. As Fig. 2 shows, there is a clear correlation in *Anemoninae* between taxonomic diversity of species groups (their "critical mass") and their capacity to establish transcontinental ranges.

The structure, diploid-polyploid nature, and distribution pattern of a group often makes it possible to characterize its status relative to historical phases of evolutionary differentiation. This can be briefly illustrated by comparing late (1), middle (2) and early (3) phases in the development of three diploid-polyploid groups of *Pulsatilla* (cf. Aichele & Schwegler 1957):

1) The relatively apomorphic and predominantly 2x *Pratensis*-group exhibits gradual eco-geographical racial differentiation combined with regional overlap and hybridization. With a distribution area extending from submediterranean-pontic S to N Europe, and only local 4x-neopolyploid populations, it conforms to a relatively recent, postglacial status.

2) The *Vulgaris*-group consists of relic 2x-elements (*P. velezensis*), but otherwise is of an exclusively 4x-allopolyploid nature. In its origin 2x-members of the *Patens*- and also of the *Pratensis*-group have participated. One can recognize two series of taxa: One (a) consisting of the disjunct *P. halleri*, *P. styriaca*, *P. slavica* and *P. taurica*, ranging from the W Alps to Crimea, obviously relics of a more *P. patens*-like pre-Würm-glacial ancestor. The other (b) with a stronger *P. pratensis* influence and a much more extensive ± continuous C, E, and N European distribution, includes the closely interrelated *P. vulgaris* and *P. grandis*, clearly resulting from postglacial differentiation/hybridization processes.

3) The *Taraoi*-group morphologically links the subgenera *Preonanthus* and *Pulsatilla* and is made up of isolated 4x-paleopolyploid populations, ranging from the Japanese Alps (*P. nipponica*) to the Kurile Islands (*P. taraoi*). This relic and

NEMOROSA-TYPE

(species groups 38-43, 46-47)

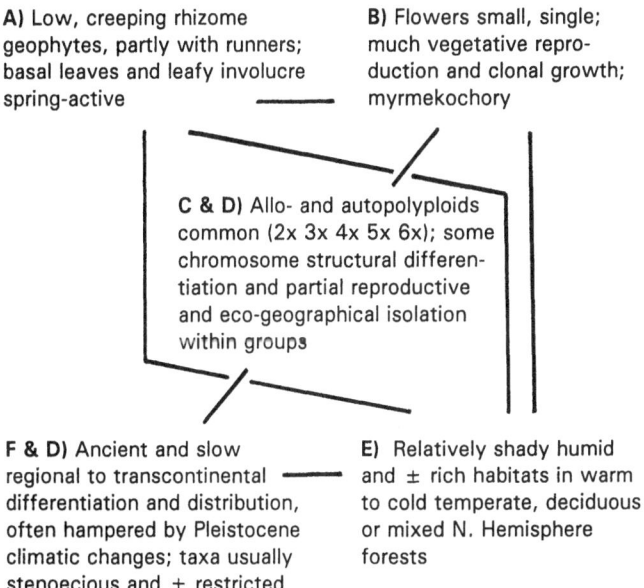

A) Low, creeping rhizome geophytes, partly with runners; basal leaves and leafy involucre spring-active

B) Flowers small, single; much vegetative reproduction and clonal growth; myrmekochory

C & D) Allo- and autopolyploids common (2x 3x 4x 5x 6x); some chromosome structural differentiation and partial reproductive and eco-geographical isolation within groups

F & D) Ancient and slow regional to transcontinental differentiation and distribution, often hampered by Pleistocene climatic changes; taxa usually stenoecious and ± restricted

E) Relatively shady humid and ± rich habitats in warm to cold temperate, deciduous or mixed N. Hemisphere forests

Fig. 3. Adaptive syndrome of the *Nemorosa*-type (from *Anemone nemorosa*) within *Anemoninae*. Aspects of diversity and differentiation are grouped in domains: A) vegetative, B) reproductive, C) cytogenetic, D) group-organizational, E) eco-geographical, F) historical; lines suggest some of the causal links connecting these domains. Species groups as in Table 1. Further explanations in the text

relatively plesiomorphic group apparently represents the most ancient pattern among the three *Pulsatilla* groups compared.

The premises discussed justify an attempt to characterize and typify adaptive syndroms within *Anemoninae* in a preliminary way. Following the domains characterized under (A)–(F) on p. 287, two of these types recognized are described in some detail and confronted in Figs. 3 and 4. Species groups included under one type are often not closely related; numbers refer to Table 1. Connecting lines between the domains should illustrate some of the causal links (functional correlations and interdependencies) which underlie these adaptive syndromes.

In the *Nemorosa*-type it is obvious that the geophytic habit, leafy involucre and reduced inflorescences are linked to a rapid development during the short spring period with its good light conditions in summer-green forests. Myrmekochory is advantageous in these forests because of the abundant ant fauna. Clonal growth – also a common trait among herbaceous forest taxa – is a prerequisite for the common occurrence and maintainance of chromosome structural differentiation and polyploidy (including anorthoploidy: 3x, 5x!), and for the usually clear isolation between species, even when their populations overlap. Many taxa belonging to the *Nemorosa*-type have evolved as elements of temperate deciduous forests, strongly exposed to the extreme climatic changes of the late Tertiary and Pleistocene, and thus reflect their ancient status and slow divergence by stenoecy and chorological

ALPINA-TYPE

(species groups 3a, 14-15, 19, 21-23, 27-28, 33, 48-50)

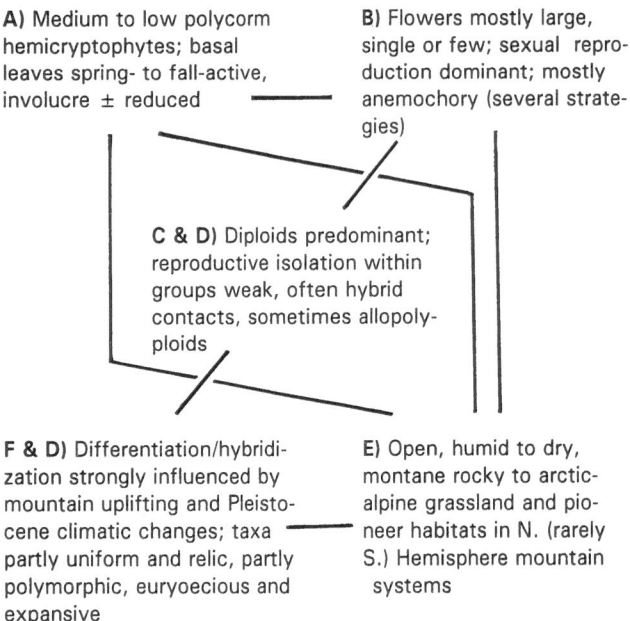

A) Medium to low polycorm hemicryptophytes; basal leaves spring- to fall-active, involucre ± reduced

B) Flowers mostly large, single or few; sexual reproduction dominant; mostly anemochory (several strategies)

C & D) Diploids predominant; reproductive isolation within groups weak, often hybrid contacts, sometimes allopolyploids

F & D) Differentiation/hybridization strongly influenced by mountain uplifting and Pleistocene climatic changes; taxa partly uniform and relic, partly polymorphic, euryoecious and expansive

E) Open, humid to dry, montane rocky to arctic-alpine grassland and pioneer habitats in N. (rarely S.) Hemisphere mountain systems

Fig. 4. Adaptive syndrome of the *Alpina*-type (from *Anemone = Pulsatilla alpina*) within *Anemoninae*. Otherwise as in Fig. 3, and with further explanations in the text

restrictions (e.g., the disjunct *A. trifolia* with scattered eastern N American and European populations).

Members of the *Alpina*-type (Fig. 4) of *Anemoninae* exhibit a strongly contrasting syndrome. They are rather slowly developing hemicryptophytic perennials with a sympodial or monopodial polycorm and medium to short flowering shoots, often with ± reduced and sometimes completely eliminated involucre. Basal leaves are active throughout the abbreviated vegetation period in higher mountains. The predominantly sexual reproduction is linked to large and often single flowers and the production of numerous achenes dispersed by anemochory, corresponding to open and wind-exposed pioneer or grassland mountain habitats. Crossing barriers between related taxa are weak, the differentiation is predominantly allopatric and on the 2x-level. As elements of the N Hemisphere mountain floras, several taxa of the *Alpina*-type were favoured by the cold periods of the Pleistocene and have reached extensive areas (e.g., the *Narcissiflora*-group), whereas others have become restricted and disjunct (e.g., the *Baldensis*-group or *Metanemone*).

Table 2 gives a survey of the 10 types of adaptive syndromes so far recognized within *Anemoninae*, including those (5, 9) already discussed. The *Rivularis*-type (1) is here regarded as the most plesiomorphic (see p. 287), characterized by a habit and growth form reminiscent of *Trautvetteria* in the *Ranunculeae*, unspecialized inflorescences, polychorous achenes, and ± open, humid and rich (sub)tropical montane to warm temperate habitats. All the other adaptive syndrome types of *Anemoninae* can be linked with this *Rivularis*-type, and can be imagined to have originated from

Table 2. Provisional survey of types 1)–10) of adaptive syndromes in *Anemoninae*. Numbers in brackets refer to species groups listed in Table 1. Some species groups hold intermediate positions: e.g., 8 between types 1) and 3), 37 between types 5) and 8); some of the types are sometimes difficult to seperate: e.g., 7) and 9). Further explanations in the text

---

1) *Rivularis*-type: tall, leafy mesophilic hemicryptophytes of ± open, humid and rich (sub)tropical montane habitats; multiflorous, polychory (1–2, 4–6, 13).

2) *Knowltonia*-type: tall, ± xerophilic hemicrypto- to chamaephytes of ± open and dry, mediterranoid S. Hemisphere habitats; polychory to endozoo-chory (7, 16–18).

3) *Crassifolia*-type: low, meso- to hygrophilic hemicryptophytes of subantarctic habitats; polychory (9–11).

4) *Dichotoma*-type: tall, leafy mesophilic root-sprouting perennials of temperate to boreal alluvial N Hemisphere habitats; multiflorous, hydrochory (44).

5) *Nemorosa*-type: low, spring-active, mesophilic creeping rhizome hemicrypto- to geophytes of ± shady and rich deciduous forests; pauciflorous, often myrmekochory (38–43, 46–47).

6) *Hepatica*-type: low, ± winter- and summer active monopodial rhizome hemicryptophytes of warm temperate deciduous forests; pauciflorous, myrmekochory (51–52).

7) *Virginiana*-type: Tall to medium sized, meso- to xerophilic, spring-to fall active, rhizome to polycorm hemicryptophytes of temperate open woodland, grassland and rocky habitats; anemochory (3, 12, 24–26, 29–31, 34).

8) *Tuberosa*-type: low, ± xerophilic and spring active, tuber-forming geophytes of open and dry, ± mediterranoid woodland to rocky habitats, N and S Hemisphere; anemochory (35–36, ?20).

9) *Alpina*-type: ± low, polycorm hemicryptophytes of ± open rocky montane to artic grassland and pioneer habitats, predominantly in N Hemisphere mountain systems; pauciflorous, anemochory (3a, 14–15, 19, 21–23, 27–28, 33, 48–50).

10) *Richardsonii*-type: low, creeping and vegetatively expanding rhizome hemicryptophytes of N Hemisphere arctic humid and snowbed habitats; polychory or anemochory (32, 45).

---

relevant ancestors. Such an evolution would have involved migrations along the classical N → S Hemisphere routes (without resorting to Gondwanland!) and divergent eco-geographical radiations into the very different habitats occupied today by members of the subtribe. Terminal points of such a radiation certainly are the groups belonging to the *Nemorosa-*, *Hepatica-*, *Tuberosa-*, *Alpina-* and *Richardsonii*-types, the latter consisting of two arctic snowbed and pioneer species, *A. parviflora* (I D: m 32) and *A. richardsonii* (II E: h 45) with remarkably similar habit but widely separate phylogeny.

Comparing eco-geographical radiation and growth form differentiation in *Anemoninae* with that in related clades of *Ranunculaceae*, we see limitations of genetic and evolutionary potentials: There are no short-lived annuals as in *Ranunculeae*, no secondarily woody lianas as in *Clematidinae*, etc. In contrast to this, most of the adaptive syndromes which were realized in *Anemoninae* (Table 2) evidently have originated not only once but independently two or more times by convergent evolution (as is obvious by comparison with Table 1). We can assume that this is due to the canalizing effect of similar environments past and present, operating on already separated but related groups, having similar genetic potentials. This has resulted in a limited number of adaptive types, often polyphyletic and with similar syndroms of structure, function, habitat, and history.

We can now return to the various difficulties in matching morphological and molecular data in *Anemoninae* systematics. First, there are those conservative elements of divergent clades which still share dominant plesiomorphic features but lack obvious apomorphies: good examples are paraphyletic taxa as sect. *Rivularidium* sensu ULBRICH (1905/06) or *Anemonidium* sensu STARODUBTSEV (1991) which include groups today placed into a "basal" position (e.g. I A: a 1, I B: c 4–11, II E: n 44).

Secondly, conspicuous morphological differences may be due to relative simple changes in major genes as recently pointed out by KADEREIT (1994). Such changes could have been responsible for the dramatic difference in achene structure (baccate versus dry) which has led to the generic separation of *Knowltonia* (I B: h 18), but which is not paralleled by major changes in cpDNA when obviously related taxa of *Anemone* s. str. (e.g., I B: g 16 or I B: c 11) are compared. Such "simple" changes of major genes may also occur easily in a parallel fashion, thus simulating phyletic affinities between clades shown to be far apart by cpDNA data. Examples from *Anemoninae* are the dense wooly achene indumentum which has prompted the artifical lumping of *Anemospermos* (I A: b 3) and certain groups of *Anemone* s. str. (I D: m 35–36) in the systems of ULBRICH (1905/06) and STARODUBTSEV (1991), or the assumption of phyletic affinities between *Clematis* and *Pulsatilla* because of their long and hairy styles.

Finally, there are the many cases of morphological homoplasies, already clearly recognized by HOOT & al. (1994), which are linked to convergent eco-geographic radiation and the origin of polyphyletic adaptive types, formerly often mistaken as "natural taxa", e.g., sect. *Anemonanthea* sensu ULBRICH (1905/06), now recognized as the *Nemorosa*-type, but including groups from very different clades (I D: m 37–43 and II E: o 46–47).

In retrospect: Evolutionary analyses of plant groups like *Anemoninae* should be as broad and multidisciplinary as possible. This applies particularly to the vegetative, reproductive, cytogenetic and molecular, group-organizational, eco-geographical, and historical aspects of diversity and differentiation. Such an approach will reconcile apparently conflicting informations (e.g., morphological and molecular), reveal various causal links between these various aspects, and make the evolutionary shaping of such plant groups in space and time better understood.

I gratefully acknowledge constructive discussions with SARA B. HOOT and technical assistance by VERONIKA MAYER.

References

AICHELE, D., SCWEGLER, H.-W., 1957: Die Taxonomie der Gattung *Pulsatilla*. – Feddes Repert. **60**: 1–230.

BAUMBERGER, H., 1970: Chromosomenzahlbestimmungen und Karyotypanalysen bei den Gattungen *Anemone*, *Hepatica* und *Pulsatilla*. – Ber. Schweiz. Bot. Ges. **80**: 17–96.

DUNCAN, T., PEREZ, P., 1979: Chromosomes of *Oreithales* (*Ranunculaceae*). – Amer. J. Bot. **66**: 989–990.

EHRENDORFER, F., 1988: Stability versus change, or how to explain evolution. – In GREUTER,

W., ZIMMER, B., (Eds): Proceedings of the XIV Intern. Bot. Congress, pp. 317–333. – Königstein/Taunus: Koeltz.

HAIR, J. B., 1963: Contributions to a chromosome atlas of the New Zealand flora, 6. – New Zealand J. Bot. **1**: 243–257.

HEDBERG, I., HEDBERG, O., 1977: Chromosome numbers of afroalpine and afromontane angiosperms. – Bot. Not. **130**: 1–24.

HOLUB, J., 1973: New names in *Phanerogamae*, 2. – Folia Geobot. Phytotax. **8**: 155–179.

HOOT, S. B., 1995: Phylogenetic relationships in *Anemone* (*Ranunculaceae*) based on DNA restriction site variation and morphology. – Pl. Syst. Evol. [Suppl.] **9**: 295–300.

– REZNICEK, A. A., PALMER, J. D., 1994: Phylogenetic relationships in *Anemone* (*Ranunculaceae*) based on morphology and chloroplast DNA. – Syst. Bot. **19**: 169–200.

HUYNH, K. L., 1965: Contribution a l'étude carylogique et embryologique des Phanérogames du Pérou. – Mém. Soc. Helvét. Sci. Nat. **85**: 1–178.

KADEREIT, J. W., 1994: Molecules and morphology, phylogenetics and genetics. – Bot. Acta **107**: 369–373.

KURITA, M., 1958: Chromosome studies in *Ranunculaceae* VIII. – Karyotype and phylogeny. – Rep. Bio. Inst., Ehime Univ. **5**: 1–14.

LOURTEIG, A., 1951: Ranunculaceas de Sudamerica Templada. – Darwiniana **9**: 397–608.

– 1956a: Ranunculaceas de Sudamerica Tropical. – Mem. Soc. Cie. Nat. La Salle **16**: 19–88, 125–223.

– 1956b: Distribution géographique des Renonculacées en Amérique du Sud. Compt. Rend. Soc. Biogéographie (Paris) **33**: 56–69.

MANEN, J. -F., NATALI, A., EHRENDORFER, F., 1994: Phylogeny of *Rubiaceae – Rubieae* inferred from the sequence of a cpDNA intergene region. – Pl. Syst. Evol. **190**: 195–211.

RASMUSSEN, H., 1979: The genus *Knowltonia* (*Ranunculaceae*). – Opera Bot. **53**: 1–44.

ROTHFELS, K., SEXSMITH, E., HEIMBURGER, M., KRAUSE, M. O., 1966: Chromosome size and DNA content of species of *Anemone* L. and related genera (*Ranunculaceae*). – Chromosoma (Berlin) **20**: 54–74.

SCHARFETTER, R., 1953: Biographien von Pflanzensippen. – Wien: Springer.

STARODUBTSEV, V. N., 1991: Wetrenitsy: sistematika i evoljutsija. – Leningrad: Akad. Nauk SSSR.

TAMURA, M., 1967: Morphology, ecology and phylogeny of the *Ranunculaceae* VII. – Sci. Rep. Osaka Univ. **16**: 21–43.

– 1993: *Ranunculaceae*. – In KUBITZKI, K., ROHWER, J. G., BITTRICH, V., (Eds): The families and genera of vascular plants, **2**, pp. 563–583. – Berlin, Heidelberg, etc.: Springer.

ULBRICH, E., 1905/06: Über die systematische Gliederung und geographische Verbreitung der Gattung *Anemone* L. Bot. Jahrb. Syst. **37**: 172–334.

Address of the author: Univ.-Prof. Dr F. EHRENDORFER, Institute of Botany, University of Vienna, Rennweg 14, A-1030 Wien, Austria.

Accepted March 9, 1995

Pl. Syst. Evol. [Suppl.] 9: 295–300 (1995)

Phylogenetic relationships in *Anemone (Ranunculaceae)* based on DNA restriction site variation and morphology

SARA B. HOOT

Received December 19, 1994

Key words: *Anemone, Ranunculaceae.* – Chloroplast DNA, restriction site variation, biogeography.

Abstract: Phylogenetic relationships of 36 species of *Anemone* (*Ranunculaceae*) and seven related genera were explored with a combination of three independent data sets: chloroplast DNA restriction sites, nuclear ribosomal DNA restriction fragments, and morphological/ cytological variation, using *Clematis* as the outgroup. In the chloroplast DNA work, 245 phylogenetically informative restriction sites were identified using ten restriction enzymes. The same ten enzymes were used to produce the nuclear ribosomal DNA restriction fragments and resulted in 16 informative characters. The conventional data set consisted of 27 characters for the same taxa used in the molecular work. The phylogeny derived from the combined data sets demonstrates that few sections of *Anemone* as established by previous classifications are monophyletic and that *Pulsatilla*, *Hepatica*, and *Knowltonia* should be subsumed within *Anemone*. Several unexpected relationships within the *Anemone* complex are discussed and hypotheses are offered to explain the interesting geographic distributions found in the genus.

Anemone L. s. str. consist of 70–90 species of perennial, low-growing herbs. The genus is characterized by a rosette of basal leaves, an involucrate peduncle bearing a single flower or compound inflorescence, a perianth of colorful sepals, and an achene fruit. There is considerable morphological variation within the genus. The perennating structures are upright stems, rhizomes, or tubers. The basal leaves vary from trilobed to several times ternate, rarely pinnate. The involucral leaves are various in number and form, sometimes reduced and bract-like. The achenes are glabrous to tomentose, few to many in number, with various style morphologies. Base chromosome numbers are either 7 or 8. Species are most numerous in the Northern Hemisphere, but also occur in mountainous and cooler regions of the Southern Hemisphere. They inhabit a wide range of habitats including arctic and alpine tundra, prairie, woodlands, Mediterranean regions, and semidesert.

Several previous classifications placed *Anemone* and *Clematis* in the same tribe, Anemoneae (LANGLET 1932, TAMURA 1967). Recent phylogenetic analyses based on molecular data also place *Anemone* and other closely related genera (*Pulsatilla*,

Knowltonia, and *Hepatica*) in a clade with *Clematis* (Johansson & Jansen 1993; Johansson 1995, this volume; Hoot 1995, this volume). The tribe *Anemoneae* are often found in a clade with the tribe *Ranunculeae* (Johansson & Jansen 1993; Johansson 1995, this volume; Hoot 1995, this volume).

The delimitation of *Anemone* has been debated for years. Early classifications often included the genera *Hepatica*, *Pulsatilla*, *Knowltonia*, *Barneoudia*, and *Oreithales* within *Anemone* (Janczewski 1892, Prantl 1891). Based mainly on differences in achene morphology, recently cited classifications exclude most of these genera (Ulbrich 1905/06; Tamura 1967, 1993).

Recently, evolutionary relationships within the *Anemone* complex (*Anemone*, *Hepatica*, *Pulsatilla*, *Knowltonia*, *Barneoudia*, and *Oreithales*) were reevaluated using restriction site variation and morphological data (Hoot & al. 1994). The reliance of past classifications on one set of characters, achene morphology, emphasized the need for this reexamination of the genus. The following presents a summary of this past work, including the phylogenetic tree based on a combination of all the data. The resulting cladogram is compared to previous classifications and questions related to biogeography are addressed.

Material and methods

Chloroplast restriction site maps for ten enzymes were constructed for each of 36 species of the *Anemone* complex and such potential outgroups as *Ranunculus* and *Clematis* (Hoot & al. 1994). The sampling included at least one species from all *Anemone* sections and subsections except for subsection *Stolonifera* (four species). To improve resolution of some species complexes and explore the potential for hybridization among species, rDNA restriction fragments were scored. The same species of *Anemone* and related genera sampled for the molecular work were scored also for 27

Fig. 1. One of the 12 shortest trees resulting from a combination of all data: chloroplast and ribosomal restriction site variation and morphology. *Clematis* was used as the outgroup. Length of each shortest tree = 667 steps, CI excluding autapomorphies = 0.56, RI = 0.85. Dotted lines indicate branches which collapsed in the strict consensus tree derived from the 12 shortest trees. Numbers above branches indicate number of characters supporting each branch. Numbers in parentheses indicate bootstrap values for 100 replications. Numbers in italics below each branch are decay indices. Numbers and symbols after taxa acronyms are base chromosome numbers and sectional designations of Ulbrich (1905/06). Brackets indicate geographical distribution. Acronyms for taxa are as follows: COCHR *Clematis ochroleuca*, CHEXA *C. hexapetala*, HAMER *Hepatica americana*, ARICH *Anemone richardsonii*, ACANA *A. canadensis*, ADICH *A. dichotoma*, AKEIS *A. keiskeana*, AFLAC *A. flaccida*, ADEMI *A. demissa*, ANARC *A. narcissiflora*, AOBTU *A. obtusiloba*, ATRUL *A. trullifolia*, ARIVU *A. rivularis*, ANUPI *A. rupicola*, AVITI *A. vitifolia*, AHUPH *A. hupehensis*, ATOME *A. tomentosa*, ACRAS *A. crassifolia*, KVESC *Knowltonia vesicatoria*, POCCI *Pulsatilla occidentalis*, PHALL *P. halleri*, PVULG *P. vulgaris*, PPATE *P. patens*, ARANU *Anemone ranunculoides*, AQUIN *A. quinquefolia*, ANEMO *A. nemorosa*, ADRUM *A. drummondii*, ALITH *A. lithophila*, APARV *A. parviflora*, AMULT *A. multifida*, ASYLV *A. sylvestris*, ARIPA *A. riparia*, AVIRG *A. virginiana*, ACYLI *A. cylindrica*, ABLAN *A. blanda*, ACORO *A. coronaria*, AHORT *A. hortensis*, AFULG *A. fulgens*, APAVO *A. pavonina*, ACARO *A. carolinana*, ATUBE *A. tuberosa*, ABERL *A. berlandieri*, AEDWA *A. edwardsiana*

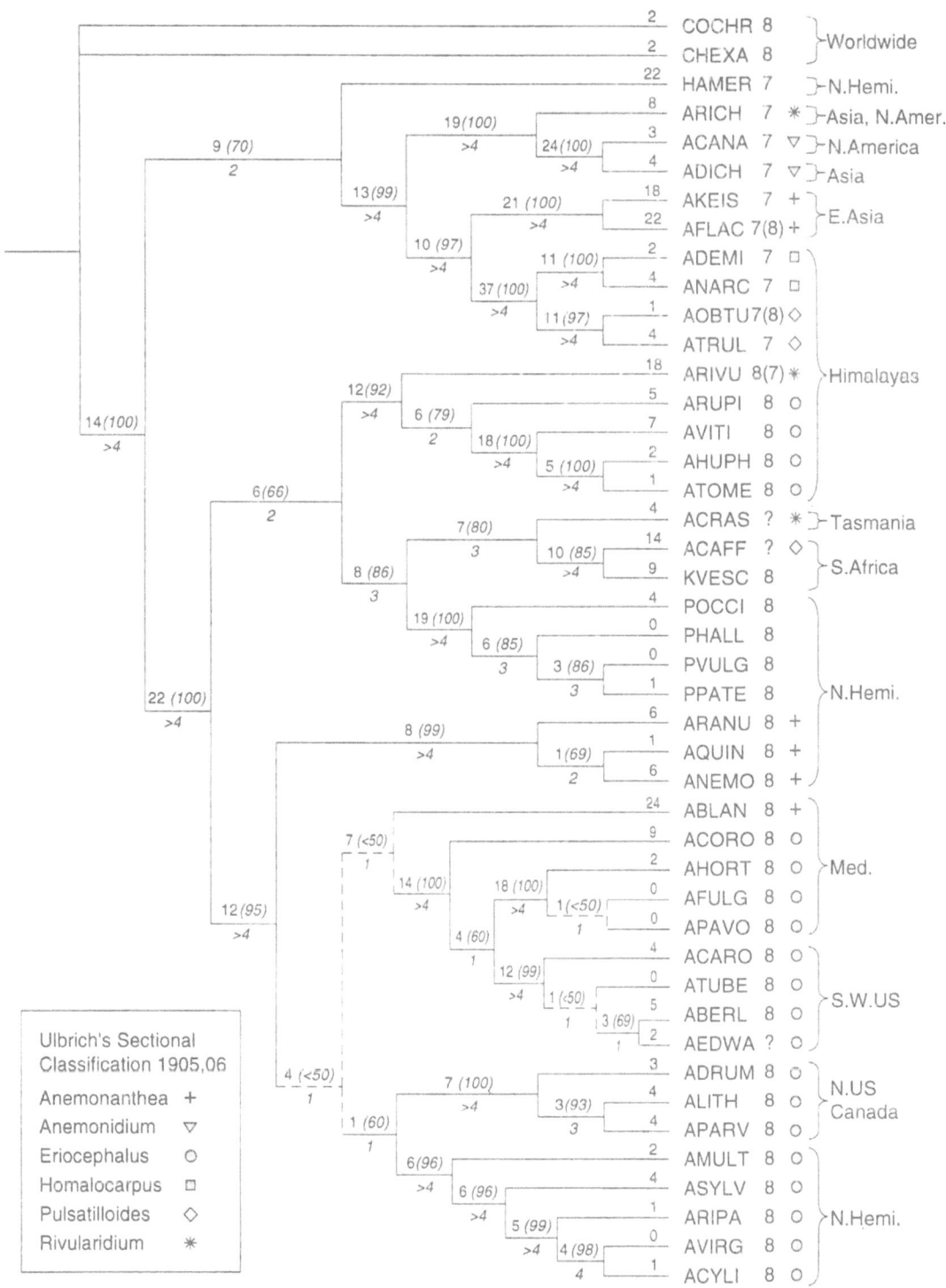

Ulbrich's Sectional
Classification 1905,06

Anemonanthea   +
Anemonidium   ▽
Eriocephalus   ○
Homalocarpus   □
Pulsatilloides   ◇
Rivularidium   ✳

morphological characters. Each data set was analysed separately (results not reported here, see Hoot & al. 1994). In addition, the three data sets from chloroplast and ribosomal DNA restriction sites and morphology were combined for an analysis based on the total evidence, using *Clematis* as the outgroup. Details of the materials and methods used in the acquisition of the restriction site and morphological data, selection of an appropriate outgroup, and subsequent data analyses can be found in Hoot & al. (1994).

Results

The matrix resulting from a combination of all the data consisted of 378 characters, 277 of which were informative. 12 shortest trees were obtained with a tree length of 667 steps; CI excluding autapomorphies was 0.56, and RI was 0.85. One of the 12 shortest trees is presented in Fig. 1. The branches with dotted lines indicate those nodes which collapse in the strict consensus tree. This tree is congruent with base chromosomes number, placing taxa with a base chromosome number of seven basal to all other members of the *Anemone* complex. Genera (*Hepatica, Knowltonia,* and *Pulsatilla*) previously placed within or closely allied with *Anemone* are found nested within the genus. Of the six sections defined by Ulbrich (1905/06), only the two small subsections *Anemonidium* and *Homalocarpus* are monophyletic (Fig. 1).

Discussion

Phylogeny. Many of the terminal clades resulting from separate analyses of the cpDNA restriction site and morphological data sets were similar to each other (trees not presented, see Hoot & al. 1994). Among the clades found by both analyses were new species alliances not recognized in previous classifications. The tree resulting from cpDNA restriction sites differed substantially from the tree based on morphological data in the degree of resolution and basal branching patterns. For example, in the cpDNA tree, the four *Pulsatilla* species are deeply nested within *Anemone* and strongly linked with their sister clade *Pulsatilloides*. In the morphological tree, however, *Pulsatilla* is next to the outgroup (in the case of one species, undifferentiated from the outgroup) and basal to all other taxa.

Some of the basal associations suggested by the tree resulting from a combination of all the data (and the cpDNA restriction site tree) were very different from both the morphological tree and past classifications (Fig. 1). These new alliances are less surprising if the traditional reliance on achene morphology is discarded. Other characters, such as base chromosome number (7 or 8), are entirely consistent with the basal split as defined by the tree (Fig. 1).

The clade consisting of taxa with x = 7 in the cpDNA tree shows some new associations (Fig. 1). One of the most striking of these is the inclusion of *A. richardsonii* Hook. f. in a clade with the two sister species, *A. canadensis* L. and *A. dichotoma* L. These are very different morphologically and had been placed in separate sections. *Anemone richardsonii* is a low-growing, rhizomatous plant, found in alpine or arctic tundra. It has glabrous achenes with long, hooked styles. *Anemone canadensis* and *A. dichotoma* are taller species found in shady, temperate habitats in which vegetative reproduction is by means of root budding. The achenes possess straight styles and wide wings. In addition to the 19 cpDNA restriction sites uniting this clade (bootstrap value 100%), these species share two unique inversions (HOOT & PALMER 1994).

The large clade consisting of taxa with base chromosome numbers of x = 8 is subdivided into two groups (Fig. 1). The first of these, including taxa segregated as *Pulsatilla* and *Knowltonia* as well as various anemones, is morphologically and geographically diverse.

The second group with x = 8 is a diverse assemblage of plants found mainly in the Northern Hemisphere (Fig. 1). These include the tuberous species (the clade consisting of *A. blanda* SCHOTT & KOTSCHY to *A. edwardsiana* Tharp), disjunct between the Mediterranean region and North and South America. These species had previously been placed in different sections and subsections (Fig. 1), but many morphological and molecular characters unite them. *Anemone blanda* is weakly associated with this clade, but geographical distribution (Mediterranean), the presence of tubers, and floral morphology (numerous linear sepals) are important characters supporting its inclusion in this tuberous clade.

The alliance of *A. quinquefolia* L., *A. nemorosa* L., and *A. ranunculoides* L., previously considered most closely related to *A. keiskeana* Ito and *A. flaccida* F. SCHMIDT, with the large clade delimited by *A. ranunculoides* and *A. cylindrica* A. GRAY (included in ULBRICH's sect. *Eriocephalus*) may never have been detected without the help of molecular data (Fig. 1). They differ markedly in morphology from other members of sect. *Eriocephalus*. They are delicate spring ephemerals found in woodlands and are characterized by few (<40/head), essentially glabrous achenes. The anemones in sect. *Eriocephalus* are found in sunny or semishady open environments and have heads with numerous wooly achenes (> 140/head). Thus the differences in achene vesture within this clade are correlated with differing environmental pressures and do not appear to be phylogenetically informative at the sectional level.

Phytogeographical implications. The phylogenetic results from a combination of molecular and morphological data for the *Anemone* complex have a geographical component (especially in regard to the terminal clades; Fig. 1) which raises some interesting questions related to past geographical distributions. For example, the analysis based on all the evidence places a Tasmanian species, *Anemone crassifolia* HOOK. f., in the same clade as the South African *A. caffra* (ECKL. & ZEYH.) Harv. and *Knowltonia vesicatoria* (L. f.) SIMS (Fig. 1). This clade is supported by seven characters, a bootstrap value of 80%, and a decay value of 2. Two explanations are possible for this distribution pattern: long-distance dispersal or the presence of the ancestors of *Anemone* throughout Gondwana when the various austral continental land masses were relatively contiguous. A more detailed discussion of these two alternate hypotheses can be found in HOOT & al. (1994).

The tuberous anemones, found in the clade delimited by *A. blanda* to *A. edwardsiana*, are disjunct among the Mediterranean region, south-western and south-central United States, and southern South America. The relatively basal position of the Mediterranean species, *A. blanda* and *A. coronaria* L., suggest a Mediterranean center of origin (Fig. 1). There are at least four possible explanations for this distribution: a former Gondwanan distribution, a spread across Europe to America when these continents were contiguous in the Eocene, dispersal across Asia to America via the Bering land bridge at this same time, or long-distance dispersal.

The above hypotheses will be tested with DNA restriction site and morphological data from additional, extant, tuberous species.

A remarkable case of disjunction is found in *A. multifida* POIR. This species was collected from six widely separated sites: Argentina, the Yukon, the mountains of Washington, Montana, Wyoming, and sand dunes in Michigan (not shown in Fig. 1, see HOOT & al. 1994). There were no restriction site differences in the cpDNA and only one restriction site difference found in the rDNA (collected from the Wyoming site). In addition, they are similar morphologically. This suggests a recent dispersal event, perhaps during the Pleistocene. To explain such disjunctions, RAVEN & AXELROD (1974) proposed a pathway along mountain ranges and mountaintops into South America during the generally cooler climate of that time.

I am grateful to Dr A. A. REZNICEK and Dr J. D. PALMER, for their valuable insights, comments, and assistance in the writing of the original paper on this topic. I thank the HORACE H. RACKHAM School of Graduate Studies for facilitating this research with a Rackham Predoctoral Fellowship awarded in 1989–1990. This work was partially supported by NSF grant BSR-8717600 to J. D. PALMER.

References

HOOT, S. B., 1995: Phylogeny of the *Ranunculaceae* based on *atp*B, *rbc*L, and 18S nuclear ribosomal DNA sequence data. – Pl. Syst. Evol. [Suppl.] **9**: 241–251.
– PALMER, J. D., 1994: Structural rearrangements, including parallel inversions, within the chloroplast genome of *Anemone* and related genera. – J. Mol. Evol **38**: 274–281.
– REZNICEK, A. A., PALMER, J. D., 1994: Phylogenetic relationships in *Anemone* (*Ranunculaceae*) based on morphology and chloroplast DNA. – Syst. Bot. **19**: 169–200.
JANCZEWSKI, M. E. DE, 1892: Études morphologiques sur le genre *Anemone* L. – Rev. Gén. Bot. **4**: 241–258.
JOHANSSON, J. T., 1995: A revised chloroplast DNA phylogeny of the *Ranunculaceae*. – Pl. Syst. Evol. [Suppl.] **9**: 253–261.
– JANSEN, R. K., 1993: Chloroplast DNA variation and phylogeny of the *Ranunculaceae*. – Pl. Syst. Evol. **187**: 29–49.
LANGLET, O., 1932: Über Chromosomenverhältnisse und Systematik der *Ranunculaceae*. – Svensk Bot. Tidskr. **26**: 381–401.
PRANTL, K., 1891: *Ranunculaceae*. – In ENGLER, A., PRANTL, K., (Eds): Die Natürlichen Pflanzenfamilien; 3, 2, pp. 43–66. – Leipzing: Engelmann.
RAVEN, P. H., AXELROD, D. I., 1974: Angiosperm biogeography and past continental movements. – Ann. Missouri Bot. Gard. **61**: 539–673.
TAMURA, M., 1967: Morphology, ecology and phylogeny of the *Ranunculaceae* VII. – Sci. Reports, Osaka Univ. **16**: 21–43.
– 1993: *Ranunculaceae*. – In KUBITZKI, K., ROHWER, J. G., BITTRICH, V., (Eds): The families and genera of vascular plants, VII, pp. 563–583. – Berlin: Springer.
ULBRICH, E., 1905/06: Über die systematische Gliederung und geographische Verbreitung der Gattung *Anemone* L. – Bot. Jahrb. Syst., Pflanzengeschichte Pflanzengeographie **37**: 172–334.

Address of the author: Dr SARA B. HOOT, Department of Biological Sciences, Lapham Hall, PO Box 413, University of Wisconsin, Milwaukee, WI 53201, USA. – email: hoot @ csd.uwm.edu, USA.

Accepted February 6, 1995

Pl. Syst. Evol. [Suppl.] 9: 301–303 (1995)

Cloning, sequencing, and phylogenetic analysis of a legumin cDNA of *Hepatica nobilis* (*Ranunculaceae*)

J. LANG and H. FISCHER

Received November 28, 1994

Key words: *Ranunculaceae*. – Legumin, molecular phylogeny.

In plant phylogenetic research, nuclear protein-encoding genes are increasingly employed. We use legumin (11–13S seed protein) gene sequences to infer phylogenies in putative basal angiosperm groups; one of the questions which we consider is the position of the *Ranunculales* which are included in the *Magnoliidae* (e.g., CRONQUIST 1988) by some authors but give name to a separate subclass *Ranunculidae* according to others (e.g., TAKHTAJAN 1980).

We have cloned and sequenced a legumin cDNA of *Hepatica nobilis* MILL. The corresponding cDNA library was constructed using mRNA of maturing *Hepatica* seeds approx. 30 days after flowering; as a probe, we used an oligonucleotide which was designed according to a *Hepatica* legumin partial sequence.

Figure 1 shows the coding nucleotide sequence of a *Hepatica* legumin cDNA. It was included in a gene tree (Fig. 2) comprising representatives of several angiosperm groups as well as some gymnosperms as an outgroup. It shows *Hepatica* clustering with *Magnolia*, both forming the sister group to the monocot representatives, whereas the remaining dicots appear as a monophyletic group for themselves. This topology remains unchanged even if sequences of more monocot and eudicot species are added. The results concerning the position of the *Ranunculaceae Hepatica* relative to *Magnolia* support the broader definition of the *Magnoliidae* and are similar to some of the analyses of rDNA data (HAMBY & ZIMMER 1992). They are in contrast, though, to the *rbcL* data derived phylogenetic analyses (CHASE & al. 1993) which place the *Ranunculidae* at the base of the Eudicots, not on the *Magnoliidae*/Monocot branch, thus advocating a natural subclass *Ranunculidae* and the tricolpate pollen condition as a possible autapomorphy of the Eudicots.

Regarding the small sample of relevant species, our results are still preliminary; some more *magnoliidae*, *Ranunculidae* and Eudicot species will be studied. Moreover, as legumins are coded by a small gene family, the data set will be extended to more genes per species.

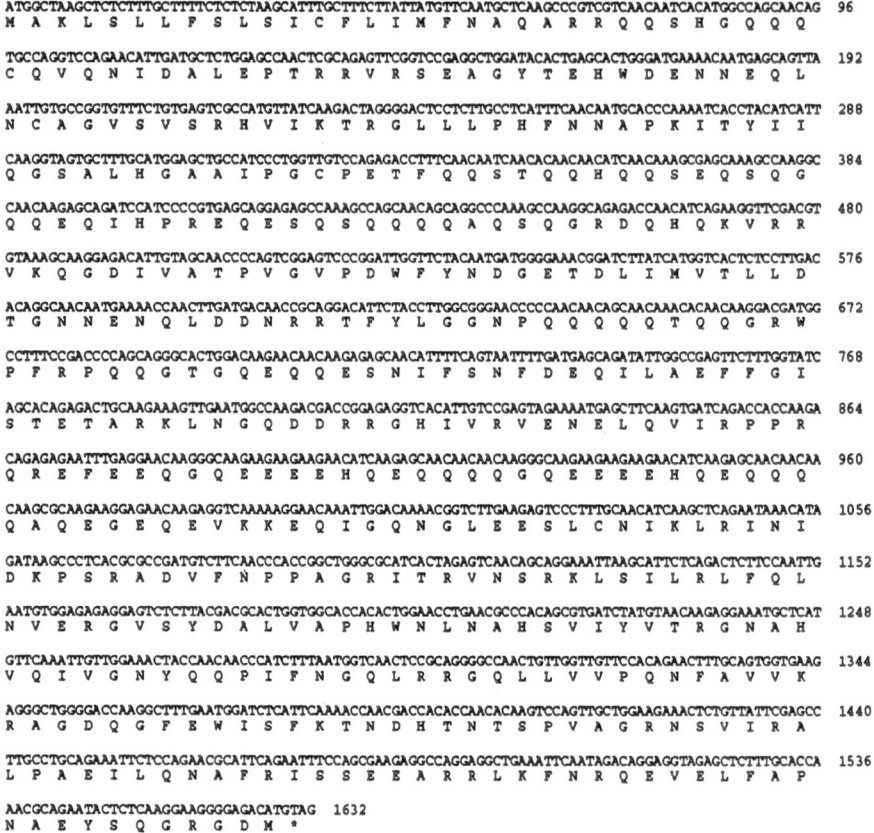

```
ATGGCTAAGCTCTCTTTGCTTTTCTCTCTAAGCATTTGCTTTCTTATTATGTTCAATGCTCAAGCCCGTCGTCAACAATCACATGGCCAGCAACAG  96
M  A  K  L  S  L  L  F  S  L  S  I  C  F  L  I  M  F  N  A  Q  A  R  R  Q  Q  S  H  G  Q  Q  Q

TGCCAGGTCCAGAACATTGATGCTCTGGAGCCAACTCGCAGAGTTCGGTCCGAGGCTGGATACACTGAGCACTGGGATGAAAACAATGAGCAGTTA  192
C  Q  V  Q  N  I  D  A  L  E  P  T  R  R  V  R  S  E  A  G  Y  T  E  H  W  D  E  N  N  E  Q  L

AATTGTGCCGGTGTTTCTGTGAGTCGCCATGTTATCAAGACTAGGGGACTCCTCTTGCCTCATTTCAACAATGCACCCAAAATCACCTACATCATT  288
N  C  A  G  V  S  V  S  R  H  V  I  K  T  R  G  L  L  L  P  H  F  N  N  A  P  K  I  T  Y  I  I

CAAGGTAGTGCTTTGCATGGAGCTGCCATCCCTGGTTGTCCAGAGACCTTTCAACAATCAACACAACAACATCAACAAAGCGAGCAAAGCCAAGGC  384
Q  G  S  A  L  H  G  A  A  I  P  G  C  P  E  T  F  Q  Q  S  T  Q  Q  H  Q  Q  S  E  Q  S  Q  G

CAACAAGAGCAGATCCATCCCCGTGAGCAGGAGAGCCAAAGCCAGCAACAGCAGGCCCAAAGCCAAGGCAGAGACCAACATCAGAAGGTTCGACGT  480
Q  Q  E  Q  I  H  P  R  E  Q  E  S  Q  S  Q  Q  Q  Q  A  Q  S  Q  G  R  D  Q  H  Q  K  V  R  R

GTAAAGCAAGGAGACATTGTAGCAACCCCAGTCGGAGTCCCGGATTGGTTCTACAATGATGGGGAAACGGATCTTATCATGGTCACTCTCCTTGAC  576
V  K  Q  G  D  I  V  A  T  P  V  G  V  P  D  W  F  Y  N  D  G  E  T  D  L  I  M  V  T  L  L  D

ACAGGCAACAATGAAAACCAACTTGATGACAACCGCAGGACATTCTACCTTGGCGGGAACCCCCAACAACAGCAACAAACACAACAAGGACGATGG  672
T  G  N  N  E  N  Q  L  D  D  N  R  R  T  F  Y  L  G  G  N  P  Q  Q  Q  Q  Q  T  Q  Q  G  R  W

CCTTTCCGACCCCAGCAGGGCACTGGACAAGAACAACAAGAGAGCAACATTTTCAGTAATTTTGATGAGCAGATATTGGCCGAGTTCTTTGGTATC  768
P  F  R  P  Q  Q  G  T  G  Q  E  Q  Q  E  S  N  I  F  S  N  F  D  E  Q  I  L  A  E  F  F  G  I

AGCACAGAGACTGCAAGAAAGTTGAATGGCCAAGACGACCGGAGAGGTCACATTGTCCGAGTAGAAAATGAGCTTCAAGTGATCAGACCACCAAGA  864
S  T  E  T  A  R  K  L  N  G  Q  D  D  R  R  G  H  I  V  R  V  E  N  E  L  Q  V  I  R  P  P  R

CAGAGAGAATTTGAGGAACAAGGGCAAGAAGAAGAAGAACATCAAGAGCAACAACAACAAGGGCAAGAAGAAGAAGAACATCAAGAGCAACAACAA  960
Q  R  E  F  E  E  Q  G  Q  E  E  E  E  H  Q  E  Q  Q  Q  Q  G  Q  E  E  E  E  H  Q  E  Q  Q  Q

CAAGCGCAAGAAGGAGAACAAGAGGTCAAAAAGGAACAAATTGGACAAAACGGTCTTGAAGAGTCCCTTTGCAACATCAAGCTCAGAATAAACATA  1056
Q  A  Q  E  G  E  Q  E  V  K  K  E  Q  I  G  Q  N  G  L  E  E  S  L  C  N  I  K  L  R  I  N  I

GATAAGCCCTCACGCGCCGATGTCTTCAACCCACCGGCTGGGCGCATCACTAGAGTCAACAGCAGGAAATTAAGCATTCTCGACTCTTCCAATTG  1152
D  K  P  S  R  A  D  V  F  N  P  P  A  G  R  I  T  R  V  N  S  R  K  L  S  I  L  R  L  F  Q  L

AATGTGGAGAGAGGAGTCTCTTACGACGCACTGGTGGCACCACACTGGAACCTGAACGCCCACAGCGTGATCTATGTAACAAGAGGAAATGCTCAT  1248
N  V  E  R  G  V  S  Y  D  A  L  V  A  P  H  W  N  L  N  A  H  S  V  I  Y  V  T  R  G  N  A  H

GTTCAAATTGTTGGAAACTACCAACAACCCATCTTTAATGGTCAACTCCGCAGGGGCCAACTGTTGGTTGTTCCACAGAACTTTGCAGTGGTGAAG  1344
V  Q  I  V  G  N  Y  Q  Q  P  I  F  N  G  Q  L  R  R  G  Q  L  L  V  V  P  Q  N  F  A  V  V  K

AGGGCTGGGGACCCAAGGCTTTGAATGGATCTCATTCAAAACCAACGACCCACACCAACACAAGTCCAGTTGCTGGAAGAAACTCTGTTATTCGAGCC  1440
R  A  G  D  Q  G  F  E  W  I  S  F  K  T  N  D  H  T  N  T  S  P  V  A  G  R  N  S  V  I  R  A

TTGCCTGCAGAAATTCTCCAGAACGCATTCAGAATTTCCAGCGAAGAGGCCAGGAGGCTGAAATTCAATAGACAGGAGGTAGAGCTCTTTGCACCA  1536
L  P  A  E  I  L  Q  N  A  F  R  I  S  S  E  E  A  R  R  L  K  F  N  R  Q  E  V  E  L  F  A  P

AACGCAGAATACTCTCAAGGAAGGGGAGACATGTAG  1632
N  A  E  Y  S  Q  G  R  G  D  M  *
```

Fig. 1. Nucleotide sequence of a *Hepatica nobilis* legumin cDNA (coding region only), together with the corresponding amino acid sequence

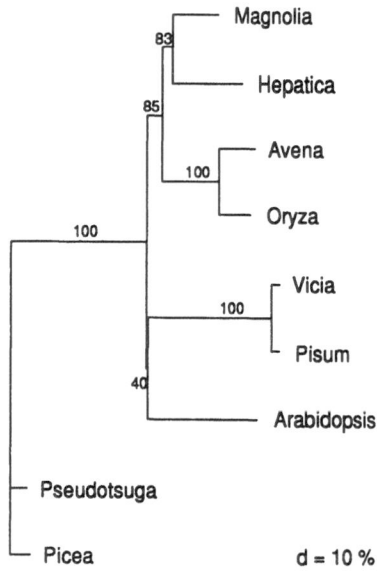

Fig. 2. Neighbor-joining (Saitou & Nei 1987) phenogram of some representative legumin gene sequences. The calculation was done on first and second codon positions only, omitting the legumin-specific non-alignable regions. Numbers at the branches refer to bootstrap values. Adding more species of the respective groups or more genes per species does not change the topology of the tree

References

Chase, M. W., Soltis, D. E., Olmstead, R. G., Morgan, D., Les, D. H., Mishler, B. D., Duvall, M. R., Price, R., Hills, H. G., Qui, Y.-L., Kron, K. A., Rettig, J. H., Conti, E.,

PALMER, J. D., MANHART, J. R., SYTSMA, K. J., MICHAELS, H. J., KRESS, W. J., DONOGHUE, M. J., CLARK, W. D., HEDRON, M., GAUT, B. S., JANSEN, R. K., KIM, K.-J., WIMPEE, C. F., SMITH, J. F., FURNIER, G. R., STRAUSS, S. H., XIANG, Q.-Y., PLUNKETT, G. M., SOLTIS, P. S., SWENSEN, S., EGUIARTE, L. E., LEARN, G. H., BARRETT, S. C. H., GRAHAM, S., DAYANANDAN, S., ALBERT, V. A., 1993: Phylogenetics of seed plants: an analysis of nucleotide sequences from the plastid gene rbcL. – Ann. Missouri Bot. Gard. **80**: 528–580.
CRONQUIST, A., 1988: The evolution and classification of flowering plants. – New York: New York Botanical Garden.
HAMBY, R. K., ZIMMER, E. A., 1992: Ribosomal RNA as a phylogenetic tool in plant systematics. – In SOLTIS, P. S., SOLTIS, D. E., DOYLE, J. J., (Eds): Molecular systematics of plants, pp. 51–91. – New York, London: Chapman & Hall.
SAITOU, N., NEI, M., 1987: The neighbour-joining method: a new method for reconstructing phylogenetic trees. – Mol. Biol. Evol. **4**: 406–425.
TAKHTAJAN, A., 1980: Outline of the classification of flowering plants (*Magnoliophyta*). – Bot. Rev. **46**: 225–359.

Address of the authors: J. LANG, H. FISCHER, Universität Bayreuth, Lehrstuhl für Pflanzenökologie und Systematik, D-95440 Bayreuth, Federal Republic of Germany.

Accepted December 5, 1994

Pl. Syst. Evol. [Suppl.] 9: 305–317 (1995)

Differentiation patterns in *Ranunculus* subgenus *Batrachium* (*Ranunculaceae*)

GERTRUD DAHLGREN

Received October 27, 1994

Key words: Angiosperms, *Ranunculaceae*, *Ranunculus*, *Batrachium*. – Karyology, polyploidy, CVA, PCA, breeding system, embryology, dioecy, confocal laser scanning, isozyme electrophoresis, flavonoid HPLC.

Abstract: The processes that have influenced speciation and evolution in *Ranunculus* subg. *Batrachium* are surveyed: polypoidy and hybridization, environmental adaptation, morphological differentiation, reproduction and the breeding system, including the embryology of the group. Preliminary results of isozyme electrophoretic studies and flavonoid HPLC-studies are presented. The rank of subgenus as well as differentiation and speciation are discussed. The new combination *Ranunculus pekinensis* (L. LIOU) G. DAHLGREN is made.

The water-buttercups are usually treated as a separate subgenus, *Batrachium* (DC.) A. GRAY of the genus *Ranunculus* L., with the following characteristics. They are aquatic during most of their life; they are heterophyllous: some taxa develop both laminate floating or aerial leaves and dissected, generally submerged leaves, while others have lost this capacity and produce only one kind of leaf; the petals are white (with one exception), not shiny, normally with a basal yellow spot and have small open nectaries at the base; the achenes are 1–2 mm long with characteristic transverse ridges.

Batrachium has its main centre in W and NW Europe but extends eastwards through Asia with another centre in China and Japan. Some species occur in C and W America and in N Africa and one in South Africa. Both annuals and perennials are known. Annual taxa grow in slow rivers and brooks, in ditches, pools and ponds etc. where *Batrachium* can form small patches or may cover most of the surface. Many localities dry up during part of the year. Perennials are found in swift rivers, the plants sometimes being quite submerged, even the flowers. The considerable biomass may completely choke the whole river.

Batrachium is a so called critical complex of c. 20 species. It is a difficult group taxonomically for three major reasons: (1) There is an extreme degree of phenotypic plasticity (COOK 1968). (2) Polyploidy is well developed with diploids, tetraploids and hexaploids as well as spontaneous triploid and pentaploid populations (DAHLGREN 1991). (3) The breeding system with a combination of various types of sexual reproduction and vegetative propagation (COOK 1968, DAHLGREN 1992 and others).

How can we reveal the evolutionary pattern of the group? Which processes have influenced differentiation in *Batrachium*? I will present a survey of the methods I used to elucidate these problems and the results to date.

Material and methods

My studies have mainly been concentrated on the Aegean archipelago and the W coast of Turkey but include material from SW England, Denmark and Sweden and some populations from China. Field studies have been carried out at intervals since 1986. Plants were sampled in the field or seeds sown for cultivation in greenhouse and outdoors at the Department of Systematic Botany in Lund.

In the Greek area occur *R. peltatus* Schrank subsp. *peltatus* (2n = 32), subsp. *boudotii* (Godron) Meikle (2n = 32), subsp. *saniculifolius* (Viv.) Cook (2n = 16, 24, 32, 40, 48), *R. tripartitus* DC. (2n = 48), *R. trichophyllus* Chaix (2n = 32, 48), *R. penicillatus* (Dumort.) Bab. subsp. *pseudofluitans* (Syme) Webster (2n = 16), *R. sphaerospermus* Boiss. & Blanche in Boiss. (2n = 16) and *R. rionii* Lagger (2n = 16).

The English material comprises *R. penicillatus* subsp. *penicillatus* (2n = 48) and subsp. *pseudofluitans* (2n = 48) and *R. fluitans* Lam. (2n = 32), the Danish *R. peltatus* subsp. *peltatus* (2n = 32) and subsp. *baudotii* (2n = 32), *R. aquatilis* L. (2n = 32), and *R. penicillatus* subsp. *penicillatus* and subsp. *pseudofluitans* (2n = 32). In Sweden material was sampled from the provinces of Skåne, Blekinge and Öland including the taxa *R. peltatus* subsp. *peltatus* and subsp. *baudotii*, *R. aquatilis*, *R. trichophyllus*, and *R. fluitans* (2n = 16). In China populations of *R. flavidus* (Hand. Mazz.) Cook (2n = 16), *R. bungei* Steudel (2n = 16), **R. pekinensis (L. Liou) G. Dahlgren, Comb. Nov.** (Basionym *Batrachium pekinense* L. Liou in Addendum Flora Republicae Popularis Sinicae, **28**, 1980: 340) (2n = 16) and *R. trichophyllus* (2n = 16) were sampled.

Plants from each locality were pressed for morphological and biometric studies. The methods used for the karyological investigation are described in Dahlgren (1991).

For the multivariate analyses of leaves and petals, see Dahlgren & Svensson (1994). A morphological analysis was also made of the whole Greek material where a minimum of five plants from each of about 100 populations were studied. The investigation included about 40 characters, some of which were binary: heterophyllous/homophyllous (the term used for taxa with only dissected leaves developed), hairy/smooth, winged/unwinged, etc., while others were treated as multistate characters like the various nectary types, or as continuous measurement variables. The data were subjected to a Principal Component Analysis (PCA) to elucidate the multivariate relationships between populations and taxa.

For the embryological investigation a confocal scanning laser microscope (CLSM) was employed, using the method described by Carlsson & Åslund (1987) and in Dahlgren (1993).

For isozyme electrophoresis leaves were ground over ice following Wyatt's extraction buffer. The starch gel was run at 90 mA for six h (system I), and then at 70 mA for four h (system 2), with bromophenol added as a front marker. The two buffer systems were modified from Wendel & Weeden (1989).

For the flavonoid analysis leaves were dried in the dark at room temperature and extracted once with boiling 80% ethanol for 20 min under reflux. Extracts were diluted (10x or 2x) for High Pressure Liquid Chromatography (HPLC). The flavonoid profile of each extract was visualized by HPLC on a nucleosil C18 reverse phase. The chromatogram was developed by a gradient of 3–35% of acetonitrile in acetic water, 0.7 ml/min; detection at 340 nm.

Results

Karyology. The karyology of the Aegean taxa has been treated in detail in a separate paper (DAHLGREN 1991). All polyploidy levels from diploids to hexaploids were found. In some species there are different chromosome numbers within a taxon. These cytotypes are genetically distinguished but cannot be separated morphologically. In the *R. peltatus* group subsp. *saniculifolius* represents the greatest variation with four cytotypes (diploids, tetraploids, pentaploids and hexaploids). In the hexaploid *R. tripartitus*, not previously found outside NW Europe, some variation in chromosome number was found between different cells of the same root tip, so-called aneusomaty. *R. penicillatus* subsp. *pseudofluitans* is diploid in Greece, while populations studied from Denmark are tetraploid (DAHLGREN 1993) and those from England hexaploid. The Chinese species studied are all diploid.

Certain karyotype differences were observed between taxa (DAHLGREN 1991) as regards size, chromosome morphology and occurrence of satellite chromosomes. Subgenus *Batrachium* is considered an advanced group of the family *Ranunculaceae* on the basis of chromosomal evidence (GOEPFERT 1974), a hypothesis supported by the fact that the group includes annuals and occurs in unstable and ephemeral habitats.

There are proportionately more diploid taxa in Greece than in the total distribution area and the diploids become more frequent eastwards. It is questionable whether subg. *Batrachium* has its origin in NW Europe, where the number of species and the variation is greatest, or in China-Tibet, where the diploid taxa dominate, and from where they may have spread and differentiated locally, developing polyploids in other parts of the world.

Morphology. Extreme adaptation to aquatic life has led to strong morphological adaptation, the strategy varying with the species. There is extensive variation in leaves and petals and the combination of character states sometimes varies to such a degree that each population seems to have its own specific appearance, particularly in the *R. peltatus* group.

Multivariate analysis (DAHLGREN & SVENSSON 1994) showed how the variation in leaf and petal was partitioned between and within populations of this group. We used Canonical Variate Analysis (CVA) and cluster analysis to investigate whether the populations clustered in groups. The population means for leaf and petal data were plotted separately for taxa, ploidy levels and islands. These diagrams showed no distinct pattern nor did they indicate that the populations form any distinct substructure. Hence, there is no support for subdividing the populations into taxa on the basis of leaf or petal characters.

In a subsequent investigation the whole Greek material was subject to Principal Component Analysis (PCA) using a wide range of vegetative and reproductive characters. It was of particular interest to examine how the total variation is partitioned according to, for example, heterophylly and species (Fig. 1). Heterophyllous and homophyllous taxa are fairly well separated regarding variation in morphological characters. For species only some groups can be seen. *R. sphaerospermus* is well separated from *R. trichophyllus*. *R. rionii* forms one more or less distinct group. *R. peltatus* subsp. *peltatus*, subsp. *baudotii* and subsp. *saniculifolius* overlap considerably but are separated from most other taxa. It is notable that *R. penicillatus* subsp. *pseudofluitans* is also found within this group.

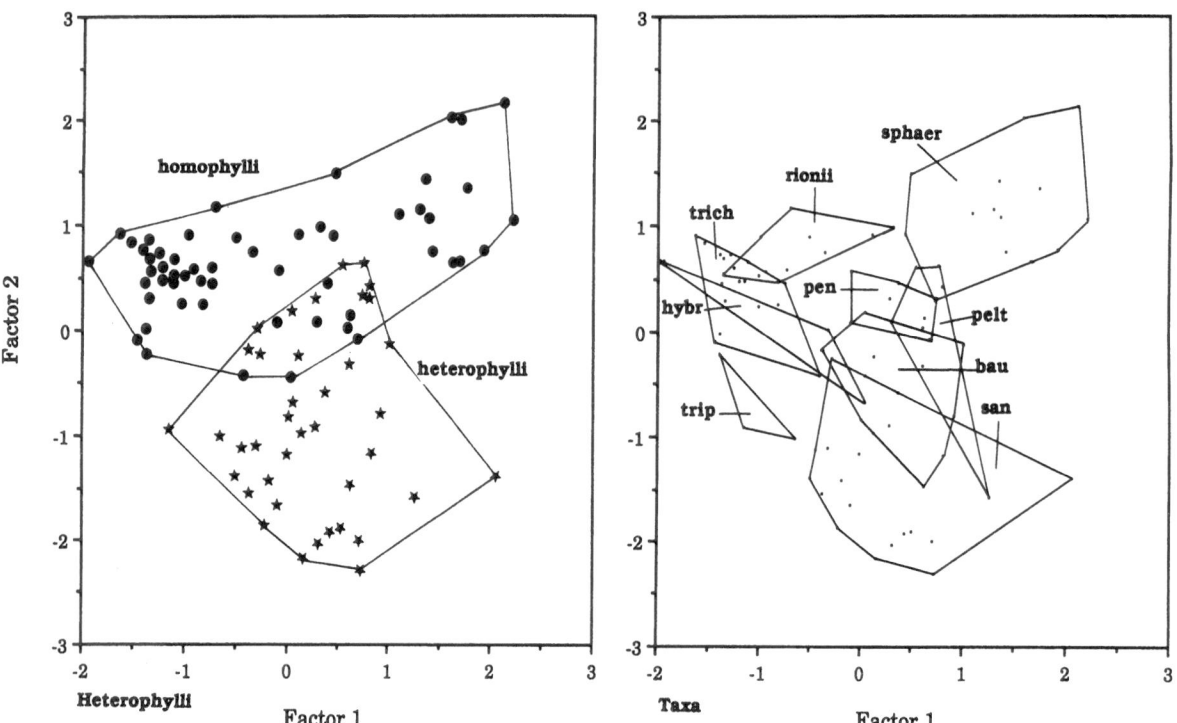

Fig. 1. Principal component analysis of the variation of Aegean taxa of *Ranunculus* subg. Batrachium, partitioned according to *a* heterophylly/homophylly and *b* species

In DAHLGREN (1992) it was shown that the appearance of the nectaries is a more useful taxonomic character than most other morphological traits and can also be used to reveal hybrid origin. In Greek populations of *Batrachium* the **lunate** type is commonest, being found in four of the eight taxa, *R. peltatus* subsp. *baudotii*, *R. tripartitus*, *R. trichophyllus* and *R. rionii*. The **cup** type (the term circular used by COOK 1966) is rare in Greek material. The **pyriform** type characterizes *R. peltatus*, a tubular type where a dorsal wall is formed. The ontogenesis of these nectary types has been studied by LEINFELLNER (1959) who demonstrated the relationship to *R. sceleratus* in the development of the nectary.

Apart from these three nectary types, described by COOK (1968) and others, several other types occur in Greek *Batrachium*. The small **ring** type is especially found in cleistogamous flowers. The **horseshoe** type is intermediate between the lunate and the pyriform, and is the dominating type in subsp. *saniculifolius*, which in other characters as well is intermediate between subsp. *peltatus* and subsp. *baudotii*. The **scale** type is attached to the petal only at the cylindrical base, and thus agrees with the nectary type characteristic of subg. *Ranunculus*. Two or more nectaries can develop side by side in populations of probable hybrid origin, and in F_1 plants after crossings in the *R. peltatus* group.

The probable evolution of the nectary types, from the scale type with its copious nectary production to the compact nectaries of ring type where no nectar is produced was discussed in DAHLGREN (1992).

Reproduction and embryology. This discussion leads up to the question of reproduction in *Batrachium*. The flower structure shows that some taxa are adapted for cross-pollination. Nevertheless, contrary to what has been written about Swedish material (HONG 1991), none of the species are self-incompatible. Other species show high levels of automatic self-pollination. Some flower under water, some are even cleistogamous and never open, but all set seed in the absence of pollinators.

In one of the outcrossing species, *R. penicillatus*, and also in some populations of other taxa, germination is extremely low even if achenes develop. In other cases the offspring is so uniform that apospory or complete self-pollination could be suspected. To throw light on the reproductive system I conducted an embryological study of the group (DAHLGREN 1993) using the so-called confocal scanning laser method, which has so far been little used for plants (FREDRIKSON & al. 1988).

The ovules are hemitropous and unitegmic (BOUMAN 1984) (Fig. 2a). Normal embryo sacs usually develop (Fig. 2b,c). Synergids and antipodals are large, the antipodals being uninucleate unlike subg. *Ranunculus* where most genera have binucleate antipodal cells. The endosperm formation is of the nuclear type and is initiated before the embryo begins to divide. It is generally possible to follow the development up to the complete embryo (Fig. 2d). In some populations of *R. penicillatus*, however, an endosperm develops but no embryo. Thus, even if the fruits appear to be normal, the plants are not necessarily fertile. Abnormalities in ovule formation and embryo sac development occur (Fig. 2e,f). The embryo sac can begin to degenerate at any stage. One, two or four large-nucleate cells can be seen in the reduced shrivelled embryo sacs but no tendency to aposporic embryo sacs has been traced, neither in the Greek nor in Scandinavian or English material.

Deviations were also observed in the male organs. For example, in some populations of *R. sphaerospermus*, the filaments of the stamens of some plants are fused while the anthers are free. In some populations of *R. sphaerospermus*, *R. penicillatus* and *R. peltatus* the filaments bear characteristic white petal-like wings 2 mm wide, interpreted as a possible link in the reduction of the stamens to staminodes.

The flowers of most populations are hermaphrodite but functional dioecy has been observed in a few populations of *R. penicillatus* and in some hybrid populations with *R. peltatus* as one of the parents. The anthers in the male flowers were twice the size of those in hermaphrodite flowers, the loculi being completely filled with well-developed pollen, while the ovaries were shrivelled, undeveloped or degenerated (DAHLGREN 1993). In other flowers the stamens were undeveloped having narrow anthers with no pollen or with small quantities of pollen with a stainability of between 20 and 30%, the ovaries being normal externally. Probably these male-sterile flowers could function as females, although seed set is very low or lacking in these populations. In populations with floral dimorphism vegetative propagation appears to dominate totally.

In all taxa vegetative propagation plays an important part side by side with the sexual reproduction. Detached branches float in the water, develop roots at the nodes and become stationary forming new plants. The same phenomenon with free-floating parts of plants, flowering or in fruit, has also been observed in the usually annual subsp. *peltatus*. There are also well-developed stolons.

The small, light and dry achenes loosen easily from the receptacle and float on the water, but in contact with water they become sticky and heavy, sink to the bottom

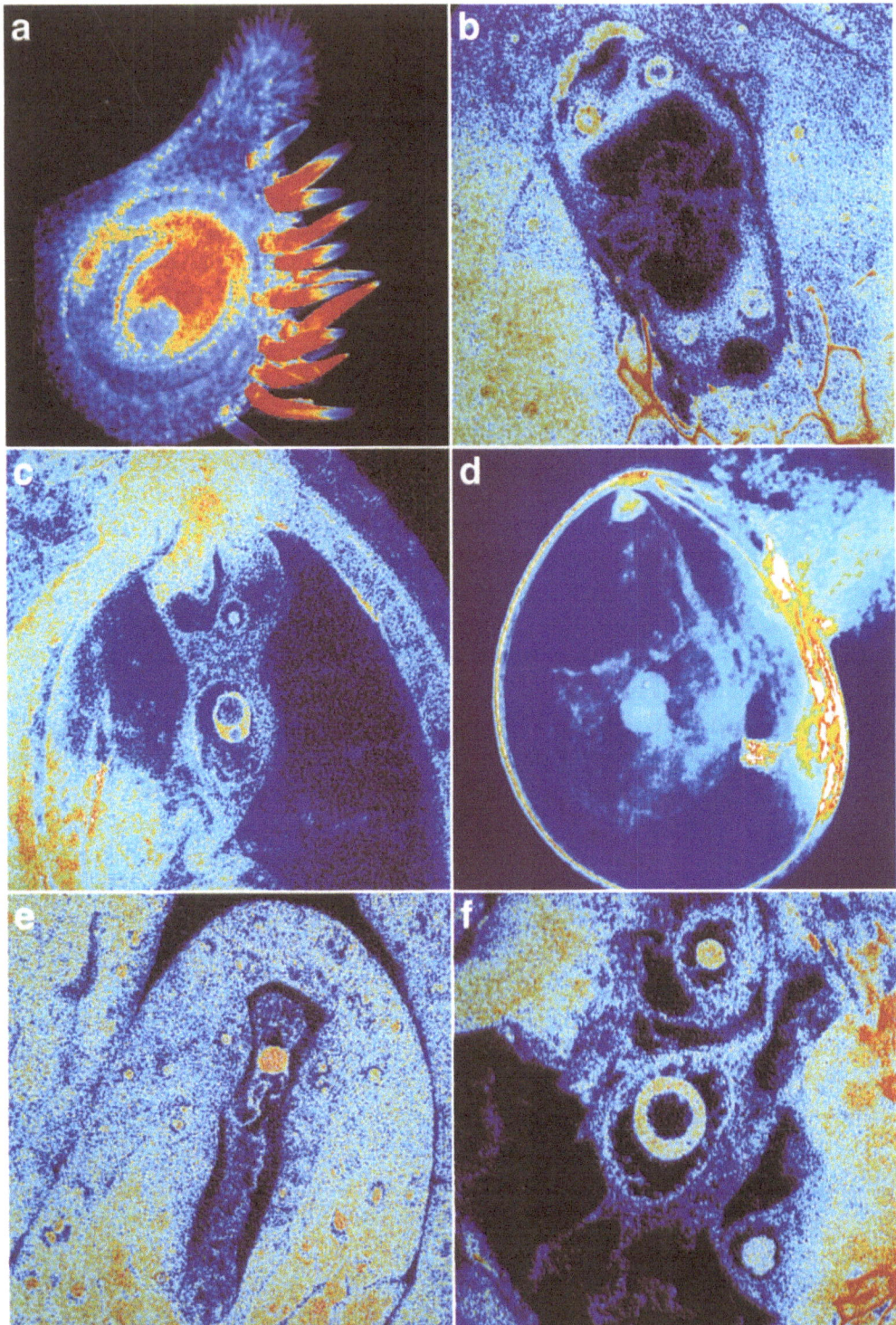

Fig. 2. Embryology, *Ranunculus penicillatus* subsp. *pseudofluitans*. *a* Hemitropous, unitegmic ovule. *b,c* Development of normal reduced embryo sacs, *b* 4-nuclear and *c* 8-nuclear,
respectively. *d* Seed with young embryo and endosperm. *e*, *f* Degenerating embryo sacs,
e with one and *f* with four unorganized large-nucleate cells in the shrivelled embryo sac. –
CSLM; pseudocolour added, weakest fluorescing parts give blue colour, then in order green,
red and white

and germinate. In plants that flower under water or plants with curved peduncles the achenes loosen immediately below the water surface. They stick easily to other plants or the feet of birds and even to sheep that come to ponds to drink, and can then be dispersed to new water systems. When dispersed in this way they examplify the so called founder effect, i.e. only part of the total variation of the parents will be transported to the newly established population, provided there is genetic variation in the parent generation.

Isozyme electrophoresis. Molecular methods should be a good complement to traditional methods to explain the variation in the subgenus. I have studied a series of populations from the Greek area and some also from England and Sweden with isozyme electrophoresis.

The laboratory work has just been finished and the results have not yet been interpreted for all enzyme systems, but some preliminary conclusions can be drawn: (1) The total amount of variation is small. (2) Some few bands in different combinations characterize *Batrachium*. (3) There is no variation between plants in a population. (4) There is no variation within the three subspecies of *R. peltatus* for some of the enzyme systems, while for others populations of subsp. *saniculifolius* show a certain degree of variation, usually having fewer bands than subsp. *peltatus* and subsp. *baudotii*, which can indicate hybrid origin. (5) Of the homophyllous species, *R. trichophyllus* and *R. aquatilis* are impossible to separate, while *R. sphaerospermus* and *R. rionii* can be distinguished from each other. (6) The electrophoretic phenotype of *R. penicillatus* (populations with 2n = 32, 48) varies between subspecies but also between populations within a subspecies. (7) In the yellow-flowered *R. flavidus* the pattern differs both from that of other Chinese taxa and of the European taxa. (8) The *Batrachium* populations of a species in each geographical area often has its specific electrophoretic phenotype.

Flavonoid HPLC-studies. Chromatographic investigations of the flavonoids of species of *Ranunculus* have been carried out by LEBRETON (1986) and WEBSTER (1991), who reached different conclusions, however, regarding the usefulness of the method. While LEBRETON interpreted the results as informative at the species and subspecies level, WEBSTER did not consider they were useful, at least for hybrid complexes. I have used the latest advanced HPLC technique for flavonoid analysis as a pilot study of populations of *R. aquatilis* and *R. trichophyllus* from the provinces Skåne and Öland in Sweden, populations of *R. sphaerospermus*, *R. rionii* and *R. peltatus* from Greece and *R. penicillatus* from England.

Quercetin is the only flavonoid present but tri- and diglycosides are often abundant. Three main chemotypes occur in the samples (I, II and III). In I the main peaks are at the beginning of the HPLC eluate, with peaks 5 (diglucosyl-3, glucosyl-7-quercetin) and 9 (its caffeyl derivative on terminal glucose in 3) dominating. *R. sphaerospermus* and *R. rionii* both belong to this type. In II the mean peaks are at the end of the diagram. They are less hydrophilic, peak 32 being the major peak. This chemotype is represented by *R. penicillatus* subsp. *pseudofluitans* from England. In III the profile is more balanced, with some variation, peaks 34–35 dominating in material of *R. penicillatus* subsp. *penicillatus*, peak 32 in *R. peltatus* and *R. aquatilis* and peak 26 in *R. trichophyllus*.

Since the chemical structure of numerous peaks has not yet been identified and no correlation can be presumed between the HPLC peaks, it is too early to draw

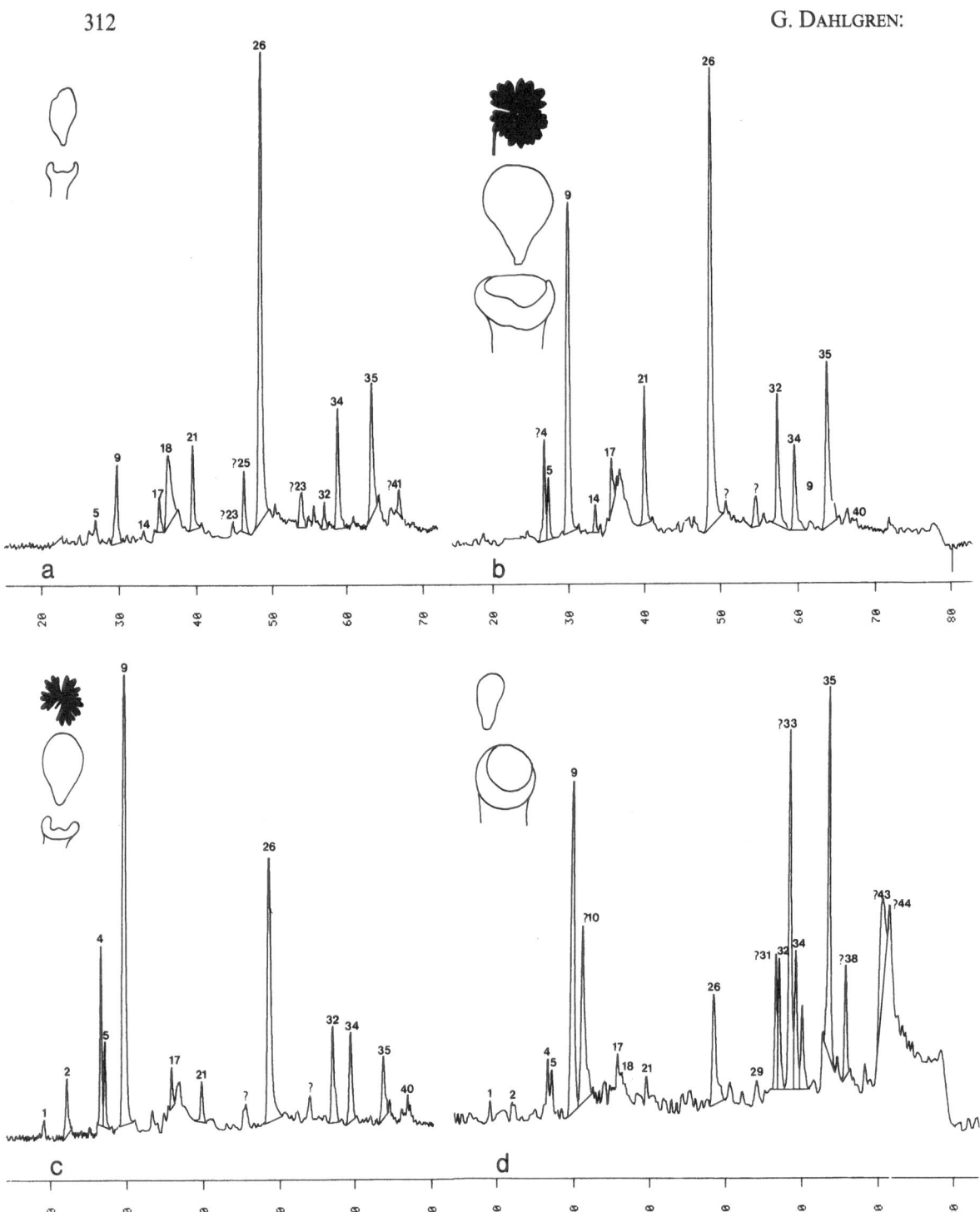

Fig. 3. HPLC-studies in the *R. aquatilis* complex. Profiles of flavonoid peaks of *a R. trichophyllus*; *b R. aquatilis*; *c,d* populations morphologically intermediate between *R. trichophyllus* and *R. aquatilis*

taxonomic conclusions from these preliminary results. Nevertheless some interesting observations can be made from the results in the *R. aquatilis* complex. The peaks 26 and 9 dominate in *R. aquatilis* populations from both Skåne and Öland, while 32 is also important in populations from Öland. The profile of the Swedish populations of *R. aquatilis* agrees well with French populations from Dombes studied by FIASSON

& FIASSON (pers. comm.) except for the increased height of 26 and 4 in the Swedish material. *R. trichophyllus* differs from *R. aquatilis* in the complete dominance of peak 26 (Fig. 3a and b). Some populations are morphologically intermediate and impossible to classify as *R. aquatilis* or *R. trichophyllus*. Two such populations from Skåne are represented in the flavonoid analyses (Fig. 3c and d). The population represented in Fig. 3c has laminate leaves, never found in *R. trichophyllus*, but lunate nectaries, characteristic of that species. The petals are of intermediate size. The flavonoid profile is most like that of *R. aquatilis*, although peak 9 is taller than 26. The population in Fig. 3d lacks laminate leaves and has small petals as is characteristic of *R. trichophyllus*, but the cup-type nectaries typical of *R. aquatilis*. The HPLC pattern differs from that in all other populations studied in the complex. Peak 26 is much lower, 35 and 9 dominate, and many additional peaks appear.

Discussion

Reproduction and species differentiation. Having studied the reproduction in the Aegean *Batrachium* as part of a greater whole we will see what this means for species differentiation in subg. *Batrachium*. *Batrachium* is rare on these islands because of the small number of suitable habitats available. Thus, they occur in small populations in one or a few places on some islands, whereas on others not at all. The genetic consequences of this small population size denote genetic draft – random changes of gene frequency from one generation to the other.

Pollination between flowers of different plants may in reality amount to pollination between rametes, i.e. the genetic effect is the same as in self-pollination or short-distance pollination, which thus increases the degree of homozygosity. The morphological similarity **within** populations and slight differences **between** populations can be explained by the fact that each isolated population has to some extent become homozygous as a result of dominating self-pollination in spite of the outcrossing structure of the flower thus explaining why the pattern of variation is the same in cross-pollinating and self-pollinating taxa.

In the *R. peltatus* group subsp. *peltatus* is outcrossing, while subsp. *baudotii* and subsp. *saniculifolius* are selfing, but both types of pollination result in seed set in all of them and series of crossings between taxa show that there are no sterility barriers at all at tetraploid level. The hybrids resemble subsp. *saniculifolius* in appearance. Subsp. *saniculifolius* found in nature differs from the other taxa in its great morphological variation. It has probably originated by hybridization between subsp. *peltatus* and subsp. *baudotii* on different occasions. The hybrids have found niches on the islands where they have succeeded better than the parents and have thus become commoner than they. The great variation probably reflects the fact that they are segregation products after hybridization or the result of back-crossings with one of the parents. The plants developed in this way reproduce vegetatively or by self-pollination and each population will maintain its characteristic appearance.

Dioecy was observed in some populations of *Batrachium*. It appears as if vegetative propagation was the dominating type of reproduction in these populations. Possibly the pollen from the male flowers might function in cross-pollination with hermaphrodite flowers. HOFFMAN (1992) described and analysed the evolution of functional dioecy in *Echinocereus* (*Cactaceae*), another polyploid complex. The hermaphrodite flowers were functionally male, and the male-sterile functionally female. The floral dimorphism was there confined to the periphery of the range of the species. Interesting questions arise when floral dimorphism evolves in polyploid complexes such as *Batrachium* and *Echinocereus*. If crosses between the dioecious plants and plants from hermaphrodite populations result in seed set and viable progeny where segregation works, the evolutionary pattern of the groups will be even more diverse.

Taxonomy. What is the taxonomic significance of the results so far? The boundaries between the three taxa of the *R. peltatus* complex can be erased in the Greek area but in the Scandinavian countries the situation is different, subsp. *peltatus* and subsp. *baudotii* being well separated while subsp. *saniculifolius* does not occur. The reason is that subsp. *peltatus* grows in freshwater, subsp. *baudotii* in brackish water. Since the Aegean islands are small, there is some influence of salt water even at some distance from the coast. When the geographic and the ecological boundaries are removed the genetic boundaries also disappear.

Thus, there are no sterility barriers at the same ploidy level in the group and there are no enzymatic differences; some populations can be distinguished morphologically but many are intermediate; ecological boundaries occur in some areas, in others not. Subspecies rank should be accorded these taxa.

R. trichophyllus belongs to a variable complex, where *R. trichophyllus* (according to COOK 1966, a tetraploid, homophyllous species with small flowers and lunate nectaries) and *R. aquatilis* L. (hexaploid, heterophyllous species with bigger flowers and cup-shaped nectaries) are often difficult to separate, especially in the Scandinavian countries and in Germany (HONG 1991, WIEGLEB & HERR 1983). Both species have the same flavonoid chemotype but with some variation in peak pattern in material studied from two provinces of Sweden, and they are not possible to distinguish in the electrophoretic phenotype. Fusion to one single species, *R. aquatilis*, has been discussed and will appear in Flora Nordica (DAHLGREN & JONSELL, in print). In Greece tetraploid, pentaploid and hexaploid populations of the complex occur, but the three cytotypes agree all morphologically with *R. trichophyllus* and no *aquatilis*-forms occur. I have so far kept species rank of *R. trichophyllus* in papers published on Greek material only.

R. sphaerospermus is largely sympatric with *R. rionii* and occurs in the same type of habitat and they are sometimes found in the same watercourse. Both are diploid. It has been possible to separate them in the investigations and the distinction between them is upheld.

R. penicillatus constitutes a complex group with considerable morphological variation and is probably of hybrid origin (COOK 1966, WIEGLEB & HERR 1983, WEBSTER 1988) with *R. peltatus*, *R. trichophyllus* and/or one diploid species, *R. fluitans*, as possible parent species. The hexaploid chromosome number dominates in the complex, but tetraploids and sterile triploids and pentaploids have also been reported (COOK 1966, 1970; TURALA-SZYBOWSKA 1978). The Greek populations of *R. penicillatus* are all diploid, and homophyllous (DAHLGREN 1991). They agree morphologically with subsp. *pseudofluitans* (according to WEBSTER 1988) and differ from diploid *R. fluitans* (not occurring in Greece but with which it is easily confused morphologically) in having leaves divided more than 6 times and thus having a greater number of leaf segments, and in the distinctly hairy receptacle. In the PCA analysis subsp. *pseudofluitans* partially fell within the variation of *R. peltatus*, however. Whether this Greek diploid may be one of the parents in the allopolyploid *R. penicillatus* complex has not yet been investigated on sufficient detail.

Finally, it can be questioned whether the subgenus rank is still the most convincing for *Batrachium*. The following exceptions are seen from the features characteristic of *Batrachium* and separating the group from other *Ranunculus*: (1) A yellow-flowered diploid species, *R. flavidus*, grows in the pools in the high plateaus of Tibet and in the Himalayas. The petals are sulphur-yellow and not shiny, however, contrary to the buttercups. (2) The morphological analysis showed that a few populations had shiny petals. (3) Free leaflike nectaries, characteristic of subg. *Ranunculus*, also occur in *Batrachium*. (4) The development of the tubelike nectaries in *Batrachium* follows the same pattern as in *R. sceleratus*. (5) The flavonoid pattern studied by HENNION & al. (1994) in a small subantarctic group of yellow ranuncles adapted to aquatic and moist biotopes agrees better with the *Batrachium* profile of flavonoids than with other *Ranunculus* species. (6) The chloroplast DNA pattern in species of *Batrachium* is similar to that of *R. hyperborus* and *R. sceleratus* in subg. *Ranunculus*.

Thus, there are no absolute limits between the two groups. Nevertheless, I consider the high degree of specialization to aquatic life and the consistency in most populations of characters distinguishing them warrant upholding subgenus rank.

Conclusion

A series of processes have influenced evolution in *Batrachium*: (1) Adaptation to different environments: to ephemeral pools or swift rivers; to freshwater or brackish habitats etc. (2) The development of both annuals and perennials. (3) Considerable morphological differentiation, partly influenced by the environment, such as leaf development, partly by various reproduction strategies, such as development of allogamous, autogamous or cleistogamous flowers. (4) A tendency to development of functionally dioecious flowers. (5) Development of petaloid stamens. (6) Reduction of the nectary types, from the scale type, resembling the nectary type in subg. *Ranunculus*, to the small ring type without nectary function, found especially in cleistogamous flowers. (7) Degeneration of embryo sacs and the sexual reproduction partly replaced by well-developed vegetative propagation. (8) Polyploidy in the form of allopolyploids, originated mainly by hybridization between taxa.

These processes in combination with geographic isolation, small population size and founder events contributed to the present-day pattern of variation in *Batrachium*.

The flavonoid HPLC-study was carried out by Dr J. -L. FIASSON and Dr K. GLUCHOFF FIASSON, Laboratoire de Biologie Micromoléculaire et Phytochimie, Université Lyon-1, Villeurbanne, France, and I am most grateful to them for stimulating cooperation.

My sincere thanks to Mrs LAURA BERGVALL, who has done the laboratory work of the isozyme analyses, to Dr MARGIT FREDRIKSON, Botanical Department, University of Göteborg for technical help with the embryological investigation, to the gardener, RUNE SVENSSON, to Dr STEFAN ANDERSSON who kindly read the manuscript and to Mrs MARGARET GREENWOOD PETERSSON for correcting the English.

The investigations have been partly financed by grants from the Royal Physiographic Society and the Crafoord Foundation.

References

Bouman, F., 1984: The ovule. – In Johri, B. M., (Ed.): Embryology of Angiosperms, pp. 124–157. – Berlin: Springer.

Carlsson, K., Åslund, N., 1987: Confocal imaging for 3 D digital microscopy. – Appl. Opt. **26**: 3232–3238.

Cook, C. D. K., 1966: A monographic study of *Ranunculus* subgenus *Batrachium* (DC.) A. Gray. – Mitt. Bot. Staatssamml. München **6**: 47–237.

– 1970: Hybridization in the evolution of *Batrachium*. – Taxon **19**: 161–166.

– 1968: Phenotypic plasticity with particular reference to three amphibious plant species. – In Heywood, V. H., (Ed.): Modern methods in plant taxonomy, pp. 97–111. London: Academic Press.

Dahlgren, G., 1991: Karyological investigations in *Ranunculus* subgenus *Batrachium* (*Ranunculaceae*) on the Aegean islands. – Pl. Syst. Evol. **177**: 193–211.

– 1992: *Ranunculus* subgenus *Batrachium* on the Aegean islands and adjacent islands. Nectary types and breeding system. – Nordic J. Bot. **12**: 299–310.

– 1993: *Ranunculus penicillatus* in Norden. – Nordic J. Bot. **13**: 593–605.

– Jonsell, B. (in print) *Ranunculus* subgenus *Batrachium*. – In Jonsell, B., (Ed.): Flora Nordica.

– Svensson, L., 1994: Variation in leaves and petals of *Ranunculus* subgenus *Batrachium* on the Aegean Islands, analysed by multivariate analysis. – Bot. J. Linn. Soc. **14**: 253–270.

Fredrikson, M., Carlsson, K., Franksson, O., 1988: Confocal scanning laser microscopy, a new technique used in an embryological study of *Dactylorhiza maculata* (*Orchidaceae*). – Nordic J. Bot. **8**: 369–374.

Goepfert, D., 1974: Karyotypes and DNA content in species of *Ranunculus* L. and related genera. – Bot. Not. **127**: 464–489.

Hennion, F., Fiasson, J. L., Gluchoff–Fiasson, K., 1994: Morphological and phytochemical relationships between *Ranunculus* species from Iles Kerguelen. – Biochem. Syst. Ecol. **22**: 533–542.

Hoffman, M. T., 1992: Functional dioecy in *Echinocereus coccineus* (*Cactaceae*): Breeding system, sex ratios, and geographic range of floral dimorphism. – Amer. J. Bot. **79**: 1382–1388.

Hong, D., 1991: A biosystematic study of *Ranunculus* subgenus *Batrachium* in S. Sweden. – Nordic J. Bot. **11**: 41–59.

Lebreton, P., 1986: Les flavonoides, marqeurs systematique chez les Ranunculacées. – Pl. Médicinales Phytothéra. **20**: 275–286.

Leinfellner, W., 1959: Über die röhrenförmige Nektarschuppe an den Nektarblättern verschiedener *Ranunculus*- und *Batrachium*-Arten. – Österr. Bot. Zeit. **106**: 88–103.

Turala-Szybowska, K., 1978: Cyto-embryological studies in self-incompatible populations of *Ranunculus penicillatus* (Dumort.) Bab. from Poland. – Acta Biol. Cracoviensis Ser. Bot. **21**: 9–21.

Webster, S. D., 1988: *Ranunculus penicillatus* (Dumort.) Bab. in Great Britain and Ireland. – Watsonia **17**: 1–22.

– 1991: A chromatographic investigation of the flavonoids of *Ranunculus* L. subgenus *Batrachium* (DC.) A. Gray (water buttercups) and selected species in subgenus *Ranunculus*. – Aquatic Bot. **40**: 11–26.

WENDEL, J. F., WEEDEN, N. F., 1989: Visualization and interpretation of plant isozymes.– In SOLTIS, D. E., SOLTIS, P. S., (Eds): Isozymes in plant biology, pp 5–45. – Portland, Oregon: Discorides Press.

WIEGLEB, G., HERR, W., 1983: Taxonomie und Verbreitung von *Ranunculus* subgenus *Batrachium* in Niedersächsischen Fliessgewässern unter besonderer Berücksichtigung des *Ranunculus penicillatus* Komplexes. – Göttinger Flor. Rundbr., Z. Arealk., Flor. Syst. **3/4**: 101–175.

Address of the author: Dr GERTRUD DAHLGREN, Department of Systematic Botany, Ö. Vallgatan 20, S-223 61 Lund, Sweden.

Accepted December 14, 1994

Pl. Syst. Evol. [Suppl.] 9: 319–323 (1995)

Phenology and pollination biology of five *Ranunculus* species in Central Germany

KATHRIN STEINBACH and GERHARD GOTTSBERGER

Received October 10, 1994

Key words: *Ranunculaceae*, *Ranunculus*, reproductive phenology, flower visitors, visitation rates, generalistic pollination syndrome, Central Europe.

Abstract: Five *Ranunculus* species, *R. acris*, *R. auricomus*, *R. bulbosus*, *R. flammula*, and *R. repens* were studied in respect to phenology and pollination biology at several places in their natural habitats around Giessen (50°35′N, 80°41′E), Central Germany, and at garden-beds of Giessen University in 1989. For *R. auricomus*, *R. bulbosus*, and *R. acris* quite distinct phenologies were observed at the two sites, with a much longer flowering period at the bed site. Flower visiting insects, belonging mainly to the *Diptera*, *Hymenoptera*, and *Coleoptera*, varied considerably in the natural habitats and at the bed sites. The great variety of flower visiting insects and the substitution of insect species at different sites are an expression of a generalistic pollination syndrome of the studied *Ranunculus* species.

Phenology

With the exception of *R. repens* L., all *Ranunculus* species studied (*R. acris* L., *R. auricomus* L., *R. bulbosus* L., *R. flammula* L., *R. repens*) showed a distinctly different phenology between the garden plots and at the natural habitat sites. These differences certainly were caused by varying environmental factors, such as different types of soil, light, wind, humidity, and the lack of competition with other plant species in the garden plots. The species studied demonstrated a broad capacity for phenological modification and adaptation.

In the garden plots, *R. bulbosus* (Fig. 1) and *R. auricomus*, usually spring-flowering species, did not only flower during spring, but also throughout summer until autumn. At this site, *R. acris*, a late spring- and early summer-bloomer, extended its flowering period until the beginning of autumn. Not only the phenological pattern was changed in the garden plots, but the longer flowering period resulted also in a higher flower and fruit production per plant. Under natural conditions *R. acris* and *R. bulbosus* employ the so-called mass flowering strategy (GENTRY 1974); in the garden plots they changed to a steady-state strategy.

In *R. auricomus*, a partly asexual (apomictic) species (see NOGLER 1984), the early peak of flowering in the garden plots corresponded with the total flowering period at the natural site. A mixture of normal flowers and pollinator-independent flowers, forming fruits either by self-pollination or apomixis caused a second flowering peak

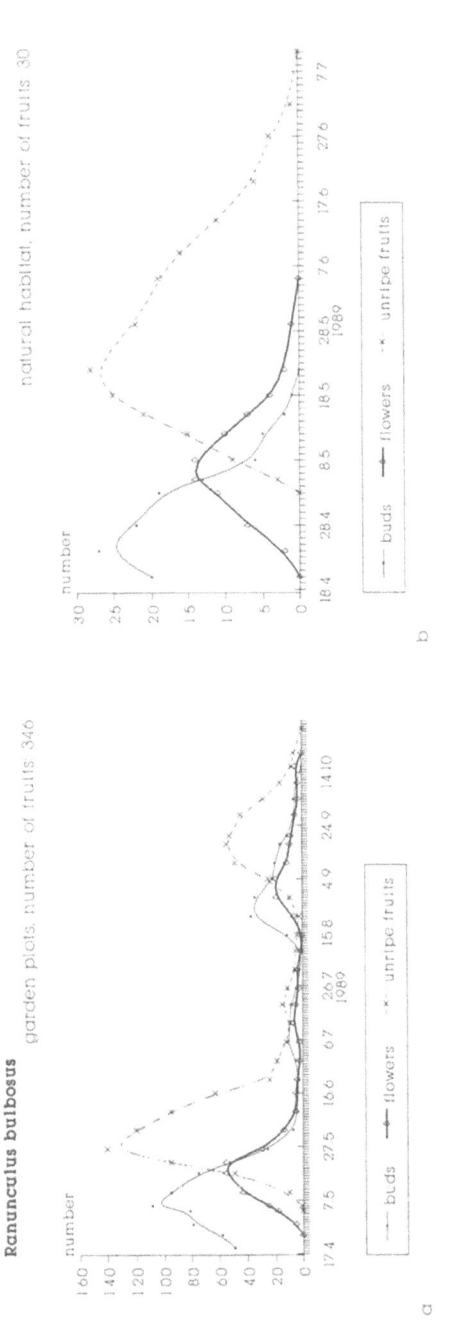

Fig. 1. Phenology of *R. bulbosus* at the garden plots (*a*) and at the natural habitat site (*b*) (10 individuals). *R. bulbosus* showed an obvious prolongation of its flowering period at the garden plot with a second flowering peak, a pattern found in 4 of the 5 species observed. With permission of the editor of Phyton (Horn, Austria).

in the garden plots in August and September. Such asexual flowers showed much reduced petals and only a few stamens. Therefore, it was almost impossible to distinguish whether these reproductive units were flowering or fruiting.

The situation in *R. repens* is somewhat different in that this species showed a similar phenological strategy at both sites. Duration of the flowering period as well as the number of flowers and fruits produced was quite similar. *R. repens* is known as a species which has a low seed production. It compensates its low generative reproduction by producing a number of creeping and rooting runners in accordance with ecological factors of the surroundings (LOVETT DOUST & al. 1990). This capacity to compensate a low generative reproduction by vegetative reproduction might be the reason for a comparably constant flowering and fruiting time at both the natural sites and the garden plots.

Pollination biology

The flower visitor spectra of all *Ranunculus* species observed comprises mainly representatives of three insect groups, *Diptera*, *Hymenoptera* and *Coleoptera*. Their proportions, however, vary between natural and garden sites as well as between the different species (Fig. 2).

The variety of flower-visiting insects and the substitution of species at different sites certainly are the expression of a generalistic pollination syndrome of the *Ranunculus* species (STEINBACH & GOTTSBERGER 1994). Moreover, several species of syrphids (KORMANN 1988), other *Diptera* (HOLLOWAY 1976), *Hymenoptera* FRISCH 1914, WESTRICH 1989), or *Coleoptera* are more adapted to a special habitat than to a flower. Therefore, at different sites, the *Ranunculus* species studied are exposed to a different pool of flower visiting insects, in accordance with the local fauna. The variety of insect species between different sites also indicates that observation of flower-visiting insects at just one site (MÜLLER 1873, KNUTH 1898) cannot reveal the whole visitor spectrum.

All *Ranunculus* species studied showed a broader flower visitor spectrum in the garden plots than at the natural habitat site. The number of flower-visiting species did not correlate with the frequency of flower visitation at the different sites. For example, groups of flower visitors with a high number of species do not necessarily represent the most frequent flower visitors (see Fig. 2 for the *Coleoptera* at the garden plots). On the other hand, a group of flower visitors may be visiting the flowers at the natural habitat site more frequently than at the garden site, while the number of species may be lower.

The large majority of the insects observed is polyphagous (KORMANN 1988, WESTRICH 1989), using a broad array of different food plants. These polylectic insects do learn flower characters, including colour of certain plant species in their surroundings (KUGLER 1951), and therefore can be as efficient as pollinators as other, more specialized groups.

The visitor spectrum comprises primitive insects such as beetles, flies, thrips, and micropterigid moths (see also KNUTH 1898), as well as derived insects such as syrphid-flies and bees. It needs to be tested whether the latter insects are primarily adapted to *Ranunculus* flowers or have learnt to exploit them secondarily.

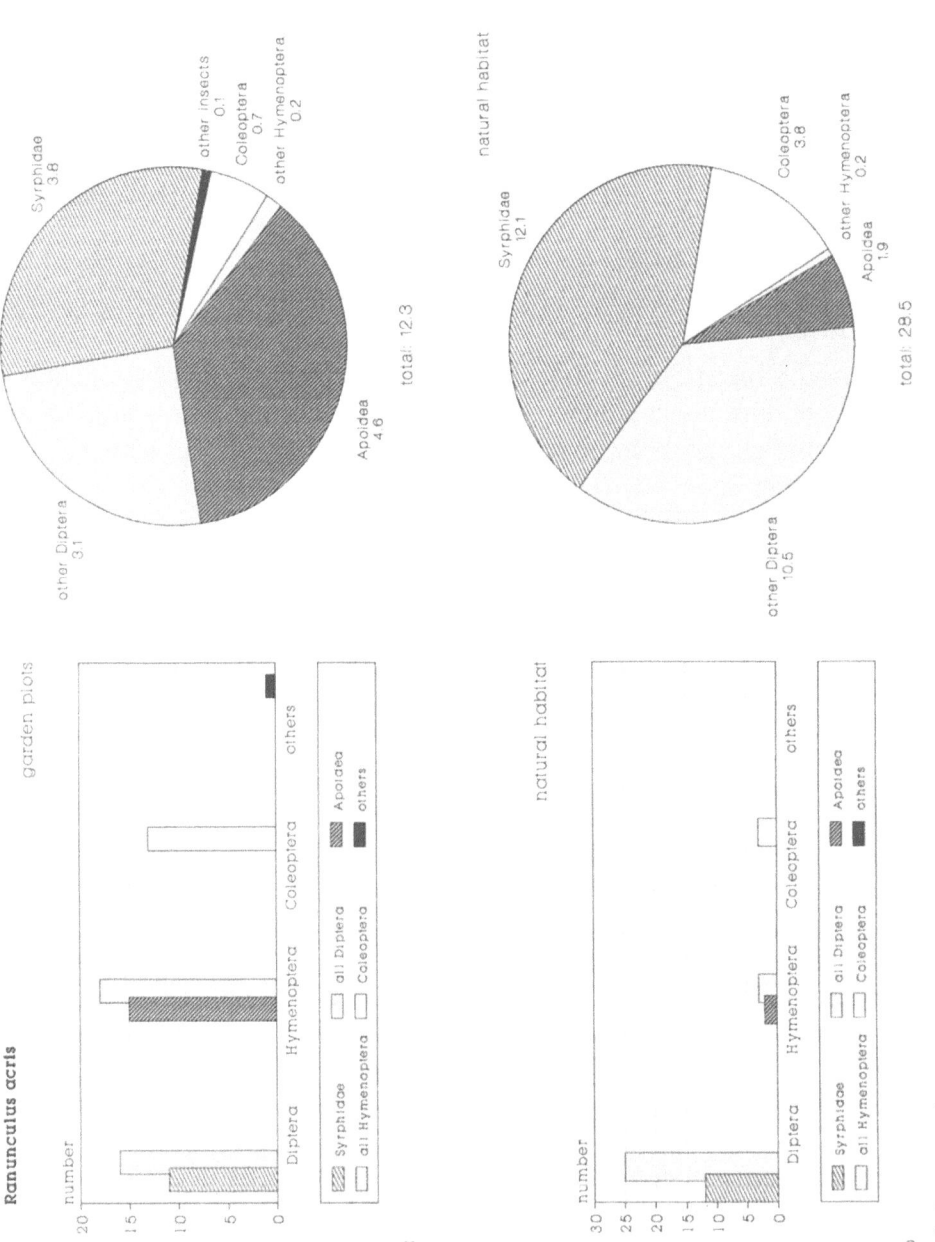

Fig. 2. *Ranunculus acris*. Left: Spectrum and number of flower visiting insect species. Right: Frequency of flower visits (per hour and 10 flowers) at the garden plots (*a*) and at the natural habitat site (*b*). *R. acris* was the species with the most diverse flower visitor spectrum. Visitation frequency was much higher at the natural habitat site than at the garden plots. With permission of the editor of Phyton (Horn, Austria).

References

FRISCH, K. VON, 1914: Der Farbensinn und Formensinn der Biene. – Zool. Jahrb., Abt. allgem. Zool. Physiol. **35**: 1–188.

GENTRY, A. H., 1974: Coevolutionary patterns in Central American *Bignoniaceae*. – Ann. Missouri Bot. Gard. **61**: 728–759.

HOLLOWAY, B. A., 1976: Pollen-feeding in hover-flies (*Diptera: Syrphidae*). – New Zealand J. Zool. **3**: 339–350.

KNUTH, P., 1898: Handbuch der Blütenbiologie, **2 (1)**. – Leipzig. Engelmann.

KORMANN, K., 1988: Schwebfliegen Mitteleuropas. – Landsberg a. L.: Ecomed.

KUGLER, H., 1951: Blütenökologische Untersuchungen mit Goldfliegen (Lucilien). – Ber. Deutsch. Bot Ges. **64**: 327–341.

LOVETT DOUST, J., LOVETT DOUST, L., GROTH, A. T., 1990: The biology of Canadian weeds. 95. *Ranunculus repens*. – Canad. J. Pl. Sci. **70**: 1123–1141.

MÜLLER, H., 1873: Die Befruchtung der Blumen durch Insekten und die gegenseitigen Anpassungen beider. – Leipzig: Engelmann.

NOGLER, G. A., 1984: Genetics of apospory in apomictic *Ranunculus auricomus*. – Bot. Helvetica **94**: 411–422.

STEINBACH, K., GOTTSBERGER, G., 1994: Phenology and pollination biology of five *Ranunculus* species in Giessen, Central Germany. – Phyton (Graz) **34**: 203–218.

WESTRICH, P., 1989: Die Wildbienen Baden-Württembergs, **1** und **2**. – Stuttgart: Ulmer.

Addresses of the authors: KATHRIN STEINBACH, Botanisches Institut I, Justus-Liebig-Universität, D-35390 Giessen, Germany. – GERHARD GOTTSBERGER, Abteilung Spezielle Botanik, Universität, D-89081 Ulm, Germany.

Accepted January 30, 1995

Pl. Syst. Evol. [Suppl.] 9: 325–326 (1995)

Pollination of *Nigella arvensis* (*Ranunculaceae*) (film presentation)

A. WEBER

Received September 22, 1994

Key words: *Ranunculaceae, Nigella*. – Floral structure, movements of floral organs, anther dehiscence, pollen presentation, pollination.

Abstract: The film (WEBER 1992) demonstrates and explains the elaborate flower structure of *Nigella arvensis* in relation to pollination. The course of anthesis (stamen and style movements in the male and female stage) is shown in rapid motion. Of special interest is the unique mode of anther dehiscence and pollen presentation. Pollination by various hymenopterans (*Apis mellifica, Pyrobombus lapidarius, Polistes dominulus, Eumenes pedunculatus*) is shown, as well as acitivites of illegitimate flower visitors (cf. *Halictus* spec., *Episyrphus* spec.).

Knowledge of the elaborate mode of pollination of *Nigella arvensis* is intimately connected with the name of C. K. SPRENGEL. In his epoch-making book "Das entdeckte Geheimnis de Natur..." SPRENGEL (1793) was the first to describe accurately the complex flower structure, to observe the movements of the stamens and the styles, and to relate them to pollination by the honey-bee. SPRENGEL's account remains the most detailed description available up to the present; later treatments of the subject are scarcely more than brief summaries of SPRENGEL's observations. When documenting the remarkable flower structure, floral dynamics and pollination (by various hymenopterans) in a film (WEBER 1992), a number of new observations were made that complete or correct those of SPRENGEL. A full presentation is given in WEBER (1993).

By invitation of the organizing committee, the film was presented at the meeting "Systematics and evolution of the *Ranunculiflorae*" at Bayreuth, Sept. 10–11, 1994. A brief summery of the topics shown in the film is given here.

a. The flower of *Nigella arvensis* is a 'round-about flower' (Umlaufblüte). Insects visiting the flower in search of nectar move around the centre of the flower and with their backs touch the anthers (male flower stage) or the styles (female stage).

b. At the onset of the male stage the numerous stamens stand ± erect. They curve outwards one by one, roughly in whorls of eight and strictly reflecting the order of initiation. When the anthers reach a horizontal position, the pollen is released. Then the stamens sink down.

c. The stamen movement is not continous, but is divided into three (four) phases. In the first phase (10–14 hours) the lower part of the filament inclines slightly while the upper part curves more strongly, so that the anther is brought into a horizontal position. After reaching this position, movement comes to a standstill (stationary phase, 10–14 hours). The third phase, in which the stamen sinks down, is much shorter than the previous ones (5–6 hours). Finally, the empty anthers curve up; this is a purely passive movement, apparently without any function, and may be referred to as a fourth phase.

d. Pollen release and exposure on the lower side of the anther is not simply by the anther opening at the lower (dorsal) side (as stated in the literature), but is brought about by a complicated mechanism. In each of the two anther locules the pollen grains stick together (by means of pollen-kitt) to form a compact package. The package itself adheres to the anther wall. The locule opens by a logitudinal ventral (!) slit, and by a very rapid twist the anther wall positions the pollen package on the lower side of the anther. There is a marked time difference in the opening of the two locules, so that each anther presents its two pollen packages in sequence.

e. Towards the end of the male stage the styles of the usually five carpels curve down and twist like a corkscrew. This ensures that in the female phase the stigmatic crests, which run down nearly the whole length of the style, touch the insect's back at several points.

f. The functionally most elaborate structures of the flowers are the nectaries. The nectar is secreted and kept in a cavity that is covered by a versatile lid. The margin of the lid and that of the cavity are studded with trichomes that operate as a filter and prevent contamination by pollen grains adhering to the proboscis of the insect. The lateral ridges bordering the cavity each end in a knob with a steep slope in front. Here the epidermal cells are completely smooth, while elsewhere they bear conspicuous cuticular folds, causing a velvety shean on the surface of the nectary. The slopes glisten like small mirrors in the sunlight and evidently act as false nectaries, simulating exposed nectar droplets. The dark transverse bands on the surface of the nectaries together form concentric rings that may serve to guide the insect in a circular course.

g. SPRENGEL regarded the honey-bee as the sole visitor and pollinator of *Nigella arvensis* because the size of its body fits perfectly into the flower. However, other insects also play an important role and must be regarded as legitimate pollinators, especially bumble-bees (*Bombus lucorum, Pyrobombus lapidarius*) and wasps (*Vespidae: Polistes dominulus, Eumenes pedunculatus; Sphecidae: Cerceris arenaria, Ammophila sabulosa*). In addition, species of *Halictus, Chrysis* and other small insects visit the flowers to drink nectar but are too small to touch the anthers or styles. Syrphids are illigetimate flower visitors too; they usually only feed on the pollen without touching the anthers with their backs.

References

SPRENGEL, C. K., 1793: Das entdeckte Geheimnis der Natur im Bau und in der Befruchtung der Blumen. – Berlin: Vieweg d. Ä.

WEBER, A., 1992: *Nigella arvensis* – Blüte und Bestäubung. – Film C 2238, Österr. Bundesinstitut f. d. Wiss. Film, Wien. – Begleitveröff. in Wiss. Film **44**: 53–60.

– 1993: Struktur, Antheseverlauf und Bestäubung der Blüte von *Nigella arvensis* (*Ranunculaceae*). – Verh. Zool.–Bot. Ges. Österreich **130**: 99–125 (in German.)

Address of the author: A. WEBER, Institute of Botany, University of Vienna, Rennweg 14, A-1030 Vienna, Austria.

Accepted December 15, 1994

Pl. Syst. Evol. [Suppl.] 9: 327–340 (1995)

Phylogeny and inflorescences of *Berberidaceae* – a morphological survey

MARTIN G. NICKOL*

Received November 7, 1994

Key words: *Berberidaceae, Achlys, Berberis, Bongardia, Caulophyllum, Diphylleia, Epimedium, Jeffersonia, Leontice, Mahonia, Nandina, Podophyllum, Ranzania, Vancouveria.* – Phylogeny, inflorescence morphology.

Abstract: The 13 genera of *Berberidaceae* are investigated morphologically and cladistically in order to examine the phylogenetic relationships among the extant taxa. Cladistic analysis of 59 morphological characters using PAUP resulted in one most parsimonious tree. In the present paper main emphasis is put on inflorescence morphological features, which contributes eight characters to the current data matrix. The analysis resulted in four major clades. *Nandina* has a basal position, and the next clade includes *Berberis* and *Mahonia*. *Ranzania* is sister genus to *Jeffersonia, Diphylleia/Podophyllum* and *Leontice/Bongardia*. These all form the sister group to a clade in which *Achlys* is sister taxon to *Epimedium/Vancouveria*. Various morphological characters suggest that *Caulophyllum* irrespective of its position in the most parsimonious tree is sister group to *Nandina*.

Berberidaceae consist of 13 genera and about 570 species. *Berberis* L. and *Mahonia* NUTT. have the largest number of species. Only these two woody genera reach the southern hemisphere, all others are concentrated in Asia and North America with only few species occurring in Europe and the northern parts of Africa. The centre of origin of the family may have been in eastern Laurasia (TAKHTAJAN 1969, WU 1983, YING 1984). Earlier analyses (MEACHAM 1980, TERABAYASHI 1985, LOCONTE & ESTES 1989) of generic relationships based on morphological characters presented different results. This was mainly due to inadequate morphological knowledge of most species of the family. Accordingly it was felt necessary to investigate various morphological features of as many specimens as possible in order to create a firm basis for an analysis, and to identify characters of high phylogenetic significance in *Berberidaceae*. The characters observed were analysed cladistically using parsimony. In the present paper the results of this cladistic analysis will be presented only briefly. Its main purpose is to utilize the resulting tree to assess inflorescence morphological features for their value in the systematics of this group. A more

* This publication is part of a doctoral thesis (D77).

detailed presentation covering other aspects of the morphology of the *Berberidaceae* is in preparation.

Inflorescence morphology of the *Berberidaceae* is known only insufficiently. Apart from a few reliable studies of individual genera like *Epimedium* (TOURN.) L. and *Vancouveria* MORR. & DECNE. (STEARN 1938), *Bongardia* C. A. MEYER (STAUFFER 1963) and *Achlys* DC. (ENDRESS 1989) and a comparative investigation of a larger number of genera (TROLL 1964, 1969; TROLL & WEBERLING 1989), no work exists comparing the inflorescence architecture of all extant genera. Complete coverage, however, should be the first step in the search for reliable phylogenetic relationships within any taxon.

Material and methods

Living plants were studied from the Botanic Gardens at Berlin, Göteborg, Kew, Mainz and München, as well as from natural habitats; herbarium material was received from K, BM, HBG, POLL, W and WU. For SEM studies material was fixed in Kew Mixture (FORMAN & BRIDSON 1991), chemically dehydrated in alcohol and FDA, critical-point dried and sputtered with gold. The preparations were viewed with a Cambridge Stereoscan Mk 250. Cladistic analysis was done using PAUP 3.1.1 (SWOFFORD 1993) with the branch-and-bound algorithm based on Wagner parsimony. To evaluate the relative robustness of clades found in the most parsimonious tree, strict consensus trees were constructed of all trees up to six steps longer than the most parsimonious tree until the consensus eventually collapsed into an unresolved bush. This procedure yields the decay index indicating the number of steps that must be added until a clade present in the minimum length tree collapses (BREMER 1988, DONOGHUE & al. 1992). *Nandina* was used to root the resulting tree. Altogether 59 morphological characters were investigated (see Appendix 1), of which 54 were informative; the data matrix is available on request. The complete matrix eventually will include about 70 characters. Character states were not polarized and were coded as unordered.

Results

Cladistic analysis of all characters. The cladistic analysis resulted in only one most parsimonious tree (Fig. 1a). This tree has a CI of 0.52 and is 160 steps long. There were nine trees one step longer and 17 trees two steps longer.

Inflorescence morphology. The monophyletic *Berberidaceae* share as an ancestral feature a terminal flower. The most basal genus in the cladistic analysis, *Nandina* THUNB., has the richest inflorescence, a many-flowered panicle. This panicle ends in a terminal flower, followed towards the base by single flowers without any further bracts, then by single flowers with sterile prophylls, and then by lateral branches with two or three flowers as shown in the diagram of Fig. 2a. The enrichment branches follow this architectural model, and the lateral branches also develop bracts and these can give rise to a flower. *Caulophyllum* MICHX., which here, as will be discussed below, is regarded as the closest relative of *Nandina*, has impoverished panicles. As in *Nandina* the lowermost enrichment branch of *Caulophyllum robustum* MAXIM. originates from the axil of a green leaf (Fig. 2b, c).

Berberis and *Mahonia* do not show differences between their individual inflorescences. These may be reduced to a single terminal flower as, e.g., in *Berberis angulosa* WALL., or may comprise a very high number of flowers, arranged as shown

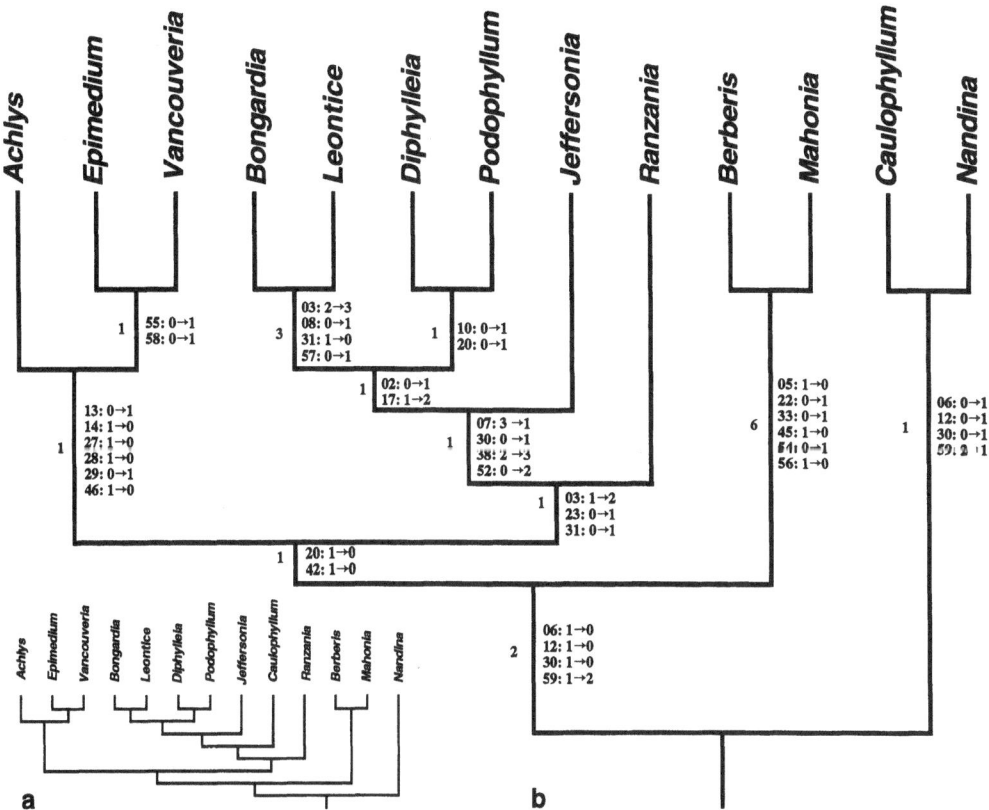

Fig. 1. Cladogram of the 13 genera of *Berberidaceae*. Numbers on the left of branches are the decay indices, character state changes are indicated on the right. For further discussion see text. *a* Most parsimonious tree. *b* Tree two steps longer with *Caulophyllum* as sister taxon to *Nandina*.

in Fig. 2 d–f. An autapomorphy of *Berberis* is the differentiation of long and short shoots. Only the short shoots as lateral axes form flowers, and the inflorescence here is terminal. *Mahonia* differs from this by having proliferating inflorescences, in which the shoot apex returns to vegetative growth after having produced floral branches in an acropetal sequence. In *Berberis* terminal flowers may be present or absent even within a single species.

Following the cladogram, the next clade comprises *Ranzania* T. Ito, *Jeffersonia* Bart., the sister genera *Podophyllum* L./*Diphylleia* Michx. and *Leontice* L./*Bongardia*. In *Ranzania* the terminal internodes of the sympodial rhizomes usually give rise to a few-flowered inflorescence of about four or five flowers. The axis ends in a terminal flower, and out of the axils of the two green leaves enrichment branches develop. The third leaf, a small bract, was sterile in every case observed. Often *Ranzania* develops only a single flower (Fig. 3a). This one-flowered inflorescence occupies a position different from the few-flowered inflorescence described above. It develops from the bud which in other genera (e.g., *Jeffersonia*) overwinters. In all such cases analysed an additional bud in the axil of one of the two leaf blades was found (Fig. 3a). *Ranzania* thus has a cymoid, consisting of a terminal flower and two mostly two- (or three-)flowered cymes. *Jeffersonia* shows some similarities with

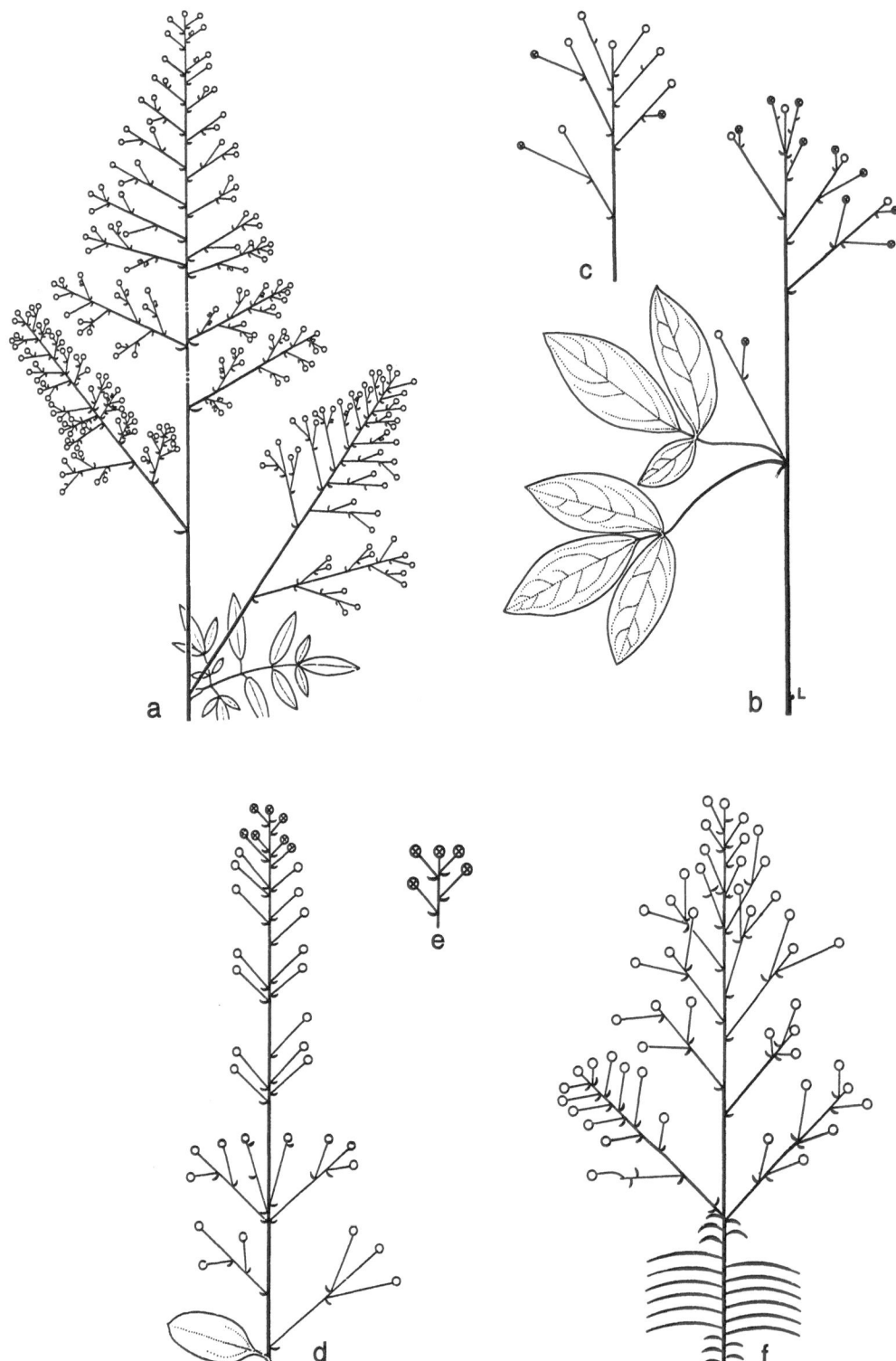

Fig. 2. Inflorescence diagrams. *a Nandina domestica. b Caulophyllum robustum. c Caulophyllum giganteum. d, e Berberis amurensis. f Berberis aggregata. – L* leaf

Ranzania. Jeffersonia diphylla Pers. (Fig. 3c) never produces more than one flower in one sympodial element. *Jeffersonia dubia* (Maxim). Baker & Moore (Fig. 3b) has only few lateral branches (paracladia) in its inflorescence and these mostly have two flowers.

Diphylleia has corymboidal thyrsoids (Fig. 3q, r). The long terminal internode is overtopped by the lateral branches. Monads are lacking, and the first branches beneath the terminal flower have at least two flowers. Because only prophyll α is fertile, the branches develop cincinnately.

The same feature is found in *Podophyllum*, but most species of this genus have fewer flowers in a single inflorescence when compared with *Diphylleia* (Fig. 3q, r). *Podophyllum hexandrum* Royle (Fig. 3d, e) from the Himalayas as well as *Podophyllum peltatum* L. (Fig. 3f–h) from North America are mainly single-flowered. Sometimes specimens of *P. peltatum* develop two flowers, a terminal flower and a monadic paracladium (Fig. 3g). Species with more flowers are concentrated in Asia: *P. delavayi* Franch. shows single flowers (monadic paracladia) beneath the terminal flower (Fig. 3i, j), and in *P. versipelle* Hance there are even branches with two flowers (dyadic paracladia, Fig. 3k–m). As in *Diphylleia* only the α-prophyll is fertile. *Podophyllum pleianthum* Hance has the most complex inflorescence (Fig. 3n–p). This is very similar to *Diphylleia* except for the corymboidal arrangement of the branches found in this genus. In *P. pleianthum* there also exist specimens with only solitary flowers beneath the terminal flower (Fig. 3p). From this it seems obvious that the two- or three-flowered branches of the few-flowered specimens are not true cymes but reduced lateral systems.

Bongardia and *Leontice* reveal common features (Fig. 4a–d). Figure 4a, b shows a few- and a many-flowered inflorescence of *Bongardia chrysogonum* (L.) Endl. The few-flowered inflorescence (Fig. 4a) appears as a raceme with a terminal flower. From this fact and from the structure of the many-flowered inflorescence it becomes obvious, however, that such a raceme is a reduced form of a paniculoid branching system (botryoid, Fig. 4b). In the distinct inhibition zone at the top of the inflorescence flowers are short-stalked or buds do not complete their development or develop very late respectively.

Leontice alberti Regel and *L. eversmanni* Bunge show a situation comparable to the two inflorescence forms of *Bongardia*. *Leontice alberti* lacks the terminal flower and all lateral branches are reduced to single flowers without any other bracts remaining (Fig. 4d). *Leontice eversmanni*, on the other hand, is completely identical with *Bongardia chrysogonum* (Fig. 4c).

In *Epimedium* one main feature of inflorescence morphology is the position of the green leaves. The primitive status seems to be the presence of cauline leaves as exemplified by the Chinese *Epimedium brevicornu* Maxim. (Fig. 4j). In the derived situation, the internode below the inflorescence is elongated and elevates the terminal inflorescence above the ground. The green leaves then arise directly from the rhizome but belong to the same sympodial element. This pattern is shown by *Epimedium alpinum* L. (Fig. 4k) and *E. pinnatum* Fisch. in DC. (Fig. 4l) which both live at high altitudes. These species develop leaves under or near the ground before climatic conditions are suitable for the exposure and full expansion of these leaves.

In *Epimedium*, inflorescence branches usually are many-flowered, and only in reduced systems does the inflorescence appear as a botryoid. This kind of inflo-

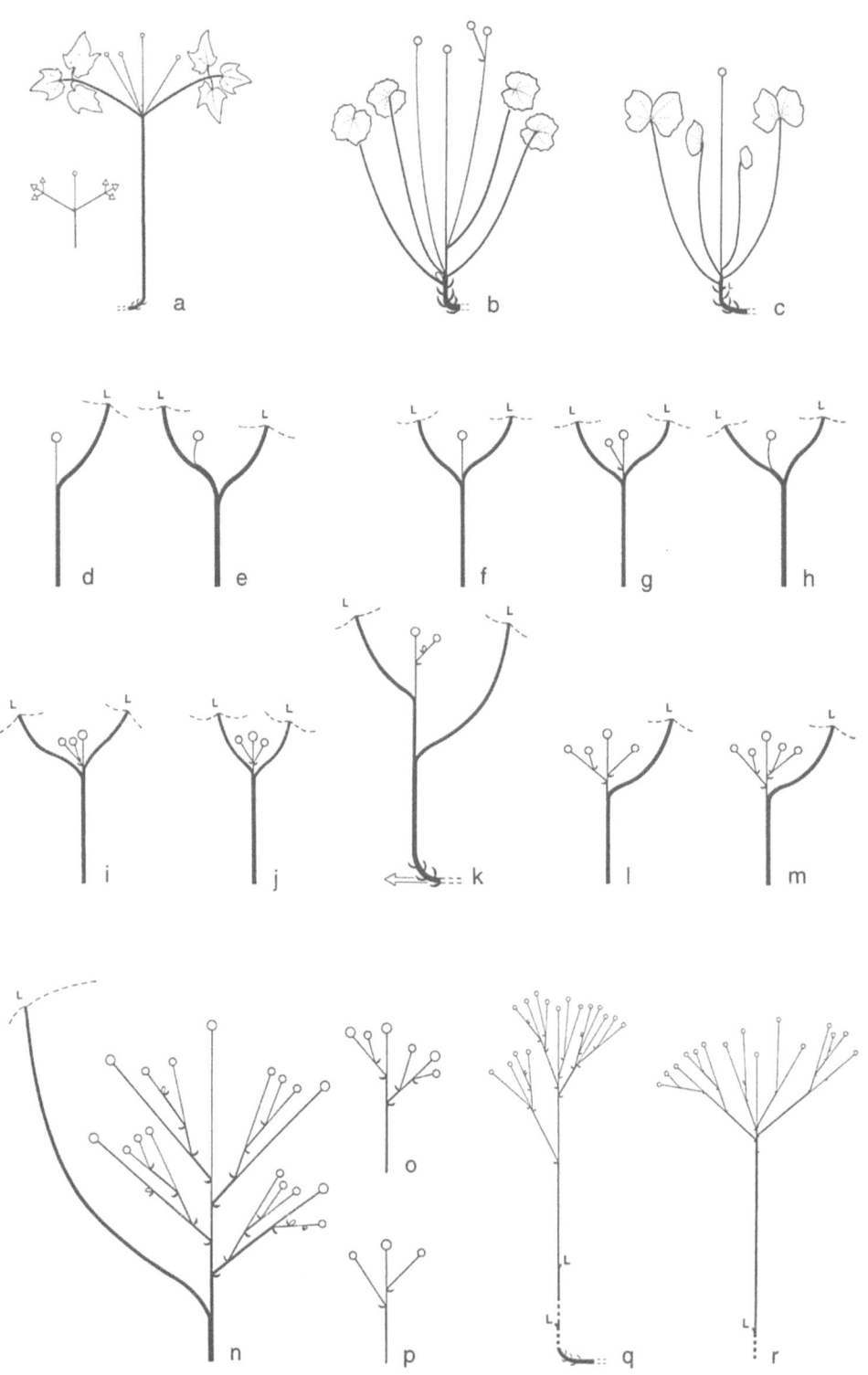

rescence seems to be a reduction of a thyrsoidal system as found, e.g., in *E. pinnatum* (Fig. 4l). This view is supported by the absence of prophylls, so that such a botryoid cannot be ancestral to a more profusely branched system but must be regarded as derived.

Vancouveria planipetala CALLONI has thyrsoidal inflorescences and the branches are excessively ramified in old plants (Fig. 4h). *Vancouveria chrysantha* GREENE and *Vancouveria hexandra* C. MORR. & DECNE. as comparably impoverished forms both have true cymes (Fig. 4g, i). The basal internodes of the lateral branches (hypopodes) are shortened. The bracts beneath the terminal flower mostly are sterile but sometimes produce a lateral flower. This suppression is homologous to the situation found in *Leontice*.

The last genus of this clade, *Achlys*, shows the most reduced inflorescence in *Berberidaceae* (Fig. 4e, f). The flowers lack a perianth and are arranged in no special order in a spike which ends with a terminal flower. The single leaf of each inflorescence belongs to that part of the sympodial rhizome which fully matures in the next season. It arises, as is the case in derived species of *Epimedium*, from below the ground. The mature inflorescence normally has no bracts, but in more than one case there are small bracts at its base. The spike-like inflorescence (a stachyoid) of *Achlys* is interpreted as a continued reduction of what we saw in *Epimedium*. From a raceme-like inflorescence, with an inhibition zone below the terminal flower, a stachyoid like that of *Achlys* can easily develop through a reduction of peduncles.

Discussion

Systematics. During work on the genera of *Berberidaceae* some systematic changes were made. *Plagiorhegma* MAXIM. was sunk into *Jeffersonia*, and *Dysosma* WOODSON (WOODSON 1928) and the illegitimate *Sinopodophyllum* T. S. YING (YING 1979) into *Podophyllum*. *Mahonia* was kept separate from *Berberis* because of its pinnate leaves, anther morphology, and the proliferation of the inflorescence apex as described above. Presumably *Mahonia* is a paraphyletic group. Whether these two genera are secondarily woody, as stated by LOCONTE & ESTES (1989) following SHEN (1954), seems doubtful. Results of BRAUN (1970), VASILEVSKAYA & OGANEZOVA (1974) and CARLQUIST (1975) do not confirm the assumption of secondary woodiness, and this is not suggested by the phylogeny presented here.

Gymnospermium SPACH was sunk into *Leontice* because their different fruit types alone give no convincing morphological reason to keep them separate at generic level. These suggestions as well as most of the morphological characters will be discussed in a future paper

Cladistics. The minimal length tree resulting from the cladistic analysis (Fig. 1a) is completely resolved. The tree was rooted with *Nandina* because of some ancestral features of this genus such as the paniculate inflorescence, the geographical range in the presumed centre of origin of the family and its woody habit with ancestral wood

◄────────

Fig. 3. Inflorescence diagrams. *a Ranzania japonica. b Jeffersonia dubia. c Jeffersonia diphylla. d, e Podophyllum hexandrum. f, g, h Podophyllum peltatum. i, j Podophyllum delavayi. k, l , m Podophyllum versipelle. n, o, p Podophyllum pleianthum. q Diphylleia cymosa* subsp. *sinensis. r Diphylleia cymosa. – L* leaf

anatomical characters (CARLQUIST 1975). The woody genera are basal to the herbaceous taxa which form two distinct clades, comprising *Achlys*, *Epimedium* and *Vancouveria* as sister group to the other herbaceous genera. The woodiness of the ancestral genera fits well into the results of other analyses (e.g., HOOT & CRANE 1995, this volume; BLACKMORE 1995, this volume), in which woody taxa like *Lardizabalaceae* and *Menispermaceae* occupy a more basal position in the *Ranunculiflorae* than *Berberidaceae*.

Although only one most parsimonious tree resulted from the present analysis I would prefer a tree two steps longer (Fig. 1b) which differs from the most parsimonious tree only by the position of *Caulophyllum* as sister group of *Nandina*. Such a tree would implicate that the herbaceous habit evolved at least twice in *Berberidaceae*: in *Caulophyllum* and all clades beyond that of *Berberis* and *Mahonia*. The close relationship of *Caulophyllum* and *Nandina* is indicated by their extremely similar mode of branching in the vegetative parts and their homologous inflorescence architecture. *Caulophyllum* looks like a *Nandina* with shortened internodes, but is identical in the construction of the branching system. The possession of usually two ovules as well as the reduced petiole are unique synapomorphies of this group (NICKOL, unpubl.).

Four groups resulted from the analysis. The relationship between *Nandina* and *Caulophyllum* was just discussed, and the close connection between *Berberis* and *Mahonia* is undoubtful. A closer relationship of *Ranzania* to this clade, as assumed by MEACHAM (1980) and LOCONTE & ESTES (1989), is not suggested by the present cladogram. Although *Ranzania* has preserved some common features with the woody genera *Berberis* and *Mahonia*, such as sensitive stamens, nectariferous glands at the base of the nectar leaves and similarities in the structure of the pollen exine (but not the actual pollen type!) (NOWICKE & SKVARLA 1981) the present analysis suggests that *Ranzania* and the other herbaceous genera of the *Berberidaceae* (except *Caulophyllum*) are derived from a more recent common ancestor. A closer relationship of *Achlys* to *Epimedium* and *Vancouveria* has been discussed since the floral analyses of TERABAYASHI (1985). It is also supported by other morphological characters, as discussed below.

The major clades of the cladogram (Fig. 1b) are supported by chromosome numbers (see also KIM & JANSEN 1995, this volume), except for *Leontice* (x = 8) which groups within an x = 6 group. *Nandina* is the only genus with x = 10 chromosomes. This is sometimes regarded as the chromosome base number of *Berberidaceae* (MIYAGJI 1930).

Inflorescence morphology. Inflorescence characters contributed only eight of the 59 characters used in the present cladistic analysis. The resulting tree with its putative phylogeny can be used in order to assess the characters of inflorescence morphology and to evaluate their systematic relevance. Lines of specialization can be seen within individual taxa of *Berberidaceae* (e.g., *Epimedium*). The ancestral inflorescence type of the family may have been a panicle with unspecialised,

Fig. 4. Inflorescence diagrams. *a, b Bongardia chrysogonum. c Leontice eversmanni. d Leontice albertii. e Achlys triphylla. f Achlys japonica. g Vancouveria chrysantha. h Vancouveria planipetala. i Vancouveria hexandra. j Epimedium brevicornu. k Epimedium pinnatum* subsp. *colchicum. l Epimedium alpinum*

allotropous flowers of medium size. Such floral organization may be regarded as primitive within angiosperms (GOTTSBERGER 1974, GOTTSBERGER & al. 1980, ENDRESS 1994a). Although it is widely accepted (WEBERLING 1992) that solitary flowers represent the primitive situation in angiosperms, it seems more likely that these solitary terminal flowers are the remnants of a reduced many-flowered inflorescence, of which only the terminal flower is developed (see also WEBERLING 1992, ENDRESS 1994b).

The first homoplastic feature of the inflorescence morphological complex to be discussed is the nature of the foliation within the inflorescence. This was used as a systematic character by, e.g., LOCONTE & ESTES (1989). Foliation can be bracteose, foliose or intermediate. Many-flowered synflorescences as found, e.g., in *Nandina* and *Bongardia*, are more variable in foliation (Figs 2a; 4a, b) than few-flowered ones. Bracteose inflorescences are found in the genera *Achlys*, *Vancouveria*, *Bongardia*, *Diphylleia*, *Podophyllum* and *Berberis*; frondo-bracteose inflorescences in species of *Caulophyllum*, *Nandina*, *Mahonia*, *Ranzania*, *Jeffersonia*, *Leontice* and *Epimedium*. The state of this character thus has changed repeatedly during evolution. This is also the case in other characters that involve only changes in the proportion of particular parts. The presence of an elongated internode below the inflorescence that raises the flowering entity above the ground is a synapomorphy of the *Achlys*-clade. On the other hand this character is homoplastic in the current tree and occurs also in *Bongardia*, *Diphylleia* and *Caulophyllum*. In all these cases a homologous internode is elongated. The parallel evolution of this character might be understandable for ecological reasons because the flowering branches have to be lifted above the ground for insect pollination and fruit dispersal. The shortening of the hypopode of the lateral branches in the inflorescences, as seen in its extreme form in *Vancouveria*, is also found in only distantly related taxa. It occurs, e.g., in species of *Podophyllum*, *Ranzania*, *Berberis* and *Mahonia*. Another case of homoplasy is the overtopping of the terminal flower by the lateral branches due to the elongation of their peduncles. It is obvious that mere morphometric features are not morphologically meaningful and are irrelevant for the systematics of *Berberidaceae*.

However, the inflorescence morphology of *Berberidaceae* provides valuable information for phylogenetic considerations. The clade of *Achlys*, *Epimedium* and *Vancouveria* is supported by the nutation of the young inflorescence as well as by the presence of an inhibition zone below the terminal flower. Other features supporting this clade are the presence of connectives with distinctly pointed tips, bracts on the rhizome (cataphylls) with a sclerotised tip, and stipules of characteristic shape which can fuse in all three genera. A character of inflorescence morphology supporting the *Ranzania*-clade is the elongation of the terminal internode. In the *Jeffersonia*-clade only axes function as storage organs. The sister group relationship of *Bongardia/ Leontice* and *Diphylleia/Podophyllum* is defined by the composition of the enrichment branches as well as the cotyledonar tube, the micromorphology of the petals and stigma morphology. The correct position of *Bongardia* in this clade is slightly doubtful because floral morphology could not yet be studied as thoroughly as in other genera due to a lack of suitable material. The shortening of the rhizome into a tuber as found in *Bongardia* as well as in *Leontice* might be the result of adaptation to similar habitats and not of common origin. The only character supporting a closer relationship of *Bongardia* to *Achlys*, which was suggested by LOCONTE & ESTES

(1989), would be the inhibition zone below the terminal flower. The large number of homologies in other inflorescence morphological features of *Leontice* and *Bongardia*, however, contradict this view.

I wish to express my sincerest thanks to Prof. JOACHIM W. KADEREIT for valuable discussions. For further stimulating comment I like to thank Prof. ALBRECHT SIEGERT and Prof. STEFAN VOGEL. Ms ANKE BERG and Mr ANDREAS HORN kindly assisted in the preparation of the figures and in laboratory work. I also want to thank the directors and staff of the Herbaria and Botanic Gardens named above for the opportunity to study their material. Prof. WILLIAM T. STEARN (Kew), Prof. ANTON WEBER (Wien), Prof. FOCKO WEBERLING (Ulm), Ms SUE ZMARZTY (Kew) and MR MARKUS BECKER (Mainz) made helpful comments on earlier drafts of the manuscript.

References

BLACKMORE, S., STAFFORD, P., PERSSON, V., 1995: Palynology and systematics of *Ranunculiflorae*. – Pl. Syst. Evol. [Suppl.] **9**: 71–82.

BRAUN, H. J., 1970: Funktionelle Histologie der sekundären Sproßachse. I. Holz. – In Handbuch der Pflanzenanatomie **9**: 1–190. – Berlin, Stuttgart: Borntraeger.

BREMER, K., 1988: The limits of amino acid sequence data in angiosperm phylogenetic reconstruction. – Evolution **42**: 795–803.

CARLQUIST, S., 1975; Ecological strategies of xylem evolution. – Berkeley, Los Angeles, London: University of California Press.

DONOGHUE, M. J., OLMSTEAD, R. G., SMITH, J. F., PALMER, J. D., 1992: Phylogenetic relationships of *Dipsacales* based on *rbc*L sequences. – Ann. Missouri Bot. Gard. **79**: 333–345.

ENDRESS, P., 1989: Chaotic floral phyllotaxis and reduced perianth in *Achlys* (*Berberidaceae*). – Bot. Acta **102**: 159–163.

– 1994a: Floral structure and evolution of primitive angiosperms: recent advances. – Pl. Syst. Evol. **192**: 79–97.

– 1994b: Diversity and evolutionary biology of tropical flowers. – Cambridge, New York, Melbourne: Cambridge University Press.

FORMAN, L., BRIDSON, D., 1991: The herbarium handbook. Rev. edn. – Kew: Royal Botanic Gardens.

GOTTSBERGER, G., 1974: The structure and function of the primitive angiosperm flower – a discussion. – Acta Bot. Neerl. **23**: 461–471.

– SILBERBAUER-GOTTSBERGER, I., EHRENDORFER, F., 1980: Reproductive biology in the primitive relic angiosperm *Drimys brasiliensis* (*Winteraceae*). – Pl. Syst. Evol. **135**: 11–39.

HOOT, S. B., CRANE, P. R., 1995: Inter-familial relationships in the *Ranunculidae* based on molecular systematics. – Pl. Syst. Evol. [Suppl.] **9**: 119–131.

KIM, Y.-D., JANSEN, R. K., 1995: Phylogenetic implications of chloroplast DNA variation in the *Berberidaceae*. – Pl. Syst. Evol. [Suppl.] **9**: 341–349.

LOCONTE, H., ESTES, J. R., 1989: Phylogenetic systematics of *Berberidaceae* and *Ranunculales* (*Magnoliidae*). – Syst. Bot. **14**: 565–579.

MEACHAM, C. A., 1980: Phylogeny of the *Berberidaceae* with an evaluation of classifications. – Syst. Bot. **5**: 149–172.

MIYAGJI, Y., 1930: Beiträge zur Chromosomenphylogenie der Berberidaceen. – Planta **11**: 650–659.

NOWICKE, J. W., SKVARLA, J. J., 1981: Pollen morphology and phylogenetic relationships of the *Berberidaceae*. – Smithsonian Contrib. Bot. **50**: 1–83.

SHEN, Y., 1984: Phylogeny and wood anatomy of *Nandina*. – Taiwania **5**: 85–92.

STAUFFER, H. U., 1963: Gestaltwandel bei Blütenständen von Dicotyledonen. – Bot. Jahrb. Syst. **82**: 216–251.

STEARN, W. T., 1938: *Epimedium* and *Vancouveria* (*Berberidaceae*), a monograph. – Bot. J. Linn. Soc. **51**: 409–535.

SWOFFORD, D. L., 1993: PAUP: Phylogenetic analysis using parsimony, ver. 3.1.1. Computer program distributed by the Illinois Natural History Survey, Champaign, Ill.

TAKHTAJAN, A. L., 1969: Flowering plants: origin and dispersal. – Edinburgh: Oliver & Boyd.

TERABAYASHI, S., 1985: The comparative floral anatomy and systematics of the *Berberidaceae*: II. Systematic considerations. – Acta Phytotax. Geobot. **36**: 1–13.

TROLL, W., 1964: Die Infloreszenzen. Typologie und Stellung im Aufbau des Vegetationskörpers, I. – Stuttgart: Fischer.

TROLL, W., 1969: Die Infloreszenzen. Typologie und Stellung im Aufbau des Vegetationskörpers, II/1. – Stuttgart, Portland: Fischer.

TROLL, W., WEBERLING, F., 1989: Infloreszenzuntersuchungen an monotelen Familien. Materialien zur Infloreszenzmorphologie. – Stuttgart, New York: Fischer.

VASILEVSKAYA, V. K., OGANEZOVA, G. G., 1974: Osbennosti formirovaniya struktury steblya drevesnykh barbarisovykh. – Vestn. Leningrad Univ., Biol. **15**: 38–45.

WEBERLING, F., 1992: Primitive Blütenstände bei primitiven Angiospermen? – Stapfia **28**: 29–51.

WOODSON, R. E., 1928: *Dysosma*: a new genus of *Berberidaceae*. – Ann. Missouri Bot. Gard. **15**: 335–340.

WU, Z., 1983: On the significance of Pacific intercontinental discontinuity. – Ann. Missouri Bot. Gard. **70**: 577–590.

YING, T.-S., 1979: On *Dysosma* WOODSON and *Sinopodophyllum* YING, gen nov. of the *Berberidaceae*. – Acta Phytotax. Sinica **17**: 15–23.

– 1984. The floristic relationship of the temperate forest regions of China and the United States. – Ann. Missouri Bot. Gard. **70**: 597–604.

Address of the author: MARTIN G. NICKOL, Institut für Spezielle Botanik und Botanischer Garten, Johannes Gutenberg-Universität, D-55099 Mainz, Germany.

Accepted January 18, 1995

Appendix 1. Characters used in cladistic analysis

Seedlings
    1 epigeous (0)/hypogeous (1)
    2 cotyledons free (0)/fused (1)

Axis
    3 internodes elongated (1)/shortened (2)/contracted (3)
    4 woody (0)/woody at the base (1)/herbaceous (2)
    5 monopodial (0)/sympodial (1)
    6 phyllotaxis 2/5 (0)/1/2 (1)
    7 storage organ: no special organ (0)/axis (1)/bracts (2) axis and bracts (3)

Leaves
    8 hypostomatic (0)/amphistomatic (1)
    9 basitonous (0)/acrotonous (1)
    10 epeltate (0)/peltate (1)
    11 not articulated (0)/articulated (1)
    12 petiole not depressed (0)/depressed (1)
    13 stipules free (0)/fused (1)

14 bracts of the rhizome with a sclerotised tip (0)/without (1)

15 wax on leave blade: no (0)/lower side (1)/upper side (2)

16 vernation: conduplicate (0)/umbraculiform (1)/involute (2)

17 petals: cuticle without (0)/with lengthwise (1)/with crosswise stripes (2)

18 sepals: cuticle without (0)/with lengthwise (1)/with crosswise stripes (2)

19 leaf margins not thickened (0)/thickened (1)

20 leaf blade nodding when emerging from the soil (0)/not (1)

Root system

21 primary root persisting (0)/adventitious root system (1)

22 no root saplings (0)/root saplings present (1)

23 roots with negative geotropism present (0)/not (1)

Inflorescence

24 terminal flower with same number of organs as lateral flowers (0)/not (1)

25 overtopping of terminal flower (0)/not (1)

26 bracteose (1) /frondose (0)/frondo-bracteose (2)

27 nutation of young inflorescence (0)/not (1)

28 inhibition zone present (0)/absent (1)

29 internode of disjunction absent (0)/internode of disjunction present (1)

30 hypopode not elongated (0)/hypopode elongated (1)

31 terminal internode not elongated (0)/elongated (1)

Floral organs

32 flower upright (0)/nodding (1) different (2)

33 fruit upright (0)/nodding (1) different (2)

34 actinomorphic (0)/zygomorphic (1) asymmetrical (2)

35 organs spirally arranged (0)/whorled (1)

36 UV-absorption by sepals (0)/petals (1)/sepals and petals (2)/filaments (3)/anthers (4)/ovarium (5)/stigma (6)/stigma and anthers (7)

Gynoecium

37 proterogynous (0)/proterandrous (1)

38 stigmatic surface convex (0)/crateriform (1) margin bent upwards (2)

39 stylar canal absent (0)/present (1)

40 whole stigmatic surface receptive (0)/only margin receptive (1)

Androecium

41 alternating (0)/obdiplostemonous (1)

42 anthers extrorse (0)/introrse (1)

43 thecae symmetrical (0)/asymmetrical (1)

44 staminodes present (0)/absent (1)

45 filaments sensitive (0)/not (1)

46 connective tip present (0)/absent (1)

47 anthers without sterile appendages (0)/with sterile appendages (1)

48 number and arrangement of stamina variable (0)/fixed (1)

49 filaments free (0)/connected with each other (1)/connected with nectary leaves (2)

50 opening of anthers valvate, valves curling inwards (0)/valvate, valves curling outwards (2)/longicidal (1)

Nectaries

51 nectar production: no nectar (0)/nectar leaves (1)/filaments (2)/gynoecium (3)

52 nectar leaves: lacking (2)/spurred (1)/without spur (0)

Sepals and petals

53 showiness: sepals (0)/petals (1)/androecium (2)/petals and androecium (3)

54 sepals not persisting (0)/persisting (1)

Fruit
    55 pericarp berry-like (0)/dry (1)/aborting (2)
    56 fruit surface waxy (0)/not (1)
Ovules
    57 anatropous (0)/atropous (1)/campylotropous (2)
    58 aril absent (0)/present (1)
    59 number of ovules: one (0)/two (1)/many (2)

Pl. Syst. Evol. [Suppl.] 9: 341–349 (1995)

Phylogenetic implications of chloroplast DNA variation in the *Berberidaceae*

Y.-D. Kim and R. K. Jansen

Received October 21, 1994

Key words: *Berberidaceae*. – Chloroplast DNA (cpDNA), *rbc*L, restriction site data.

Abstract: In spite of extensive studies using chromosome number, floral anatomy, pollen morphology, and serology, substantial disagreement still remains in the systematics of the *Berberidaceae*. We employed two different molecular systematic approaches, gene sequencing and restriction site analysis, to test competing hypotheses about intergeneric relationships in the family. Both *rbc*L gene sequence and restriction site data are congruent in recognizing four major chromosomal groups ($x = 10$, 8, 7, and 6). Fragmentation of the *Berberidaceae* into smaller families, especially separation of *Nandina* as a distinct family was not supported. The large woody genera, *Berberis* and *Mahonia*, are related to the monotypic herbaceous genus, *Ranzania*. *Bongardia* is nested within the $x = 6$ clade, and has a remote relationship to the *Leontice* group with which the genus has been placed for long time in most previous classifications. Although there are some major incongruences, our molecular systematic studies support the morphological classification proposed by Loconte & Estes (1989).

Berberidaceae sensu lato are a heterogeneous assemblage of 13–17 genera which exhibit much morphological divergence. Because of the diverse morphology several different classification systems have been proposed for the family (Table 1). In spite of extensive systematic studies based on floral anatomy (Terabayashi 1977, 1978, 1979, 1981, 1983a, 1983b, 1983c, 1985a, 1985b), serology (Jensen 1973), pollen morphology (Kosenko 1980, Nowicke & Skvarla 1981), and chromosome number (Kuroki 1965, 1968, 1970), substantial disagreement still remains regarding phylogenetic relationships among the genera.

One major systematic issue of *Berberidaceae* is fragmentation of the family into smaller coherent families (Airy Shaw 1973, Hutchinson 1973, Takhtajan 1969). The separation of the monotypic genus *Nandina* as the *Nandinaceae* is the most well known controversy (Nakai 1936, Shen 1954, Kumazawa 1938, Cronquist 1981). Another major disagreement among the classification systems (Table 1) is the phylogenetic affinity of the distinct woody genera *Berberis* and *Mahonia* to the herbaceous genera. In Airy Shaw's system these genera are placed with *Epimedium* and *Vancouveria*, and the four genera are segregated as the separate family *Ber-*

Table 1. Different classification systems proposed for the *Berberidaceae*. MEACHAM (1980) recognized four subfamilial groups of *Berberidaceae* (excluding *Nandina*) without taxonomic treatment. Numbers in LOCONTE & ESTES (1989) system denote base chromosome numbers

AIRY SHAW (1973)	TERABAYASHI (1985)	MEACHAM (1980)	LOCONTE & ESTES (989)
Nandinaceae	*Nandinoideae*	*Nandinaceae*	*Nandinoideae*
Nandina	*Nandina*	*Nandina*	*Nandina* (10)
			Berberidoideae
Berberidaceae	*Berberidoideae*	*Berberidaceae*	*Leoticeae*
Mahonia	*Berberideae*		*Caulophyllum* (8)
Berberis	*Mahonia*	*Mahonia*	Leontice (8)
Epimedium	*Berberis*	*Berberis*	*Gymnospermium* (8)
Vancouveria	*Ranzanieae*	*Ranzania*	*Berberideae*
	Ranzania		*Berberidinae*
Leonticaceae	*Epimediae*		*Mahonia* (7)
Caulophyllum	*Epimediinae*	*Caulophyllum*	*Berberis* (7)
Leontice	*Epimedium*	*Leontice*	*Ranzania* (7)
Gymnospermium	*Vancouveria*	*Gymnospermium*	*Epimediinae*
Bongardia	*Jeffersonia*	*Bongardia*	*Epimedium* (6)
	Plagiorhegma		*Vancouveria* (6)
Podophyllaceae	*Achlys*	*Epimedium*	*Jeffersonia* (6)
Ranzania	*Leoticinae*	*Vancouveria*	*Plagiorhegma* (6)
Jeffersonia	*Caulophyllum*	*Jeffersonia*	*Achlys* (6)
Plagiorhegma	*Leontice*	*Plagiorhegma*	*Bongardia* (6/7)
Achlys	*Gymnospermium*	*Achlys*	*Podophyllum* (6)
Podophyllum	*Bongardia*		*Dysosma* (6)
Dysosma	*Podophylleae*		*Sinopodophyllum* (6)
Diphylleia	*Podophyllum*	*Podophyllum*	*Diphylleia* (6)
	Dysosma	*Dysosma*	
	Diphylleia	*Diphylleia*	

beridaceae sensu stricto. However, cladistic analyses of morphological data (MAECHAM 1980, LOCONTE & ESTES 1989) suggested that the Japanese monotypic genus *Ranzania* is more closely associated with the woody genera. A recent palynological study by KOSENKO (1980) also casted doubt on the phylogenetic position of *Bongardia*, which includes the single species *B. chrysogonum*. The species was originally described as *Leontice chrysogonum* (LINNAEUS 1753) and has been placed with *Caulophyllum*, *Gymnospermiun*, and *Leontice* in most classification systems. LOCONTE & ESTES (1989), however, argued that a superficial resemblance between the *Bongardia* and *Leontice* group is the result of parallel evolution, and placed the genus with *Achlys*.

To test competing hypotheses about the intergeneric relationships of *Berberidaceae* mentioned above, we employed two different molecular systematic approaches, gene sequencing and restriction site studies of the chloroplast genome.

Material and methods

*rbc*L **sequencing.** We carefully selected nine ingroup genera representing indisputable groups in most classification systems (Table 1). Excluding the sequence of *Mahonia*, the *rbc*L sequences of the rest genera of the *Berberidaceae* are new reports (Y.-D. KIM & R. K. JANSEN, unpubl.). The sequence data are available from the first author upon request. The sequences of *Mahonia* and genera from related families including *Caltha*, *Xanthorhiza*, *Cocculus*, *Akebia*, *Sanguinaria*, and *Papaver* were gathered from GenBank (see CHASE & al. 1993 for accession number). The sequences of *Hydratis*, *Glaucidium*, and *Tinospora* were kindly provided by Dr SARA HOOT.

Total DNA was isolated from fresh leaves using the CTAB method (DOYLE & DOYLE 1987), followed by purification in cesium chloride-ethidium bromide gradients. The DNA region (ca. 1.5 kb) containing the *rbc*L gene was amplified using Polymerase Chain Reaction (PCR) (OLMSTEAD & al. 1992). To remove unused amplifying primers and deoxynucleotide triphosphates, the PCR product was electrophoresed and recovered from the agarose gel by the geneclean kit (Bio 101) according to the manufacturer's instruction. Purified double stranded PCR product was sequenced by the snap chill method (WINSHIP 1989) using the Sequenase version 2.0 kit (United States Biochemical Corporation), S^{35}-labelled ATP, and a series of forward and reverse sequencing primers (obtained from G. ZURAWSKI).

A total of 1402 bp of *rbc*L sequences from 18 genera was analyzed by maximum parsimony method using PAUP version 3.1 (SWOFFORD 1993). To identify minimum length trees and multiple islands of equally parsimonious trees (MADDISON 1991) we used heuristic search and MULPARS options with 100 random entries of the taxa, TBR branch swapping, and steepest descent. Bootstrap (FELSENSTEIN 1985) and decay analyses (BREMER 1988, DONOGHUE & al. 1992) were performed to evaluate the degree of support for given clades.

Restriction site analyses. Based on information from JOHANSSON & JANSEN (1993), we selected 10 restriction endonucleases (*Ava*I, *Bam*HI, *Ban*I, *Bgl*II, *Eco*RV, *Hinc*II, *Hind*III, *Nsi*I, *Xba*I, and *Xho*I) for total DNA digestion. Agarose gel electrophoresis, bidirectional transfer of DNA onto Zetabind (AMF Cuno), P^{32} labelling of probe DNA by nick-translation, filter hybridization (at 58 °C for 16 h), and autoradiography were performed according to methods described by PALMER (1986). The *Nicotiana tabacum* L. cpDNA clone bank (SUGIURA & al. 1986) including 43 subclones (OLMSTEAD & PALMER 1992) were used for the filter hybridizations.

We constructed restriction site maps of chloroplast genome for 16 species representing 16 genera of the *Berberidaceae* and two species from two putatively related families. The map of *Bongardia* is incomplete at this point, thus the genus is excluded from the analyses. A detailed restriction site map of *Mahonia higginsae* cpDNA (KIM & JANSEN 1994), which was constructed by single and double digestion procedure, was utilized as a reference to determine relative positions of the restriction sites in other cpDNAs. Phylogenetic analyses were performed with the same options used for *rbc*L sequence data. Chloroplast DNA restriction site maps for 10 enzymes and the data matrix used in this study are available from the first author upon request.

Results

*rbc*L **sequence analyses.** Sequences of *rbc*L were compared for 18 taxa including nine genera from five allied families. All *rbc*L sequences were complete except the sequence of *Papaver* where ca. 200 bp were missing at 3′ end. Parsimony analyses with equal character weighting resulted in a single minimum length tree (Fig. 1) of 610 steps with a consistency index (CI) of 0.47 excluding autapomorphic characters

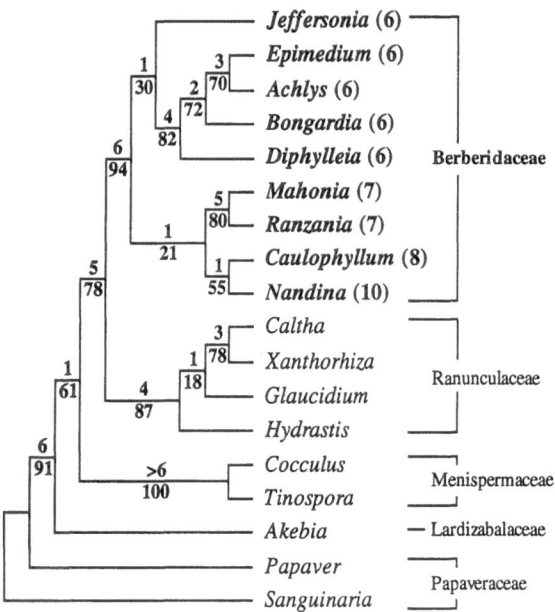

Fig. 1. Single most parsimonious tree based on *rbc*L sequence data (length = 610 steps, CI = 0.47, RI = 0.44). Genera of *Berberidaceae* are indicated in bold. Bootstrap and decay values are indicated below and above the nodes, respectively. Numbers in the parentheses of berberidaceous genera denote base chromosome numbers

and a retention index (RI) of 0.44. Based on information from much a broader study by CHASE & al. (1993) the tree was rooted by *Papaveraceae* (i.e., *Sanguinaria* and *Papaver* in this data matrix).

Although many clades are weakly supported as indicated by low bootstrap and decay values, several important phylogenetic implications are evident in the *rbc*L gene tree. (1) *Ranunculaceae* (including *Glaucidium* and *Hydrastis*) are the sister group to the *Berberidaceae* with a moderate bootstrap value (78%) and a high decay index of 5. (2) *Berberidaceae* s.l. form a strong monophyletic groups as indicated by high bootstrap (94%) and decay (6) values. (3) The basal position or sister group relationship of *Nandina* to the other genera of Berberidaceae is not supported because the genus is grouped with *Caulophyllum*. (4) The Japanese herbaceous genus *Ranzania* is related to *Mahonia*, and the latter genus is remotely related with *Epimedium*. (5) *rbc*L data support the mnophyly of the x = 6 group (LOCONTE & ESTES 1989), although the clade is weakly supported (bootstrap value of 30% and decay index of 1). This clade consists of two lineages, *Jeffersonia* and the four genera *Epimedium*, *Achlys*, *Bongardia*, and *Diphylleia*, which are strongly held together with a bootstrap value of 82% and a decay index of 4. (6) In contrast to many traditional classifications of the *Berberidaceae*, the monotypic genus *Bongardia* is evidently nested within the x = 6 group.

Restriction site data analyses. Chloroplast genomes of the genera included in this study showed no major structural rearrangements except the extension of the inverted repeat in *Berberis* and *Mahonia* (KIM & JANSEN 1994). Several regions, especially the area corresponding to tobacco cpDNA probes 10–12 (OLMSTEAD & PALMER 1992), were excluded from the analyses because of considerable length variation. A total of 545 restriction site mutations was mapped, 467 of which were phylogenetically informative for the 18 genera compared.

Phylogenetic analyses produced two equally parsimonious Wagner trees with 949 steps, a CI of 0.55, and a RI of 0.73. A strict consensus tree is shown in Fig. 2. The

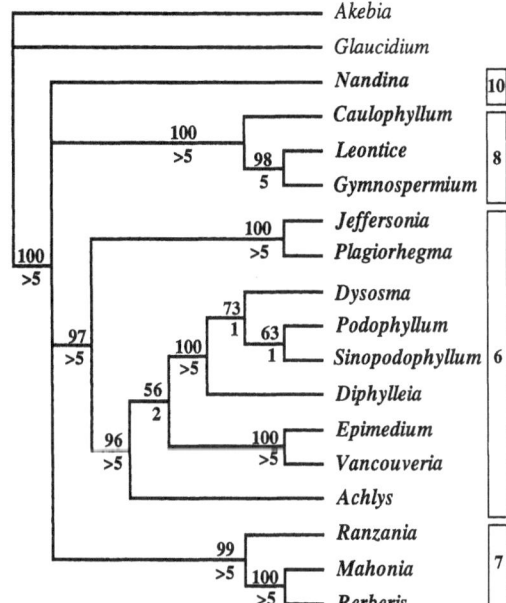

Fig. 2. Strict consensus tree of two equally parsimonious Wagner trees based on cpDNA restriction site data (length = 949 steps, CI = 0.55, RI = 0.73). Genera of *Berberidaceae* are indicated in bold. Bootstrap and decay values are indicated below and above the nodes, respectively. Numbers in the boxes denote base chromosome numbers

only difference between the two equally parsimonious trees is the relationship of the *Berberis* group and *Nandina* (*Nandina* is basal in one tree, whereas the *Berberis* group is the basal in the other). One major outcome of these trees is the strong support for four chromosomal groups (x = 10, 8, 7, 6). The first chromosomal group (x = 10 group) includes the single genus, *Nandina*, which is often viewed as basal in traditional classifications. The x = 8 group, which has a high bootstrap value (100%) and decay index (over 5), consists of three genera, *Caulophyllum*, *Leontice*, and *Gymnospermium*. Within the x = 8 group, *Caulophyllum* is the sister group to the *Gymnospermium* and *Leontice* group. Two closely related woody genera, *Berberis* and *Mahonia*, formed a strong monophyletic group (the x = 7 group) with their herbaceous sister genus *Ranzania*. The largest chromosomal group of the family (x = 6) is also well supported (97% bootstrap value and decay index of over 5). Within the x = 6 group, four evident lineages are observed: (1) *Achlys*, (2) *Epimedium* and *Vancouveria*, (3) *Diphylleia*, *Podophyllum*, *Dysosma*, and *Sinopodophyllum*, (4) *Jeffersonia* and *Plagiorhegma*. Among the four lineages, it is also evident that *Jeffersonia-Plagiorhegma* form a sister group to the other three lineages (core genera of the x = 6 group) which are strongly held together. However, relationships among the chromosomal groups are not resolved in this analysis.

Discussion

Sister group relationship and monophylyl of the *Berberidaceae* s.l. Although the major intention of this study was to test competing hypotheses about the intergeneric relationships within the *Berberidaceae*, attention was also paid to the monophyly and sister group relationships of the family. Our results, as shown in *rbc*L tree (Fig. 1), provide substantial support for the monophyly of the *Berberidaceae* s.l. Paraphyly of the family in the big *rbc*L tree by CHASE & al. (1993) is probably due to insufficient taxon sampling. *Ranunculaceae* appear to form sister group to the *Berberidaceae* as suggested by CRONQUIST (1981). A close phylogenetic relationship between *Lardizabalaceae* and *Berberidaceae* (LOCONTE & ESTES 1989) is not supported by the *rbc*L data. *Hydrastis* and *Glaucidium*, the ranunculaceous genera which have often been considered as closely related to the members of *Berberidaceae* or segregated as distinct families (KUMAZAWA 1930, DAHLGREN 1980, TOBE & KEATING 1985, THORNE 1992), occupy basal positions in the *Ranunculaceae*.

Fragmentation of the *Berberidaceae* s.l. Because of high levels of morphological divergence, few systematists have included all genera of *Berberidaceae* sensu lato in a single family (Table 1). The monotypic genus *Nandina* has most frequently been separated from the *Berberidaceae*. If included in the family the genus is considered as the sister group or basal to the remaining genera of the family based on the base chromosome number (x = 10), unusual floral morphology (TERABAYASHI 1983c), and distinct wood anatomy (SHEN 1954). Separation of *Nandina* as a distinct family or the basal position of genus is not supported by either *rbc*L sequence or restriction site data. The genus is associated with *Caulophyllum* in the *rbc*L tree. The basal position of *Nandina* is observed (with weak support) only in one of the two equally parsimonious trees from the restriction site data. It is interesting that the close relationship between *Nandina* and *Caulophyllum*, which is suggested for the first time in this study, is also independently proposed by NICKOL (1995, this volume) and ADACHI & al. (1995, this volume) based on their morphological and molecular data analyses, respectively. Segregation of other smaller families such as *Podophyllaceae*, *Leonticaceae*, and *Berberidaceae* sensu stricto is not warranted by our molecular data simply because the segregation leads to recognition of paraphyletic or polyphyletic groups.

Intergeneric relationships. The most important finding from our molecular systematic study is that the cpDNA data support the recognition of four chromosomal groups in the *Berberidaceae*. This is also one of major congruent features between the recent morphological tree (LOCONTE & ESTES 1989) and our molecular tree. However, neither tree exhibits robust phylogenetic relationships among the four chromosomal groups. MIYAJI (1930) postulated that there was a phylogenetic reduction of the base chromosome number (from 8 to 7 to 6) in the evolution of the *Berberidaceae*. However, MOORE (1963) opposed MIYAJI's idea and hypothesized that karyotypes of *Ranzania* (x = 7) and *Caulophyllum* (x = 9) are derived from those of *Epimedium* (x = 6) through chromosome misdivisions. Critical reevaluation of the chromosomal characters including more sampling of representative genera will provide valuable insights into the evolution and phylogenetic relationships among the four chromosomal groups.

The x = 8 group, corresponding to tribe *Leonticeae*, includes three genera. Among the three genera *Leontice* and *Gymnospermium* form a monophyletic group and they are the sister group to *Caulophyllum*. The same relationships have been proposed by LOCONTE & ESTES (1989). *Bongardia*, a genus previously placed with those three genera, is excluded from this group, and forms a sister group to *Achlys* and *Epimedium* in the x = 6 group.

Both *rbc*L sequence and restriction site data strongly support a clade consisting of *Ranzania* and *Berberis* & *Mahonia* group. Although *Ranzania* differs markedly from *Berberis* & *Mahonia* in gross morphology and habit, phylogenetic affinities among the genera were implicated by their sharing sensitive stamens (KUMAZAWA 1937), pollen wall structure (NOWICKE & SKVARLA 1981), and a base chromosome number of 7 (LOCONTE & ESTES 1989). This clade was also recognized in recent classifications of MEACHAM (1980) and LOCONTE & ESTES (1989).

The placement of *Jeffersonia* and *Plagiorhegma* in the $X = 6$ group is supported by both *rbc*L and restriction site data. Additional information from Internal Transcribed Spacer (ITS) sequences of nuclear ribosomal repeat also supports this position (Y.-D. KIM & R. K. JANSEN, unpubl.). The position of these genera in morphological trees of MAECHAM (1980) and LOCONTE & ESTES (1989) is incongruent with our result. However, features of the wood anatomy of *Jeffersonia* suggest that the genus may have diverged relatively early from other *Berberidaceae* (S. CARLQUIST, pers. comm.). Excluding *Jeffersonia* and *Plagiorhegma*, the rest of the genera of the $x = 6$ clade form a core group. Within this group in the *rbc*L tree, *Diphylleia* formed a strongly supported clade with *Podophyllum* and allied genera, *Sinopodophyllum* and *Dysosma*. *Achlys* is grouped with *Epimedium*, with *Bongardia* as sister taxon. In contrast to the *rbc*L tree, restriction site data do not support the *Achlys* & *Epimedium* clade. Relationships within the core group are also weakly supported in *rbc*L tree. The ITS sequence data of the $x = 6$ group, however, provide relatively strong support of the relationships observed in the *rbc*L tree (Y.-D. KIM & R. K. JANSEN, unpubl.). Inclusion of *Bongardia* in restriction site analyses will enable us to directly compare the resulting tree with the *rbc*L and ITS trees for the $X = 6$ group. Mapping of the chloroplast genome of *Bongardia* is in progress.

We thank JAN JOHANSSON, LES MEHRHOFF, TOM PHILLBRICK, and JUN WEN for collecting plant material, the Arnold Arboretum, Kew, New York, Rancho Santa Ana, and Edinburgh Botanical Gardens for access to their living collections, GERALD ZURAWSKI for *rbc*L sequencing primers, SARA HOOT for sequence data and helpful comments, J. D. PALMER for providing tobacco cpDNA probes and V. ALBERT and J. PALMER for comments on an earlier draft of the manuscript. This paper represents a portion of the senior authors Ph.D. thesis. The research was supported by NSF grant DEB-9318279 RKJ.

References

ADACHI, J., KOSUGE, K., DENDA, T., WATANABE, K., 1995: Phytogenetic relationships of the *Berberidaceae* based on partial sequences of the *gap*A gene. – Pl. Syst. Evol. [Suppl.] **9**: 351–353.

AIRY SHAW, H. K., 1973: In WILLIS, J. C., (Ed.): A dictionary of the flowering plants and ferns, 8th edn. – Cambridge: Cambridge University Press.

BREMER, K., 1988: The limits of amino acid sequence data in angiosperm phylogenetic reconstruction. – Evolution **42**: 795–803.

CHASE, W. M., SOLTIS, D. E., OLMSTEAD, R. G., MORGAN, D., LES, D. H., MISHLER, B. D., DUVALL, M. R., PRICE, R. A., HILIS, H. G., QIU, Y.-L., KRON, K. A., RETTIG, J. H., CONTI, E., PALMER, J. D. MANHART, J. R., SYTSMA, K. J., MICHAELS, H. J., KRESS, W. J., KAROL, K. G., CLARK, W. D., HEDRÉN, M., GAUT, B. S., JANSEN, R. K., KIM, K.-J., WIMPEE, C. F., SMITH, J. F., FURNIER, G. R., STRAUSS, S. H., XIANG, Q.-Y., PLUNKETT, G. M. SOLTIS, P. S., SWENSEN, S. M., WILLIAMS, S. E., GADEK, P. A., QUINN, C. J., EGUIARTE, L. E., GOLENBERG, E., LEARN, G. H., Jr., GRAHAM, S. W., BARRETT, S. C. H., DAYANANDAN, S., ALBERT, V. A., 1993: Phylogenetics of seed plants: an analysis of nucleotide sequences from the plastid gene *rbc*L. – Ann. Missouri Bot. Gard. **80**: 528–580.

CRONQUIST, A., 1981: An integrated system of classification of flowering plants. – Cambridge: Cambridge University Press.

DAHLGREN, R. M. T., 1980: A revised system of classification of the angiosperms. – J. Linn. Soc. Bot. **80**: 91–124.

DONOGHUE, M. J., OLMSTEAD, R. G., SMITH, J. F., PALMER, J. D., 1992: Phylogenetic relationships of *Dipsacales* based of *rbcL* sequences. – Ann. Missouri. Bot. Gard **79**: 333–345.

DOYLE, J. J., DOYLE, J. L., 1987: A rapid DNA isolation procedure for small quantities of fresh leaf tissue. – Phytochem. Bull. **19**: 11–15.

FELSENSTEIN, J., 1985: Confidence limits on phylogenies: an approach using the bootstrap. – Evolution **39**: 783–791.

HUTCHINSON, J., 1973: The families of flowering plants, 3rd edn. – Oxford: Oxford University Press.

JENSEN, U., 1973: The interpretation of comparative serological results: Novel symposium 25. – In BENDZ, G., SANTESSON, J., (Eds): Chemistry in botanical classification, pp. 217–227. – New York: Academic Press.

JOHANSSON, J. T., JANSEN, R. K., 1993: Chloroplast DNA variation and phylogeny of the *Ranunculaceae*. – Pl. Syst. Evol. **187**: 29–49.

KIM, Y.-D., JANSEN, R. K., 1994: Characterization and phylogenetic distribution of a chloroplast DNA rearrangement in the *Berberidaceae*. – Pl. Syst. Evol. **193**: 107–114.

KOSENKO, V. N., 1980: Comparative palynomorphological study of the family *Berberidaceae*: 2. Morphology of the pollen grains of the genera *Gymnospermium*, *Leontice*, *Caulophyllum*, *Bongardia*, *Epimedium*, *Vancouveria*, *Achlys*, and *Jeffersonia*. – Bot. Ž (Moscow, Leningrad) **65**: 1412–1423.

KUMAZAWA, M., 1930: Morphology and biology of *Glaucidium palmatum* SIB. et ZUCC. with notes of affinities to the allied genera *Hydrastis*, *Podophyllum*, and *Diphylleia*. – J. Fac. Sci. Univ. Tokyo, Sect. 3, Bot. **2**: 345–380.

– 1937: *Ranzania japonica*, its morphology, biology and systematic affinities. – Japan. J. Bot. **9**: 55–70.

– 1938: Systematic and phylogenetic consideration of the *Ranunculaceae* and *Berberidaceae*. – Bot. Mag. (Tokyo) **52**: 9–15.

KUROKI, Y., 1865: Chromosome study in three species of *Berberidaceae*. – Mem. Ehime Univ., Sect. II, Sci., Ser. B, Biol. **5**: 19–24.

– 1968: Chromosome study in five species of *Berberidaceae*. – Mem. Ehime Univ., Sect. II, Sci., Ser. B, Biol. **6**: 11–16.

– 1970: Chromosome study in five species of *Berberidaceae*. – Mem. Ehime Univ., Sect. II, Sci., Ser. B, Biol. **6**: 63–69.

LINNAEUS, C., 1953: Species plantarum. 2 vols. – Stockholm.

LOCONTE, H., ESTES, J. R., 1989: Phylogenetic systematics of *Berberidaceae* and *Ranunculales* (*Magnoliidae*). – Syst. Bot. **14**: 565–579.

MADDISON, D. R., 1991: The discovery and importance of multiple islands of most parsimonious trees. – Syst. Zool. **40**: 315–328.

MEACHAM, C. A., 1980: Phylogeny of the *Berberidaceae* with an evoluation of classifications. – Syst. Bot. **5**: 149–172.

MIYAJI, Y., 1930: Beiträge zur Chromosomenphylogenie der Berberidaceen. – Planta **11**: 650–659.

MOORE, R. J., 1963: Karyotype evolution in *Caulophyllum*. – Canad. J. Genet. Cytol. **5**: 384–388.

NAKAI, T., 1936: Flora Sylvatica Koreana. 21. – Keijo.

NICKOL, M. G., 1995: Phylogeny and inflorescences of *Berberidaceae* – a morphological survey. Pl. Syst. Evol. [Suppl.] **9**: 327–340.

NOWICKE, J. W., SKVARLA, J. J., 1981: Pollen morphology and phylogenetic relationships of the *Berberidaceae*. – Smithson. Contrib. Bot. **50**: 1–83.

OLMSTEAD, R. G., PALMER, J. D., 1992: A chloroplast phylogeny of the *Solanaceae*: subfamilial relationships and character evolution. – Ann. Missouri Bot. Gard. **79**: 346–360.

– MICHAELS, H. J., SCOTT, C. M., PALMER, J. D., 1992: Monophyly of the *Asteridae* and identification of their major lineages inferred from DNA sequences of *rbc*L. – Ann. Missouri Bot. Gard. **79**: 249–265.

PALMER, J. D., 1986: Isolation and structural analysis of chloroplast DNA. – Meth. Enzymol. **118**: 167–186.

SHEN, Y., 1954: Phylogeny and wood anatomy of *Nandina*. – Taiwania **5**: 85–92.

SUGIURA, M., SHINOZAKI, K., ZAITA, N., KUSUDA, M., KUMANO, M., 1986: Clone bank of the tobacco (*Nicotiana tabacum*) chloroplast genome as a set of overlapping restriction endonuclease fragments: mapping of eleven ribosomal protein genes. – Pl. Sci. **44**: 211–216.

SWOFFOR, D. L., 1993: PAUP. Phylogenetic analysis using parsimony, version 3.1 for Macintosh. Computer package. – Champaign, Illinois Nat. Hist. Survey.

TAKHTAJAN, A., 1969: Flowering plants: original and dispersal. Transl. by C. JEFFREY. – Washington, D.C.: Smithsonian Institution Press.

TERABAYASHI, S., 1977: Studies in the morphology and systematics of *Berberidaceae*. I. Floral anatomy of *Ranzania japonica*. – Acta Phytotax. Geobot. **28**: 45–57.

– 1978: Studies in the morphology and systematics of *Berberidaceae*. II. Floral anatomy of *Mahonia japonica* and *Berberis thunbergii*. – Acta Phytotax. Geobot. **29**: 106–118.

– 1979: Studies in the morphology and systematics of *Berberidaceae*. III. Floral anatomy of *Epimedium grandiflorum* ssp. *sempervirens* and *Vancouveria hexandrum*. – Acta Phytotax. Geobot. **30**: 153–168.

– 1981: Studies in the morphology and systematics of *Berberidaceae*. IV. Floral anatomy of *Palgiorhegma dubia*, *Jeffersonia diphylla*, and *Achlys triphylla* ssp. *japonica*. – Bot. Mag. (Tokyo) **94**: 141–157.

– 1983a: Studies in the morphology and systematics of *Berberidaceae*. V. Floral anatomy of *Caulophyllum*, *Leontice*, *Gymnospermium*, and *Bongardia*. – Mem. Fac. Sci. Kyoto Univ., Ser. Biol. **8**: 197–217.

– 1983b: Studies in the morphology and systematics of *Berberidaceae*. VI. Floral anatomy of *Diphylleia*, *Podophyllum*, and *Dysosma*. – Acta Phytotax. Geobot. **34**: 27–47.

– 1983c: Studies in the morphology and systematics of *Berberidaceae*. VII. Floral anatomy of *Nandina domestica*. – J. Phytogeogr. Taxon. **31**: 16–21.

– 1985a: The comparative floral anatomy and systematics of the *Berberidaceae*. I. Morphology. – Mem. Fac. Sci. Kyoto Univ., Ser. Biol. **10**: 73–90.

– 1985b: The comparative floral anatomy and systematics of the *Berberidaceae*. II. Systematic considerations. – Acta Phytotax. Geobot. **36**: 1–13.

THORNE, R. F., 1992: An updated phylogenetic classification of the flowering plants. – Aliso **13**: 365–389.

TOBE, H., KEATING, R. C., 1985: The morphology and anatomy of *Hydrastis* (*Ranunculaceae*): systematic reevaluation of the genus. – Bot. Mag. (Tokyo) **98**: 291–316.

WINSHIP, P. R., 1989: An improved method for direct sequencing of PCR amplified material using dimethyl sulphoxide. – Nucleic Acids Res. **17**: 1266

Address of the authors: YOUNG-DONG KIM* and ROBERT K. JANSEN, Department of Botany, University of Texas, Austin, TX 78713–7640, USA. *Current address: Department of Biology, Hallym University, Chunchon, Kangwon Province, South Korea.

Accepted February 20, 1995

Pl. Syst. Evol. [Suppl.] 9: 351–353 (1995)

Phylogenetic relationships of the *Berberidaceae* based on partial sequences of the *gap*A gene

J. Adachi, K. Kosuge, T. Denda, and K. Watanabe

Received October 27, 1994

Key words: *Berberidaceae, Nandina. – gap*A, phylogeny, sequencing.

Abstract: A Phylogenetic tree of six species in the *Berberidaceae* based on partial sequences of the nuclear gene for glyceraldehyde-3-phosphate dehydrogenase (*gap*A) shows that the *Berberidaceae* is a distinctly monophyletic group, and divided into three clades:*Nandina-Caulophyllum, Mahonia-Ranzania, Epimedium-Diphylleia.*

The *Berberidaceae* comprise 16 genera with about 650 species, distributed throughout temperate Eurasia and North America. This family is characterized by pseudomonomerous pistils, anthers opening by valves, trimerous flowers and isoquinoline alkaloids. Intergeneric relationships, however, are poorly known due to distinct morphologies which are difficult to interpret in a phylogenetic context, specifically the systematic position of the monotypic genus *Nandina* Thunb.

In order to clarify the systematic position of *Nandina* and some other genera, we have analyzed partial sequences of the *gap*A gene. Total DNA was extracted from six species in the *Berberidaceae* and two species in the *Ranunculaceae* with the CTAB (Cetyl-trimethyl-ammonium bromide) method (David & Steven 1990). Voucher specimens are deposited at the Kobe University. The *gap*A fragment was amplified from total DNA using the polymerase chain reaction (PCR) and sequenced directly. The phylogenetic tree was constructed based on 661 bp of the partial sequences of exon 4 and 5 using the neighbour-joining (NJ) method (Fig. 1).

Phylogenetic analysis

In the NJ tree, *Nandina* and other five species of the *Berberidaceae* form an independent clade separated from the *Ranunculaceae* by relatively high bootstrap support values (93%). This clade is divided into three lineages, (1) *Nandina-Caulophyllum* Michaux., (2) *Mahonia* Nutt. – *Ranzania*, T. Ito, (3) *Epimedium* L. – *Diphylleia* Michaux.

Nandina and *Caulophyllum* form a clade with relatively low bootstrap support: 54%. Kumazawa (1938 a, b) recognized *Nandina* as the monotypic family *Nandinaceae* based on the ovular structure and type of anther dehiscence. Meacham

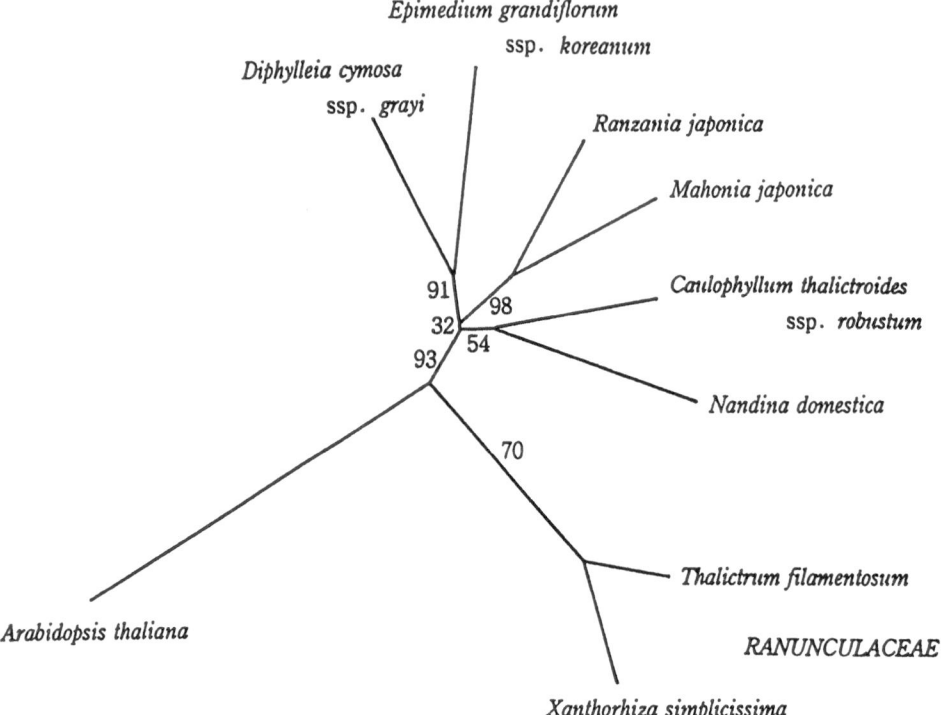

Fig 1. Molecular phylogenetic tree of *gap*A segments in the *Berberidaceae* using the neighbour-joining method (the Clustal V program constructed by HIGGINS 1991). The sequence for *Arabidopsis thaliana* (L.) HEYNH. *gap*A was obtained from SHIH & al. (1991). The numbers on the nodes indicated the proportion of 100 bootstrap replicates

(1980) also excluded *Nandina* from the *Berberidaceae* based on the longitudinal anther dehiscence and distinct chromosome number. Alternatively, TERABAYASHI (1982, 1983) treated it as the monotypic subfamily *Nandinoideae* in the *Berberidaceae*. LOCONTE & ESTES (1989) divided this family into two subfamilies. *Nandina* was considered to be the most primitive genus of the *Berberidaceae* because of the relative plesiomorphy of woody habit and the longitudinal anther dehiscence and was placed by these authors in a monotypic subfamily. *Caulophyllum* was placed in a different subfamily, *Berberidoideae*. In contrast, KIM & JANSEN (1995, this volume) suggested that *Nandina* and *Caulophyllum* form a clade based on *rbc*L sequence and RFLP data. In spite of various morphological dissimilarities, *gap*A sequence data suggests that *Nandina* is included in the *Berberidaceae*, related to *Caulophyllum*.

The *Mahonia-Ranzania* and *Epimedium-Diphylleia* clades are supported with high bootstrap support values: 98% and 91%, respectively. TERABAYASHI (1977, 1982) suggested that *Ranzania* was placed as an intermediate between arborescent *Berberis* and *Mahonia* and herbaceous *Epimedium*, *Achlys* DC. and allied genera. LOCONTE & ESTES (1989) proposed that *Ranzania* and *Berberis-Mohonia* formed a clade, and recognized the sister group relationship between *Ranzania* group and subtribe *Epimediinae* (including *Epimedium* and *Diphylleia*). GapA sequence data also supports that *Ranzania* is closely related to *Mahonia* and the sister group relationship between *Mahonia-Ranzania* and *Epimedium-Diphylleia*.

We are indebted to Dr YOUNGDONG KIM and Mr TODD BARKMAN for carefully reading and correcting the English text, and to Dr SUSUMU TERABAYASHI and Prof. KEN INOUE for providing plant materials.

References

DAVID, M. W., STEVEN, J. K., 1990 : DNA Extraction from a previously recalcitrant plant genus. – Pl Mol. Biol. Rep. **8** (3): 180–185.

HIGGINS, D. G., 1991 : Clustal V Documentation. – EMBL.

KIM, Y.-D., JANSEN, R. K., 1995: Phylogenetic implications of chloroplast DNA variation in the *Berberidaceae*. – Pl. Syst. Evol. [Suppl.] **9**: 341–349.

KUMAZAWA, M., 1938a : Systematic and phylogenetic consideration of the *Ranunculaceae* and *Berberidaceae*. – Bot. Mag. (Tokyo) **52**: 9–15.

– 1938b : On the ovular structure in the *Ranunculaceae* and *Berberidaceae*. – Japan. J. Bot. **14**: 10–25.

LOCONTE, H., ESTES, J. R., 1989: Phylogenetic systematics of *Berberidaceae* and *Ranunculales* (*Magnoliidae*). – Syst. Bot. **14**: 565–579.

MEACHAM, C. A., 1980: Phylogeny of the *Berberidaceae* with an evaluation of classifications. – Syst. Bot. **5**: 149–172.

SHIH, M.-C., HEINRICH, P. C., GOODMAN, H. M., 1991 : Cloning and chromosomal mapping of nuclear genes encoding chloroplast and cytosolic glyceraldehyde-3-phosphate dehydrogenase from *Arabidopsis thaliana*. – Gene **104**: 133–138.

TERABAYASHI, S., 1977: Studies in the morphology and systematics of *Berberidaceae*. I. Floral anatomy of *Ranzania japonica*. – Acta Phytotax. Geobot. **28**: 45–57.

– 1982: Systematic consideration of the *Berberidaceae*. – Acta Phytotax. Geobot. **33**: 355–370. (In Japanese.)

– 1983: Studies in the morphology and systematics of *Berberidaceae*. VII. Floral anatomy of *Nandina domestica*. – J. Phytogeogr. Taxon. **31**: 16–21.

Address of the authors : JUNKO ADACHI, Dr KEIKO KOSUGE, TESTSUO DENDA, Prof. KUNIAKI WATANABE, Department of Biology, Faculty of Science, Kobe University, Rokkodai Nada-ku, Kobe 657, Japan.

Accepted January 25, 1995

Pl. Syst. Evol. [Suppl.] 9: 355–361 (1995)

Combination of data in phylogenetic analysis

H.-J. Bandelt

Received September 19, 1994

Key words: Phylogenetic analysis, combination of data, consensus techniques, networks.

Abstract: In phylogenetic analysis, when faced with different data sets from different sources, two approaches, "taxonomic congruence" and "total evidence", compete with one another in seeking a best-fitting hypothesis. In either approach one deliberately ignores the notorious uncertainty typically associated with the estimate of a best tree. It is proposed here to represent the partly conflicting information contained in the data by specific networks with reticulations. This allows more flexibility in evaluating and eventually combining data.

Should data sets be pooled – to strive for "total evidence" (MIYAMOTO 1985, KLUGE 1989) – or be analyzed separately in order to seek for "taxonomic congruence" (MICKEVICH 1978)? This controversy is not only a philosophical debate but also has a practical impact on current phylogenetic studies (e.g., HEDGES & BEZY 1993 versus CROTHER & PRESCH 1992). One is not really forced to chose between these two extreme alternatives since a less stereotyped phylogenetic analysis would anyway incorporate features of either approach; cf. SWOFFORD (1991), DE QUEIROZ (1993), BULL & al. (1993), and HUELSENBECK & al. (1994). The latter alliance of authors emphasizes the importance of detecting character "heterogeneity" (that may reflect different phylogenetic histories). Separate analyses of data sets or even parts thereof can attribute to this goal. Standard aggregation of the partial information is not without problem. "If we compare only the optimal ⋯ trees for each data set, we effectively ignore the uncertainty associated with the estimate of the tree" (SWOFFORD 1991). This caveat, in particular, applies to the way of combination favoured by DOYLE (1992), BAUM (1992), RAGAN (1992); for further discussion of those related methods, see RODRIGO (1993), BAUM & RAGAN (1993), and WILLIAMS (1994). Any estimation of trees can be regarded as a drastic reduction of the data by tentatively discarding irrelevant or misleading information while enhancing only a small set of consistent signals. A somewhat less extreme reduction, leading to reticulate networks, could be more appropriate in capturing the essential features of data sets in intermediate stages of the phylogenetic analysis.

Networks

Ambiguity In many phylogenetic studies, no single dichotomous tree (as a best hypothesis) would receive overwhelming support from the data. Then "an ambigu-

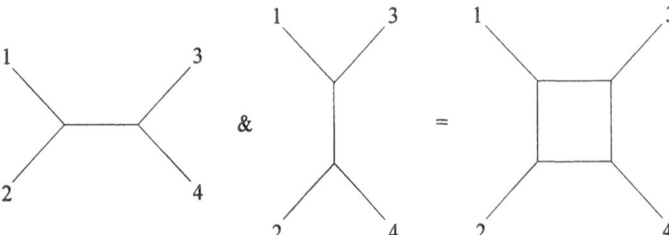

Fig. 1. Network expressing ambiguity between two alternative quartet trees

ous solution that contains the truth is, in many situations, preferable to an unambiguous solution that is wrong" (SWOFFORD 1991). There are however different degrees of ambiguity, even in the case of four taxa. Assume that the three alternative (unrooted) trees for taxa no. 1, 2, 3, 4 are almost equally likely. This lack of resolution between the alternatives would usually be expressed by a polytomous tree. In other situations there may only be a conflict between two alternatives, the third one being considerably less likely. A polytomy could not express this partial ambiguity appropriately. Instead, the two solutions can be integrated in a single network with a reticulation; see Fig. 1. Collapsing the rectangle in an either vertical or horizontal direction then reproduces the two incorporated trees.

Splits. An unrooted tree which displays the relationships among n taxa can be described by its system of splits (i.e., partitions of the taxa into two complementary clusters): each link of the tree joins exactly two subtrees that connect the taxa of either cluster. Given a fixed reference taxon, each split can be coded as a binary character, viz., all taxa in the cluster containing the reference taxon receive state 0, while the taxa of the complementary cluster have state 1. The system of all tree splits coded in this way is a "clique", that is, it has the following characteristic property, referred to as "compatibility" (MEACHAM & ESTABROOK 1985): no four taxa can be discriminated by any single pair of characters, i.e., for any two characters not all combinations 00, 01, 10, 11 of states are present among the taxa.

Weak compatibility. Relaxing the compatibility requirement allows to express partially ambiguous relationships among taxa. According to BANDELT & DRESS (1992a), three splits are called "weakly compatible" if – when coded as binary characters – not all four of the combinations 000, 011, 101, 110 occur among the taxa, i.e., for any four taxa at least one of the three alternative (quartet) trees has no support from the splits. A system of splits (or binary characters) that are (triplewise) weakly compatible is briefly called a "weak clique". Whereas a clique on n taxa can include at most $2n-3$ distinct splits, a weak clique may contain up to $\binom{n}{2}$ distinct members. The most intriguing feature of weak compatibility is that every weak clique of weighted splits can be reconstructed from the matrix of distances between taxa (i.e., total weights of mismatches between the associated character strings). A condition that is even stronger than weak compatibility requires that no three splits be pairwise incompatible. This condition is met by the system of splits obtained from a collection of cliques by applying the "one-third minority" rule: select those splits which occur in more than one third of the given cliques. In particular, the union of two cliques is of this kind. Pooling together the splits from two (but not more) trees always results in a weak clique.

Graphical representation. Cliques of splits are conveniently represented by their corresponding trees. It is therefore desirable to realize weak cliques by networks generalizing trees in a natural way. Such an approach has been proposed by BANDELT & DRESS (1992 b). The employed networks are ("distance-preserving") parts of high-dimensional Boolean cubes, and are not difficult to construct by hand (BANDELT 1992, 1994). In contrast to the case of trees, a split of such a network is associated with normally more than one link. These links, all drawn in parallel by convention, join the two subnetworks carrying the complementary clusters of the corresponding split. All nodes, those representing the taxa and the intermediate ones, can uniquely be described by 0–1 strings (of length equal to the number of distinct splits), once the character string of any reference node is prescribed. The network is distance-preserving in the sense that for any two nodes there is a connecting path of length equal to the number of mismatches of their character strings. In the case of weighted splits, the length of the links corresponding to a split is proportional to the weight of that split.

Inference of splits

Split decomposition. Typically, even binary data (e.g., restriction sites) would rarely constitute a weak clique, so that it becomes necessary to reduce the information by inferring a number of splits that do not necessarily fit in a tree but still obey weak compatibility. A natural way to achieve this is via split decomposition (BANDELT & DRESS 1992 a,b), that was originally designed only for distance data, but can readily be adjusted to process sequence data as well (BANDELT & DRESS 1993). First, one would choose an optimality criterion (such as maximum parsimony, minimum length, or maximum likelihood) in order to discriminate between the three alternative quartet trees. Then, given the data and the criterion, each of the alternative quartet trees has a certain level of support. This is expressed numerically as the length (either in sequence or distance space) taken negative or the log-likelihood. Subtracting for any four taxa the minimum value from each of the three levels of support yields the (nonnegative) "excess of support" for the particular alternative. An algorithm (running in polynomial time) then finds all those splits for which all quartet trees supported by that split have positive excess of support from the data. Necessarily, these splits form a weak clique since at least one quartet tree among the three alternative ones receives zero excess of support.

In the case that one departs from a distance matrix d, the minimum length criterion would be invoked: for each quartet tree, one determines the smallest possible total length subject to the constraint that the given distances never exceed the estimated distances along the quartet tree. A little computation shows that this minimum length is bounded below by

$$\frac{1}{2}\left(\max\left(d_{12}+d_{34}, d_{13}+d_{24}, d_{14}+d_{23}\right)+d_{12}+d_{34}\right)$$

for the tree with pairs 1, 2 and 3, 4 of neighbours (i.e., the first tree in Fig. 1). If the distances satisfy the triangle inequality (which can be achieved by adding a sufficiently large constant to all entries of the matrix d except the diagonal ones), this lower bound is actually attained. Therefore, the excess of support for this quartet tree equals

$$\tfrac{1}{2}(\max(d_{12} + d_{34}, d_{13} + d_{24}, d_{14} + d_{23}) - d_{12} - d_{34}).$$

Given sequence data, the excess of support for this tree according to the parsimony criterion equals the number of informative sites supporting this tree minus the minimum number of supporting informative sites among the three alternative trees for taxa 1, 2, 3, 4 (BANDELT & DRESS 1993, cf. BANDELT & al., unpubl.).

Bootstrap minority splits. A convenient way to arrive at a weighted weak clique is to employ any conventional tree-building method combined with the bootstrap or the jackknife procedure. As to bootstrapping, one would estimate trees from resampled data matrices in, say, 1000 experiments and record the frequencies of splits occurring in those 1000 trees (cf. SWOFFORD & OLSEN 1990). The "bootstrap minority" splits, appearing as tree splits in more than one third of the replications, then from a weak clique. The corresponding bootstrap percentages (or suitably scaled values) can serve as split weights.

Spectral analysis. For certain invertible models of sequence evolution, Hadamard conjugation provides a system of weighted splits (the "conjugate spectrum"), the weights being referred to as estimated "edge lengths"; see HENDY & PENNY (1993) and STEEL & al. (1992, 1994). From this spectrum one may extract a weak clique either using a least squares procedure or simply by greedy selection: as long as weak compatibility can be maintained, successively sample the splits of largest weight from the spectrum.

Consensus

Threshold rules. Feasible consensus rules for (weighted) weak cliques can be designed in analogy to the case of cliques. For instance, choose any threshold q with $\tfrac{2}{3} \leqslant q < 1$. A "$q$-majority coalition" in a family $\mathscr{F}$ of weighted weak cliques C is any subfamily $\mathscr{G}$ comprising more than q of the total weight. The "q-majority consensus" of the family $\mathscr{F}$ collects all splits σ that belong to the strict consensus of at least one q-majority coalition. (In the case of unit weights, these "q-majority splits" are exactly the splits appearing in more than q members of $\mathscr{F}$.) If the splits in each weak clique have their own weights, then one may compute the "q-majority weight" of a split σ as follows: for each q-coalition $\mathscr{G}$, determine the smallest weight of σ in the weak cliques from $\mathscr{G}$, and then take the maximum of all these weights. The q-majority splits are therefore precisely the splits with positive q-majority weight.

Union-inference methods. The procedure proposed by BAUM (1992) and RAGAN (1992) for combining cliques has its counterpart for weak cliques. Given a family of weighted weak cliques, one first takes the union by pooling together all splits that occur in these weak cliques, thereby adding up weights. This pooled system of splits can be regarded as a binary data matrix (with weighted characters) and thus may be subjected to any of the above procedures (split decomposition, etc.) for inferring a single weak clique.

An illustration

Conflicting data. BARRETT & al. (1991) presented two artificial data sets that were designed in order to assist their arguments against consensus. A quite similar example was constructed by CHIPPENDALE & WIENS (1994). Either one of the former

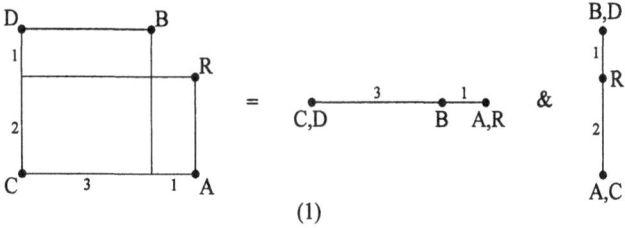

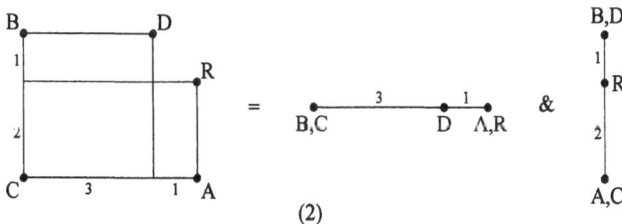

Fig. 2. Network representation of the two artificial data sets (1) and (2) considered by BARRETT & al. (1991); R denotes the hypothetical ancestor with character states 0, and numbers along links indicate weights of the associated splits

two sets consists of four taxa A, B, C, D and four polarized binary characters with different weights (altogether summing up to 7). The two data sets only differ in that the roles of taxa B and D are interchanged, which then leads to a conflict between the separately inferred trees. Polarity is realized here by adding a fifth taxon R representing the hypothetical ancestor with character states 0. As is seen from Fig. 2, either individual data set is ambiguous, being the disjoint union of the unique optimal clique (having weight 4) and the suboptimal clique (with weight 3). If these were real data, one should simply reject them as being unsuitable for phylogeny reconstruction. It would be of interest then to find out what really caused this ambiguity: hybrid nature of taxon C, or systematical bias, or random noise? Such problematic data may actually arise in reality; see BANDELT (1994) for an example.

Consensus versus pooling. Intuition would tell us that a combination of the two data sets (1) and (2) from Fig. 2 could not provide a clearer picture. The strict consensus of the participating weighted splits yields three splits with total weight 4, two of which are incompatible (Fig. 3a). The pooled system of splits is no longer a weak clique, as all three alternative splittings of taxa A, B, C, D are induced. So, split decomposition would come into action, returning three trivial splits (supporting terminal links with respective weights 2, 2, and 3) and two incompatible splits having weights 3 and 4, respectively. The corresponding network (Fig. 3b) equally testifies to the ambiguity in the data. Observe that split decomposition slightly favours the strict consensus tree of the two optimal trees over the common suboptimal tree, whereas for the strict consensus of splits (Fig. 3a) the order is reversed. Either way of data combination confirms that these data do not reliably support any single dichotomous tree, and moreover, helps to pin-point the ambiguity.

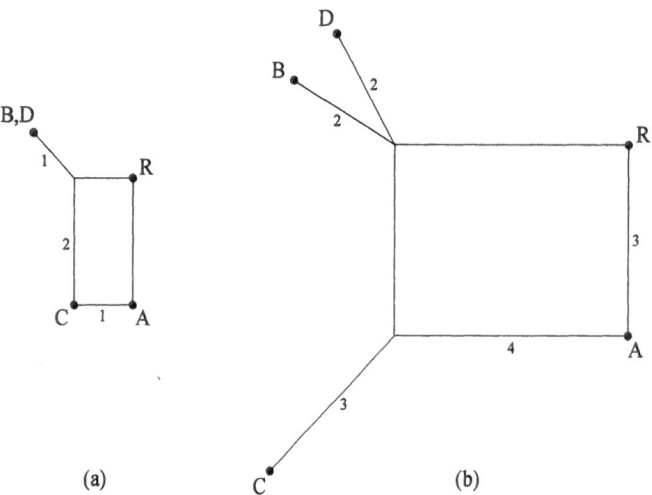

Fig. 3. Combination of data sets (1) and (2) from Fig. 2: (a) strict consensus of weighted splits; (b) split decomposition applied to the pooled data

Conclusion

This paper presents a framework for the combination of molecular (and other) data sets by systematically employing networks with reticulations rather than trees. A network expresses (to some extent) the uncertainty associated with phylogeny reconstruction and constitutes a more realistic picture of the data.

Phylogenetic analysis of different data sets for the same group of taxa is suggested to proceed in several stages. In the first step, each individual data set is thoroughly analyzed in order to detect potential systematic biases and to evaluate the information content. These single analyses then assist in adjudicating which parts of the total data may be pooled together (being regarded as rather homogeneous) and which portions should be discarded as being misleading or randomized. At the next stage, networks with weighted links are inferred from the homogeneous parts of the total data. The weights of links indicate the strengths of signals attributed to the corresponding splits. The information obtained from the separate parts can be compared and eventually aggregated in consensus networks by employing different strategies. It may be reasonable then to additionally weight the different parts of the full data set. The thus weighted union of weighted weak cliques can be subjected to the same estimation procedure for inferring a single weak clique that was already used for the homogeneous parts of the data. More focussed analyses may then explore particular areas of reticulation or low resolution in the resulting network.

References

BANDELT, H.-J., 1992: Generating median graphs from Boolean matrices. – In DODGE, Y., (Ed.): L_1-statistical analysis and related methods, pp. 305–309. – Amsterdam: North-Holland.
– 1994: Phylogenetic networks. – Verhandl. Naturwiss. Vereins Hamburg **34**: 51–71.
– DRESS, A. W. M., 1992 a: A canonical decomposition theory for metrics on a finite set. – Advances Math. **92**: 47–105.

– – 1992 b: Split decomposition: a new and useful approach to phylogenetic analysis of distance data. – Molec. Phyl. Evol. **1**: 242–252.

– – 1993: A relational approach to split decomposition. – In OPITZ, O., LAUSEN, B., KLAR, R., (Eds.): Information and classification, pp. 123–131. – Berlin: Springer.

BARRETT, M., DONOGHUE, M. J., SOBER, E., 1991: Against consensus. – Syst. Zool. **40**: 486–493.

BAUM, B. R., 1992: Combining trees as a way of combining data sets for phylogenetic inference, and the desirability of combining gene trees. – Taxon **41**: 3–10.

– RAGAN, M. A., 1993: Reply to A. G., RODRIGO's "A comment on BAUM's method for combining phylogenetic trees". – Taxon **42**: 637–640.

BULL, J. J., HUELSENBECK, J. P., CUNNINGHAM, C. W., SWOFFORD, D. L., WADDELL, P. J., 1993: Partitioning and combining data in phylogenetic analysis. – Syst. Biol. **42**: 384–397.

CHIPPENDALE, P. T., WIENS, J. J., 1994: Weighting, partitioning, and combining characters in phylogenetic analysis. – Syst. Biol. **43**: 278–287.

CROTHER, B. I., PRESCH, W. F., 1992: The phylogeny of Xantusiid lizards: the concern for analysis in the search for a best estimate of phylogeny. – Molec. Phyl. Evol. **1**: 289–294.

DE QUEIROZ, A., 1993: For consensus (sometimes). – Syst. Biol. **42**: 368–372.

DOYLE, J. J., 1992: Gene trees and species trees: molecular systematics as one-character taxonomy. – Syst. Bot. **17**: 144–163.

HEDGES, S. B., BEZY, R. L., 1993: Phylogeny of Xantusiid lizards: concern for data and analysis. – Molec. Phyl. Evol. **2**: 76–87.

HENDY, M. D., PENNY, D., 1993: Spectral analysis of phylogenetic data. – J. Classification **10**: 5–24.

HUELSENBECK, J. P., SWOFFORD, D. L., CUNNINGHAM, C. W., BULL, J. J., WADDELL, P. J., 1994: Is character weighting a panacea for the problem of data heterogeneity in phylogenetic analysis? – Syst. Biol. **43**: 288–291.

KLUGE, A. G., 1989: A concern for evidence and a phylogenetic hypothesis of relationships among *Epicrates* (*Boidae, Serpentes*). – Syst. Zool. **38**: 7–25.

MEACHAM, C. A., ESTABROOK, G. F., 1985: Compatibility methods in systematics. – Annu. Rev. Ecol. Syst. **16**: 431–446.

MICKEVICH, M. F., 1978: Taxonomic congruence. – Syst. Zool. **27**: 143–158.

MIYAMOTO, M. M., 1985: Consensus cladograms and general classifications. – Cladistics **1**: 186–189.

RAGAN, M. A., 1992: Phylogenetic inference based on matrix representation of trees. – Molec. Phyl. Evol. **1**: 53–58.

RODRIGO, A. G., 1993: A comment on BAUM's method for combining phylogenetic trees. – Taxon **42**: 631–636.

SWOFFORD, D. L., 1991: When are phylogeny estimates from molecular and morphological data incongruent? – In MIYAMOTO, M. M., CRACRAFT, J., (Eds): Phylogenetic analysis of DNA sequences, pp. 295–333. – New York: Oxford University Press.

– OLSEN, G. J., 1990: Phylogeny reconstruction. – In HILLIS, D. M., MORITZ, C., (Eds): Molecular systematics, pp. 411–501. – Sunderland: Sinauer.

STEEL, M. A., HENDY, M. D., PENNY, D., 1994: A discrete Fourier analysis for evolutionary trees. – Proc. Natl. Acad. Sci. USA **91**: 3339–3343.

– – SZÉKELY, L. A., PENNY, D., 1992: Spectral analysis and a closest tree method for genetic sequences. – Appl. Math. Letters **5**: 63–67.

WILLIAMS, D. M., 1994: Combining trees and combining data. – Taxon **43**: 449–453.

Author's address: Prof. Dr H.-J. BANDELT, Mathematisches Seminar, Universität Hamburg, Bundesstrasse 55, D-20146 Hamburg, Federal Republic of Germany.

Accepted February 9, 1995

Springer-Verlag
and the Environment

SpringerBiology

Hermann Meusel, Arndt Kästner

Lebensgeschichte der Gold- und Silberdisteln

Monographie der mediterran-mitteleuropäischen Compositen-Gattung *Carlina*

Band I: Merkmalsspektren und Lebensräume der Gattung

1990. 140 text figures, 32 plates. 294 pages.
Cloth DM 198,–, öS 1386,–
ISBN 3-211-82214-3

Band II: Artenvielfalt und Stammesgeschichte der Gattung

In collaboration with Ernst Vitek, Klaus Werner, Friedrich Ehrendorfer
1994. 179 text figures, 32 colour plates. 657 pages.
Soft cover DM 380,–, öS 2500,–
ISBN 3-211-86558-6

The two-volume monograph of the Mediterranean-Middle European genus *Carlina* is based on multidisciplinary analyses. Besides morphological, anatomical, karyological, embryological and phytochemical descriptions it includes detailed informations on the ecologically relevant aspects of the taxa as growth form, distribution pattern and phytocoenological amplitude. On the basis of all these features a reconstruction of the differentiation and evolution is presented: from its origin in the Mediterranean basin during subtropical periods of the Upper Tertiary up to the postglacial expansion into temperate Europe and W. Siberia.

Along with the authors' original research the volumes incorporate contributions by various experts. Each volume includes a summary in English. The volumes are amply illustrated with drawings and fotos especially in respect to diagnostic characters. Legends as well as table titles are translated into English.

SpringerWienNewYork

P.O.Box 89, A-1201 Wien • New York, NY 10010, 175 Fifth Avenue
Heidelberger Platz 3, D-14197 Berlin • Tokyo 113, 3-13 Hongo 3-chome, Bunkyo-ku

SpringerBiology

Peter K. Endress, Else Marie Friis (eds.)

Early Evolution of Flowers

1994. 130 partly colored figures. V, 229 pages.
Cloth DM 220,–, öS 1540,–
Reduced price for subscribers to "Plant Systematics and Evolution":
Cloth DM 198,–, öS 1386,–
ISBN 3-211-82599-1
Plant Systematics and Evolution / Supplement 8

The recent discovery of a large number of excellently preserved fossil flowers, studied with new techniques, and the comparative study of the flowers of extant basal clades of the angiosperms revolutionized the conception of early flower evolution. This volume brings together contributions of 17 palaeo- and neobotanists including critical reviews on the origin of flowers and the homologies of angiosperm floral organs, on the relationships of floral traits of magnoliids to those of lower eudicots (ranunculids, hamamelidids, dilleniids) and monocots, and articles on particular new Cretaceous fossils and new results on extant groups. These studies also influence the notion of early angiosperm evolution and of the macrosystematics of flowering plants.

SpringerWienNewYork

P.O.Box 89, A-1201 Wien • New York, NY 10010, 175 Fifth Avenue
Heidelberger Platz 3, D-14197 Berlin • Tokyo 113, 3-13 Hongo 3-chome, Bunkyo-ku